A FIRST COURSE IN

PROBABILITY AND STATISTICS WITH APPLICATIONS

A FIRST COURSE IN

PROBABILITY AND STATISTICS WITH APPLICATIONS

Peggy Tang Strait

Queens College of the City University of New York

HARCOURT BRACE JOVANOVICH, INC.

New York   San Diego   Chicago   San Francisco   Atlanta
London   Sydney   Toronto

ISBN: 0-15-527520-8
Library of Congress Catalog Card Number: 82-83468
Printed in the United States of America

In memory of my parents, Doris and Paul Tang (1902–1982)

PREFACE

This book was written primarily to answer the growing need for a one-semester course in probability and statistics for students with some background in calculus. Two current developments have led to the demand for this type of course: first, more and more students who are not mathematics majors are taking calculus; second, probability and statistics are becoming increasingly important in both the natural and the social sciences. For these reasons, a growing number of students—non-mathematics majors as well as mathematics majors not intending to become mathematical statisticians—have been seeking a course in which they can acquire a working knowledge of both probability and statistics and their applications in a single semester, or in two quarters. In many colleges, it has been the practice for such students to take only the first semester of the two-semester sequence. Unfortunately, in so doing they miss entirely the important area of statistical inference, which is usually covered in the second half of the course. Clearly, we are not adequately meeting the legitimate needs of these students.

This text, based on notes successfully used at Queens College for a number of years, was developed with just such students in mind. A great many of them are not mathematics majors. At Queens, they tend to be in computer science, economics, biology, engineering, and pre-medical studies, and often there are majors in philosophy, history, geology, accounting, physical education, and home economics. It is my hope that this book will stimulate the development of one-semester courses in probability and statistics for this expanding group of students.

The book is written with the realization that concepts in probability and mathematical statistics—even though they often appear deceptively simple—are in fact difficult to comprehend. Every basic concept and method is therefore explained in full, in a language that is easily understood. To emphasize applications, the explanations are accompanied by a wealth of illustrative examples and exercises. Concepts that are hard to grasp are given special attention. For example, there are discussions of the difference between probability one and certainty, between probability zero and impossibility, between independence and mutual exclusiveness, and between confidence levels and probabilities.

Understandably, in a one-semester text it is impossible to discuss in depth every relevant topic. I have included the following: elementary probability (including conditional probability and independence), discrete and continuous random variables and their distributions, mathematical expectations and applications (including Monte Carlo methods and decision problems), sums of random variables (but without proofs for a number of theorems), and statistical inference (including estimation, hypothesis testing, regression, correlation, and non-parametric methods). There is also a thorough discussion of the implications of the strong law of large numbers and the central limit theorem, even though the proofs are not given (a proof of the central limit theorem for random variables with moment generating functions is given in the appendix). The following topics are not included in the main body of the text: product moments, moment generating functions, and derivations of sampling distributions. However, the standard presentation of moment generating functions and the derivation of some of the sampling distributions are given in the appendix.

How to use this book. For a one-semester course in probability and statistical inference, I recommend that Chapters 1, 3–6, and 10–12 be studied first, and then as much other material as time permits. Chapter 3 (conditional probability) may be omitted if the instructor wishes to spend more time on topics of statistical inference, and selected topics from Chapter 2 (combinatorial methods) will enhance the understanding of discrete probability. For a one-semester course in probability, Chapters 1–8 and the appendix on moment generating functions offer a good introduction to its theory and applications.

The book affords great flexibility in the choice of topics. It is possible to omit entire chapters without loss of continuity, since explanations of important concepts are often repeated in several chapters, making it unnecessary for students to refer to previous chapters. For example, although the main discussion of independence of random variables appears in Chapter 7, the topic is introduced again in Chapters 8 and 10, enabling the instructor to proceed to topics of statistical inference immediately after the study of Chapter 6. The following list gives the prerequisites for the study of each chapter in the book.

Chapter	Topic	Prerequisites
1	Probability	Intermediate algebra
2	Combinatorial Methods	Chapter 1
3	Conditional Probability	Chapter 1
4	Discrete Random Variables	Chapter 1
5	Continuous Random Variables	Chapters 1, 4, and calculus
6	Mathematical Expectation	Chapters 1, 4, 5, and calculus
7	Multivariate Distributions	Chapters 1, 3–6, and calculus
8	Sums of Random Variables	Chapters 1, 4, 5, and 6
9	Descriptive Statistics	Intermediate algebra
10	Elements of Statistical Inference	Chapters 1, 4, 5, and 6
11	Estimation	Chapters 1, 4, 5, 6 and 10
12	Understanding Testing of Statistical Hypotheses	Chapters 1, 4, 5, 6, and 10
13	Some Standard Tests of Statistical Hypotheses	Chapters 1, 4, 5, 6, 10 and 12
14	Regression and Correlation	Chapters 1, 4, 5, 6, and 10
15	Nonparametric Methods	Chapters 1, 4, 5, 6, and 10

Answers to most of the exercises appear at the end of the book. A Solutions Manual is available with complete solutions to all exercises.

Acknowledgments

I would like to express my deep appreciation to acquisition editors Marilyn Davis and Richard Wallis, manuscript editor Liana Beckett, production editor Marji James, production supervisors Sue Crosier and Diane Polster, designer Nancy Shehorn, and art editor Avery Hallowell—the team at Harcourt Brace Jovanovich who worked with great care and dedication to ensure the publication of a book of the highest quality; to Ralph D'Agostino, Cyrus Derman, Murray Eisenberg, Morris Hamburg, Raymond F. Heiser, Clarence Jones, Valerie Miké, Fred W. Morgan, John D. Neff, Ronald Rothenberg, Harold Shane, Frederick Stern, James Vick, and Sol Weintraub, who reviewed the manuscript and offered many helpful suggestions which were incorporated into the text; to my student Sarah Hecht, who was invaluable as proofreader of the original manuscript and as assistant in the preparation of answers to the exercises; to Rebecca Amann, Miriam Green, and Phyllis Rubin for their expert typing; and most of all to my husband, Roger, a professor in the social sciences, who not only graciously endured the hardships of being an author's spouse, but who took upon himself the task of reading the entire manuscript to assure its clarity and readability.

CONTENTS

Chapter 6

Mathematical Expectation   166

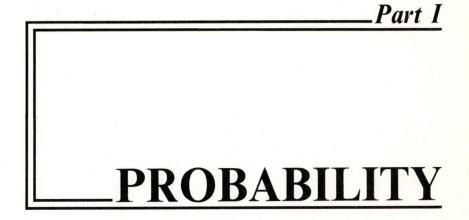

Part I

PROBABILITY

Chapter **1**

PROBABILITY

1.1   Introduction

There is something about the unpredictable that brings out the probabilist in each of us. Will there be another world war, will I be drafted, can I expect a decent grade in this course? What are the chances, what are the probabilities? Such questions bring us to the topics discussed in Part I of this book, which is divided into two major segments. Part I introduces the basic concepts and applications of probability; Part II, building on the structure of Part I, examines the methods and applications of statistical inference. Before entering into a formal discussion of the mathematics of probability, however, let us briefly consider the meaning and historical evolution of the term **probability**.

1.2   What Is Probability?

What do we mean when we say that the probability of a particular event is some number such as 1/6? To a certain extent, the answer depends on the nature of the event and on the nature of our philosophical conception of uncertainty. It is not surprising, then, that not one, but several commonly used definitions of probability have developed since the seventeenth century. Generally, they are referred to as the **classical**, **empirical**, **subjective**, and **axiomatic** definitions of probability. Following the standard modern approach, this book will develop the mathematics of probability according to the axiomatic definition, a detailed discussion of which is given in Section 1.4. Problems of application, however, will often be interpreted in terms of the other, more intuitive, definitions. A brief description of each definition follows.

1.2.1   Classical Probability

Historically, the first recorded discussion of probability dealt with games of chance. In the middle of the seventeenth century the Chevalier de Méré, a French nobleman and inveterate gambler, consulted the mathematician Blaise Pascal (1623–1662) about a problem concerning a game of dice (ref. 8 and Exercise 6, Section 2.2 of Chapter 2). The theory that evolved concerned games with a finite number of equally likely outcomes. If the event under consideration was the occurrence of any one of n possible outcomes among the total number N of all possible outcomes, then the probability of that event was defined to be n/N. For example, in rolling a balanced die, there are six equally likely outcomes. Thus, according to the classical definition, the probability of obtaining an even number on the die is 3/6, there being three out of a total of six outcomes that are even numbers. A formal statement of this theory of probability was given by Laplace in his classic work on the subject (ref. 5).

   An important observation made by several founders of classical probability—Pascal, Fermat, Huygens, and J. Bernoulli—was that the relative frequency ratio of a given event of certain games of chance tends to converge to a definite value when the game is repeated a great number of times. If was, in fact, this phenomenon that led them to postulate the classical theory of probability. In games of chance, events that can be objectively determined to be equally likely are, in fact, observed to occur with equal frequency when the game is repeated many times. Thus, according to classical theory, they are events with equal probabilities.

To illustrate this point, the author's 10-year-old son threw a die 600 times and a throw-by-throw record of the experiment was kept. Figure 1.1 summarizes the results with a graphic display of the phenomenon of a convergence to a stable relative frequency ratio of 1/6 for the outcomes 1 and 2. (Illustration of the same phenomenon for the outcomes of 3, 4, 5, and 6 was omitted to permit a less cluttered diagram.) This confirms the classical theory that events that can be objectively determined to be equally likely—in this case the events 1, 2, 3, 4, 5, and 6 in a throw of a die—are, in fact, observed to occur with equal frequency when the game is repeated many times.

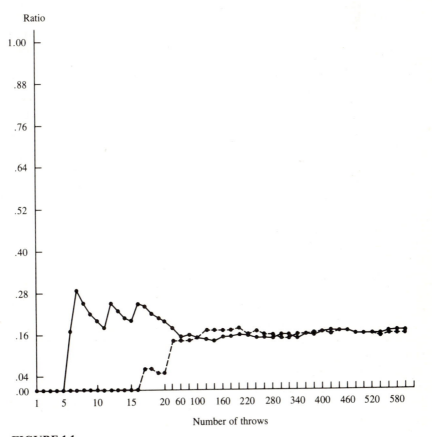

FIGURE 1.1
Relative frequency ratios of ones (solid line) and twos (broken line) in a sequence of throws with a die.

The classical theory can be extended beyond games of chance to any chance situation involving a finite number of equally likely outcomes. There are many problems, applied and theoretical, that fall into this category. Especially interesting and challenging problems of this sort are connected with combinatorial methods, a topic discussed in Chapter 2. Nevertheless, classical probability has one major limitation. The restriction to a finite number of equally likely outcomes excludes the consideration of many chance phenomena of interest.

1.2.2 Empirical Probability: The Relative Frequency Interpretation

In the eighteenth and nineteenth centuries, observers of natural phenomena noted again and again that the long-run stability of relative frequency ratios was not limited to games of chance, but was to be found in various demographic and experimental data. One of the first recorded instances was in connection with the sex of newborn infants at a certain city hospital in England. It was noted that the long-run relative frequency ratio of male births tended toward 1/2. This seemed to imply that the event of a male birth can be looked upon as a chance phenomenon, much as is the event of heads in a toss of a coin, and that the probability of this event was 1/2. Observations of this sort eventually led to the formation of a definition of probability called empirical probability—a relative frequency interpretation of probability. This theory of probability was developed mainly by R. A. Fisher and R. von Mises (refs. 1 and 7). The definition of empirical probability may be briefly stated as follows.

If $\mathscr{E}$ is an experiment with unpredictable outcomes, and if A is one of the possible outcomes of $\mathscr{E}$, then we say that the (empirical) probability of the outcome A is the number p if, in repeated performances of $\mathscr{E}$, the long-run relative frequency ratio of occurrences of A tends to the number p. That is, if n denotes the number of repetitions of $\mathscr{E}$, and n_A denotes the number of times A occurs, and $p = \lim_{n \to \infty}(n_A/n)$, then the (empirical) probability of A is the number p. Because, in practice, it is not possible to repeat an experiment an infinite number of times, the general rule is to accept the relative frequency ratio n_A/n as the probability of A if n is a large number.

The empirical definition of probability is the most widely accepted definition of probability. It is the definition that most frequently comes to mind when we are confronted with a probability statement. For example, when the weather bureau predicts a .9 probability of rain, we assume that conclusion to be the result of many observations of days with identical weather conditions, 90% of which days were rainy. In other words, .9 is an empirical probability of the event of rain.

However, in spite of its enormously persuasive properties, the empirical definition of probability is problematic when a rigorous development of the mathematics of probability is desired. We list a few of the problems here.

1. It is not possible, in practice, to repeat an experiment an infinite number of times to obtain the probability $p = \lim_{n \to \infty} (n_A/n)$.

2. Even if it were possible to repeat an experiment an infinite number of times, it is conceivable that a different infinite sequence of performances of the same experiment could produce a different value for p. For example, it is within the realm of possibility that one sequence of tosses of a coin will produce only heads, while a second sequence of tosses will produce only tails. We do not consider such happenings to be even remotely likely, but they must be admitted as possible.

3. If we define p to be the ratio n_A/n for n large, it is not clear how large n should be before we are certain that p is close to the limiting value of n_A/n as $n \to \infty$.

These limitations notwithstanding, almost everyone would still insist that, although the empirical definition or relative frequency interpretation of probability is not a rigorous one, nevertheless, any useful definition of probability must contain the basic feature of empirical probability—that if p is the probability of an event A, then $\lim_{n \to \infty} (n_A/n) = p$, if not always, then at least with the greatest likelihood. As we shall see in the chapters to follow, it is a triumph of the axiomatic definition of probability that it offers this possibility. (See Chapter 6 on the strong law of large numbers and Section 6.10.2.)

1.2.3   Subjective Probability

As the name suggests, subjective probability is a personal evaluation of the likelihood of chance phenomena. It is an important element in many decision-making processes and is a basic ingredient of Bayesian decision theory, to be discussed in a later chapter. Early proponents of subjective probability theory were Keynes and Jeffreys (refs. 3 and 2). As in all other definitions, the subjective probability of an event is defined to be a number ranging from 0 to 1. If it is the judgment of an individual that an event A is less likely to occur than an event B, then a smaller probability would be assigned to A than to B. Exactly how much smaller a probability would be assigned is, again, a personal matter. If the probabilities can be related somehow to betting odds, however, the monetary commitment involved in such situations can be used to determine the probabilities. For example, if Tom Smith is willing to bet $100 against $50 that interest rates will increase by 3% within 6 months, we could say that Tom Smith assigns 100 chances ($100) against a total of $100 + 50 = 150$ chances ($150) to the event that interest rates will increase by 3% within 6 months. This implies an assignment of a subjective probability of $100/(100 + 50) = 2/3$ to the event of a 3% increase in interest rates against a subjective probability of $50/(100 + 50) = 1/3$ to the event that there will not be a 3% increase in interest rates.

In major business decision-making processes, operations research consultants are often called upon to advise management. These analysts are aware that high-level corporate executives are astute judges of business situations. Therefore, their judgments—which are sometimes expressed in terms of subjective probabilities—are entered as input into the operations research analysis. Since betting odds are second nature to most business executives, they are often used as a guide to the determination of subjective probabilities.

1.2.4   Axiomatic Definition of Probability

The name sounds high-powered and frightening—yet, the axiomatic definition of probability is perhaps the simplest of all the definitions and is certainly the least controversial. Essentially, it is a definition based on a set of axioms that are statements of minimal requirements for a definition of probability. The advantage of the axiomatic approach is that it allows for a rigorous development of the mathematics of probability; yet, because of the careful choice of axioms, it remains essentially compatible with the other definitions of probability. This approach was first introduced by the Russian mathematician A. N. Kolmogorov and is now universally accepted by all probabilists and mathematical statisticians. A detailed discussion of the axioms is given in Section 1.4.1.

Exercises

In each of the following examples, a probability is defined. Is it classical, empirical, or subjective?

(a)   A box contains 10 balls, 3 of which are red. A ball is drawn at random from the box. The probability of drawing a red ball is .3.

(b)   A card is drawn at random from a standard deck of 52 cards. The probability of drawing the Queen of Hearts is 1/52. The probability of drawing a heart is 1/4.

(c)   One group of 1,000 mice were fed diet A plus a fixed amount of chemical X. A second group of 1,000 mice were fed only diet A. None of the mice in this second group developed cancer. Thirty of the other mice developed cancer. From these results, one can infer that for mice on diet A, the addition of chemical X increases the probability of cancer.

(d)   Madam X, the palm reader, told Karen that she has a 9 to 1 chance of meeting her future husband within the year. The probability that Karen will meet her future husband within the year is $9/(9 + 1) = .9$.

(e)   In a certain town in Switzerland, a train arrives at the station 25 times a day. Records kept over 20 years show that fewer than 5 trains per week are late. The probability that a train will be late at this station is less than .03.

(f)   Two thousand tosses of a certain coin resulted in 1,200 heads. Therefore, the probability of heads in a toss of this coin is .6.

(g)   The probability of obtaining two heads in two tosses of a coin is 1/4.

(h)   David has a certain appearance of confidence. You think that his probability of winning first prize in the judo match is .75.

(i)   The leader of a nuclear project feels that the odds are 7 to 5 that the project will be completed on schedule. Therefore he assigns a probability of 5/12 to the event of an on-schedule completion of the project.

1.3 Elements of Probability

Regardless of our choice of definition of probability—be it classical, empirical, subjective, or axiomatic—it is necessary to define certain terms and concepts so that the topic may be developed in an orderly and unambiguous fashion.

1.3.1 Experiments, Outcomes, Sample Spaces

Applied problems in probability and statistics are often stated within the context of an experiment. For such purposes, the definition of an experiment is the following.

Definition 1.1 An **experiment** is a specific set of actions the results of which cannot be predicted with certainty.

For example, the process of rolling a pair of dice to observe the values on the top sides of the dice is an experiment, as is the process of subjecting laboratory animals to certain treatments to determine the effects of the treatments, and as is the process of introducing a new product to the market to determine its appeal.

A well-designed experiment should include a specification of the data that will be recorded. For example, in designing the experiment of rolling a pair of dice, we may include instructions for recording the value on the top side of each die. For such experiments, each possible set of recorded data defines a result which is called a **simple outcome** of the experiment.

EXAMPLE 1.1

Let $\mathscr{E}$ be the following experiment: roll a red die and a green die, and record the value on the top side of each die. Use mathematical notation to list all the possible simple outcomes of this experiment.

SOLUTION

The specified data of the experiment are the values on the die. If the value on the red die is a 2 and the value on the green die is a 5, then a simple mathematical notation for this pair of values is $(2, 5)$. Speaking in more general terms, if the value on the red die is i and the value on the green die is j, then a simple mathematical notation for this pair of values is (i, j). Thus, each (i, j), for $i = 1, 2, 3, 4, 5, 6$ and $j = 1, 2, 3, 4, 5, 6$, represents a distinct simple outcome of the experiment. Conversely, each distinct simple outcome of the experiment is represented by a pair of values (i, j).

This example demonstrates the use of mathematical notation to represent the simple outcomes of an experiment. The set of all simple outcomes of an experiment defines the sample space of the experiment.

Definition 1.2   A **sample space** is a nonempty set of elements. A set S is a sample space for an experiment $\mathscr{E}$ if each element of S represents a unique simple outcome of $\mathscr{E}$, and if each simple outcome of $\mathscr{E}$ is represented by a unique element of S. The elements of S are called **sample points**.

In Example 1.1, each pair (i, j), for $i = 1, 2, 3, 4, 5, 6$ and $j = 1, 2, 3, 4, 5, 6$, represents a distinct simple outcome of the experiment. Furthermore, each distinct simple outcome is represented by a pair of values (i, j). Thus the set S of all pairs (i, j) with $i = 1, 2, \ldots,$ or 6 and $j = 1, 2, \ldots,$ or 6 is called the sample space of the experiment. Using the standard notation for sets, the notation for this particular sample space is

$$S = \{(i, j)\,|\, i = 1, 2, 3, 4, 5, 6; \quad j = 1, 2, 3, 4, 5, 6\}$$

(Readers who are unfamiliar with the various methods for specifying the elements in a set should refer to the discussion on sets in Section 1.3.2.) Figure 1.2 contains a graphic illustration of the sample space for this experiment of rolling a pair of dice.

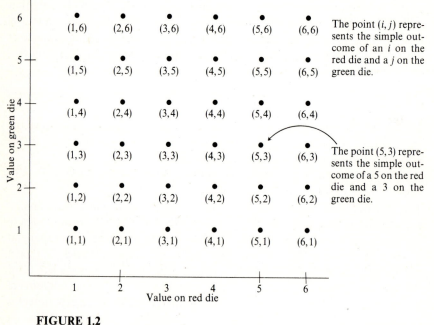

FIGURE 1.2
Sample Space for Example 1.1
The sample space for the experiment of rolling two dice is the set

$$S = \{(i, j)\,|\, i = 1, 2, 3, 4, 5, 6; \quad j = 1, 2, 3, 4, 5, 6\}$$

EXAMPLE 1.2

Define a sample space for each of the following experiments.

(a)   Toss a coin and record the face on the top side of the coin.

(b)   Draw a card from a standard deck of 52 cards and record the suit of the card drawn.

(c)   The heights in inches of the children Mary, John, Paul, and Susan are 48, 52, 49, and 50, respectively. Select a child at random from this group of children, then measure and record the child's height.

(d)   Toss a coin and then roll a die. Record the face on the coin and the value on the top side of the die.

(e)   Select a number at random from the interval $(0, 1)$ of real numbers. Record the value of the number selected.

   SOLUTION

We will use the following standard notation for sets: if the set A consists of the elements $a_1, a_2, \ldots, a_n$, then we write

$$A = \{a_1, a_2, \ldots, a_n\}$$

(a)   Let h and t represent the simple outcomes of heads and tails, respectively. A sample space for the experiment is the set

$$S = \{h, t\}$$

(b)   Let the symbols $\heartsuit$, $\diamondsuit$, $\spadesuit$, $\clubsuit$ represent the simple outcomes of hearts, diamonds, spades, and clubs, respectively. A sample space is

$$S = \{\heartsuit, \diamondsuit, \spadesuit, \clubsuit\}$$

(c)

$$S = \{48, 52, 49, 50\}$$

(d)   Let h and t represent the simple outcomes of heads and tails, respectively. A sample space for the experiment is

$$S = \begin{Bmatrix} (h, 1), (h, 2), (h, 3), (h, 4), (h, 5), (h, 6) \\ (t, 1), (t, 2), (t, 3), (t, 4), (t, 5), (t, 6) \end{Bmatrix}$$

The number in the second coordinate of each pair of values (,) is the value on the die.

(e)

$$S = \{x \mid x \text{ is a number in } (0, 1)\}$$

The expression on the right of the equals sign stands for the set of all elements x such that x is a number in the interval $(0, 1)$.

Sample spaces are classified according to the number of elements they contain. A sample space is said to be a **finite sample space** if it consists of a finite number of elements. It is a **discrete sample space** if the elements can be placed in a one-to-one correspondence with the positive integers, or if it is finite. Sample spaces that do not satisfy either of the preceding criteria are called **nondiscrete sample spaces**. These distinctions become important when we speak of events (Section 1.3.3) and random variables (Chapters 3 and 4).

Summary

A sample space is a **finite sample space** if it consists of a finite number of sample points. It is a **discrete sample space** if the sample points can be placed in a one-to-one correspondence with the positive integers, or if it is finite. It is a **nondiscrete sample space** if it satisfies neither of these criteria.

Given an experiment $\mathscr{E}$ and a sample space S for $\mathscr{E}$, generally there will be results that cannot be clearly defined by single elements in S. For example, in rolling two dice each of the sample points $(1, 6)$, $(2, 5)$, $(3, 4)$, $(4, 3)$, $(5, 2)$, and $(6, 1)$ represents a sum of 7. We cannot single out one of these sample points to represent the general result of "a sum of 7." A convenient method of representing this result is to use the entire subset

$$\{(1, 6), (2, 5), (3, 4), (4, 3), (5, 2), (6, 1)\}$$

Generalizing from this example, we shall use subsets of sample spaces to represent results of experiments. The understanding is that if A is a subset of a sample space S, then the result represented by A occurs if any of the simple outcomes represented by sample points in A occur. Results of experiments that are represented by subsets of the sample space are called **outcomes**. Furthermore, the subsets themselves are called outcomes. A **simple outcome** is an outcome that is represented by a subset consisting of a single sample point.

EXAMPLE 1.3

Consider the experiment of rolling two dice (Example 1.1).

   (a)   Let B represent the outcome of a sum of 11. List the sample points in B.

   (b)   Express in words the outcomes represented by the subsets

$$D = \{(1, 1), (2, 2), (3, 3), (4, 4), (5, 5), (6, 6)\}$$

and

$$E = \{(5, 6), (6, 5), (6, 6)\}$$

   (c)   What subset C of the sample space represents the outcome of a 1 on the red die and a 1 on the green die?

(d)   Let A be the outcome of a sum of 7. Let B, C, D, and E be the outcomes described above. If the simple outcome (1, 1) occurs, then which of the outcomes A, B, C, D, and E occur?

(e)   Draw a line around those sample points in Figure 1.2 that are contained in the subsets A, B, and C.

SOLUTION

(a)
$$B = \{(5,6), (6,5)\}$$

(b) D represents the outcome of a "double." E represents the outcome of "a sum greater than or equal to 11" or "a sum of 11 or 12."

(c)
$$C = \{(1,1)\}$$

This illustrates the fact that simple outcomes represented by a single sample point can also be expressed as a subset consisting of the single sample point.

(d)   The sample point (1, 1) is an element in both C and D. If the simple outcome (1, 1) occurs, then the outcome C occurs and the outcome D occurs.

(e)   See Figure 1.3.

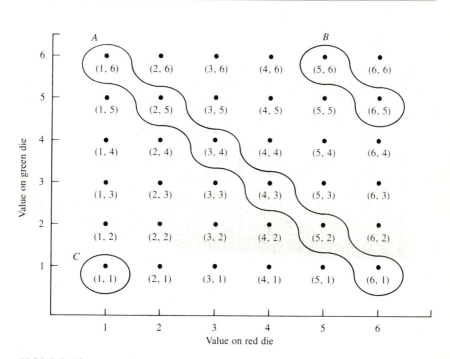

FIGURE 1.3

Outcomes of experiments are represented by subsets of the sample space.

$A = \{(1,6), (2,5), (3,4), (4,3), (5,2), (6,1)\}$ = outcome of a sum of 7

$B = \{(5,6), (6,5)\}$ = outcome of a sum of 11

$C = \{(1,1)\}$ = outcome of a 1 on each die

Summary

Outcomes are results of experiments that are represented by subsets of sample spaces. If A is a subset of a sample space, then the outcome represented by A occurs if any of the simple outcomes represented by sample points in A occur. The subsets representing outcomes are also called outcomes.

1.3.2   Set Operations Applied to Outcomes

Outcomes of experiments are represented by subsets of sample spaces. The notations, definitions, and operations of set theory take on special meaning when applied to subsets representing outcomes of experiments.

1.   Sets

There are two common methods for specifying the elements in a set. If the set has a finite number of elements, we may simply list the elements and enclose the list within braces. For example, the set A consisting of elements $a_1, a_2, \ldots, a_n$ is written

$$A = \{a_1, a_2, \ldots, a_n\}$$

The set I of integers from 1 to 100 is written

$$I = \{1, 2, \ldots, 100\}$$

If the number of elements in a set is large or infinite, then a better method is to include a statement of qualification concerning the elements. For example, the set I of positive integers may be written

$$I = \{i \mid i \text{ is a positive integer}\}$$

The vertical bar ($\mid$) is equivalent to the phrase "such that," so that the expression to the right of the equals sign may be translated as "the set of all elements i, such that i is a positive integer." This method of specifying the elements in a set was used in Figure 1.2 and Example 1.2(e).

2.   Subsets

Set A is said to be a **subset** of a set B if every element in A is an element in B. If A is a subset of B, then we write

$$A \subset B$$

The meaning of subset for outcomes   Recall that outcomes are represented by subsets of the sample space, and that we say that an outcome C occurs if any of the simple outcomes represented by sample points in C occur. Now, suppose that an outcome A is a subset of an outcome B, and that neither A nor B are empty sets. That is, suppose that $A \subset B$ and $A \neq \emptyset$. This means that every sample point in A is also a sample point in B. Furthermore, this means that if the outcome A occurs (that is, if a simple outcome represented by a sample point s in A occurs), then the outcome B also occurs (because the sample point s is also in B). In other words, if A is a subset of B, then the occurrence of the outcome A *implies* the occurrence of the outcome B. Conversely, if the occurrence of an outcome A implies the occurrence of an outcome B, then A is a subset of B. We can therefore conclude that the statement "A is a subset of B" is equivalent to the statement "the occurrence of the outcome A implies the occurrence of the outcome B," and we write

$$[(\text{set } A) \subset (\text{set } B)] \equiv [(\text{outcome } A) \text{ implies } (\text{outcome } B)]$$

provided that A and B are not empty sets. (The symbol "$\equiv$" stands for the phrase "is equivalent to.") Consider, for example, the experiment of rolling two dice (Examples 1.1 and 1.3). If B is the outcome of "a sum of 11" (i.e. $B = \{(5,6), (6,5)\}$), and E is the outcome of "a sum of at least 11" (i.e., $E = \{(5,6), (6,5), (6,6)\}$), then $B \subset E$ and the occurrence of the outcome B (i.e., a sum of 11) *implies* the occurrence of the outcome E (i.e., a sum of at least 11).

3.   Union

The **union** of a set A with a set B, written $A \cup B$, is the set of all elements belonging to A or to B. The word "or" is used here and elsewhere in the book in the inclusive sense, so that elements belonging to both A and B are also in the union. See the Venn diagram of $A \cup B$ in Figure 1.4(a).

The meaning of union for outcomes   Let A and B be outcomes. If the outcome $A \cup B$ occurs, then a simple outcome represented by one of the sample points in $A \cup B$ occurs. A sample point is in $A \cup B$ if it is in A or in B. Thus, the outcome $A \cup B$ occurs if and only if the outcome A occurs *or* the outcome B occurs. In terms of outcomes, then, the operation $\cup$ can be translated to the word "or." Shown symbolically

$$[(\text{set } A) \cup (\text{set } B)] \equiv [(\text{outcome } A) \text{ or } (\text{outcome } B)]$$

Consider, for example, the experiment of rolling two dice. If A is the outcome of "a sum of 7" (i.e., $A = \{(1,6), (2,5), (3,4), (4,3), (5,2), (6,1)\}$) and B is the outcome of "a sum of 11" (i.e., $B = \{(5,6), (6,5)\}$), then

$$A \cup B = \{(1,6), (2,5), (3,4), (4,3), (5,2), (6,1), (5,6), (6,5)\}$$

is the outcome of "a sum of 7" *or* "a sum of 11."

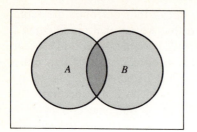

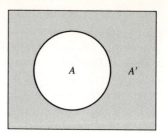

(a) $A \cap B$ = dark shaded region   **(b)** A' = shaded region
           $A \cup B$ = dark and light shaded
               regions

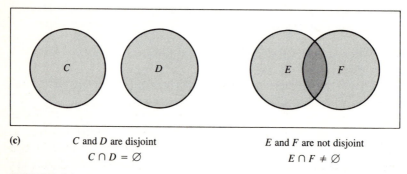

(c) C and D are disjoint E and F are not disjoint
                   $C \cap D = \varnothing$ $E \cap F \neq \varnothing$

FIGURE 1.4
Venn Diagrams of Operations on Sets and Outcomes

4.   Intersection

The **intersection** of a set A with a set B, written $A \cap B$ (or simply AB) is the set of all elements belonging to both A and B. See the Venn diagram of $A \cap B$ in Figure 1.4(a).

   The meaning of intersection for outcomes   Let A and B be outcomes. A sample point is in $A \cap B$ if and only if it is in both A and B. Thus, the outcome $A \cap B$ occurs if and only if both A *and* B occur. In terms of outcomes, then, the operation $\cap$ can be translated to the word "and."

$$[(\text{set } A) \cap (\text{set } B)] \equiv [(\text{outcome } A) \text{ and } (\text{outcome } B)]$$

Consider, for example, the experiment of rolling two dice. If D is the outcome of a double (i.e., $D = \{(1,1),\ (2,2),\ (3,3),\ (4,4),\ (5,5),\ (6,6)\}$) and F is the outcome of "a sum of 4" (i.e., $F = \{(1,3),\ (2,2),\ (3,1)\}$), then $D \cap F = \{(2,2)\}$ is the outcome of "a double" *and* "a sum of 4." In other words, $D \cap F$ is the outcome of a 2 on each of the two dice. See Figure 1.5.

5.   Complementation

Complementation is a set operation applicable when dealing with a universal set consisting of all elements under consideration. If S is the universal set, then

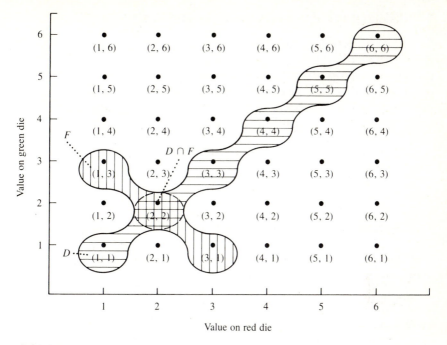

FIGURE 1.5

D is the outcome of a double; F is the outcome of a sum of 4; $D \cap F$ is the outcome of a double and a sum of 4; therefore, $D \cap F = \{(2, 2)\}$.

the complement of A, written A', is the set of all elements in S that are not in A. See the Venn diagram of A' in Figure 1.4(b).

The meaning of complementation for outcomes With respect to outcomes, the sample space S is considered the universal set. If A is an outcome, then the outcome A' occurs if and only if a simple outcome represented by one of the sample points in A' occurs. A sample point is in A' if and only if it is *not* in A. Therefore, A' is the outcome when A does *not* occur.

$$[(\text{set } A)'] \equiv [\text{not (outcome } A)]$$

6. The empty set $\emptyset$

The set consisting of no elements is called the **empty set** or **null set**. We use the symbol $\emptyset$ to denote the empty set.

Since S' is the set of all elements in S (the universal set) that are not in S, it is clear that S' and $\emptyset$ are identical sets. That is

$$\emptyset = S'$$

The meaning of $\emptyset$ for outcomes The outcome $\emptyset$ is the outcome S'. The outcome S' is the outcome when S does not occur. Therefore, $\emptyset$ is the outcome that none of the simple outcomes represented by sample points in the sample space occur.

7.   Disjoint sets and mutually exclusive outcomes

Sets A and B are said to be **disjoint** if none of the elements in A are elements in B. In other words, A and B are disjoint if $A \cap B = \emptyset$.

   The meaning of disjoint sets for outcomes   If A and B are outcomes and $A \cap B = \emptyset$, then none of the sample points in A are sample points in B. See Figure 1.4(c). Therefore, A and B cannot occur simultaneously. Such outcomes are said to be **mutually exclusive**.

$$[A \cap B = \emptyset] \equiv [\text{outcomes } A \text{ and } B \text{ are mutually exclusive}]$$

   A sequence $A_1, A_2, A_3, \ldots$ of outcomes is said to be mutually exclusive if $A_i \cap A_j = \emptyset$ for $i \neq j$, and $i = 1, 2, 3, \ldots; j = 1, 2, 3, \ldots$.

8.   Other useful facts concerning sets and outcomes

The following true statements concerning sets and outcomes may also be verified with Venn diagrams:

(a)   The commutative law $A \cup B = B \cup A$; $A \cap B = B \cap A$

(b)   The associative law $A \cup (B \cup C) = (A \cup B) \cup C$;
$A \cap (B \cap C) = (A \cap B) \cap C$

Because of these two statements, we can use the simpler notations $A \cup B \cup C$ and $A \cap B \cap C$ without fear of ambiguity when referring to $A \cup (B \cup C)$ and $A \cap (B \cap C)$, respectively.

(c)   The distributive law $A \cup (B \cap C) = (A \cup B) \cap (A \cup C)$;
$A \cap (B \cup C) = (A \cap B) \cup (A \cap C)$

(d) $S' = \emptyset$; $\emptyset' = S$; $(A')' = A$

(e) $A \cup S = S$; $A \cap \emptyset = \emptyset$ for all subsets A (outcomes) of S

(f) $A \cup A' = S$; $A \cap A' = \emptyset$

(g)   de Morgan's laws $(A \cup B)' = A' \cap B'$; $(A \cap B)' = A' \cup B'$

EXAMPLE 1.4

Use Venn diagrams to verify the statement $(A \cup B)' = A' \cap B'$.

   SOLUTION

Figure 1.6 contains a Venn diagram of $(A \cup B)'$ and of $A' \cap B'$. In the Venn diagram for $(A \cup B)'$ the region covered by both horizontal and vertical lines represents the set $(A \cup B)'$. In the Venn diagram for $A' \cap B'$, the region covered by horizontal lines represents A' and the region covered by vertical lines represents B'. Thus, the region covered by both horizontal and vertical lines represents $A' \cap B'$.

   Observe that the regions covered by both horizontal and vertical lines in the two Venn diagrams are identical. Therefore, $(A \cup B)' = A' \cap B'$.

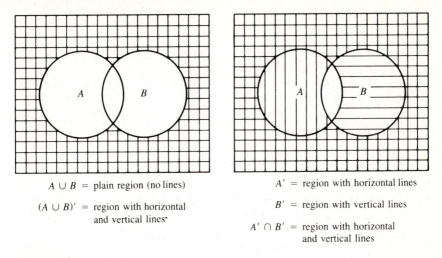

$A \cup B$ = plain region (no lines)

$(A \cup B)'$ = region with horizontal
and vertical lines·

A' = region with horizontal lines

B' = region with vertical lines

$A' \cap B'$ = region with horizontal
and vertical lines

FIGURE 1.6
Venn Diagram Proving $(A \cup B)' = A' \cap B'$

1.3.3   Events

In the formal language of probability and mathematical statistics, the word
event is reserved for subsets belonging to a special class of subsets of a sample
space. For applied problems in probability and statistics, if the sample space is
discrete, then all subsets are events; if the sample space is an interval or a union
of intervals of numbers on the line of real numbers, then events are either
subintervals of the sample space, or else they are unions, intersections, or
complementations of subintervals of the sample space. (It is necessary to be
restrictive in defining events for nondiscrete sample spaces because certain
subsets, called nonmeasurable sets, are problematic in a rigorous definition of
probability.) In all cases, the sample space S and the empty set $\emptyset$ are events.
According to common usage, results and outcomes represented by subsets
that are events are also called events. In this book we shall restrict our
consideration of results and outcomes to those that are represented by subsets
that are events; thus we may use the terms results, outcomes, and events
interchangeably.

Summary

Certain subsets of sample spaces are called **events**—as are results and
outcomes of experiments represented by subsets that are events. If a sample
space is discrete, then all subsets are events. If a sample space is an interval
on the line of real numbers, then events are subintervals, or are unions,
intersections, or complements of subintervals of the sample space.

Exercises

1. Define a sample space *S* for each of the following experiments.
    (a)   Submit five brands of coffee, labeled A, B, C, D, and E to a panel of judges for appraisal. A best-tasting and worst-tasting brand are to be chosen and the labels are to be recorded.
    (b)   Randomly arrange the three letters *a*, *t*, and *c* in a row. Record the arrangement of the three letters.
    (c)   The possible scores on a certain test are H, P, and F, for High Pass, Pass, and Fail, respectively. Give this test to two students and record their scores on the test.
    (d)   Toss a coin three times and record the outcome after each toss.
    (e)   Toss a coin until a head appears, and record the number of tosses required.
    (f)   Toss a coin three times and record the number of heads.
    (g)   Susan has a penny, a nickel, a dime, and a quarter in her coin purse. She randomly selects a coin from the purse and records its value (in cents). She randomly selects another coin from the purse and records the value of the second coin also.
    (h)   Suppose that instead of recording the values of both coins, Susan records only the sum of the values of the two coins.
    (i)   Assume you are to give a certain new medication to three patients suffering from a particular disease. After the treatment, a doctor is to determine and record the condition of each patient as greatly improved (G), moderately improved (M), or not at all improved (N).

2. Consider the experiments and sample spaces of Exercise 1. List the sample points in the following outcomes.
    (a)   In 1(a), Brand A is judged to be the best-tasting coffee.
    (b)   In 1(a), Brand A is judged to be neither the best- nor the worst-tasting coffee.
    (c)   In 1(c), neither of the students received a score of High Pass.
    (d)   In 1(d), there were at least 2 heads.
    (e)   In 1(f), there were at least 2 heads.
    (f)   In 1(g) one of the coins selected was a penny.
    (g)   In 1(g), Susan selected a nickel and a dime.
    (h)   In 1(h), Susan selected a nickel and a dime.
    (i)   In 1(i), one patient was greatly improved, one moderately improved, and one not at all improved.

3. A certain corporation has 5 sedans and 3 station wagons for company use. The number of cars in use at any moment is unpredictable and therefore constitutes a simple outcome of an experiment. Let (i, j) represent the simple outcome that *i* sedans and *j* station wagons are in use.
    (a)   Draw a diagram showing the points of the sample space.
    (b)   List the sample points in the following outcomes:
       Outcome *A*:   All sedans are in use.

Outcome B: All station wagons are in use.
Outcome C: None of the cars are in use.

(c)   Find an expression in terms of outcomes A, B, and C for each of the following outcomes:

Outcome D: At least one car is in use.
Outcome E: All cars are in use.

(d)   Describe in as few words as possible the outcomes represented by the following sets:

$F = \{(5, 3)\}$
$G = \{(5, 0), (4, 1), (3, 2), (2, 3)\}$
$H = \{(5, 1), (4, 1), (3, 1), (2, 1), (1, 1), (0, 1)\}$

4.   The possible but unpredictable states of 3 devices are "On" or "Off." Let 1 denote "On" and 0 denote "Off." Let $(0, 1, 0)$ denote the simple outcome of "Off," "On," and "Off," for devices 1, 2, and 3, respectively.
(a)   Using these notations, define a sample space for the possible simple outcomes.
(b)   Draw a diagram showing the sample points of the sample space.
(c)   How many sample points are there in the sample space?
(d)   List the sample points in the following outcomes.

Outcome A: Exactly one device is "On."
Outcome B: Exactly two devices are "On."
Outcome C: At least one device is "On."

5.   Consider the outcomes B, C, and D of Example 1.3. Let S be the sample space.
(a)   Does the occurrence of C imply the occurrence of D?
(b)   Does the occurrence of S imply the occurrence of B?
(c)   Does the occurrence of E imply the occurrence of B?

6.   Let A, B, and C be three events of a sample space S. Find the simplest expression for each of the following events:
(a) A did not occur.
(b)   Both A and B occurred.
(c)   None of the events A, B, or C occurred.
(d)   At least one of the events A, B, C occurred.
(e)   At most one of the events A, B, C occurred.

Use set notation to make the following statements:

(f)   If A occurred, then B occurred.
(g)   Events B and C cannot occur simultaneously.
(h)   At least one of the events A, B, or C must occur.

7.   Toss a coin until a head appears and record the number of tosses required. Let $S = \{1, 2, 3, 4, \ldots\}$ be the sample space. Let A be the event that an even number of tosses is required. Let B be the event that fewer than 11 tosses are required. Let C be the event that more than 9 tosses are required.
(a)   List the sample points in each of the events A, B, and C.

(b)   Find expressions for the following events in terms of A, B, and C.

(1) It requires an odd number of tosses.
(2) It requires 10 tosses.
(3) It requires more than 10 tosses.

8. A survey has been conducted to determine the opinion of residents of a certain county concerning a proposal to build a nuclear plant in the area. Let A be the event that, when surveyed, a majority of residents have indicated their opposition to the proposal only if the plant is to be located within 5 miles of their residences. Let B be the event that when surveyed, a majority of residents have indicated their opposition to the proposal no matter where in the county a nuclear plant is to be located. Let C be the event that, when surveyed, a majority of residents have indicated favoring the proposal provided it can be determined that building the plant would reduce the energy costs by 50%. Which, if any, of the events A, B, C, A', B', and C' are certain to have occurred if the survey indicates that

(a)   at most, 50% of residents are presently opposed to the proposal;
(b)   75% of residents are presently opposed to the proposal;
(c)   90% of residents are in favor of the proposal if the plant is to be more than 5 miles away from their residences and if it can be determined that the cost of energy will be reduced by 50%;
(d)   60% of residents are presently in favor of the proposal.

9. Use Venn diagrams to verify the following statements:
(a) $(A \cap B)' = A' \cup B'$
(b) $(A \cap B) \cup (A \cap C) = A \cap (B \cup C)$
(c) $A \cup (B \cap C) = (A \cup B) \cap (A \cup C)$
(d) $A' \cap B' \cap C' = (A \cup B \cup C)'$

10. In which of the following are events A and B mutually exclusive?
(a)   Toss a coin twice. A is the event of a head on the first toss, and B is the event of a head on the second toss.
(b)   Roll two dice. A is the event of a sum of 7. B is the event of a double (i.e., same value on both dice).
(c)   Roll two dice. A is the event of a 2 on at least one of the dice. B is the event of a 3 on one of the dice.
(d)   Draw five cards from a deck of cards. A is the event of drawing at least one spade. B is the event of drawing no aces.

11. In each of the following experiments, determine the size of the appropriate sample space. Is it finite, infinite but discrete (i.e., countably infinite), or nondiscrete?
(a)   Measure the amount of water used by the city of New York on a randomly selected day.
(b)   Randomly select 5 members of the student body to meet with the president of the college.
(c)   Measure the boiling point of a given liquid.
(d)   Continue tossing a coin until heads occurs. Count the required number of tosses.

(e)   Select 9 consumers at random and ask each of them if they find a certain new product pleasing or not pleasing.

(f)   Record the Dow Jones Industrial Averages for the next 10 business days.

(g)   Record the number of days in the coming month when the Dow Jones Industrial Average is above 1,000 points.

1.4   Rules of Operation for Probability of Events

1.4.1   The Axioms of Probability

An **axiom** is a statement that is assumed to be true; whereas a **theorem** is a statement that can be deduced either from axioms or from previously proved theorems. In the axiomatic definition of probability, several simple statements concerning probability are assumed to be true. These statements are called the **axioms of probability**. All other statements concerning probability are deduced from these axioms, and they are called the **theorems of probability**. Such axioms and theorems provide the rules of operation for the mathematics of probability.

When we speak of the probability of an event, we refer to a number between 0 and 1 (inclusive). In other words, we assign numbers called probabilities to sets called events. **Probabilities**, therefore, are values of a set function P that assigns numbers $p = P(A)$ to events A of a sample space S. The axiomatic definition of a probability set function follows.

Definition 1.3   A set function P that assigns numbers $p = P(A)$ to events A of a sample space S is called a **probability set function** or a **probability measure**, if the following axioms are satisfied.

AXIOM 1

$$0 \leq P(A) \leq 1 \qquad \text{for all events } A \text{ of the sample space } S$$

AXIOM 2

$$P(S) = 1$$

AXIOM 3
   (a)   If A and B are mutually exclusive events, then

$$P(A \cup B) = P(A) + P(B)$$

   (b)   If $A_1, A_2, A_3, \ldots$ is a finite or infinite sequence of mutually exclusive events of S, then

$$P(A_1 \cup A_2 \cup A_3 \cup \cdots) = P(A_1) + P(A_2) + P(A_3) + \cdots$$

If P assigns the value $p = P(A)$ to the event A, then we say that p, or $P(A)$, is the probability of the event A.

Observe that the axioms are indeed reasonable assumptions. Regardless of our convictions concerning classical, empirical, and subjective probabilities, we would all agree that:

1. A probability is a number between 0 and 1 (inclusive). This is Axiom 1.
2. In performing an experiment, we assume that the result will be one of the simple outcomes. In other words, we assume that the event S will occur. Therefore it is reasonable to assume that $P(S) = 1$. This is Axiom 2.
3. If $A_1, A_2, A_3, \ldots$ are events that cannot occur simultaneously, then the probability that one among them will occur is the sum of their probabilities. For example, the probability of either a 1 or a 2 in a roll of a die is $1/6 + 1/6 = 2/6$. This is Axiom 3.

1.4.2   The Elementary Theorems

The following intuitively correct statements are theorems derived from the axioms of probability.

Theorem 1.1

$$P(\varnothing) = 0$$

Proof   From 8(e) of Section 1.3.2, we have

$$\varnothing \cup S = S$$

Therefore

$$P(\varnothing \cup S) = P(S) \tag{1}$$

Applying Axiom 2 to the term on the right, we have

$$P(S) = 1 \tag{2}$$

and applying Axiom 3, followed by Axiom 2, to the term on the left in Equation (1), we have

$$P(\varnothing \cup S) = P(\varnothing) + P(S) = P(\varnothing) + 1 \tag{3}$$

Combining the results of Equations (1), (2), and (3), we have

$$P(\varnothing) + 1 = P(\varnothing) + P(S) = P(\varnothing \cup S) = P(S) = 1$$

Therefore

$$P(\varnothing) + 1 = 1$$

and

$$P(\varnothing) = 0$$

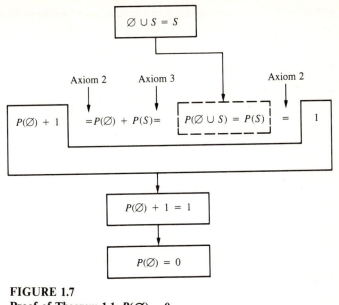

FIGURE 1.7
Proof of Theorem 1.1: $P(\varnothing) = 0$

See Figure 1.7 for a diagram of the flow of equations in the proof of this theorem.

Theorem 1.1 tells us that if we accept the axioms of probability, then we must accept the consequences that $P(\varnothing) = 0$. But this is easy to accept. The event $\varnothing$ is the event that none of the simple outcomes represented by sample points in the sample space occur. We would all agree that the probability of this event should be zero.

Theorem 1.2

$$P(A') = 1 - P(A)$$

Proof   From 8(f) of Section 1.3.2, we have

$$A \cup A' = S$$

Therefore

$$P(A \cup A') = P(S) \tag{1}$$

Applying Axiom 2 to the term $P(S)$, we have

$$P(S) = 1 \tag{2}$$

Events A and A' are mutually exclusive. Therefore, we may apply Axiom 3 to obtain

$$P(A \cup A') = P(A) + P(A') \tag{3}$$

Combining the results of Equations (1), (2), and (3), we have

$$P(A) + P(A') = P(A \cup A') = P(S) = 1$$

Therefore

$$P(A) + P(A') = 1$$

and

$$P(A') = 1 - P(A)$$

This completes the proof of Theorem 1.2. See Figure 1.8 for a diagram of the flow of equations in the proof.

The statement of Theorem 1.2 is that if the probability of occurrence of an event A is $P(A)$, then the probability of nonoccurrence of the event A is $P(A') = 1 - P(A)$. Again, this conforms with our intuitive notion of probability. For instance, if the probability of a sum of 7 in 2 rolls of a die is 1/6, then the probability of a sum that is not 7 is $1 - (1/6)$, or 5/6.

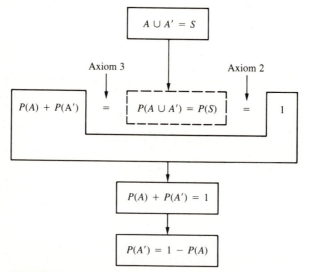

FIGURE 1.8
Proof of Theorem 1.2: $P(A') = 1 - P(A)$

Theorem 1.3

$$P(A \cup B) = P(A) + P(B) - P(A \cap B)$$

Proof   We shall use the abbreviated notation $AB = A \cap B$. The following statements are evident from the Venn diagram of Figure 1.9.

Statement 1

$$A \cup B = AB' \cup AB \cup A'B$$

Therefore

$$P(A \cup B) = P(AB' \cup AB \cup A'B)$$

Statement 2

$$A = AB' \cup AB$$

Therefore

$$P(A) = P(AB' \cup AB)$$

Statement 3

$$B = AB \cup A'B$$

Therefore

$$P(B) = P(AB \cup A'B)$$

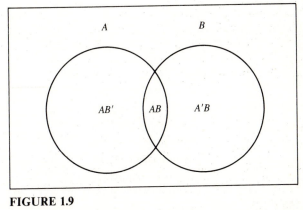

FIGURE 1.9

$A \cup B = AB' \cup AB \cup A'B;$ $A = AB' \cup AB;$ $B = AB \cup A'B.$

Using Statements 1, 2, 3, and Axiom 3, we obtain

(Statement 1)
↓
$$P(A \cup B) = P(AB' \cup AB \cup A'B)$$

(Axiom 3)
↓
$$= P(AB') + P(AB) + P(A'B)$$

(add and subtract $P(AB)$)
↓
$$= P(AB') + P(AB) + P(A'B) + P(AB) - P(AB)$$

(Axiom 3)
↓
$$= P(AB' \cup AB) \quad + P(A'B \cup AB) \quad - P(AB)$$

(Statements 2 and 3)
↓
$$= P(A) \qquad\qquad + P(B) \qquad\qquad - P(AB)$$

Therefore

$$P(A \cup B) = P(A) + P(B) - P(AB)$$

This completes the proof of Theorem 1.3. See Figure 1.10 for a diagram of the flow of equations in the proof.

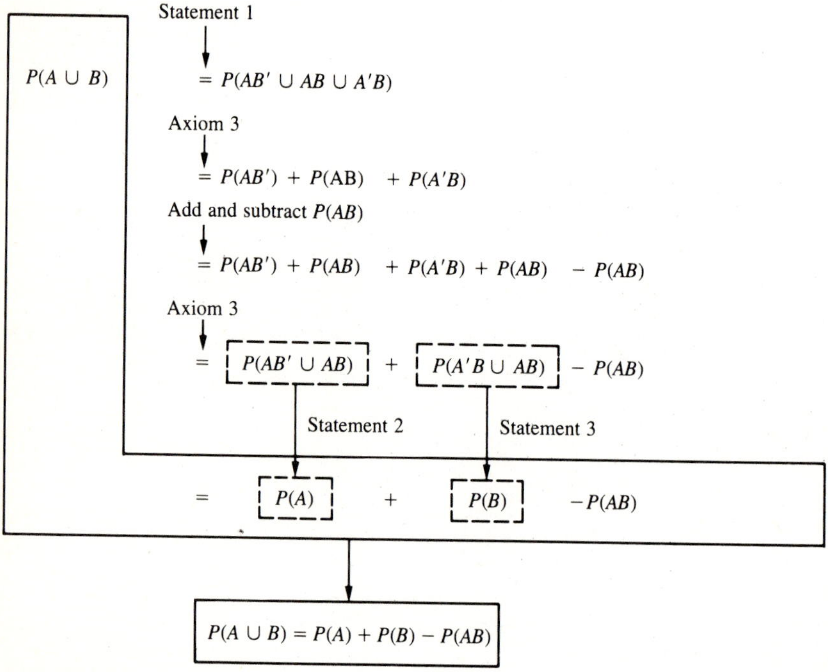

FIGURE 1.10
Proof of Theorem 1.3: $P(A \cup B) = P(A) + P(B) - P(A \cap B)$

EXAMPLE 1.5

Events A, B, and C have probabilities $P(A) = 1/2$, $P(B) = 1/3$, and $P(C) = 1/4$. Furthermore, $A \cap C = \emptyset$, $B \cap C = \emptyset$, and $P(A \cap B) = 1/6$. Use the axioms and theorems of probabilities as well as the facts concerning sets and events to find: **(a)** $P[(A \cap B)']$, **(b)** $P(A \cap B')$, **(c)** $P[(A \cup B)']$, **(d)** $P(A' \cap B')$, and **(e)** $P(A \cup B \cup C)$.

SOLUTION

(a) Applying Theorem 1.2, we have

$$P[(A \cap B)'] = 1 - P(A \cap B) = 1 - (1/6) = 5/6$$

(b) Use the relationship $(A \cap B') \cup (A \cap B) = A$ that was established in the proof of Theorem 1.3. Then

(Axiom 3)

$$1/2 = P(A) = P[(A \cap B') \cup (A \cap B)] \overset{\downarrow}{=} P(A \cap B') + P(A \cap B)$$
$$= P(A \cap B') + (1/6)$$

Therefore

$$1/2 = P(A \cap B') + (1/6)$$

and

$$P(A \cap B') = 2/6 = 1/3$$

(c) Applying Theorem 1.2, we have

$$P[(A \cup B)'] = 1 - P(A \cup B)$$

From Theorem 1.3, we have

$$P(A \cup B) = P(A) + P(B) - P(A \cap B)$$
$$= (1/2) + (1/3) - (1/6) = 4/6 = 2/3$$

Therefore

$$P[(A \cup B)'] = 1 - (2/3) = 1/3$$

(d) The first of de Morgan's laws (Section 1.3.2, 8(g)) states that

$$A' \cap B' = (A \cup B)'$$

Therefore

$$P(A' \cap B') = P[(A \cup B)'] = 1/3$$

(e)   Since $A \cap C = \varnothing$ and $B \cap C = \varnothing$, it follows that $(A \cup B) \cap C = \varnothing$. In other words, the events $(A \cup B)$ and C are mutually exclusive events. Therefore, Axiom 3 and the probabilities $P(A \cup B) = 2/3$ (see (c)) and $P(C) = 1/4$ give us the following results:

$$P(A \cup B \cup C) = P[(A \cup B) \cup C] = P(A \cup B) + P(C)$$
$$= 2/3 + 1/4 = 11/12$$

Theorem 1.4   *If the sample space S consists of a finite number N of sample points, and if each subset consisting of a single sample point is an event with probability $1/N$, then for any event A in S*

$$P(A) = N_A/N$$

where N_A is the number of sample points in the event A.

Proof   Let $s_1, s_2, \ldots, s_{N_A}$ be the sample points in A, and let E_i be the event consisting of the single sample point s_i, $i = 1, 2, \ldots, N_A$. The events $E_1, E_2, \ldots, E_{N_A}$ are clearly mutually exclusive and $A = E_1 \cup E_2 \cup \cdots \cup E_{N_A}$. Therefore

$$P(A) = P(E_1 \cup E_2 \cup \cdots \cup E_{N_A})$$
$$\text{(Axiom 3)}$$
$$\downarrow$$
$$= P(E_1) + P(E_2) + \cdots + P(E_{N_A})$$
$$= (1/N) + (1/N) + \cdots + (1/N)$$
$$= N_A/N$$

In conclusion

$$P(A) = N_A/N$$

This completes the proof of Theorem 1.4.

This theorem tells us that if we are dealing with a finite sample space with equally probable sample points, then the probability of any event A may be computed by simply counting the number of sample points in A and dividing that number by the number of points in the sample space. This coincides exactly with the classical definition of probability. In other words, the axiomatic definition of probability is compatible with the classical definition when we are dealing with finite sample spaces with equally probable sample points.

EXAMPLE 1.6

In rolling 2 dice we assume that all 36 outcomes are equally probable.

Consider the events $A, B, C, D,$ and E of Example 1.3 and let F be the event of "a sum of 4." Find: $P(A)$, $P(A')$, $P(D \cap E)$, $P(C \cup D)$, and $P(D \cap F)$.

   SOLUTION

There are 36 points in the sample space. Therefore, $N = 36$. By simply counting the number of sample points in each of the sets $A, A', D \cap E, C \cup D$, and $D \cap F$, we obtain $N_A = 6$, $N_{A'} = 30$, $N_{D \cap E} = 1$, $N_{C \cup D} = N_D = 6$, and $N_{D \cap F} = 1$. Therefore

$$P(A) = N_A/N = 6/36 = 1/6$$
$$P(A') = N_{A'}/N = 30/36 = 5/6$$
$$P(D \cap E) = N_{D \cap E}/N = 1/36$$
$$P(C \cup D) = N_{C \cup D}/N = N_D/N = 6/36 = 1/6$$
$$P(D \cap F) = N_{D \cap F}/N = 1/36$$

1.4.3   A List of the Axioms and Theorems

The axioms of Section 1.4.1 and the theorems of Section 1.4.2 provide the basic rules of operation for the mathematics of probability. They are listed here for easy reference.

   Axiom 1: $0 \le P(A) \le 1$ for all events A of the sample space S

   Axiom 2: $P(S) = 1$

   Axiom 3:   If $A_1, A_2, A_3, \ldots$ is a finite or infinite sequence of mutually exclusive events, then $P(A_1 \cup A_2 \cup \cdots) = P(A_1) + P(A_2) + \cdots$.

   Theorem 1.1: $P(\emptyset) = 0$

   Theorem 1.2: $P(A') = 1 - P(A)$

   Theorem 1.3: $P(A \cup B) = P(A) + P(B) - P(A \cap B)$

   Theorem 1.4:   If S consists of a finite number of equally probable sample points, then for any event A in S, $P(A) = N_A/N$ where N_A is the number of sample points in A, and N is the number of sample points in S.

1.4.4   The Assignment of Probabilities to Key Events

Although the axioms and theorems of probability provide the operating rules for the mathematics of probability, they cannot assign numerical values to probability of events, other than $P(\emptyset) = 0$ or $P(S) = 1$, unless such events are in a mathematical equation involving other events whose probabilities are known. For instance, in Example 1.5 the probabilities of events A, B, and C were given. We then applied the axioms and theorems to these numerical values to find the probabilities of events $(A \cap B)'$, $A \cap B'$, $(A \cup B)'$, $A' \cap B'$, and $A \cup B \cup C$. In Example 1.6 we assumed a finite sample space with equally probable sample points. This is equivalent to an assignment of numerical values to probabilities of events consisting of single sample points. The probabilities of more-complex events are then obtained through application of Theorem 1.4. In other words, if probabilities of events are to be computed by using the axioms and theorems of probability, there must be some initial assignment of numerical values to probabilities of key events of the sample space. How do we make such initial assignments? In some instances, it can be done through objective evaluations that result in a classical definition and assignment of probabilities. Such was the case for Example 1.5. In other instances, it can be done through the use of empirical data that result in an empirical definition and assignment of probabilities. Still other instances may require a personal evaluation of events leading to a subjective definition and assignment of probabilities. As long as the probabilities assigned are consistent with the axioms of probability, all the theorems that follow from the axioms may be utilized to solve the more complex problems of probability.

EXAMPLE 1.7

Helen is quite competent in biology and calculus, but she feels that she has no chance of getting A's in any of her other courses. With respect to biology and calculus, however, she feels that the odds against getting an A in biology are 1 to 3, and the odds against getting an A in calculus are 1 to 4. She feels that the odds against getting A's in both courses are 2 to 3. Assuming that Helen's evaluation of the situation is correct, how would you answer the following questions?
   (a)   What is the probability that Helen will get at least one A?
   (b)   What is the probability that she will get no A's?
   (c)   What is the probability that she will get only one A?

   SOLUTION

The subjective probabilities assigned by Helen to event B, that she will receive an A in biology, to the event C, that she will receive an A in calculus, and to the event BC, that she will receive A's in both subjects, are

$$P(B) = 3/4, \qquad P(C) = 4/5, \quad \text{and} \quad P(BC) = 3/5$$

The solutions to (a), (b), and (c) may now be obtained by applying the axioms and theorems of probability to these numerical values.
   (a)   The event of at least one A is the event $B \cup C$. Therefore

$$P(\text{at least one A}) = P(B \cup C)$$

(Theorem 1.3)
↓
$$= P(B) + P(C) - P(BC) = (3/4) + (4/5) - (3/5)$$
$$= 19/20 = .95$$

(b)
$$P(\text{no A's}) = 1 - P(\text{at least one A}) = 1 - .95 = .05$$

(c)
$$P(\text{only one A}) = P(\text{at least one A}) - P(\text{A's in both courses})$$
$$= P(B \cup C) - P(BC) = .95 - .6 = .35$$

EXAMPLE 1.8

A recent survey at a museum showed that, of one million visitors to the museum, 800,000 visited the European galleries and 500,000 visited the Egyptian galleries. Use these empirical data to estimate the probabilities that a visitor to the museum will visit (a) the European galleries, (b) the Egyptian galleries, or (c) both galleries. Use these probabilities and the applicable axioms and theorems to find the probability that a visitor will visit at least one of these two galleries.

SOLUTION

Let A and B be the events of a visit to the European galleries and a visit to the Egyptian galleries, respectively. Let N, N_A, N_B, and N_{AB} be the number of visitors to the museum, the European galleries, the Egyptian galleries, and both galleries, respectively. Under the empirical definition of probability, $P(A) = N_A/N$, $P(B) = N_B/N$, and $P(AB) = N_{AB}/N$. Now, an analysis of the data shows that $300,000 \leq N_{AB} \leq 500,000$.

Therefore

$$\frac{300,000}{1,000,000} \leq \frac{N_{AB}}{N} \leq \frac{500,000}{1,000,000}$$

Consequently

$$P(A) = .8, \qquad P(B) = .5, \quad \text{and} \quad .3 \leq P(AB) \leq .5$$

The probability that a visitor will visit at least one of the two galleries is given by the expression $P(A \cup B)$. From Theorem 1.3 we have $P(A \cup B) = P(A) + P(B) - P(AB)$. Therefore

$$.8 \leq P(A \cup B) \leq 1$$

EXAMPLE 1.9

Mr. Thomas feels the probability of the Eagles' winning their first game to be 2/3, of winning their second game 3/4, and of winning both games 1/3. Is this set of probabilities consistent with the axioms?

SOLUTION

Let A be the event of an Eagles first-game win. Let B be the event of a second-game win. Mr. Thomas assigned the following probabilities:

$$P(A) = 2/3, \qquad P(B) = 3/4, \qquad P(AB) = 1/3$$

Applying Theorem 1.3 we conclude that

$$P(A \cup B) = P(A) + P(B) - P(AB) = (2/3) + (3/4) - (1/3) = 1.08$$

The last statement is not consistent with Axiom 1 which requires that probabilities of events be numbered between 0 and 1. Therefore, the set of probabilities is not consistent with the axioms.

1.4.5 The Probabilities Zero and One

The probabilities zero and one deserve a special note because they are often subjects of misinterpretation. Within the context of an experiment, if A is an event that is certain to occur, then $P(A) = 1$. By the same token, if B is an event that cannot occur, then $P(B) = 0$. Exercise 25 provides an elaboration on these two statements. However, if events C and E have probabilities $P(C) = 1$ and $P(E) = 0$, it does not mean that event C is certain to occur and it does not mean that event E cannot occur. An example will clarify this point. Consider a hypothetical experiment in which a number is selected at random from the unit interval $[0, 1]$ (i.e., the set of all real numbers from 0 to 1, including 0 and 1). Clearly, the probability of obtaining a number between 0 and 1 (inclusive of 0 and 1) is 1. The probability of obtaining a number between $\frac{1}{4}$ and $\frac{3}{4}$ is $1/2$, since the length of the interval from $\frac{1}{4}$ to $\frac{3}{4}$ is one-half the length from 0 to 1. The probability of obtaining a number between $\frac{3}{8}$ and $\frac{5}{8}$ is $1/4$, since the length of the interval from $\frac{3}{8}$ to $\frac{5}{8}$ is one-fourth the length from 0 to 1. The probability of obtaining a number between $\frac{7}{16}$ and $\frac{9}{16}$ is $1/8$. Continuing in this manner, the argument forces us to conclude that the probability of obtaining the number $\frac{1}{2}$ is zero. Now, let E be the event of obtaining the number $\frac{1}{2}$. We have $P(E) = 0$. But surely event E can occur, because the number $\frac{1}{2}$ is one of the numbers in the interval $[0, 1]$. Thus, a probability of zero does not imply an impossible event. On the other hand, suppose we let $C = E'$. Then

$$P(C) = P(E') = 1 - P(E) = 1$$

Now, suppose that in performing the experiment we had selected the number $\frac{1}{2}$. Selection of the number $\frac{1}{2}$ is occurrence of event E. Event C is the complement of E. Thus, selection of the number $\frac{1}{2}$ implies that C does not occur. Hence, C is an event with probability one, but it is not an event that is certain to occur. In the language of probability and mathematical statistics, we say that if $P(C) = 1$, then event C is almost sure to occur. We cannot say that it is sure to occur.

Exercises

1. Let A and B be mutually exclusive events with $P(A) = .35$, and $P(B) = .15$. Find (a) $P(A \cup B)$; (b) $P(A')$; (c) $P(A \cap B)$; (d) $P(A' \cup B')$.

2. Let A and B be events with $P(A) = .25$, $P(B) = .40$, and $P(A \cap B) = .15$. Find (a) $P(A')$; (b) $P(A \cup B)$; (c) $P(A' \cap B')$; (d) $P(A \cap B')$; (e) $P(A' \cap B)$.

3. The probability that a person stopping at a certain bank will cash a check is .25, the probability that he or she will make a deposit into a savings

account is .39, and the probability that he or she will do both is .09. What is the probability that a person stopping at this bank will
(a)   do at least one of the two things mentioned above;
(b)   neither cash a check nor make a deposit into a savings account;
(c)   cash a check but not make a deposit into a savings account;
(d)   cash a check or make a deposit into a savings account, but not both;
(e)   not cash a check.

4.   A box of 10 balls contains 2 red, 3 blue, and 5 green balls. If a ball is drawn at random from the box in such a manner that each ball has equal probability of being drawn, and if R, B, and G are the events of drawing a red, blue, and green ball, respectively, then what is (a) $P(R)$; (b) $P(B')$; (c) $P(R \cup B)$; (d) $P(R \cup B \cup G)$; (e) $P(B \cap G)$; (f) $P(R' \cap B')$; (g) $P(B' \cup G')$.

5.   The probability that a student will miss the first meeting of a class is .15, the probability that he will miss the second meeting is .10, and the probability that he will miss both meetings is .05.
(a)   What is the probability that a student will miss at least one of the first two meetings of the class?
(b)   What is the probability that a student will attend both the first and second meetings of the class?
(c)   What is the probability that a student will miss exactly one of the first two meetings of the class?

6.   A sample space S consists of 25 equally probable sample points. If $N_A = 10$, $N_B = 15$, $N_C = 8$, $N_{AB} = 5$, $N_{AC} = 3$, $N_{BC} = 5$, $N_{ABC} = 1$, find:

$$P(A), \qquad P(AB'), \qquad P(A'B), \qquad P(A'B'), \qquad P(A \cup B \cup C), \qquad P(A')$$

(*Hint:*   draw a Venn diagram and count sample points.)

7.   A certain restaurant has two cooks—the chief cook and his assistant On any given day, the probability that the chief cook will show up for work is .98, the probability that the assistant cook will show up for work is .97, and the probability that at least one of the two cooks will show up for work is .99. Find the probabilities that on any given day the restaurant will have
(a)   both cooks;   (d)   only the chief cook;
(b)   neither of the two cooks;   (e)   only the assistant cook.
(c)   only one of the two cooks;

8.   A card is drawn at random from a standard deck of 52 playing cards so that each card has an equal probability of being drawn. Find the probability of drawing
(a)   a black queen;
(b)   a 7, 8, or 9;
(c)   a red card;
(d)   a black ace or a red queen;
(e)   a card that is neither red nor a queen;
(f)   a card that is neither a spade nor a king;
(g)   a card that is not black.

9. Which of the sample spaces of Exercise 1, Section 1.3, would you consider to be sample spaces of equally probable sample points?

10. If each sample point of the sample spaces of Exercise 2(d) and Exercise 2(g) of Section 1.3 is equally probable, what is the probability of each of the events described in those exercises?

11. If each sample point of the sample space of Exercise 3 of Section 1.3 is equally probable, what is the probability of each of the events A, B, C, D, E, F, G, and H?

12. If each sample point of the sample space of Exercise 4 of Section 1.3 is equally probable, what is the probability of each of the events $A, B,$ and C?

13. Let $A, B,$ and C be events. Prove that

$$P(A \cup B \cup C) = P(A) + P(B) + P(C) - P(A \cap B) - P(A \cap C)$$
$$- P(B \cap C) + P(A \cap B \cap C)$$

(*Hint:* let $D = B \cup C$. Apply Theorem 1.3.)

14. Of 10,000 automobiles manufactured by a certain company, 25 have faulty transmissions, 20 have faulty steering wheels, 30 have faulty windshield wipers, 9 have faulty transmissions and steering wheels, 8 have faulty transmissions and windshield wipers, 7 have faulty windshield wipers and steering wheels, and 5 have faulty transmissions, steering wheels, and windshield wipers. What are the probabilities that an automobile randomly selected from these 10,000 will have
    (a) a faulty transmission;
    (b) a faulty windshield wiper;
    (c) a faulty transmission and a faulty steering wheel;
    (d) a faulty transmission, a faulty steering wheel, and faulty windshield wipers;
    (e) neither a faulty transmission nor faulty windshield wipers;
    (f) neither faulty windshield wipers nor a faulty steering wheel;
    (g) faulty windshield wipers but not a faulty transmission;
    (h) at least one of the three parts mentioned above in perfect working condition;
    (i) all of the three parts mentioned above in perfect working condition;
    (j) exactly one of the three parts mentioned above in perfect working condition;
    (k) no more than one of the three parts mentioned above in perfect working condition;
    (l) exactly two of the three parts mentioned above in perfect working condition.

15. It can be shown that if S is a finite sample space consisting of the mutually exclusive outcomes $A_1, A_2, \ldots, A_N$, then P is a probability set function which is consistent with the axioms of probability if $0 \leq P(A_i) \leq 1$ for each $i = 1, 2, \ldots, N$, and $P(A_1) + P(A_2) + \cdots + P(A_N) = 1$. Let $\mathscr{E}$ be an experiment with exactly 5 mutually exclusive outcomes

A, *B*, *C*, *D*, and *E*. Check whether the following assignments of probability are permissible.

(a) $P(A) = .28$, $P(B) = .14$, $P(C) = .13$,
       $P(D) = .31$, $P(E) = .14$

(b) $P(A) = .25$, $P(B) = -.05$, $P(C) = .55$,
       $P(D) = .10$, $P(E) = .15$

(c) $P(A) = .20$, $P(B) = .14$, $P(C) = .13$,
       $P(D) = .21$, $P(E) = .12$

(d) $P(A) = 1/2$, $P(B) = 1/4$, $P(C) = 1/4$,
       $P(D) = 1/3$, $P(E) = 1/5$

16.   If Allen is willing to bet $4 against $1, but not $5 against $1, that the school team will win the next basketball game, what does this tell us about Allen's subjective probability concerning the school team's ability to win the next basketball game?

17.   A recent survey at a fast-food restaurant showed that, of one million customers, 800,000 ordered hamburgers and 300,000 ordered milk shakes. Use these empirical data to estimate the probability that a customer will order both hamburgers and milk shakes.

18.   A project leader speculates that projects A and B will be completed on schedule with probabilities .75 and .90, respectively. Furthermore, she speculates that there is a probability of .99 that at least one of these two projects will be completed on schedule. If these subjective probabilities are correct, what are the probabilities that
(a)   both projects will be completed on schedule;
(b)   only one of the projects will be completed on schedule;
(c)   neither project will be completed on schedule.

19.   Addition of a certain nutrient to the soil produced an increase in foliage for 35% of the plants, an increase in height for 25% of the plants, and an increase in both foliage and height for 20% of the plants. Use these empirical data to estimate the probabilities that addition of this nutrient to the soil will result in
(a)   an increase in at least one of the measurements of foliage and height for plants of this type;
(b)   neither an increase in foliage nor an increase in height;
(c)   no increase in height;
(d)   an increase in height but no increase in foliage.

20.   Jason's parents have promised him a trip to Paris if he receives A's in both chemistry and English and a trip to Canada if he receives only one A. Jason feels that the odds against his getting an A in chemistry are 1 to 3, the odds against his getting an A in English 2 to 3, and the odds against his getting A's in both subjects 1 to 1. Translating these odds into subjective probabilities, what is the probability that
(a)   Jason will get a trip to Paris?
(b)   Jason will get a trip to Paris or to Canada?
(c)   Jason will get a trip to Canada?

21. Prove that if $A \subset B$, then $P(A) \leq P(B)$ and $P(A'B) = P(B) - P(A)$.
    (*Hint:*   draw a Venn diagram of A and B to see that if $A \subset B$, then $B = A \cup A'B$. Events A and $A'B$ are mutually exclusive.)

22. Mrs. Smith predicts that if it rains tomorrow it is bound to rain the day after tomorrow. She also claims that the chance of rain tomorrow is 1/2 and that the chance of rain the day after tomorrow is 1/3. Are the subjective probabilities defined by Mrs. Smith consistent with the axioms and theorems of probability?

23. Prove that Theorem 1.1 is really a corollary of Theorem 1.2. (i.e., that Theorem 1.1 is a consequence of Theorem 1.2).

24. Explain why each of the following assignments of probabilities is inconsistent with the axioms and theorems of probability.
    (a)   John feels that the probability of his getting an A in physics is .3, and the probability of his not getting an A in physics is .4
    (b)   Sarah is elated about her prospects for attending the prom. She feels there is a probability of .8 that Jim will ask her, .9 that Mark will ask her, and .5 that both will.
    (c)   The Yankees are scheduled to play 2 games with the Royals. Steve feels that the probability is .5 that the Yankees will win only one game, and .6 that they will win both games.
    (d)   Marilyn feels that if inflation exceeds 10% this year, it will exceed 15% next year. Furthermore, she feels that the probability of inflation exceeding 10% this year is .6, and that the probability of inflation exceeding 15% next year is .5.

25. Let $\mathscr{E}$ be an experiment. Let S be a sample space for $\mathscr{E}$.
    (a)   Show that if A is an event that is certain to occur, then $A = S$. Consequently, $P(A) = 1$.
    (b)   Show that if B is an event that cannot occur, then $B = \varnothing$. Consequently, $P(B) = 0$.

26. Which of the following are true statements?
    (a)   If $P(A \cup B \cup C) = 1$, then events A, B, and C are mutually exclusive events.
    (b)   If $P(A) + P(B) + P(C) = 1$, then events A, B, and C are mutually exclusive events.
    (c)   If the occurrence of event A implies the occurrence of event B, then $P(A) \geq P(B)$.

References

1. Fisher, R. A. 1921. On the mathematical foundations of theoretical statistics. *PTRS* 222:309.
2. Jeffreys, H. 1939. *Theory of probability*. Oxford: Clarendon.
3. Keynes, J. M. 1921. *A treatise on probability*. London: Macmillan and Co., Ltd.
4. Kolmogorov, A. N. 1933. *Grundbegriffe der Wahrscheinlichkeitsrechnung*. Berlin.

5. Laplace, P. S. de. 1956. *Foundations of probability theory*. 2d English ed., translated from German. New York: Chelsea Publishing Co.
6. Laplace, P. S. de. 1951. *A philosophical essay on probabilities* (translation with an introduction by E. T. Bell). New York: Dover Publications, Inc.
7. Mises, R. von. 1941. On the foundation of probability and statistics. *AMS* 12:191.
8. Ore, Oystein. 1960. Pascal and the invention of probability theory. *American Mathematical Monthly* 67(5):409–19.

Chapter **2**

COMBINATORIAL METHODS

A Basic Counting Procedure
Permutations
Combinations: The Binomial Coefficient
The Multinomial Coefficient
Tree Diagrams

This chapter is concerned with the description of methods for counting sample points in events and sample spaces and with their application to problems involving sample spaces with equally probable simple outcomes. By using the formula $P(A) = N_A/N$ from Theorem 1.4, we can compute the probability of an event A once we have counted the number N_A of sample points in A and the number N of sample points in the sample space S.

2.1   A Basic Counting Procedure

The following basic counting procedure (also called the multiplication principle) can be described in terms of the process of selecting one element from each of k sets.

Let us begin with A and B, two sets with 3 and 4 elements, respectively.

$$A = \{a_1, a_2, a_3\}$$
$$B = \{b_1, b_2, b_3, b_4\}$$

Suppose we were interested in counting the distinct pairs (a_i, b_j) of elements that can be formed by selecting first an element from A and then an element from B. There are 3 ways to select an element from A. For each of them there are 4 ways to select an element from B. Hence there are $3 \times 4 = 12$ distinct pairs (a_i, b_j) of elements selected, each pair containing one element from A and one from B. See Figure 2.1 for a visual representation of these pairs of elements. (See also Figure 2.5; the tree diagram is another method for counting the distinct pairs of elements.)

Let us generalize to k sets, $A_1, A_2, \ldots, A_k$; with $n_1, n_2, \ldots, n_k$ elements, respectively. Let us count the distinct ways of selecting first an element from A_1, then an element from $A_2, \ldots,$ and finally an element from A_k. By using the basic counting procedure employed in the previous example, we note that the set A_1 contains n_1 elements. Hence there are n_1 ways of selecting an element from A_1. For each way of selecting an element from A_1 there are n_2 ways of selecting an element from A_2. Therefore there are $n_1 \cdot n_2$ ways of selecting first an element from A_1 and then an element from A_2. For each of these $n_1 \cdot n_2$ ways of selecting elements from A_1 and A_2 there are n_3 ways to select an element from A_3. Hence there are $n_1 \cdot n_2 \cdot n_3$ ways to select the elements from A_1, A_2, and A_3. Continuing in this manner, we conclude that there are $n_1 \cdot n_2 \cdots n_k$ ways of selecting 1 element from each of $A_1, A_2, \ldots, A_k$, respectively. This result is stated in Theorem 2.1.

Theorem 2.1   *Let $A_1, A_2, \ldots, A_k$ be sets consisting of $n_1, n_2, \ldots, n_k$ elements, respectively. There are $n_1 \cdot n_2 \cdots n_k$ ways to select first an element from A_1, then an element from A_2, then an element from $A_3, \ldots,$ and finally an element from A_k.*

Elements of A

	a_1	a_2	a_3
b_1	(a_1, b_1)	(a_2, b_1)	(a_3, b_1)
b_2	(a_1, b_2)	(a_2, b_2)	(a_3, b_2)
b_3	(a_1, b_3)	(a_2, b_3)	(a_3, b_3)
b_4	(a_1, b_4)	(a_2, b_4)	(a_3, b_4)

Elements of B

FIGURE 2.1

EXAMPLE 2.1

At a certain Chinese restaurant a fixed-price dinner for two allows a choice of one dish from column A and one from column B.

(a) If there are 6 items in column A and 8 in column B, what is the number of combinations of dishes possible under this dining plan?

(b) If there are 3 seafood dishes in column A and 4 seafood dishes in column B, what is the probability that a random selection of dishes will result in a combination of 2 seafood dishes?

SOLUTION

(a) There are 6 ways of selecting an item from column A. For each of them there are 8 ways of selecting an item from column B. In total, there are $6 \cdot 8 = 48$ combinations of 2 dishes. This is an application of Theorem 2.1.

(b) For this problem the sample space S is the set of all combinations of 2 dishes, chosen 1 from A and 1 from B. From (a), we have

$$\text{(the number of sample points in } S) = N = 48$$

Let E be the event of a combination of 2 seafood dishes. There are 3 seafood dishes in A and 4 in B. Therefore the number of ways to choose 1 seafood dish from A and 1 from B is $3 \cdot 4 = 12$. In other words, the number of sample points in event E is

$$N_E = 12$$

Applying Theorem 1.4, we have

$$P(E) = N_E / N = 12/48 = 1/4$$

EXAMPLE 2.2

Mr. Levine has 5 jackets, 10 shirts, 6 ties, and 7 pairs of slacks.

(a) How many different outfits of a jacket, a shirt, a tie, and a pair of slacks are possible for him?

(b)   If 2 jackets, 3 shirts, 2 ties, and 2 pairs of slacks are blue, what is the probability that a random selection will result in an all-blue outfit?

SOLUTION

(a)   An outfit is a combination of 1 each from 5 sets consisting of 5, 10, 6, and 7 items, respectively. According to Theorem 2.1, there are $5 \cdot 10 \cdot 6 \cdot 7 = 2{,}100$ different possible outfits.

(b)   There are 2 blue jackets, 3 blue shirts, 2 blue ties, and 2 pairs of blue slacks. Therefore, there are $2 \cdot 3 \cdot 2 \cdot 2 = 24$ different all-blue outfits.

The sample space for this problem is the set of 2,100 outfits of (a). Since an outfit is being selected at random, we can assume that each of the sample points in the sample space has equal probability. Thus we may apply Theorem 1.4 to compute $P(B)$ where B is the event of a completely blue outfit. The answer is

$$P(B) = N_B/N = 24/2{,}100 = .0114$$

EXAMPLE 2.3

Count the possible outcomes of **(a)** 5 flips of a coin; **(b)** 4 throws of a die.

SOLUTION

(a)   Coins can land either heads or tails. In other words, there are two ways for a coin to land in each flip of the coin. Therefore, in 5 flips of a coin there are $2 \cdot 2 \cdot 2 \cdot 2 \cdot 2 = 2^5 = 32$ possible outcomes.

(b)   Dice can land in any one of six ways. Therefore, in 4 throws there are $6 \cdot 6 \cdot 6 \cdot 6 = 6^4 = 1{,}296$ possible outcomes.

Exercises

For the exercises in this section assume that sample spaces consist of equally probable sample points so that the formula $P(A) = N_A/N$ of Theorem 1.4 is applicable.

1.   Define a sample space and count the number of sample points in the sample space for each of the following experiments.
     (a)   Roll 6 dice and record the value on each die.
     (b)   Flip a coin 20 times and record the "heads" or "tails" result of each flip.
     (c)   Answer 15 true-or-false questions by flipping a coin. Mark "true" if heads and "false" if tails.
     (d)   Ten patients with the same disease are to be examined after one week on a certain medical program. Each patient is to be recorded as showing "improvement" or "no improvement."
     (e)   Four abandoned buildings are to be demolished, renovated, or left as a slum.

2. Consider experiments (a), (b), (c), (d), and (e) of Exercise 1.
    (a)   Find the probability of getting the same value on each of the 6 dice in experiment (a).
    (b)   Find the probability of getting either all heads or all tails in experiment (b).
    (c)   Find the probability of getting only correct answers in experiment (c).
    (d)   Find the probability that after the medical program, none of the patients in experiment (d) showed improvement.
    (e)   Find the probability of renovating all 4 buildings in experiment (e).

3. Again, consider experiments (a), (b), (c), (d), and (e) of Exercise 1.
    (a)   Find the probability of getting a 6 on exactly one of the dice in experiment (a).
    (b)   Find the probability of getting only even numbers on each of the 6 dice in experiment (a).
    (c)   Find the probability of getting a sum of 7 or less in experiment (a).
    (d)   Find the probability of getting exactly 1 head in experiment (b).
    (e)   Find the probability of getting only correct answers or only incorrect answers in experiment (c).
    (f)   Find the probability of getting only one correct answer in experiment (c).
    (g)   Find the probability of at least one patient showing improvement in experiment (d).
    (h)   Find the probability of either demolishing or renovating each of the four buildings in experiment (e).

4. Define a sample space and count the number of sample points in the sample space for each of the following experiments:
    (a)   Roll a die and toss a coin 6 times. Record the value on the die and the face on each toss of the coin.
    (b)   Select at random 1 physics book, 1 math book, and 1 English book from a collection of 3 physics, 5 math, and 7 English books.
    (c)   Select at random 1 income tax return from each of the groups A, B, and C. There are 100 returns in group A, 200 in group B, and 150 in group C.
    (d)   Select at random 1 calculator from each of shipments A, B, and C containing 100, 150, and 125 calculators, respectively.

5. Consider experiments (a), (b), (c), and (d) of Exercise 4.
    (a)   Find the probability of getting a 1 on the die and exactly 1 head in the 6 tosses of the coin in experiment (a).
    (b)   If 2 of the physics, 3 of the math, and 6 of the English books in experiment (b) are at the college level, what is the probability of selecting only books at the college level?
    (c)   If 30%, 20%, and 10% of the returns in groups A, B, and C, respectively, are fraudulent, what is the probability of selecting only fraudulent returns in experiment (c)?
    (d)   If 2, 1, and 3 calculators in shipments A, B, and C, respectively, are damaged, what is the probability of selecting at least one damaged calculator in experiment (d)?

2.2   Permutations

A permutation of r elements is an ordered set of r elements. There is a first element, a second element, ..., and an rth element. The usual notation for a permutation of r elements is a list of the elements in the proper order enclosed by parentheses. For example, $(1, 3, 5)$ is a permutation of the elements 1, 3, and 5 with the understanding that 1 is the first element, 3 is the second element, and 5 is the third element. The permutation $(1, 3, 5)$ is different from the permutation $(3, 1, 5)$ because the arrangement of the elements is different.

We wish to develop a formula for counting the distinct permutations of r elements from a set of n elements. Let us consider first a case where $r = 3$ and $n = 4$.

EXAMPLE 2.4

(a) List the permutations of 3 elements from the set $\{1, 2, 3, 4\}$ of 4 elements.

(b) Count the number of permutations of 3 elements from a set of 4 elements

SOLUTION

(a) The distinct permutations are given in Figure 2.2.

(b) By counting the number of permutations in Figure 2.2 we see that they total 24. We could have arrived at this number by using the basic counting procedure of Section 2.1. Following the basic procedure, we would observe that we began with four elements in the set; hence there are four ways to select the first element in the permutation. After this is done, three elements remain in the set; hence there are three ways to select the second element in the permutation. Therefore, there are $4 \cdot 3 = 12$ ways to select the first and second elements in the permutation. Finally, two elements remain in the set; hence there are two ways to select the third element in the permutation. This means that in total there are $4 \cdot 3 \cdot 2 = 24$ ways to form a permutation of three elements from the set of four elements.

$(1, 2, 3)$	$(1, 3, 4)$	$(1, 2, 4)$	$(2, 3, 4)$
$(1, 3, 2)$	$(1, 4, 3)$	$(1, 4, 2)$	$(2, 4, 3)$
$(2, 1, 3)$	$(3, 1, 4)$	$(2, 1, 4)$	$(3, 2, 4)$
$(2, 3, 1)$	$(3, 4, 1)$	$(2, 4, 1)$	$(3, 4, 2)$
$(3, 1, 2)$	$(4, 1, 3)$	$(4, 1, 2)$	$(4, 2, 3)$
$(3, 2, 1)$	$(4, 3, 1)$	$(4, 2, 1)$	$(4, 3, 2)$

The distinct permutations of 3 elements from the set $\{1, 2, 3, 4\}$ of 4 elements.

FIGURE 2.2

Consider now the case of permutations of r elements from a set of n elements, with $r \leq n$. Again we employ the basic counting procedure. There are n ways to select an element from the set to be the first element in the permutation. Once we have made the selection, there remain $n - 1$ elements, so that for each way of selecting the first element there are $n - 1$ ways to select the second element. Hence there are $n(n - 1)$ ways to select the first and second elements. There remain $n - 2$ elements for the third choice. Thus, there are $n(n - 1)(n - 2)$ ways to select the first, second, and third elements. Continuing in this manner, we conclude that there are $n(n - 1)(n - 2)\cdots(n - r + 1)$ permutations of r elements from a set of n elements. (*Note:* in case the reader wonders why the rth (and last) factor in the expression $n(n - 1)(n - 2)\cdots(n - r + 1)$ is $(n - r + 1)$, consider the following line of thought. The **1**st factor in the expression is $n = n - 1 + 1$, the **2**nd factor is $n - 1 = n - 2 + 1$, the **3**rd factor is $n - 2 = n - 3 + 1, \ldots,$ and so on. Therefore, the rth (and last) factor is $n - r + 1$.) The notation for the number of permutations of r elements from a set of n elements is P_r^n. We have verified the following theorem and corollary.

Notation and Definition P_r^n denotes the number of distinct permutations or r elements that can be obtained from a set of n elements. P_r^r denotes the number of distinct permutations of r elements.

Theorem 2.2

$$P_r^n = n(n - 1)(n - 2)\cdots(n - r + 1)$$

for positive integers r and n with $r \leq n$.

Corollary 2.1

$$P_r^r = r(r - 1)(r - 2)\cdots 1 = r!$$

for positive integers r.

Notation and Definition   For positive integers n

$$n! = n(n - 1)(n - 2)\cdots 1$$
$$0! = 1$$

The symbol $n!$ is pronounced "n factorial" and represents the quantity $n(n - 1)(n - 2)\cdots 1$.

EXAMPLE 2.5

A club has 100 members. A president, a vice president, a secretary, and a treasurer are to be selected from the membership. All members are eligible for the offices, but each member can hold at most one office.

(a)   How many distinct choices of officers are there?

(b)   If the officers are selected at random from the membership, what is the probability that Mr. Lee, Ms. Jones, Mr. Williams, and Ms. Schwartz will be selected as president, vice president, secretary, and treasurer, respectively?

(c)   What is the probability that these 4 members will hold offices?

SOLUTION

(a)   A slate consists of 4 offices. If we designate the president, vice president, secretary, and treasurer as the first, second, third, and fourth person on the slate, respectively, then a distinct choice of offices is a permutation of 4 persons from the membership of 100 people. Therefore, the answer to the problem is

$$P_4^{100} = 100 \cdot 99 \cdot 98 \cdot 97 = 94{,}109{,}400$$

(b)   The sample space consists of $N = 94{,}109{,}400$ sample points. Let A be the event of selecting Mr. Lee, Ms. Jones, Mr. Williams, and Ms. Schwartz as president, vice president, secretary, and treasurer, respectively. Then $N_A = 1$. Therefore the answer to the problem is

$$P(A) = \frac{N_A}{N} = \frac{1}{94{,}109{,}400} = .106 \times 10^{-7}$$

(c)   Each permutation of the 4 names—Lee, Jones, Williams, and Schwartz—is a different slate. Therefore, if B is the event that these 4 people will hold the 4 offices, then $N_B = 4!$ Thus

$$P(B) = \frac{N_B}{N} = \frac{4!}{94{,}109{,}400} = .255 \times 10^{-6}$$

Exercises

For the exercises in this section assume that sample spaces consist of equally probable sample points so that the formula $P(A) = N_A/N$ of Theorem 1.4 is applicable.

1.   Define a sample space and count the number of sample points in the sample space for each of the following experiments:

(a)   Select at random a catcher, pitcher, first baseman, and second baseman from a group of 15 baseball players.

(b)   Select at random a chairman and a deputy chairman from a department of 40 professors.

(c)   Select at random a best-tasting, a second-best-tasting, and a worst-tasting wine from a selection of 20 different types of wine.

(d)   Select at random three mutual funds from a list of 20 mutual funds for an investment of $3,000, $10,000, and $15,000, respectively.

2. Consider experiments (a), (b), (c), and (d) of Exercise 1.

(a)   Find the probability that the position of catcher, pitcher, first baseman, and second baseman will be filled by Jackson, Williams, Gonzales, and Jacobs, respectively, in experiment 1(a).

(b)   Find the probability that the 2 positions in experiment (b) will be filled by Professors Lee and Weiss.

(c)   Find the probability that the 3 choices in experiment (c) will be from the 6 cheapest types of wine.

(d)   Find the probability that the monies in experiment (d) will be invested in the three most profitable mutual funds.

3. Find the probability of obtaining a different value on each of the 6 dice in experiment (a) of Exercise 1 of Section 2.1.

4. Define a sample space and count the number of sample points in the sample space for each of the following experiments.

(a)   Successively draw 4 cards from a deck of 52 cards. Record the value and suit of the first, second, third, and fourth card drawn.

(b)   Draw 3 raffle tickets from a box containing 100 raffle tickets numbered 1 through 100. The first, second, and third tickets drawn receive first, second, and third prizes, respectively.

5. Consider experiments (a) and (b) of Exercise 4.

(a)   Find the probability of drawing the Ace, King, Queen, and Jack of Spades and in that order in experiment (a).

(b)   Find the probability of drawing the Ace, King, Queen, and Jack of Spades in any order in experiment (a).

(c)   Find the probability of drawing an ace, a king, a queen, and a jack of any combination of suits and in any order in experiment (a).

(d)   Find the probability of having only even-numbered winning tickets in experiment (b).

(e)   Find the probability of no prizes going to tickets numbered over 50 in experiment (b).

(f)   Find the probability of having the first, second, and third prizes going to tickets numbered n, $n + 1$, and $n + 2$, respectively, for some integer n in experiment (b).

6.   A Problem of Pascal.   A letter from Pascal to Fermat on July 29, 1654, contains the following passage:

> M. de Méré told me that he had found a fallacy in the theory of numbers, for this reason: If one undertakes to get a six with one die, the advantage in getting it in 4 throws is as 671 to 625. If one undertakes to throw 2 sixes with two dice, there is a disadvantage in undertaking it in 24 throws. And nevertheless 24 is to 36 (which is the number of pairings of the faces of two dice) as 4 is to 6 (which is the number of faces of one die). This is what made him so indignant and made him say to one and all that the propositions were not consistent and Arithmetic was self-contradictory...

If you were Pascal, how would you deal with de Méré's problem (aside from writing to your friend Fermat)? Suggestions:

(a)   Show that the number of possible outcomes in 4 throws of a die is 1,296.

(b)   Show that the number of possible outcomes in 4 throws of a die, in which at least one of the results is a six, is 671.

(c)   Show that the (classical) probability of getting at least one six in 4 throws of a die is $(671/1,296) = .5177$.

(d)   Show that the number of possible outcomes in 24 throws of a pair of dice is $36^{24} = 2.24522 \times 10^{37}$.

(e)   Show that the number of possible outcomes in 24 throws of a pair of dice, in which at least one of the results is a pair of sixes, is $36^{24} - 35^{24} = 1.10331 \times 10^{37}$.

(f)   Show that the (classical) probability of getting at least one pair of sixes in 24 throws of a pair of dice is $(1.10331/2.24522) = .4914$.

(g)   Show that the chances of getting at least one six in 4 throws of a die are indeed greater than the chances of getting at least one pair of sixes in 24 throws of a pair of dice.

2.3   Combinations: The Binomial Coefficient

A combination of r elements is an unordered set of r elements. We use the notation $\{a_1, a_2, \ldots, a_r\}$ to designate a combination of the elements $a_1, a_2, \ldots, a_r$. The notation is identical to the notation for a set because a combination is just another name for a set. We refer to a set of elements as a combination of elements if we wish to emphasize the unordered nature of the elements. Observe that the combination $\{1, 3, 5\}$ is identical to the combination $\{3, 1, 5\}$.

The number of distinct combinations of r elements that can be formed from a set of n elements is an important quantity in the mathematics of probability. There are two special notations for this quantity.

Notation   The notation for the number of combinations of r elements from a set of n elements is C_r^n, or $\binom{n}{r}$. The quantity C_r^n or $\binom{n}{r}$ is also called a binomial coefficient because of its appearance in the binomial expansion of $(a + b)^n$ given by the formula

$$(a + b)^n = \sum_{r=0}^{n} C_r^n a^{n-r} b^r$$

(A proof of this formula is considered in Exercise 10.)

In developing a formula for counting the distinct combinations of r elements that can be formed from a set of n elements ($n \geq r$), let us begin with a case where $r = 3$ and $n = 4$.

EXAMPLE 2.6

(a) List the combinations of 3 elements that can be formed from the set $\{1,2,3,4\}$ of 4 elements.

(b) Count the number of combinations of 3 elements that can be formed from a set of 4 elements.

SOLUTION

(a) The combinations of 3 elements from the set $\{1,2,3,4\}$ of 4 elements are $\{1,2,3\}, \{1,3,4\}, \{1,2,4\}, \{2,3,4\}$.

(b) There are 4 combinations of 3 elements from a set of 4 elements.

There is an interesting relationship between the permutations listed in Figure 2.2 of Example 2.4 and the combinations listed in Example 2.6(a). The permutations in the *first* column of Figure 2.2 are permutations of the elements of the first combination, $\{1,2,3\}$, of Example 2.6. The permutations in the *second* column of Figure 2.2 are permutations of the elements of the second combination, $\{1,3,4\}$, of Example 2.6. The permutations in the third and fourth columns of Figure 2.2 are permutations of the elements of the third and fourth combinations, respectively, of Example 2.6. Figure 2.3 illustrates this relationship. Moving from left to right in Figure 2.3, we can see that a set of 4 elements produces 4 combinations of 3 elements each. Each of these combinations of 3 elements produces 6 permutations of the same 3 elements. In total there are $4 \cdot 6 = 24$ permutations of 3 elements from a set of 4 elements. In other words, Figure 2.3 establishes the following relationship between permutations and combinations:

$$P_3^4 = C_3^4 \cdot P_3^3$$
$$\downarrow \quad \downarrow \quad \downarrow$$
$$24 = 4 \ \cdot \ 6$$

This relationship can easily be extended to arbitrary values for r and n, as we can see from the following argument. A set of n elements produces, by definition, C_r^n combinations of r elements. Each of these combinations of r elements produces, according to Corollary 2.1, P_r^r permutations of the r elements. Therefore a set of n elements produces $C_r^n \cdot P_r^r$ permutations of r elements. In other words

$$P_r^n = C_r^n \cdot P_r^r$$

By solving for the term C_r^n in this equation, we obtain the following theorem:

Theorem 2.3

$$\binom{n}{r} = C_r^n = \frac{P_r^n}{P_r^r} = \frac{n(n-1)(n-2)\cdots(n-r+1)}{r!} = \frac{n!}{r!(n-r)!}$$

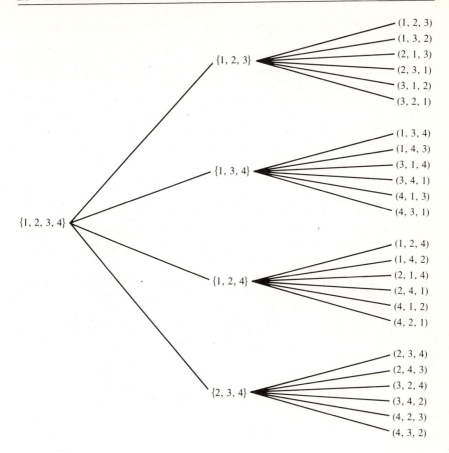

FIGURE 2.3

EXAMPLE 2.7

A poker hand is a set of 5 playing cards.

   (a)   Find the number of distinct poker hands of 5 cards.

   (b)   If 5 cards are dealt at random from a standard deck of 52 cards, what is the probability that all the cards have values lower than 10? (Consider an ace as a card with a value of 1.)

   (c)   What is the probability that all 5 cards are of the same suit?

   SOLUTION

   (a)   A poker hand of 5 cards is a combination of 5 cards from the set of 52 cards. Therefore, the answer is

$$C_5^{52} = \frac{52 \cdot 51 \cdot 50 \cdot 49 \cdot 48}{5!} = 2,598,960$$

(b)   Let A be the event of selecting 5 cards with values lower than 10. There are 9 cards in each suit with values lower than 10. (Consider an ace to have a value of 1.) Therefore there are $4 \times 9 = 36$ cards in a deck that have values lower than 10. Any 5 cards from this set of 36 represent a sample point in event A. Hence

$$N_A = \binom{36}{5}$$

and

$$P(A) = \frac{N_A}{N} = \frac{\binom{36}{5}}{\binom{52}{5}} = .145$$

(c)   Let B be the event of selecting 5 cards from the same suit. There are 4 suits. Each suit has 13 cards. There are $\binom{13}{5}$ ways to select 5 cards from each suit. Hence

$$N_B = 4 \cdot \binom{13}{5}$$

and

$$P(B) = \frac{N_B}{N} = \frac{4 \cdot \binom{13}{5}}{\binom{52}{5}} = .00198$$

EXAMPLE 2.8

A certain carton of eggs has 3 bad and 9 good eggs.

(a)   If an omelette is made of 3 eggs randomly chosen from the carton, what is the probability that there are no bad eggs in the omelette?

(b)   What is the probability of having at least 1 bad egg in the omelette?

(c)   What is the probability of having exactly 2 bad eggs in the omelette?

SOLUTION

(a)   Let A be the event of having no bad eggs in the omelette. There are $\binom{12}{3}$ ways of choosing 3 eggs from a carton of a dozen eggs. Therefore, the number of sample points in the sample space is

$$N = \binom{12}{3} = 220$$

There are 9 good eggs in the carton. Therefore, there are $\binom{9}{3}$ ways of choosing

3 eggs that are not bad. Hence

$$N_A = \binom{9}{3} = 84$$

The answer is

$$P(A) = \frac{N_A}{N} = \frac{\binom{9}{3}}{\binom{12}{3}} = \frac{84}{220} = .38$$

(b)   The event of having at least 1 bad egg is the complement of the event of having no bad eggs. Therefore the answer is

$$P(A') = 1 - P(A) = .62$$

(c)   Let B be the event of having exactly 2 bad eggs in the omelette. Let us count the number of sample points in B. We shall use the basic counting procedure and Theorem 2.3. There are 3 bad eggs and 9 good eggs in the carton. Therefore there are $\binom{3}{2}$ ways to choose 2 bad eggs from the carton to go into the omelette and there are $\binom{9}{1}$ ways to choose the 1 good egg to go into the omelette. Hence there are $\binom{3}{2}\binom{9}{1}$ ways to choose 2 bad eggs and 1 good egg. In other words

$$N_B = \binom{3}{2}\binom{9}{1} = 27$$

The answer is

$$P(B) = \frac{N_B}{N} = \frac{\binom{3}{2}\binom{9}{1}}{\binom{12}{3}} = \frac{27}{220} = .12$$

EXAMPLE 2.9

(a)   Let $A_1, A_2, \ldots, A_k$ be sets with $n_1, n_2, \ldots, n_k$ elements, respectively. Use the basic counting procedure and binominal coefficients to show that there are $\binom{n_1}{r_1} \cdot \binom{n_2}{r_2} \cdots \binom{n_k}{r_k}$ ways to select a combination of r_1 elements from A_1, then a combination of r_2 elements from $A_2, \ldots$, and finally a combination of r_k elements from A_k.

(b)   Use this formula to find the number of ways a disc jockey can select 3 rock, 4 jazz, 2 blues, and 3 country records from a collection of 8 rock, 10 jazz, 7 blues, and 5 country records.

SOLUTION

(a) The binomial coefficients $\binom{n_1}{r_1}, \binom{n_2}{r_2}, \ldots, \binom{n_k}{r_k}$ provide the formulas for the number of ways to select $r_1, r_2, \ldots, r_k$ elements from the sets $A_1, A_2, \ldots, A_k$, respectively. Applying the basic counting procedure, we see that there are $\binom{n_1}{r_1}$ ways to select r_1 elements from A_1. For each of these $\binom{n_1}{r_1}$ ways to select r_1 elements from A_2, there are $\binom{n_2}{r_2}$ ways to select r_2 elements from A_2. Hence there are $\binom{n_1}{r_1} \cdot \binom{n_2}{r_2}$ ways to select r_1 elements from A_1 and r_2 elements from A_2. Continuing in this fashion we conclude that there are $\binom{n_1}{r_1} \cdot \binom{n_2}{r_2} \cdots \binom{n_k}{r_k}$ ways to select r_1 elements from A_1, r_2 elements from $A_2, \ldots$, and finally r_k elements from A_k.

(b) We are selecting 3 elements (rock records) from a set of 8 elements, 4 elements (jazz records) from a set of 10 elements, 2 elements (blues records) from a set of 7 elements, and finally, 3 elements (country records) from a set of 5 elements. The answer is

$$\binom{8}{3} \cdot \binom{10}{4} \cdot \binom{7}{2} \cdot \binom{5}{3} = 2{,}469{,}600$$

Exercises

For the exercises in this section assume that sample spaces consist of equally probable sample points so that the formula $P(A) = N_A/N$ of Theorem 1.4 is applicable.

1. Define a sample space and count the number of sample points in the sample space for each of the following experiments.
    (a) Select at random a set of 5 questions from a set of 15 questions.
    (b) Form by random selection a committee of 5 from a membership of 20.
    (c) Select at random a set of 4 cards from a standard deck of 52 cards.
    (d) From a group of 25 failing small businesses select 5 at random to receive federally insured loans.

2. Consider experiments (a), (b), (c), and (d) of Exercise 1.
    (a) Find the probability of selecting only questions numbered 1 through 10 in experiment (a).
    (b) If 8 members are over 60 years old, find the probability of obtaining a committee that consists entirely of members over 60 in experiment (b).
    (c) Find the probability of getting only black cards in experiment (c).
    (d) If 4 of the failing small businesses in the group are family owned, find the probability of selecting all 4 of them to receive the loans.

3. Can you use the sample space of Exercise 1(c) to find the probability in any of the Section 2.2 Exercises 5(a), 5(b), or 5(c)? If so, find the probability wherever possible and compare with the answer in Section 2.2 Exercises 5(a), 5(b), or 5(c). If the answer is "no" for any of the cases, explain.

4. Consider experiments (a), (b), (c), and (d) of Exercise 1.
    ✓ (a) Find the probability of selecting 3 odd-numbered questions and 2 even-numbered questions in experiment (a).
    ✓ (b) If 8 members are women and 12 members are men, find the probability of selecting a committee of 1 woman and 4 men in experiment (b).
    (c) Find the probability of selecting 3 hearts and 1 diamond in experiment (c).
    (d) Find the probability of selecting no more than 2 family-owned businesses in experiment (d). (Assume a total of 4 such businesses.)

5. What is the probability of answering exactly 5 questions correctly in Exercise 1(c) of Section 2.1?

6. What is the probability of answering fewer than 5 questions correctly in Exercise 1(c) of Section 2.1?

7. Evaluate $\binom{n}{0}$ and show that it is equal to 1.

8. Show that $\binom{n}{r} = \binom{n}{n-r}$.

9. **Pascal's Triangle.** Pascal discovered the following arrangement of binomial coefficients.

$$
\begin{array}{ccccccccccc}
 & & & & & 1 & & 1 & & & \\
 & & & & 1 & & 2 & & 1 & & \\
 & & & 1 & & 3 & & 3 & & 1 & \\
 & & 1 & & 4 & & 6 & & 4 & & 1 \\
 & 1 & & 5 & & 10 & & 10 & & 5 & & 1 \\
1 & & 6 & & 15 & & 20 & & 15 & & 6 & & 1 \\
\end{array}
$$

· · · · · · · · ·

Each row begins with a 1, ends with a 1, and each of the other entries is the sum of the nearest two entries in the row immediately above. The rth entry of the nth row is the binomial coefficient $\binom{n}{r-1}$.

(a) Read off the values of $\binom{5}{2}$ and $\binom{6}{3}$ from the triangle. Compare these values with the values obtained from the formula

$$
\binom{n}{r} = \frac{n!}{r!(n-r)!}
$$

(b) Construct the next two rows of the triangle. Read off the values of $\binom{7}{3}$ and $\binom{8}{2}$.

10. The expansion of the binomial expression $(a + b)^n$ is given by the formula

$$(a + b)^n = \sum_{r=0}^{n} C_r^n a^{n-r} b^r$$

where a and b are real numbers, and n is a positive integer.

(a)   Verify this formula for the case of $n = 2$.

(b)   Verify this formula for the case of $n = 3$.

(c)   Verify this formula for the general case of any positive integer n. (*Hint:* if we multiply out $(a + b)^n$, term by term, the coefficient of $a^{n-r}b^r$ is equal to the number of ways in which r factors of b and $n - r$ factors of a can be selected, one from each of the n factors of $(a + b)$ in the expression $(a + b)^n$.)

11. A certain record collection consists of 10 records of chamber music, 15 records of vocal music, and 20 records of symphonic music.

(a)   Find the number of ways 2 records may be selected from each of the 3 groups of records.

(b)   Suppose that 3 of the chamber music records, 4 of the vocal music records, and 7 of the symphonic music records consist of compositions by Mozart. What, then, is the probability that a random selection of 2 records from each of the 3 groups will result in 6 records of compositions by Mozart?

(c)   What is the probability that a random selection of 2 records from each of the 3 groups will result in no compositions by Mozart?

12. An inspector randomly selects 2 typewriters from a shipment of 15, 3 radios from a shipment of 20, and 2 televisions from a shipment of 10. If 1 typewriter, 2 radios, and 2 televisions in the shipments have defective parts, what is the probability that the inspector's random selection includes all 5 of these defective items?

2.4   The Multinomial Coefficient

A very useful formula for counting sample points in events and sample spaces is the multinomial coefficient. We shall derive this formula through an example.

Suppose a set A contains n elements. We wish to count the ways we can select n_1 of these elements to put into group 1, n_2 of these elements to place into group 2,..., and finally n_k of these elements to place into group k. Assume that $n_1 + n_2 + \ldots + n_k = n$. On application of the basic counting procedure, we see that there are $\binom{n}{n_1}$ ways to select n_1 elements to put into group 1. After this selection, there remain $n - n_1$ elements in A. Hence, there are $\binom{n - n_1}{n_2}$ ways of selecting n_2 elements for group 2. In total there are $\binom{n}{n_1} \cdot \binom{n - n_1}{n_2}$

ways of selecting n_1 elements from A for group 1, and then n_2 elements from A for group 2. Continuing in this fashion, we conclude that there are

$$\binom{n}{n_1}\binom{n-n_1}{n_2}\binom{n-n_1-n_2}{n_3}\cdots\binom{n-n_1-n_2-\cdots-n_{k-1}}{n_k}$$

ways to select n_1 elements from A for group 1, n_2 elements from A for group 2, ..., and finally n_k elements from A for group k.

This long expression is equivalent to the factorial expression

$$\frac{n!}{n_1!n_2!\cdots n_k!}$$

The following computation verifies this equivalence:

$$\binom{n}{n_1}\binom{n-n_1}{n_2}\binom{n-n_1-n_2}{n_3}\cdots\binom{n-n_1-n_2-\cdots-n_{k-1}}{n_k}$$

$$= \frac{n!}{n_1!(n-n_1)!}\cdot\frac{(n-n_1)!}{n_2!(n-n_1-n_2)!}\cdots\frac{(n-n_1-n_2-\cdots-n_{k-1})!}{n_k!(n-n_1-n_2-\cdots-n_k)!}$$

$$= \frac{n!}{n_1!n_2!\cdots n_k!}$$

Theorem 2.4 *With $n_1 + n_2 + \cdots n_k = n$, there are*

$$\frac{n!}{n_1!n_2!\cdots n_k!}$$

ways to divide a set of n elements into k groups so that the first group contains n_1 element, the second group contains n_2 elements, ..., and the kth group contains n_k elements.

EXAMPLE 2.10

How many ways can we divide 100 students into 4 groups so that there are 30, 20, 35, and 15 students in groups 1, 2, 3, and 4, respectively?

SOLUTION

Direct application of Theorem 2.4 yields the answer

$$\frac{100!}{30!20!35!15!} = 1.070253614 \times 10^{55}$$

The formula of Theorem 2.4 can also be used to count the number of permutations of objects that are not all distinct. Example 2.11 shows how this is done. The result is summarized in Corollary 2.2.

EXAMPLE 2.11
Count the distinct permutations of the letters in the word *Mississippi*.

SOLUTION

The word consists of 11 letters; therefore it occupies 11 spaces. There are 4 *i*'s, 4 *s*'s, 2 *p*'s, and 1 *M*. Think of selecting 4 spaces for *i*'s (the same as placing 4 elements into group 1), then 4 spaces for *s*'s (4 elements into group 2), then 2 spaces for *p*'s (2 elements into group 3), and finally 1 space for *M* (1 element into group 4). Therefore the number of distinct permutations is

$$\frac{11!}{4!4!2!1!} = 34{,}650$$

Corollary 2.2   *There are*

$$\frac{n!}{n_1!n_2!\cdots n_k!}$$

distinct permutations of n elements of which n_1 are of one kind, n_2 are of a second kind,..., n_k are of a kth kind, and $n_1 + n_2 + \cdots + n_k = n$.

Exercises

1. Fourteen cars of the same style, of which 4, 3, and 7 are white, black, and blue, respectively, are randomly parked in a row. How many distinct arrangements are possible?

2. One hundred calculus students are randomly assigned to take a test: 30 go to room A, 20 to room B, 35 to room C, and 15 to room D. If Professors Green, Silva, Jackson, and Chin are calculus instructors who have, respectively, 30, 20, 35, and 15 of the students, what is the probability that students who have the same professor will be assigned to the same room?

3. In 6 throws of a die, what is the probability of getting exactly 2 ones, 3 twos, and 1 six?

4. The eight shoes of four different pairs are randomly distributed two to each of four closets. What is the probability that matching pairs will be together?

5. In Exercise 2, what is the probability that all of Professor Green's students are together, all of Professor Jackson's student are together, but Professor Silva's students are divided equally into two rooms?

6. In how many ways can 10 salespersons be assigned so that 2 are assigned to district A, 3 to district B, and 5 to district C? If 5 of the salespersons are men and 5 are women, what is the probability that a random assignment of 2 salespersons to district A, 3 to district B, and 5 to district C will result in a segregation of the salespersons by sex? What is the probability that a random assignment will assign at least one female salesperson to each of the three districts?

2.5 Tree Diagrams

Tree diagrams are often used to count the number of outcomes in a sequence of actions. The following examples illustrate the technique.

EXAMPLE 2.12

Alice and Mary are engaged in a chess tournament. The first person to win either 2 games in a row or a total of 3 games will be declared the winner. The tree diagram in Figure 2.4 gives the possible outcomes of the tournament.

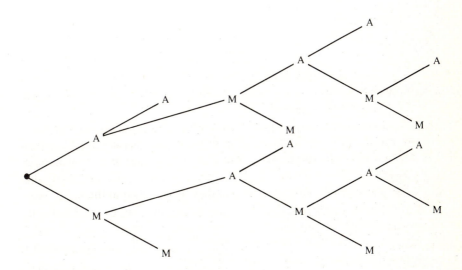

FIGURE 2.4

A denotes a win for Alice; M denotes a win for Mary. The possible outcomes are obtained by following the branches of the tree. They are: AA, AMAA, AMM, AMAMA, AMAMM, MM, MAMM, MAA, MAMAM, MAMAA.

EXAMPLE 2.13

Set A consists of the elements a_1, a_2, a_3. Set B consists of the elements b_1, b_2, b_3, b_4.

The tree diagram of Figure 2.5 illustrates the process of selecting first an element from A and then an element from B to form a pair (a_i, b_j) of elements.

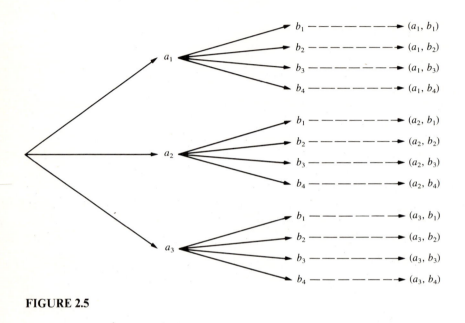

FIGURE 2.5

Exercises

1. One can travel from Chicago to New York by bus, train, or airplane. One can travel from New York to London by ship or airplane. Draw a tree diagram to show the different ways one can travel from Chicago to London via New York.

2. A certain student can spend his evening, before his parents are home, watching television, reading comic books, loafing, or studying. After that, he can either study until bedtime or pretend he is tired and go to bed early. Draw a tree diagram to show the different ways the student can spend his evening.

3. Toss a coin until either 2 heads occur or until the coin has been tossed 4 times. Draw a tree diagram of the possible outcomes.

4. Joe decides to play chess with Ron until he either wins one game or has played a total of 5 games. Draw a tree diagram of the possible outcomes.

5. Players A, B, and C are engaged in a series of games of chance. The series begins with A and B as opponents. The winner of each game continues the series by playing the player who was not a participant in the most current game. The series continues until one player has won 3 games. Draw a tree diagram of the possible outcomes.

√ 6. If each of the branches of the tree in Exercise 1 is equally probable, what is the probability of traveling by airplane at least part of the way from Chicago to London via New York?

7. If each of the branches of the tree in Exercise 2 is equally probable, what is the probability that the student will not study at all?

Supplementary Exercises

√ 1. Roll 3 dice. What is the probability of obtaining 3 different values on the dice?

2. One hundred raffle tickets were sold and Jill bought 4. What is the probability of her winning at least one of 3 prizes? What is the probability of her winning none of the 3 prizes? What is the probability of winning exactly one of the 3 prizes?

√ 3. Joseph knows that the last 4 digits of Sylvia's phone number is some permutation of the digits 7, 2, 6, and 4. If he dials a random permutation of the 4 digits what is the probability of his getting Sylvia's number?

√ 4. Five double agents have infiltrated a group of secret agents to form a group of 25. A commission of 10 agents is to be randomly selected from the group of 25 for an important mission. What is the probability of having at least one double agent on the commission? What is the probability of having no double agents on the commission? What is the probability of having exactly 2 double agents on the commission? (*Hint:* see Example 2.8 on bad eggs in the omelette.)

5. Define a sample space and count the number of sample points in the sample space for each of the following experiments.
   (a) Select at random a bridge hand of 13 cards from a standard deck of 52 cards.
   (b) From a group of 25 string players, select at random a first violinist, a second violinist, a violist, and a cellist for a string quartet.
   (c) Form a committee of 2 freshmen, 2 sophomores, 2 juniors, and 2 seniors by random selection from a student body of 1,000 freshmen, 980 sophomores, 850 juniors, and 840 seniors.
   (d) Select at random an appetizer, an entree, and a dessert from a menu listing 5 choices of appetizer, 10 choices of entree, and 6 choices of dessert.

6. Consider experiments (a), (b), (c), and (d) of Exercise 5.

(a)   Find the probability that all cards in experiment (a) are from the same suit.

(b)   Find the probability that the position of first violinist, second violinist, violist, and cellist in experiment (b) will be filled by the oldest, second-oldest, third-oldest, and youngest player, respectively.

(c)   If 400 freshmen, 350 sophomores, 300 juniors, and 300 seniors are women, what is the probability that all members of the committee in experiment (c) are women?

(d)   Find the probability of selecting either the first or last item in each category in experiment (d).

7.   In selecting 8 cards at random from a deck of 52 cards, what is the probability of selecting exactly 2 cards from each suit?

8.   In a random selection of 13 cards from a deck of 52 cards, what is the probability of selecting exactly 2 aces, 3 kings, 1 queen, 4 jacks, 2 tens, and 1 nine?

9.   Referring to Exercise 1 of Section 2.4,

(a)   what is the probability that a random parking of the cars in a row will result in the first 4 cars, the next 3 cars, and the last 7 cars being white, black, and blue, respectively;

(b)   what is the probability that a random parking of these cars in a row will result in a grouping of the cars by color?

Supplementary Exercises on Combinatorial Methods

Permutations

1.   In how many ways can 10 students line up at the bursar's window?

2.   How many 4-digit numbers can be formed from the integers 1 through 9 if no integer can be used more than once? How many can be formed if the number begins with the integer 1?

3.   In how many ways can 9 books be arranged on a shelf?

4.   How many 3-letter words (i.e., permutations of 3 letters) can be formed from the letters of the alphabet if no letter can be used more than once?

5.   In how many ways can 4 books be chosen from 10 different books and arranged in 4 spaces on a shelf?

6.   In how many ways can 5 dice be thrown so that no 2 dice have the same value?

Combinations

1.   In how many ways can a committee of 4 be chosen from 10 people?

2.   In how many ways can a child choose 5 toys from a collection of 15 different toys?

3. In how may ways can a student choose 3 books from a set of 5 different books?

4. In how many ways can a committee consisting of 2 men and 3 women be chosen from 6 men and 8 women?

5. In a certain examination, a student must answer 3 out of 5 questions for part A, 2 of 3 questions for part B, and 4 of 6 questions for part C. How many different groups of 9 questions can be chosen?

Multinomial coefficients

1. Count the distinct permutations of all the letters in the word *abracadabra*.

2. How many different 12-digit numbers can be formed by arranging the digits in the number 222,445,557,777?

3. In how many ways can 8 toys be divided among 3 children if the youngest gets 3 toys and the oldest gets 2?

4. In how many ways can a teacher divide her class of 20 students into 3 reading groups so that there are 7 in group 1, 7 in group 2, and 6 in group 3?

References

The reader may wish to refer to the following books for additional examples on the use of combinatorial methods in problems of probability.

1. Feller, William. 1968. *An introduction to probability theory and its applications*, Vol. I, 3d ed. New York: John Wiley & Sons, Inc.
2. Lipschutz, S. 1968. *Probability*, Schaum's Outline Series. New York: McGraw-Hill Book Co.
3. Ross, Sheldon. 1976. *A first course in probability*. New York: Macmillan Publishing Co., Inc.

Chapter 3

CONDITIONAL PROBABILITY AND INDEPENDENCE

3.1 Conditional Probability

One of the most important and practical concepts of probability is the conditional probability. As the name suggests, the conditional probability of an event is the probability that an event will occur, subject to the existence of certain conditions. The exact definition follows.

3.1.1 Definition of Conditional Probability

Definition 3.1   Let A and B be events with $P(B) > 0$. The **conditional probability** of the occurrence of A, given the occurrence of B, is defined to be the quantity

$$P(A \mid B) = \frac{P(AB)}{P(B)}$$

Instead of using the longer statement given in the formal definition above, we generally call $P(A \mid B)$ "the conditional probability of A given B."

Note: throughout this book we use the notation AB to represent the event $A \cap B$. The notation $A_1 A_2 \cdots A_n$ represents the event $A_1 \cap A_2 \cap \cdots \cap A_n$.

3.1.2 Use of the Conditional Probability

The following example illustrates the appropriate use of the conditional probability.

EXAMPLE 3.1

A box contains 10 balls of which 6 are black and 4 are white. Four of the black balls have smooth surfaces and 2 have rough surfaces. One of the white balls has a smooth surface and 3 have rough surfaces. (Figure 3.1 depicts the contents of the box.) Consider an experiment in which a ball is selected at random from the box in such a manner that each ball has an equal probability of being selected. Let B, W, and R denote the events of selecting a black, white, and rough ball, respectively.

(a)   Find $P(B)$, $P(W)$, $P(R)$, $P(BR)$, and $P(WR)$.
(b)   If the ball selected is rough, what are the chances that it is white?
(c)   Find the conditional probability of W given R.
(d)   Compare $P(W \mid R)$ with the answer to (b).

SOLUTION

(a)   There are 10 balls in the box. Hence, the number of points in the sample space is $N = 10$. The number of black, white, rough, black and rough, and white and rough are $N_B = 6$, $N_W = 4$, $N_R = 5$, $N_{BR} = 2$, and $N_{WR} = 3$,

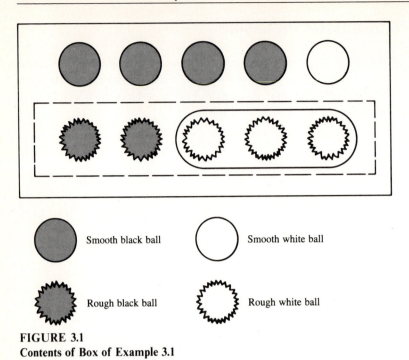

Smooth black ball

Smooth white ball

Rough black ball

Rough white ball

FIGURE 3.1
Contents of Box of Example 3.1

respectively. Therefore the desired probabilities are:

$$P(B) = N_B/N = 6/10 = .6$$
$$P(W) = N_W/N = 4/10 = .4$$
$$P(R) = N_R/N = 5/10 = .5$$
$$P(BR) = N_{BR}/N = 2/10 = .2$$
$$P(WR) = N_{WR}/N = 3/10 = .3$$

(b)   In considering question (b) it should be noted that if we are to use the sample space of (a), it is impossible to define in a meaningful way an event A for which $P(A)$ is the answer to the question. Nevertheless, we have no trouble answering the question because the generally accepted answer is 3/5. The argument is that there are 5 rough balls, of which 3 are white. (See Figure 3.1.) In other words the answer to the question is 3/5 since

$$N_{WR}/N_R = 3/5$$

(c)   By definition, the conditional probability of W given R is the quantity

$$P(W \mid R) = \frac{P(WR)}{P(R)} = \frac{N_{WR}/N}{N_R/N} = \frac{N_{WR}}{N_R} = \frac{3}{5}$$

(d)   The quantity $P(W \mid R)$ is identical to the answer for (b). The agreement is not only in the numerical value of 3/5, but also in terms of the concept expressed by the term $(N_{WR})/(N_R)$. In other words, the conditional probability of W given R is the same as the chances of obtaining a white ball if it is known that the ball selected is rough; that is, the conditional probability $P(W \mid R)$ is the probability of selecting a white ball from the reduced sample space of rough balls.

Example 3.1 illustrates the use of the conditional probability. In other words, for the experiment of drawing balls from a box, $P(A \mid B)$ is the answer to the question, "If the event B occurs, then what is the probability that the event A occurs?" This use of the conditional probability may be extended to the general case.

Summary

If the event B occurs, then the probability that the event A occurs is $P(A \mid B)$.

EXAMPLE 3.2

In rolling 2 balanced dice, if the sum of the two values is 7, what is the probability that one of the values is 1?

SOLUTION

Let A be the event that one of the values is 1. Let B be the event that the sum is 7. If we use the standard sample space and notation for the experiment of rolling 2 dice, then the event AB consists of the sample points $(1, 6)$ and $(6, 1)$. The event B consists of the sample points $(1, 6)$, $(2, 5)$, $(3, 4)$, $(4, 3)$, $(5, 2)$, and $(6, 1)$. Thus, $N_{AB} = 2$, $N_B = 6$, and

$$P(AB) = N_{AB}/N = 2/36$$
$$P(B) = N_B/N = 6/36$$

Therefore, the answer to the question is

$$P(A \mid B) = \frac{P(AB)}{P(B)} = \frac{2/36}{6/36} = 1/3$$

EXAMPLE 3.3

Robert feels that the probability that he will get an A in the first physics test is 1/2 and the probability that he will get A's in the first and second physics tests is 1/3. If Robert is correct, what is the conditional probability that he will get an A in the second test, given that he gets an A in the first test?

SOLUTION

Let A_1 and A_2 be the events of an A in the first and second tests, respectively.

Robert claims that $P(A_1) = 1/2$ and $P(A_1 A_2) = 1/3$. If he were correct, then

$$P(A_2 \mid A_1) = \frac{P(A_1 A_2)}{P(A_1)} = \frac{(1/3)}{(1/2)} = 2/3$$

EXAMPLE 3.4

Suppose that 5 cards are to be drawn at random from a standard deck of 52 cards. If all the cards drawn are red, what is the probability that all of them are hearts?

SOLUTION

Let A be the event that all five cards are hearts and let B be the event that all five cards are red. We wish to find $P(A \mid B)$. The simplest way to find $P(A \mid B)$ is to follow the ideas of Example 3.1. In other words, if we are dealing with a sample space with equally probable sample points then

$$P(A \mid B) = \frac{P(AB)}{P(B)} = \frac{N_{AB}/N}{N_B/N} = \frac{N_{AB}}{N_B}$$

In this example, we are interested in sets of 5 red cards. N_B is the number of distinct sets of 5 red cards and N_{AB} is the number of distinct sets of 5 red cards that are also hearts. Therefore

$$N_B = \binom{26}{5}, \qquad N_{AB} = \binom{13}{5}, \qquad \text{and} \qquad P(A|B) = \frac{N_{AB}}{N_B} = \frac{\binom{13}{5}}{\binom{26}{5}} = .0196$$

Summary

For sample spaces with equally probable sample points

$$P(A \mid B) = N_{AB}/N_B$$

where N_{AB} and N_B are the number of sample points in the events AB and B, respectively.

Exercises

1. A box contains 10 red, 10 white, and 5 blue balls. A ball is chosen at random from the box. If the ball chosen is not blue, what is the probability that it is white?

2. In two tosses of a coin, what is the conditional probability of two heads if the first toss results in heads?

3. Consider the rolling of two balanced dice.
   (a) What is the conditional probability that at least one lands on 2, given that the dice land on different numbers?
   (b) What is the conditional probability that at least one lands on an even number, given that the dice land on different numbers?

4. One-third of the applicants for a certain job are college graduates, and one-fourth of them are college graduates with at least one year of experience in their field. If an applicant is selected at random, what is the probability that if the person is a college graduate he or she also has at least one year of experience?

5. Kevin, a computer science student, feels that he has a fifty-fifty chance of completing the programming portion of his project within a week but that the odds against his completing the programming and getting a chance to try it on the computer within the week are 5 to 1. If these odds are correct, what is the probability that Kevin will get a chance to try his program on the computer within the week if he completes the programming within the week?

6. Five cards each are to be dealt to two players. If one player has the Ace of Spades, what is the probability that the other player has the Ace of Hearts?

7. Five cards each are to be dealt to two players. If one player has only hearts, what is the probability that the other player has only diamonds?

8. A certain system has 5 components. A failure in the system is caused 35%, 30%, 20%, 10%, and 5% of the time by a failure in components A, B, C, D and E, respectively. Assume that simultaneous failures in more than one component are so rare that we may consider the probability of their occurrence to be zero.
   (a) If, on a given occasion, a failure in the system is not caused by a failure in component A, what is the probability that it is caused by a failure in component B?
   (b) If, on a given occasion, a failure in the system is caused by neither a failure in component A, nor a failure in component B, what is the probability that it is caused by a failure in either component C or D?

3.2 Bayes' and Other Theorems on Conditional Probabilities

The following theorems greatly enhance the use of conditional probabilities in problem solving.

Theorem 3.1

$$P(A' \mid B) = 1 - P(A \mid B)$$

Proof

$$P(A' \mid B) = \frac{P(A'B)}{P(B)} = \frac{P(B) - P(AB)}{P(B)} = \frac{P(B)}{P(B)} - \frac{P(AB)}{P(B)}$$

$$= 1 - P(A \mid B)$$

Theorem 3.2

(a)

$$P(AB) = P(A \mid B)P(B) \qquad \text{if} \quad P(B) > 0$$
$$P(AB) = P(A)P(B \mid A) \qquad \text{if} \quad P(A) > 0$$

(b)

$$P(ABC) = P(A)P(B \mid A)P(C \mid AB) \qquad \text{if} \quad P(AB) > 0$$

(c)

$$P(A_1 A_2 \cdots A_n) = P(A_1)P(A_2 \mid A_1)P(A_3 \mid A_1 A_2) \cdots P(A_n \mid A_1 A_2 \cdots A_{n-1})$$
$$\text{if} \quad P(A_1 A_2 \cdots A_{n-1}) > 0$$

Proof   We shall prove only (a) of this theorem. The other parts will be considered in Exercise 13.

By the definition of conditional probability we have

$$P(A \mid B) = \frac{P(AB)}{P(B)}$$

Therefore

$$P(AB) = P(A \mid B)P(B)$$

Interchange A with B to obtain $P(AB) = P(A)P(B \mid A)$.

EXAMPLE 3.5

Suppose that 60% of adults over 30 in a certain community are college graduates. Furthermore, suppose that 80% of the college graduates over 30 have incomes over $15,000.

   (a)   If a college graduate over 30 is selected at random from the community, what is the probability that he or she has an income of $15,000 or less?

   (b)   What percent of adults over 30 in the community are college graduates and have incomes over $15,000?

   SOLUTION

   (a)   Consider an experiment in which an adult over 30 is selected at random from the community. Let A be the event of selecting a person with an income over \$15,000. Let B be the event of selecting a college graduate. Then, $P(A \mid B) = .80$ and $P(B) = .60$. The answer to (a) is

$$P(A' \mid B) = 1 - P(A \mid B) = 1 - .80 = .20.$$

(b)

$$P(AB) = P(A \mid B)P(B) = (.80)(.60) = .48$$

Therefore, the answer to (b) is 48%.

EXAMPLE 3.6

Two defective fuses have been mixed up with three good ones. The fuses are tested, one by one, until both defective ones are found.
  (a) What is the probability that the testing of exactly two fuses is required?
  (b) What is the probability that the testing of exactly three fuses is required?

  SOLUTION

  (a) Let G_i be the event of finding a good fuse on the ith test, and let D_i be the event of finding a defective fuse on the ith test, for $i = 1, 2, 3, 4, 5$. The event that the testing of exactly two fuses is required is $D_1 D_2$, and the probability of that event, upon application of Theorem 3.2, is

$$P(D_1 D_2) = P(D_1)P(D_2 \mid D_1)$$

The probability $P(D_1)$ is 2/5 because in choosing a fuse for the first test there are 5 fuses of which 2 are defective. The conditional probability $P(D_2 \mid D_1)$ is 1/4 because if the first fuse tested is defective, there remain 4 fuses of which 1 is defective. Thus

$$P(D_1 D_2) = P(D_1)P(D_2 \mid D_1) = (2/5)(1/4) = .10$$

  (b) The event that the testing of exactly 3 fuses is required is $G_1 D_2 D_3 \cup D_1 G_2 D_3 \cup G_1 G_2 G_3$. (Note that if the first three fuses tested are good, the remaining two must be defective.) Thus the solution is

$$
\begin{aligned}
P(G_1 D_2 D_3 &\cup D_1 G_2 D_3 \cup G_1 G_2 G_3) \\
&= P(G_1 D_2 D_3) + P(D_1 G_2 D_3) + P(G_1 G_2 G_3) \\
&= P(G_1)P(D_2 \mid G_1)P(D_3 \mid G_1 D_2) + P(D_1)P(G_2 \mid D_1)P(D_3 \mid D_1 G_2) \\
&\quad + P(G_1)P(G_2 \mid G_1)P(G_3 \mid G_1 G_2) \\
&= (3/5)(2/4)(1/3) + (2/5)(3/4)(1/3) + (3/5)(2/4)(1/3) = .3
\end{aligned}
$$

Theorem 3.3   *Let S be a sample space. If $B_1, B_2, \ldots, B_n$ are mutually exclusive events such that*

$$S = B_1 \cup B_2 \cup \cdots \cup B_n \quad \text{and} \quad P(B_i) > 0 \qquad \text{for} \quad i = 1, 2, \ldots, n$$

then for any event A of S

   (a)

$$P(A) = P(AB_1) + P(AB_2) + \cdots + P(AB_n) \quad \text{and}$$

   (b)

$$P(A) = P(A \mid B_1)P(B_1) + P(A \mid B_2)P(B_2) + \cdots + P(A \mid B_n)P(B_n)$$

   Proof

   (a)   Let us begin with an informal verification of the statement. Figure 3.2 illustrates the conditions for the theorem, given a case of $n = 4$. From the figure it is clear that event A is the union of the mutually exclusive events AB_1, AB_2, AB_3, and AB_4. Therefore, $P(A) = P(AB_1) + P(AB_2) + P(AB_3) + P(AB_4)$.

   A formal proof of (a) follows:

$$S = B_1 \cup B_2 \cup \cdots \cup B_n$$

Therefore

$$A = A \cap S = A \cap (B_1 \cup B_2 \cup \cdots \cup B_n)$$
$$= AB_1 \cup AB_2 \cup \cdots \cup AB_n$$

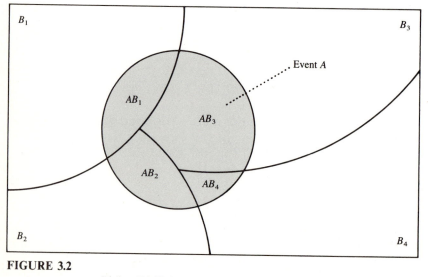

FIGURE 3.2

$$P(A) = P(AB_1) + P(AB_2) + P(AB_3) + P(AB_4)$$

Clearly, events $AB_1, AB_2, \ldots, AB_n$ are mutually exclusive. Therefore

$$P(A) = P(AB_1 \cup AB_2 \cup \cdots \cup AB_n)$$
$$= P(AB_1) + P(AB_2) + \cdots + P(AB_n)$$

(b)   By applying Theorem 3.2 to Theorem 3.3(a) we obtain

$$P(A) = P(AB_1) + P(AB_2) + \cdots + P(AB_n)$$
$$= P(A \mid B_1)P(B_1) + P(A \mid B_2)P(B_2) + \cdots + P(A \mid B_n)P(B_n)$$

EXAMPLE 3.7

Suppose it is known that carriers of a certain dread disease constitute 1% of the population. Furthermore, suppose that there is a simple blood test that is 85% accurate in detecting both presence and absence of this condition. It follows that if a person is a carrier, the probability that the test will be positive is .85; if the person is not a carrier, the probability that the test will be negative is also .85.

(a)   What is the probability that the blood test of a person selected at random from the population will be positive?

(b)   If a person's test results are positive, what is the probability that he or she is a carrier of the dread disease?

SOLUTION

(a)   Let C, P, and N, respectively, denote the events that a person selected at random from the population will be a carrier, will test positive, and will test negative. The given probabilities are $P(C) = .01$, $P(P \mid C) = .85$, and $P(N \mid C') = .85$. Therefore, the answer to the question is

(Theorem 3.3)
$$\downarrow$$
$$P(P) = P(P \mid C)P(C) + P(P \mid C')P(C')$$

(Theorem 3.1)
$$\downarrow$$
$$= P(P \mid C)P(C) + [1 - P(N \mid C')]P(C')$$
$$= (.85)(.01) + (1 - .85)(1 - .01) = .157$$

(b)   We are interested in the conditional probability $P(C \mid P)$. The computation follows:

(Definition 3.1) (Theorem 3.2)
$$\qquad\quad\downarrow \qquad\quad\downarrow$$
$$P(C \mid P) = \frac{P(CP)}{P(P)} = \frac{P(P \mid C)P(C)}{P(P)} = \frac{(.85)(.01)}{(.157)} = .054$$

EXAMPLE 3.8

The following table shows the proportion of families in a certain community who live in privately owned single-family houses, rental apartments, or cooperative apartments, and who own one or more cars.

	Single-family houses	Rental apartments	Cooperative apartments
One car	.40	.20	.10
More than one car	.20	.02	.08

If a family selected at random from this community owns more than one car, what is the probability that the family lives in a single-family house?

SOLUTION

Let A_1, A_2, and A_3 be the respective events that the family selected lives in a privately owned single-family house, rental apartment, and cooperative apartment. Let B_1 and B_2 be the respective events that the family selected owns one car and more than one car. Then

P(the family lives in a privately owned single-family house, given that the family owns more than one car)

$$= P(A_1 \mid B_2) = P(A_1 B_2)/P(B_2)$$

$$= \frac{P(A_1 B_2)}{P(A_1 B_2) + P(A_2 B_2) + P(A_3 B_2)}$$

$$= \frac{.20}{.20 + .02 + .08} = 2/3$$

Theorem 3.4　Bayes' Theorem　*If the conditions of Theorem 3.3 are satisfied, then for any $k = 1, 2, \ldots, n$*

$$P(B_k \mid A) = \frac{P(A \mid B_k)P(B_k)}{P(A \mid B_1)P(B_1) + P(A \mid B_2)P(B_2) + \cdots + P(A \mid B_n)P(B_n)}$$

Proof　This theorem is a consequence of the definition of conditional probability, of Theorem 3.2, and of Theorem 3.3.

(Definition 3.1) (Theorem 3.2)
$$\downarrow \qquad \qquad \downarrow$$

$$P(B_k \mid A) = \frac{P(AB_k)}{P(A)} = \frac{P(A \mid B_k)P(B_k)}{P(A)}$$

(Theorem 3.3)
$$\downarrow$$

$$= \frac{P(A \mid B_k)P(B_k)}{P(A \mid B_1)P(B_1) + P(A \mid B_2)P(B_2) + \cdots + P(A \mid B_n)P(B_n)}$$

EXAMPLE 3.9   CONDITIONAL PROBABILITIES OF CAUSES

Bayes' Theorem is sometimes referred to as a formula for finding the conditional probabilities of causes. The following problem illustrates this point.

Suppose that at a certain accounting office 30%, 25%, and 45% of the statements are prepared by Ms. Jones, Mr. Lane, and Ms. Brown, respectively. These employees are very reliable. Nevertheless, they are in error some of the time. Suppose that .01%, .005%, and .003% of the statements prepared by Ms. Jones, Mr. Lane, and Ms. Brown, respectively, are in error. If a statement from the accounting office is in error, what is the probability that it was prepared (caused) by Ms. Jones?

SOLUTION

Consider an experiment in which a statement is selected at random from those prepared by the accounting office. Let B_1, B_2, and B_3 be the events that the statement was prepared by Ms. Jones, Mr. Lane, and Ms. Brown, respectively. Let E be the event of an error in the statement, and let the conditional probabilities of error, given that the statement was prepared by Ms. Jones, Mr. Lane, and Ms. Brown, be $P(E \mid B_1)$, $P(E \mid B_2)$, and $P(E \mid B_3)$, respectively. We can use Bayes' Theorem to find the answer as follows:

$$P(B_1 \mid E) = \frac{P(E \mid B_1)P(B_1)}{P(E \mid B_1)P(B_1) + P(E \mid B_2)P(B_2) + P(E \mid B_3)P(B_3)}$$

$$= \frac{(.0001)(.30)}{(.0001)(.30) + (.00005)(.25) + (.00003)(.45)} = .5357$$

EXAMPLE 3.10   WHODUNNIT?

The following problem, albeit somewhat whimsical, also illustrates the use of Bayes' Theorem in finding the conditional probability of causes.

Mrs. Vandergill has a cat and a butler. The butler hates the cat, and the chances that he would kill it, if left alone with it, are 1/2. However, the butler is a busy man, so the chances that the cat would end up alone with the butler when it is not with its mistress are only 1/5. Then, again, this cat is rather stupid; if not attended to by its mistress, the chances that it will have an accident and die are 1/4. Recently, Mrs. Vandergill was called away on an emergency and left her cat unattended. When she returned, the cat was dead. Did the butler do it?

(Assume that if the cat was with the butler when it died, death was not accidental. Also, assume that no one else is after the cat.)

SOLUTION

Let us assume there are only two possible causes of death, namely, murder or accident. This means that if the cat dies, either the butler killed it or the cat had a fatal accident. But this means that if the cat dies, the butler killed the cat if and only if the cat was alone with the butler. In other words, we are looking for the conditional probability of the event that the butler is alone with the cat given the event that the cat has died.

Let D represent the event that the cat has died. Let B represent the event that the butler was alone with the cat. We are given the following values:

$$P(D\,|\,B) = 1/2, \quad P(B) = 1/5, \quad P(D\,|\,B') = 1/4, \quad P(B') = 4/5$$

Therefore the answer to the problem is

$$P(B\,|\,D) = \frac{P(D\,|\,B)P(B)}{P(D\,|\,B)P(B) + P(D\,|\,B')P(B')}$$

$$= \frac{(1/2)(1/5)}{(1/2)(1/5) + (1/4)(4/5)} = 1/3$$

In other words, under the conditions of the problem, the chances that the butler killed the cat are $1/3$, whereas the chances that the cat had a fatal accident are $2/3$. (Moral of the story: if you are a stupid cat, you don't need any enemies!)

EXAMPLE 3.11   UPDATING PROBABILITIES: A MATTER OF STRATEGY

Bayes' Theorem is sometimes referred to as a formula for updating probabilities as new information becomes available. The following problem illustrates this point.

General X is in the process of determining the overall strategy for the army. Will the next attack from the enemy be at locations 1, 2, or 3? After consultations with his staff, he feels that the chances of attack at locations 1, 2, and 3 are .5, .3, and .2, respectively. Suddenly a report arrives at army headquarters indicating that the enemy is taking certain actions which we shall here label as action A. A new round of consultations with his staff leaves the general with the feeling that—depending on whether the enemy plans to attack at location 1, 2, or 3—the probabilities of action A are .6, .2, and .1, respectively. With this information, how might the general use Bayes' Theorem to revise his initial probabilities concerning the location of the next attack from the enemy?

SOLUTION

Let A be the event of action A by the enemy, and let B_1, B_2, and B_3 be the events of an attack by the enemy at locations 1, 2, and 3, respectively. The general's initial probabilities concerning the events B_1, B_2, and B_3, are $P(B_1) = .5$, $P(B_2) = .3$, and $P(B_3) = .2$. His conditional probabilities concerning the event of A are $P(A|B_1) = .6$, $P(A|B_2) = .2$, and $P(A|B_3) = .1$. Upon application of Bayes' Theorem, we obtain

$$P(B_k|A) = \frac{P(A|B_k)P(B_k)}{P(A|B_1)P(B_1) + P(A|B_2)P(B_2) + P(A|B_3)P(B_3)}$$

for $k = 1, 2$, and 3.

The denominator of the expression given above is equal to

$$(.6)(.5) + (.2)(.3) + (.1)(.2) = .38$$

Therefore

$$P(B_1|A) = \frac{(.6)(.5)}{.38} = .79$$

$$P(B_2|A) = \frac{(.2)(.3)}{.38} = .16$$

$$P(B_3|A) = \frac{(.1)(.2)}{.38} = .05$$

If the general is a Bayesian probabilist, he may, with the new information concerning the enemy, wish to update his set of probabilities concerning an attack from the enemy at locations 1, 2, and 3 from .5, .3, and .2, respectively, to .79, .16, and .05, respectively. The initial set of probabilities are called **prior** probabilities, and the revised set are called **posterior** probabilities. The method used in revising the probabilities is called Bayesian because of the use of Bayes' Theorem.

Exercises

1. Joan feels that her probability of scoring a high mark in the MCAT examination is 1/2, and the probability of her being accepted by a prestigious medical school if she does score a high mark is 2/3. If Joan is correct, what is the probability that she will score a high mark in the MCAT examination and be accepted by a prestigious medical school?

2. Mr. Westman applied for a bank loan. The probability that it will be approved is 2/3. If the loan is approved, the probability that his business venture will succeed is 3/4. However, if the loan is not approved, the

probability that his business venture will succeed is only 1/4. What is Mr. Westman's probability of success in his business venture?

3. John is undecided as to whether to call Teresa or Lisa for a date for Saturday. Although he prefers Teresa's company, John estimates that if he asks Teresa for a date, the probability that she will say yes is only 1/3, whereas the probability that Lisa will say yes, if asked, is 1/2. John decides to base his decision on the flip of a coin. If John's subjective probabilities are correct,

(a)   what is the probability that he will get a date with Teresa for Saturday?

(b)   what is the probability that he will get a date with one of the two young women? Assume that if John is rejected on the first call he will be too distraught to try another.

4. Three cards are to be randomly selected, without replacement, from a standard deck of 52 cards. If the second and third cards selected are hearts, what is the probability that the first card selected is also a heart?

5. Let A, B, and C be events with $P(A) = .4$, $P(B) = .5$, $P(C) = .6$, $P(AB) = .1$, $P(BC) = .2$, $P(AC) = .3$, and $P(ABC) = .02$. Find (a) $P(A \mid B)$; (b) $P(A \mid BC)$;   (c) $P(AB \mid C)$;   (d) $P(A \cup B \mid C)$;   (e) $P(C \mid A \cup B)$; (f) $P(A' \mid C)$; (g) $P(B \mid C')$; (h) $P(ABC \mid A \cup B)$; (i) $P(A \cup B \cup C \mid AB)$; (j) $P(A' \cup B' \mid C)$.

6. A certain construction company buys 20%, 30%, and 50% of their nails from hardware suppliers A, B, and C, respectively. Suppose it is known that .05%, .02%, and .01% of the nails from A, B, and C, respectively, are defective.

(a)   What percent of the nails purchased by the construction company are defective?

(b)   If a nail purchased by the construction company is defective, what is the probability that it came from supplier C?

7. The supervisor of a certain large group of factory workers knows from past experience that a worker who lives within 5 miles of the factory has a probability of .95 of arriving to work on time, while a worker who lives more than 5 miles from the factory has a probability of only .85 of arriving to work on time.

(a)   If 75% of the workers live more than 5 miles from the factory, what is the probability that a worker will arrive to work on time?

(b)   Use the answer to (a) to find the conditional probability of a worker living within 5 miles of the factory if he arrives to work on time.

8. During the holiday season the salespersons handling telephone orders at a certain large department store are so busy that on the average only 60% of the customers are able to speak to a salesperson immediately. The other 40% are asked to hold the line until a salesperson is available. About 30% of the time a customer who is asked to wait becomes impatient and hangs up the phone before a salesperson becomes available.

(a)   What percent of telephone customers who call this department store during the holiday season actually speak to a salesperson?

(b)   If a telephone customer speaks to a salesperson, what is the probability that he or she was able to do so immediately?

9.   Consider 3 boxes. Box A contains 2 pennies and 4 nickels; box B contains 6 pennies and 6 nickels; and box C contains 1 penny and 3 nickels. If one coin is selected at random from each box, what is the probability that a penny was selected from box A, given that exactly 2 pennies were selected?

10.   Prove (b) and (c) of Theorem 3.2. (*Hint:* for proof (b), replace BC with D in $P(ABC)$ and then apply (a) to $P(AD)$. Explain why (c) follows from (b).)

11.   A pathologist is certain that the cause of death is due to one of three poisons. Up to this point she feels that the test results indicate probabilities of .6, .3, and .1 that the death was caused by poisons 1, 2, and 3, respectively. She conducts another test. She was startled by the result, because the probability of this result if the cause of death is poison 1, 2, or 3, is .1, .7, or .2, respectively.

(a)   How might the pathologist use Bayes' Theorem to update her subjective probabilities concerning the cause of death?

(b)   What are the prior probabilities?

(c)   What are the posterior probabilities?

12.   Consider Example 3.6.

(a)   What is the probability that it requires testing exactly four fuses?

(b)   Explain why the sum of the answers to (a) and (b) of Example 3.6 with the answers to (a) of this exercise should be 1?

13.   (a)   Prove Theorem 3.2(b). (*Hint:* let $D = AB$ and use Theorem 3.2(a).)

(b)   Prove Theorem 3.2(c) for the case of $n = 4$.

3.3   Independent Events

Many concepts in probability have names that bring to mind familiar nonmathematical connotations. The concept of independence of events is an example. Of course, the names were adopted because of some connection with the usual meaning. One must be careful, however, not to jump to conclusions concerning the exact mathematical meaning of these concepts. A case in point is the definition of independence of events.

3.3.1   Definition of Independence

Definition 3.2   Independence of Two Events   Events A and B are said to be independent if

$$P(AB) = P(A)P(B)$$

Otherwise, the events are said to be dependent.

Definition 3.3   Independence of Several Events   Events $A_1, A_2, ..., A_n$ are said to be independent if the probability of the intersection of any 2, 3, ..., or n of these events equals the product of their respective probabilities.

3.3.2   Meaning of Independence

The formal definition of independence does not give much of a clue as to the intuitive meaning of independence. One may initially associate independence with separation and conclude that independence implies mutual exclusiveness. A more careful analysis shows that, on the contrary, mutual exclusiveness, except for the trivial case of zero probabilities, precludes the possibility of independence. (See Figure 3.3.)

Observation   *If events A and B are independent, $P(A) > 0$, and $P(B) > 0$, then $A \cap B \neq \emptyset$.*

Proof   If A and B are independent, then by definition

$$P(AB) = P(A)P(B)$$

Since $P(A) > 0$ and $P(B) > 0$, it follows that

$$P(A)P(B) > 0$$

Therefore

$$P(AB) = P(A)P(B) > 0$$

This implies that $A \cap B \neq \emptyset$, since $P(\emptyset) = 0$.

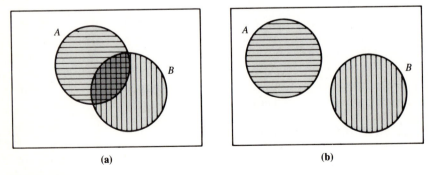

(a)   (b)

FIGURE 3.3
Question: If A and B are independent events with positive probabilities, which of these figures cannot be correct?

Answer: Figure 3.3(b) is incorrect.

But what is the intuitive meeting of independence? The following theorem provides the answer.

Theorem 3.5   *If events A and B have nonzero probabilities, then they are independent if and only if*

$$P(A \mid B) = P(A) \quad \text{and} \quad P(B \mid A) = P(B)$$

Proof   If A and B are independent, then

$$P(A \mid B) = \frac{P(AB)}{P(B)} = \frac{P(A)P(B)}{P(B)} = P(A)$$

and

$$P(B \mid A) = \frac{P(BA)}{P(A)} = \frac{P(B)P(A)}{P(A)} = P(B)$$

On the other hand, if $P(A \mid B) = P(A)$ and $P(B \mid A) = P(B)$, then

$$P(A) = P(A \mid B) = \frac{P(AB)}{P(B)}$$

and

$$P(B) = P(B \mid A) = \frac{P(BA)}{P(A)}$$

This implies that

$$P(A)P(B) = P(AB)$$

which is the condition of independence of events A and B.

Consequences of Theorem 3.5 and the Definition of Independence
   (a)   *If events A and B are independent, then events A' and B' are independent.*
   (b)   *If events A and B have nonzero probabilities that are less than 1, then they are independent if and only if*

$$P(A \mid B') = P(A) \quad \text{and} \quad P(B \mid A') = P(B)$$

The proofs of these consequences are considered in Exercises 4 and 5 of Section 3.3.

Theorem 3.5 and its consequence tell us that (a) events A and B with positive probabilities are independent if and only if the probability of the occurrence of A is not affected by the occurrence or nonoccurrence of B (i.e., $P(A \mid B) = P(A)$ and $P(A \mid B') = P(A)$) and (b) the probability of the occurrence of B is not affected by the occurrence or nonoccurrence of A (i.e., $P(B \mid A) = P(B)$ and $P(B \mid A') = P(B)$). This is the intuitive meaning of independence.

Intuitive Meaning of Independence   If events A and B have nonzero probabilities, then they are independent if and only if the probability of the occurrence of each is not affected by the occurrence or nonoccurrence of the other.

EXAMPLE 3.12   EXAMPLES OF INDEPENDENT EVENTS

(a)   Show that in two tosses of a fair coin, if A is the event of heads on the first toss and B is the event of tails on the second toss, then A and B are independent.

(b)   Show that in three throws of a balanced die, if A is the event of a 1 on the first throw, B is the event of a 4 on the second throw, and C is the event of an even number on the third throw, then events A, B, and C are independent.

SOLUTION

(a)   We need to show that $P(AB) = P(A)P(B)$.

Let $S = \{(t, t), (t, h), (h, t), (h, h)\}$ be the usual sample space for the experiment of two tosses of a fair coin. In terms of this sample space the events A, B, AB, and the number of sample samples N_A, N_B, N_{AB} are as follows:

$$
\begin{aligned}
A &= \{(h, t), (h, h)\} & N_A &= 2 \\
B &= \{(h, t), (t, t)\} & N_B &= 2 \\
AB &= \{(h, t)\} & N_{AB} &= 1
\end{aligned}
$$

Therefore,

$$
\begin{aligned}
P(A) &= N_A/N = 2/4 = 1/2 \\
P(B) &= N_B/N = 2/4 = 1/2 \\
P(AB) &= N_{AB}/N = 1/4
\end{aligned}
$$

and

$$P(A)P(B) = (1/2)(1/2) = 1/4 = P(AB)$$

(b)   To show that events A, B, and C are independent we need show that $P(AB) = P(A)P(B)$, $P(BC) = P(B)P(C)$, $P(AC) = P(A)P(C)$,   and $P(ABC) = P(A)P(B)P(C)$.

Let $S = \{(i, j, k) \mid i = 1, 2, \ldots, 6; j = 1, 2, \ldots, 6; k = 1, 2, \ldots, 6\}$ be the usual sample space for the experiment of three throws of a die. The sample space consists of $N = 216$ equally probable sample points. The number of sample points in the events under consideration are $N_A = 36$, $N_B = 36$, $N_C = 108$, $N_{AB} = 6$, $N_{BC} = 18$, $N_{AC} = 18$, and $N_{ABC} = 3$. The probabilities of the events under consideration are $P(A) = 1/6$, $P(B) = 1/6$, $P(C) = 1/2$, $P(AB) = 1/36$, $P(BC) = 1/12$, $P(AC) = 1/12$, and $P(ABC) = 1/72$. It is now clear that the events A, B, and C are independent because

$$P(AB) = 1/36 = (1/6)(1/6) = P(A)P(B)$$
$$P(BC) = 1/12 = (1/6)(1/2) = P(B)P(C)$$
$$P(AC) = 1/12 = (1/6)(1/2) = P(A)P(C)$$
$$P(ABC) = 1/72 = (1/6)(1/6)(1/2) = P(A)P(B)P(C)$$

If we generalize from the two cases of Example 3.12, we can easily see that if $A_1, A_2, \ldots, A_n$ are events whose occurrence or nonoccurrence are completely determined by the performances of separate and distinct experiments $\mathscr{E}_1, \mathscr{E}_2, \ldots, \mathscr{E}_n$, respectively, then the events $A_1, A_2, \ldots, A_n$ are independent. For instance, the experiments $\mathscr{E}_1, \mathscr{E}_2$, and $\mathscr{E}_3$ may be the first, second, and third throws of a die; the events A_1, A_2, and A_3 may be the events of a 1 on the first throw, a 4 on the second throw, and an even number on the third throw, respectively.

EXAMPLE 3.13

A certain experiment is measured only in terms of its success or failure. Suppose that the probability of success is p and the probability of failure is $q = 1 - p$. If the experiment is repeated 4 times under identical conditions, what are the probabilities of obtaining no success, exactly 1 success, exactly 2 successes, exactly 3 successes, and exactly 4 successes?

SOLUTION

Let S_1, S_2, S_3, and S_4 be the events of success on the first, second, third, and fourth performances of the experiment, respectively, and let F_1, F_2, F_3, and F_4 be the events of failure on the first, second, third, and fourth performances of the experiment, respectively. Using the property of independence of events, we obtain

$$P(0 \text{ success}) = P(F_1 F_2 F_3 F_4) = P(F_1)P(F_2)P(F_3)P(F_4) = q^4$$
$$P(1 \text{ success}) = P[(S_1 F_2 F_3 F_4) \cup (F_1 S_2 F_3 F_4) \cup (F_1 F_2 S_3 F_4)$$
$$\cup (F_1 F_2 F_3 S_4)]$$
$$= P(S_1 F_2 F_3 F_4) + P(F_1 S_2 F_3 F_4)$$
$$+ P(F_1 F_2 S_3 F_4) + P(F_1 F_2 F_3 S_4)$$
$$= P(S_1)P(F_2)P(F_3)P(F_4) + \cdots$$
$$+ P(F_1)P(F_2)P(F_3)P(S_4)$$
$$= 4pq^3$$

$$P(2 \text{ successes}) = P[(S_1 S_2 F_3 F_4) \cup (S_1 F_2 S_3 F_4) \cup (S_1 F_2 F_3 S_4)$$
$$\cup (F_1 S_2 S_3 F_4) \cup (F_1 S_2 F_3 S_4) \cup (F_1 F_2 S_3 S_4)]$$
$$= P(S_1 S_2 F_3 F_4) + \cdots + P(F_1 F_2 S_3 S_4)$$
$$= P(S_1) P(S_2) P(F_3) P(F_4) + \cdots$$
$$+ P(F_1) P(F_2) P(S_3) P(S_4)$$
$$= 6p^2 q^2$$

$$P(3 \text{ successes}) = 4p^3 q \qquad \text{(See Exercise 7.)}$$
$$P(4 \text{ successes}) = p^4 \qquad \text{(See Exercise 7.)}$$

Exercises

✓ 1. Use the property of independence to find the probability of getting 10 heads in 10 tosses of a fair coin.

2. Use the property of independence to find the probability of getting only even values in 5 rolls of a balanced die.

✓ 3. If a coin is tossed, a die is rolled, and a card is drawn at random from a standard deck of 52 cards, what is the probability of obtaining a head on the coin, a 6 on the die, and an ace on the card?

4. Prove that if A and B are independent events
  ✓ (a) A and B' are independent,
      (b) A' and B' are independent.

5. Prove that if events A and B have nonzero probabilities, they are independent if and only if $P(A \mid B') = P(A)$ and $P(B \mid A') = P(B)$.

✓ 6. Let events A and B be independent with $P(A) = .5$ and $P(B) = .8$. Find the probability that neither of the events A nor B occurs.

7. Find the probabilities of 3 successes and of 4 successes in Example 3.13.

✓ 8. Find the probabilities of obtaining no success, exactly 1 success, exactly 2 successes, and exactly 3 successes in 3 performances of an experiment in which probability of success is p and probability of failure is $1 - p$. (*Hint:* see Example 3.13.)

✓ 9. Consider three rolls of a balanced die. Let A, B, and C be the events of a 1 on the first roll, a sum of 7 in the first two rolls, and a 6 on the third roll, respectively. Which of the following sets of events are independent? (a) A and B; (b) A and C; (d) A, B, and C.

3.4   Additional Applications and Examples

These applications and examples illustrate the combined use of the concepts of conditional probability and independent events.

EXAMPLE 3.14 A PROBLEM OF PASCAL

On each play of a game, one of two players scores a point. Both players have equal chances of making a point in each play, and three points are required to win. If for some unforeseen reason the game must be terminated when one player has 2 points and the other has 1 point, how should the stakes be divided? Pascal said that the stakes should be split 3 to 1 in favor of the player who is ahead. Do you agree? How did Pascal arrive at his answer?

PASCAL'S SOLUTION

Suppose that player A has 1 point and player B has 2 points. Then

$$P(A \text{ wins} \mid B \text{ already has 2 points}) = P(A \text{ wins the next two plays})$$

(by independence)

$$\downarrow$$
$$= P(A \text{ wins the next play}) P(A \text{ wins the play after the next play})$$
$$= (1/2) \cdot (1/2) = 1/4$$

By using Theorem 3.1 on the conditional probability of the complement of an event we obtain

$$P(B \text{ wins} \mid B \text{ already has 2 points})$$
$$= 1 - P(A \text{ wins} \mid B \text{ already has 2 points}) = 1 - (1/4) = 3/4$$

This means that if player B already has 2 points, he or she is 3 times as likely to win as player A. It is reasonable, therefore, to split the stakes 3 to 1 in favor of the player who was ahead.

EXAMPLE 3.15 A TRICKY CARD GAME

Only an unscrupulous person would trick someone with this game!

Begin with 3 cards. Card I is red on both sides, Card II is black on both sides, and Card III is red on one side and black on the other. The game follows:

Step 1: Player A mixes the 3 cards in a box.

Step 2: Player B draws a card from the box and places it on a table without revealing the underside of the card.

Step 3: Player A guesses the color on the underside of the card.

Step 4: If player A guesses correctly, player B pays $1; otherwise player B receives $1 from player A.

If you were player A, how should you do your guessing?

SOLUTION

Let R_1 and B_1 be the events that the topside of the card on the table is red and black, respectively. Let I, II, and III be the events that the card on the table is card I, card II, and card III, respectively. Let us find the conditional probability of black on the underside given red on the topside. Note that this is equivalent to $P(III \mid R_1)$. Upon application of Bayes' Theorem we obtain

$$P(III \mid R_1) = \frac{P(R_1 \mid III)P(III)}{P(R_1 \mid I)P(I) + P(R_1 \mid II)P(II) + P(R_1 \mid III)P(III)}$$

$$= \frac{(1/2)(1/3)}{(1)(1/3) + (0)(1/3) + (1/2)(1/3)} = \frac{1}{3}$$

By symmetry, the conditional probability of red on the underside, given black on the topside, is also 1/3.

This lesson in conditional probability tells us that if player A consistently guesses the same color on the underside as that on the topside, in the long run this strategy will yield a correct response approximately two-thirds of the time.

EXAMPLE 3.16   A TECHNIQUE OF CONDITIONING

The technique of conditioning used to solve the following problem may be applied to more complex situations. The problem below was one of the questions in a Graduate Record Examination given a number of years ago.

Consider an infinite sequence of actions where the odd steps consist of tossing a coin and the even steps consist of rolling a die. In this sequence of action what is the probability that a head on a coin will appear before a 5 or 6 on a die?

SOLUTION

Let A be the event of a head on a coin before a 5 or 6 on a die. Let H_n, T_n, and B_n be the events of a head, a tail, and a 5 or 6, on the nth step of the sequence of action. Then

$$P(A) = P(A \mid H_1)P(H_1) + P(A \mid T_1)P(T_1) \tag{1}$$
$$= (1)(1/2) + P(A \mid T_1)(1/2) = (1/2) + (1/2)P(A \mid T_1)$$

Now, because T_1 and B_2 are independent events, the statement that follows is true (see Exercise 3 for justification).

$$P(A \mid T_1) = P(A \mid T_1 B_2)P(B_2) + P(A \mid T_1 B_2')P(B_2') \tag{2}$$
$$= (0)(1/3) + P(A)(2/3)$$

In Equation (2), we were able to replace $P(A \mid T_1 B_2')$ with $P(A)$ because if the first and second steps in the sequence of actions yielded tails and neither 5 nor 6, we can behave as though we were again at the beginning of the sequence.

To complete the computation, substitute the value for $P(A \mid T_1)$ from Equation (2) into Equation (1), to obtain

$$P(A) = (1/2) + (1/2)P(A \mid T_1) = (1/2) + (1/2)[(2/3)P(A)]$$

Solve the above equation for $P(A)$. Then

$$P(A) = 3/4$$

Exercises

1. Suppose that in Example 3.14 four points are required to win.
   (a) If the game were terminated when one player has 3 points and the other has 1 point, how should the stakes be divided?
   (b) If the game were terminated when one player has 3 points and the other has 2 points, how should the stakes be divided?
   (c) If the game were terminated when one player has 2 points and the other has 1 point, how should the stakes be divided? (*Hint:* for (c) refer to Example 3.16.)

2. Three boxes have identical exteriors. Box I contains two pennies, box II contains two nickels, and box III contains a penny and a nickel. A box is selected at random, then a coin is selected at random from the box. If the coin selected is a penny, what is the probability that the other coin in the box is also a penny? (*Hint:* see Example 3.15.)

3. Prove that if B and C are independent events, for any event A, $P(A \mid B) = P(A \mid BC) P(C) + P(A \mid BC') P(C')$. (*Hint:* replace each conditional probability in the equation with its definition in terms of probabilities.)

4. A coin is tossed, a die is rolled, and then a card is drawn from a standard deck of 52 cards. If this sequence of actions is repeated *ad infinitum*, then what is the probability of obtaining a head on a coin before a 6 on a die or an ace in a card? (*Hint:* see Example 3.16.)

3.5 Tree Diagrams

The solutions of problems involving conditional probabilities often is greatly simplified by using tree diagrams. Consider the following example.

EXAMPLE 3.17

Samples arriving at a certain laboratory may be given a number of tests to determine the presence of trace amounts of Chemical X. If the result of test A is positive, then the sample is subjected to test B. If the result of test B is positive, then it is subjected to test C. If the result of test C is positive, then it is assumed that the sample contains Chemical X. Suppose that the probabilities of false positives for tests A, B, and C, are .05, .10, and .15, respectively, and the probabilities of false negatives are .01, .02, and .03, respectively. Furthermore, suppose that 10% of samples arriving at the laboratory contain trace amounts of Chemical X.

(a)   Draw a tree diagram of the probabilities given in this example.

(b)   Find the probability that the laboratory will falsely assume the presence of Chemical X in a sample.

(c)   Find the probability that the laboratory will falsely assume the absence of Chemical X in a sample.

(d)   What percent of samples tested by the laboratory are reported to contain Chemical X?

(e)   What percent of samples tested by the laboratory are reported to contain no trace of Chemical X?

SOLUTION

(a)   A tree diagram of the probabilities is given in Figure 3.4. In the diagram, X_+ and X_- denote, respectively, the presence and absence of trace amounts of Chemical X in the sample. The subscripts of $+$ and $-$ with the letters A, B, and C are used to denote, respectively, the result of positive and the result of negative in the respective tests.

(b)   Reading off the appropriate probabilities from the tree diagram we find that the answer to (b) is

P(the laboratory will falsely assume the presence of Chemical X)

$= P(A_+B_+C_+ \mid X_-) = P(A_+ \mid X_-)P(B_+ \mid A_+X_-)P(C_+ \mid A_+B_+X_-)$

$= (.05)(.10)(.15) = .00075$

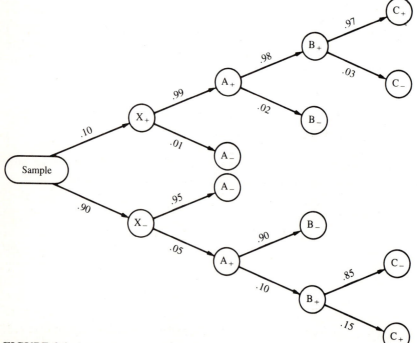

FIGURE 3.4
Example 3.17: Tree Diagram of Probabilities

(c) Reading off the appropriate probabilities from the tree diagram we find that the answer to (c) is

P(the laboratory will falsely assume the absence of Chemical X)

$$= P(A_- \,|\, X_+) + P(A_+ B_- \,|\, X_+) + P(A_+ B_+ C_- \,|\, X_+)$$
$$= P(A_- \,|\, X_+) + P(A_+ \,|\, X_+) P(B_- \,|\, A_+ X_+)$$
$$\quad + P(A_+ \,|\, X_+) P(B_+ \,|\, A_+ X_+) P(C_- \,|\, A_+ B_+ X_+)$$
$$= (.01) + (.99)(.02) + (.99)(.98)(.03) = .05891$$

(d) Reading off the appropriate probabilities from the tree diagram we find that the answer to (d) is

P(the laboratory will report the presence of Chemical X in a sample)

$$= P(A_+ B_+ C_+)$$
$$= P(A_+ B_+ C_+ \,|\, X_+) P(X_+) + P(A_+ B_+ C_+ \,|\, X_-) P(X_-)$$
$$= P(X_+) P(A_+ \,|\, X_+) P(B_+ \,|\, A_+ X_+) P(C_+ \,|\, A_+ B_+ X_+)$$
$$\quad + P(X_-) P(A_+ \,|\, X_-) P(B_+ \,|\, A_+ X_-) P(C_+ \,|\, A_+ B_+ X_-)$$
$$= (.10)(.99)(.98)(.97) + (.90)(.05)(.10)(.15) = .09478$$

In other words, the laboratory will report the presence of Chemical X in approximately 9.5% of the samples.

(e) Reading off the appropriate probabilities from the tree diagram we find that the answer to (e) is

P(the laboratory will report the absence of Chemical X in a sample)

$$= P(A_- \cup A_+ B_- \cup A_+ B_+ C_-)$$
$$= P(X_+) P(A_- \cup A_+ B_- \cup A_+ B_+ C_- \,|\, X_+)$$
$$\quad + P(X_-) P(A_- \cup A_+ B_- \cup A_+ B_+ C_- \,|\, X_-)$$
$$= (.10) \cdot [(.01) + (.99)(.02) + (.99)(.98)(.03)]$$
$$\quad + (.90) \cdot [(.95) + (.05)(.90) + (.05)(.10)(.85)]$$
$$= .9052$$

In other words, the laboratory will report the absence of Chemical X in approximately 90.5% of the samples. This answer may, of course, be also obtained by referring to (d).

Exercises

1. Applicants for employment at a certain large firm are initially interviewed by any one of three personnel officers. If an interview is favorable, the applicant is given a written test. If the applicant's test result is favorable, he

or she is interviewed by a technical staff member. A favorable report from the technical staff member places the applicant on a waiting list for the next job opening. The following statistics are known concerning the application process at this firm: personnel officers A, B, and C interview 40%, 35%, and 25%, respectively, of the applicants. These personnel officers give favorable reports to 60%, 75%, and 80%, respectively, of the applicants. Only 10% of the applicants who take the written test receive favorable results, and of these applicants only 30% receive favorable reports from the interview with a technical staff member.

(a) Draw a tree diagram of the empirical probabilities defined by these statistics.

(b) What percentage of all applicants are eventually placed on the waiting list for employment?

(c) What percentage of applicants who receive favorable results on their initial interviews with the personnel officer are eventually placed on the waiting list for employment?

2. Alice and Mary are engaged in a chess tournament. The first person to win either 2 games in a row or a total of 3 games will be declared the winner.

(a) If Mary has a probability of .6 of winning each of the individual games, what is the probability that Mary will win the tournament?

(b) What is the probability that Mary will win the tournament after fewer than 4 games? (*Hint:* draw a tree diagram of the pertinent probabilities. This problem was considered in Example 2.12 of Chapter 2.)

3. Players A, B, C are engaged in a series of games of chance. The series begins with A and B as opponents. The winner of each game continues the series by playing against the player who was not a participant in the most recent game. The series continues until one player has won 3 games.

(a) Draw a tree diagram of the possible outcomes with the associated probabilities.

(b) What is the probability that player A will win the series of games?

(c) What is the probability that player B will win the series of games? (*Note:* this problem was considered in Exercise 5 of Section 2.5 of Chapter 2.)

Supplementary Exercises

1. Suppose that 5% of males and 2.5% of females are color blind. If a color-blind person is chosen at random, what is the probability that the person is female? Assume that there are an equal number of males and females.

2. Perhaps not all politicians are liars, but candidates A, B, and C are—at least some of the time. Suppose that candidates A, B, and C lie 5%, 10%, and 15% of the time, respectively; tell half-truths 40%, 45%, and 50% of the time, respectively; and are truthful 55%, 45%, and 35% of the time, respectively. Furthermore, suppose that candidate A issues twice as

many statements as candidate B and three times as many statements as candidate C.

(a)   If a lie was issued by one of these candidates, what is the probability that candidate A was responsible?

(b)   What percent of the combined statements issued by these candidates are half-truths?

 3.   Suppose that each child born to a couple is equally likely to be a boy or a girl, independently of the sex of the other children in the family. If a couple has four children, then what are the probabilities of the following events:

(a)   All four children are girls.

(b)   All four children are of the same sex.

(c)   Exactly two are boys.

(d)   The eldest is a girl.

(e)   There is at least one boy.

 4.   Players A and B take turns rolling a die. The first player to get a 6 wins the game. If player A has the first turn, what is the probability that he will be the winner. (*Hint:* see Example 3.16.)

 5.   Suppose that in Exercise 4 there are three players.

(a)   What is the probability that player A will be the winner if he has the first turn?

(b)   What is the probability that player B will be the winner if he has the second turn?

 6.   Twenty items, 6 of which are defective and 14 not defective, are inspected one after the other. If these items are chosen at random, what is the probability that:

(a)   The first three items inspected are defective?

(b)   The first three items inspected are not defective?

(c)   Among the first three items inspected, there is one defective and two not defective?

 7.   In a nail factory, machines A, B, and C manufacture 20%, 30%, and 50% of the total output, respectively. Of their outputs, 4%, 5%, and 1%, respectively, are defective nails. A nail is chosen at random and found to be defective. What is the probability that the nail came from machine A? B? C?

 8.   Suppose that in the production of a certain item, defects of one type occur with probability .05 and defects of a second type occur with probability .01. If the occurrence or nonoccurrence of one type of defect does not affect the occurrence or nonoccurrence of the other, what is the probability that, with respect to these types of defects,

(a)   an item of this type is defective?

(b)   an item of this type has both kinds of defects?

(c)   an item of this type has only one type of defect, given that it is defective?

 9.   A system consists of components A and B. The probability that

component A fails is .01, that component B fails is .02, and that both
components fail is .001.

(a)   Are the events of failure of components A and B independent
events?

(b)   What is the conditional probability of a failure in component A,
given that there is a failure in component B?

(c)   What is the conditional probability of a failure in component B,
given that there is a failure in component A?

(d)   Does the event of a failure in one component increase or decrease
the likelihood of failure in the other component?

(e)   What is the probability of a failure in at least one of the components?

(f)   What is the probability of a failure in exactly one of the
components?

10.   Let S be a sample space. Let C be an event in S with $P(C) > 0$. Prove

(a) $0 \le P(A|C) \le 1$ for all events A in S;

(b) $P(S|C) = 1$;

(c) $P(A \cup B|C) = P(A|C) + P(B|C)$ for mutually exclusive events A
and B in S (i.e., prove that if P_C is a set function that is defined by the
relation $P_C(A) = P(A|C)$, then P_C is a probability set function.)

Chapter *4*

DISCRETE RANDOM VARIABLES AND PROBABILITY FUNCTIONS

4.1   Introduction

Results of experiments are often summarized in terms of numerical values. For example, in a game of dice, the result may be summarized in terms of the sum of the values of the two dice; in testing the mathematical aptitude of business executives, the result may be summarized in terms of the average test score; and in a survey of voters, the result may be summarized in terms of the proportion of respondents who prefer a certain candidate. The sum of the values of two dice, the average test score, and the proportion of respondents are numerical quantities. However, the specific value of each of these quantities depends on the outcomes of the experiments—in the throw of the dice, in the performance of the business executives, and in the preference of the voters. In other words, the specific values of these variable quantities cannot be predicted with certainty before the experiment is performed. For this reason, these variable quantities (e.g., the sum of the values of two dice, the average test score, and the proportion of respondents) are called **random variables**. Random variables and their distributions play a fundamental role in the mathematics of probability and statistics—so much so, in fact, that one may say the key to the understanding of probability and statistics is an understanding of random variables and their distributions. The precise definition of a random variable for experiments with a discrete sample space follows. Random variables for experiments with a continuous sample space are discussed in Chapter 5.

4.2   Discrete Random Variables

In each of the examples in Section 4.1, the informally defined random variable is a function that assigns a real number to each outcome of the experiment. For instance, consider the game of dice. The numerical quantity under consideration is the sum of the values of the two dice. Thus, if we let X denote the random variable that is the sum of the values of the two dice, then X assigns to the outcome $(1, 3)$ the real number $1 + 3 = 4$; it assigns to the outcome $(3, 5)$ the real number $3 + 5 = 8$; and, in the general case, it assigns to the outcome (i, j) the real number $i + j$. Using the notation of functions (such as $f(x) = 2x + 3$), we may write

$$X(1, 3) = 1 + 3 = 4$$
$$X(3, 5) = 3 + 5 = 8$$

and

$$X(i, j) = i + j$$

for $i = 1, 2, \ldots, 6$ and $j = 1, 2, \ldots, 6$. Thus, X is a function that assigns the real number $i + j$ to each outcome (i, j) of the experiment. In other words, X is a function that assigns the real number $i + j$ to each sample point (i, j) of the sample space for the experiment.

Another way of saying that X assigns the number 4 to the outcome $(1, 3)$, that it assigns the number 8 to the outcome $(3, 5)$, and that it assigns the number $i + j$ to the outcome (i, j), is to say that X is equal to 4 when the outcome is $(1, 3)$, that X is equal to 8 when the outcome is $(3, 5)$, and that X is equal to $i + j$ when the outcome is (i, j). Figure 4.1 is a table of values for the random variable X for corresponding outcomes (i.e., sample points) of the experiment. It illustrates the assignment of numbers by the random variable X.

In the experiment of testing for the mathematical aptitude of business executives, suppose that tests are given to three randomly selected business executives, and that (i, j, k) denotes the outcome of scores of i, j, and k, respectively, for the first, second, and third executive chosen. The numerical quantity under consideration, in this case, is the average test score. Thus if we let X denote the random variable that is the average of the three test scores, then X assigns to the outcome $(60, 95, 88)$ the real number $(60 + 95 + 88)/3 = 81$. It assigns to the outcome $(98, 47, 89)$ the real number $(98 + 47 + 89)/3 = 78$ and, in the general case, it assigns to the outcome (i, j, k) the real number $(i + j + k)/3$. Using the notations of functions, we may write $X(i, j, k) = (i + j + k)/3$. Thus, again, the random variable X is a function that assigns a real number to each sample point in the sample space of the experiment.

The formal definition of a discrete random variable follows.

Definition 4.1   Let S be a discrete sample space with a probability set function P. Let X be a function that assigns a real number to each sample point in the sample space S. X is called a random variable. In particular, it is called a **discrete random variable** because S is a discrete sample space.

Sample point	Value of X
(1, 1)	2
(1, 2)	3
(1, 3)	4
⋮	⋮
(1, 6)	7
(2, 1)	3
(2, 2)	4
⋮	⋮
(6, 1)	7
⋮	⋮
(6, 6)	12

FIGURE 4.1
Assignment of Numbers by the
Random Variable X

EXAMPLE 4.1

Consider a throw of two dice (one red, one green). Use the notations of functions to define the following variables.
   (a) X is the absolute value of the difference in values of the two dice.
   (b) Y is the value of the red die plus the square of the value of the green die.

   SOLUTION

Let (i, j) denote the outcome of an i on the red die and a j on the green die.
   (a)

$$X(i, j) = |i - j| \quad \text{for} \quad i = 1, 2, \ldots, 6 \quad \text{and} \quad j = 1, 2, \ldots, 6$$

   (b)

$$Y(i, j) = i + j^2 \quad \text{for} \quad i = 1, 2, \ldots, 6 \quad \text{and} \quad j = 1, 2, \ldots, 6$$

EXAMPLE 4.2

Consider 3 flips of a coin. Use a table of values to define the following random variables.
   (a) X is the number of heads in the 3 flips.
   (b) Y is the number of heads minus the number of tails.

   SOLUTION

See Figure 4.2.

Exercises

1. Consider a throw of two dice (one red, one green). Use the notation of functions to define the following random variables.
    (a) X is the value of the red die minus the value of the green die.
    (b) Y is the value of the red die times the value of the green die. (*Hint:* see Example 4.1.)

Sample point	Value of X	Value of Y
(h, h, h)	3	3
(h, h, t)	2	1
(h, t, h)	2	1
(t, h, h)	2	1
(h, t, t)	1	−1
(t, h, t)	1	−1
(t, t, h)	1	−1
(t, t, t)	0	−3

FIGURE 4.2
Assignment of Numbers by the Random
Variables X and Y

2. Consider 3 flips of a coin. Use a table of values to define the following random variables.

(a) X is the number of heads times the number of tails.

(b) Y is the absolute value of the difference in the number of heads and tails. (*Hint:* see Example 4.2 and Figure 4.2.)

3. The possible ratings on a certain music test are 1, 2, and 3. Suppose that three students selected at random take this test and X is their average test score.

(a) Use the notation of functions to define the random variable X. (*Hint:* let (i, j, k) denote the outcome of the first, second, and third student receiving ratings of $i, j,$ and k, respectively.)

(b) Use a table of values to define the random variable X.

4.3   Probability Functions

Because the value of a random variable is determined by the outcome of the experiment, associated with each possible value of a discrete random variable there is a probability that the random variable will take on that value. The following example illustrates this point.

EXAMPLE 4.3

Consider 3 flips of a coin. Let X be the number of heads in the 3 flips. The possible values of X are 3, 2, 1, and 0 (see Figure 4.2). Find the probabilities of X being equal to 3, 2, 1, and 0, respectively.

SOLUTION

The experiment of the example is identical to the experiment of Example 4.2. Therefore, we may use the data of Figure 4.2 to compute the following probabilities:

$$P(X = 3) = P[\{(h, h, h)\}] = 1/8$$
$$P(X = 2) = P[\{(h, h, t), (h, t, h), (t, h, h)\}] = 3/8$$
$$P(X = 1) = P[\{(h, t, t), (t, h, t), (t, t, h)\}] = 3/8$$
$$P(X = 0) = P[\{(t, t, t)\}] = 1/8$$

It is important to understand that when we write $P(X = 3)$, the expression $X = 3$ refers to the event of the random variable X being equal to 3 (that is, the event of the random variable X taking on the value 3). The expression does not imply that X is equivalent to the number 3. In the above example, the event $X = 3$ is equivalent to the event of 3 heads in 3 flips of the coin, the event $X = 2$ is the event of 2 heads in 3 flips of the coin, the event $X = 1$ is the event of 1 head in 3 flips of the coin, and the event $X = 0$ is the event of 0 heads in 3 flips of the coin. We saw that for each possible value x of the random variable X we could find the probability $P(X = x)$. Thus, we could define a function $f(x) = P(X = x)$. This function is called the probability function or the probability distribution of the random variable X. The formal definition follows.

Definition 4.2   Let X be a discrete random variable defined on a sample space S with probability set function P. The **probability function** or probability distribution of X is the function f defined by

$$f(x) = P(X = x)$$

for each possible value x of the random variable X.

EXAMPLE 4.4

Consider a throw of two dice. Let X be the sum of the values of the two dice.
  (a)   Find the probability function of X.
Use the probability function to find the probabilities of obtaining
  (b)   a sum of 7 or 11;
  (c)   a sum that is greater than 8;
  (d)   a sum that is greater than 3 but less than 9.

  SOLUTION

  (a)   The possible values of X are the integers $2, 3, \ldots, 12$. The values of the probability function f are as follows:

$$f(2) = P(X = 2) = P[\{(1, 1)\}] = 1/36$$
$$f(3) = P(X = 3) = P[\{(1, 2), (2, 1)\}] = 2/36$$
$$f(4) = P(X = 4) = P[\{(1, 3), (2, 2), (3, 1)\}] = 3/36$$
$$\vdots$$
$$f(12) = P(X = 12) = P[\{(6, 6)\}] = 1/36$$

Figure 4.3 gives the values of f in tabular form.

x	$f(x)$
2	1/36
3	2/36
4	3/36
5	4/36
6	5/36
7	6/36
8	5/36
9	4/36
10	3/36
11	2/36
12	1/36

FIGURE 4.3
The Probability Function of X

(b)

$$P(\text{a sum of 7 or 11})$$
$$= P[(X = 7) \quad \text{or} \quad (X = 11)]$$
$$= P(X = 7) + P(X = 11)$$
$$= f(7) + f(11) = (6/36) + (2/36) = 2/9$$

(c)

$$P(\text{a sum that is greater than } 8) = P(X > 8)$$
$$= P(X = 9) + P(X = 10) + P(X = 11) + P(X = 12)$$
$$= f(9) + f(10) + f(11) + f(12)$$
$$= (4/36) + (3/36) + (2/36) + (1/36) = 5/18$$

(d)

$$P(\text{a sum that is greater than 3 but less than 9})$$
$$= P(3 < X < 9) = f(4) + f(5) + f(6) + f(7) + f(8)$$
$$= 23/36$$

It is easy to see that if x is a possible value of a discrete random variable X, then $f(x)$ is a number that is greater or equal to 0 and less than or equal to 1. This is because, for each fixed value of x, $f(x)$ is a probability. Also, it is easy to see that if you add the values of $f(x)$ over all possible values of x, then the sum would be 1. This is because the result is identical to the sum of the probabilities of all sample points in the sample space. These 2 properties of the probability function are formally stated in the following theorem.

Theorem 4.1 *Let $f(x)$ be the probability function of a discrete random variable X. Then*
   (a)

$$0 \le f(x) \le 1 \quad \text{for each possible value } x \text{ of } X$$

   (b)

$$\sum_x f(x) = 1$$

(The summation is over all possible value x of the random variable X.)

EXAMPLE 4.5

Suppose that $f(x)$ is a function that is defined only for values $x = -1, 0, 1,$ and 3. Furthermore, suppose that $f(-1) = 1/2$, $f(0) = 1/4$, $f(1) = 1/8$, and $f(3) = 1/16$. Can $f(x)$ serve as the probability function of a random variable?

SOLUTION

The function $f(x)$ satisfies (a) of Theorem 4.1, but it does not satisfy (b) of the theorem. Therefore, $f(x)$ cannot be the probability function of a random variable.

Statements (a) and (b) of Theorem 4.1 are often referred to as the **defining properties** of a probability function. In other words, if a function f has the properties of (a) and (b), then f is the probability function of a random variable. The following example shows why this is true in a simple case. The reader can easily extend the result to the general case.

EXAMPLE 4.6

Suppose that f is a function that is defined for only the values of $x = -3, 0, 1$, and 4. Furthermore, suppose that $f(-3) = 1/10$; $f(0) = 3/10$, $f(1) = 4/10$, and $f(4) = 2/10$, so that conditions (a) and (b) of Theorem 4.1 are satisfied. Find a random variable X for which f is the probability function.

SOLUTION

Consider a box of ten cards: one has the number -3, three have the number 0, four have the number 1, and two have the number 4 written on them. Let X be the number on the card selected at random from the box, then the function f given above is clearly the probability function of X.

Exercises

1. Find the probability functions of X and Y of Exercise 1 of Section 4.2.
2. Find the probability functions of X and Y of Exercise 2 of Section 4.2.
3. Suppose that Adam, Bernie, Charles, and Dan have annual allowances of $1,000, $500, $200, and $1,300, respectively. If 2 of these young men are chosen at random to be roommates in room 101A, and if X is the average of their annual allowances, then what is the probability function of X?
4. The IQ's of a certain group of four students are 100, 150, 125, and 100, respectively. If three of the students are chosen at random to represent the school and X is the average IQ of the representatives, what is the probability function of X?
5. Consider the random variable of Example 4.4. Use the probability function to find the probabilities of obtaining
    (a) a sum that is less than 4;
    (b) a sum that is greater or equal to 3 but less than or equal to 9;
    (c) a sum that is divisible by 3.
6. A box contains a penny, two nickels, and a dime. If two coins are selected at random from the box and X is the sum of values of the two coins, what is the probability function of X?

7. Let X be a random variable whose probability function is defined by the values $f(-2) = 1/10$, $f(0) = 2/10$, $f(4) = 4/10$, and $f(11) = 3/10$. Find $P(-2 \le X < 4)$, $P(X > 0)$, and $P(X \le 4)$.

8. Check whether the following functions satisfy the conditions of a probability function. Explain your answers.
   (a) $f(x) = 1/4$ for $x = -3, 0, 10, 20$
   (b) $f(x) = 1/x$ for $x = 1, 2, 3, 4$
   (c) $f(x) = 1 - x$ for $x = 0, 1/2, 3/2$
   (d) $f(x) = (1/2)^x$ for $x = 1, 2, 3, 4, \ldots$
   (*Hint:* $\sum_{n=0}^{\infty} \cdot t^n = 1/(1 - t)$ (the geometric series))

9. If $f(x) = cx$ for $x = 2, 4, 6, 8$ is a probability function, what must be the value of c?

4.4 Probability Histograms and Bar Charts

The two standard conventions for depicting probability functions graphically are the **probability histogram** and the **bar chart**. Figure 4.4 is a histogram of the probability function for the sum of two dice (the random variable of Example 4.4); Figure 4.5 is a bar chart of the same probability function. In the histogram, the height of each rectangle equals the probability that the random variable X will take on the value that is the midpoint of its base. (Histograms are often adjusted so that it is the *area* of the rectangle rather than its height that represents the probability of the midpoint of the base of the rectangle.) In the bar chart, the height of each bar equals the probability of the corresponding value of the random variable.

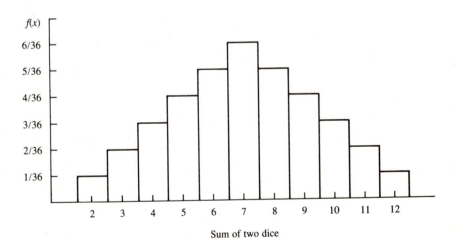

Sum of two dice

FIGURE 4.4
Probability Histogram

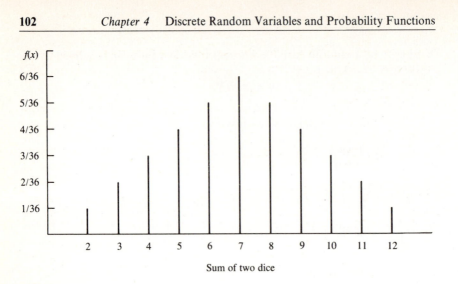

FIGURE 4.5
Bar Chart of Probability Function

Exercises

1. Draw a histogram of the probability function of the random variable X in Exercise 2 of Section 4.2.

2. Draw a bar chart of the probability function of the random variable Y in Exercise 2 of Section 4.2.

3. Draw a histogram and a bar chart of the random variable X in Example 4.3.

4.5   Several Important Discrete Distributions

A number of discrete random variables and their probability functions are used so frequently that they deserve special mention.

4.5.1   The Discrete Uniform Distribution

Consider the experiment of selecting a number at random from the set consisting of the numbers $x_1, x_2, \ldots, x_n$. Let X be the value of the number selected. If we assume that each of the numbers $x_1, x_2, \ldots,$ and x_n has an equal chance of being selected, then

$$f(x_i) = P(X = x_i) = 1/n \quad \text{for} \quad i = 1, 2, \ldots, n$$

This probability function is called the discrete uniform probability function or the discrete uniform probability distribution. (Recall that a probability function is also called a probability distribution.)

Definition 4.3   The function

$$f(x_i) = 1/n$$

for positive integer n and $i = 1, 2, \ldots, n$, is called the **discrete uniform probability function**.

A random variable X is said to have a discrete uniform distribution, and is referred to as a discrete uniform random variable if its probability function is the discrete uniform probability function.

4.5.2   The Bernoulli Distribution

An experiment consisting of only two possible outcomes, either success or failure, is called a Bernoulli trial. If X is a random variable that takes on the value 1 when the Bernoulli trial is a success and the value 0 when the Bernoulli trial is a failure, then X is called a Bernoulli random variable. If p is the probability of success and $q = 1 - p$ is the probability of failure, clearly the probability function of X is

$$f(x) = \begin{cases} p & \text{if } x = 1 \\ q & \text{if } x = 0 \end{cases}$$

This probability function is called the Bernoulli probability function or the Bernoulli probability distribution.

Definition 4.4   The function

$$f(x) = \begin{cases} p & \text{if } x = 1 \\ q & \text{if } x = 0 \end{cases}$$

where $0 < p < 1$ and $p + q = 1$, is called the **Bernoulli probability function**.

A random variable X is said to have a Bernoulli distribution, and is referred to as a Bernoulli random variable, if its probability function is the Bernoulli probability function.

4.5.3   The Binomial Distribution

Consider n Bernoulli trials. That is, consider n performances of an experiment consisting of only two possible outcomes. Suppose that the outcomes are labeled success and failure, and that their respective probabilities are p and $q = 1 - p$. If S_n is the total number of successes in the n Bernoulli trials, what is the probability function of S_n? In other words, what is $P(S_n = x)$ for

$x = 0, 1, 2, \ldots, n$? (This problem, for the case $n = 4$, was solved in Example 3.13 of Chapter 3.)

Let us approach this problem by considering first an easier, but related, problem. Let us find the probability of success on the first x Bernoulli trials followed by probability of failure on the remaining $n - x$ Bernoulli trials. To solve this problem, suppose we let s_i denote the event of success and f_i the event of failure on the ith Bernoulli trial. Then, the probability we seek is $P(s_1 s_2 \cdots s_x f_{x+1} f_{x+2} \cdots f_n)$. Now, because we are dealing with repeated performances of identical experiments (Bernoulli trials) the events $s_1, s_2, \ldots, s_x, f_{x+1}, f_{x+2}, \ldots, f_n$ are independent (see Section 3.3.2 of Chapter 3). Therefore

$$P(s_1 s_2 \cdots s_x f_{x+1} f_{x+2} \cdots f_n) = P(s_1)P(s_2) \cdots P(s_x)P(f_{x+1})P(f_{x+2}) \cdots P(f_n)$$
$$= p^x q^{n-x}$$

We see, then, that the probability of getting x successes and $n - x$ failures in the one specific order designated above is $p^x q^{n-x}$. In fact, the probability of getting x successes and $n - x$ failures in any one specific order is $p^x q^{n-x}$. Thus, the probability of getting exactly x successes in n Bernoulli trials is the number of ways in which x successes can occur among the n trials multiplied by the quantity $p^x q^{n-x}$. Now, the number of ways in which x successes can occur among n trials is the same as the number of distinct arrangements of n items of which x are of one kind (success) and $n - x$ are of another kind (failure). We saw in Chapter 2 that this number is the binomial coefficient $\binom{n}{x} = n!/(x!(n - x)!)$. In conclusion, the probability function for S_n, the number of successes in n Bernoulli trials, is

$$P(S_n = x) = \binom{n}{x} p^x q^{n-x} \qquad \text{for} \quad x = 0, 1, 2, \ldots, n$$

This probability function is called the **binomial probability function** because of the role of the binomial coefficient in deriving the exact expression of the function. It is also called the binomial probability distribution, or simply, the binomial distribution. In view of its importance in probability and statistics there is a special notation for this function, as we will see in the following formal definition.

Definition 4.5   Let n be a positive integer, and let p and q be probabilities, with $p + q = 1$. The function

$$b(x; n, p) = \binom{n}{x} p^x q^{n-x} \qquad \text{for} \quad x = 0, 1, 2, \ldots, n$$

is called the **binomial probability function**.

A random variable is said to have a binomial distribution, and is referred to as a **binomial random variable**, if its probability function is the binomial

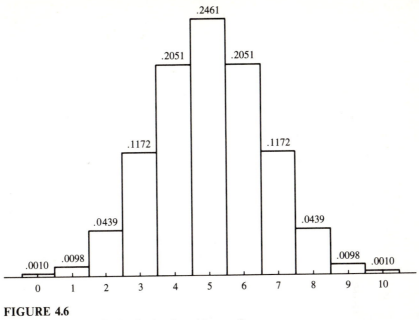

FIGURE 4.6
Histogram: Binomial Distribution $(n = 10, p = .5)$

probability function. In this book we reserve the notation S_n for the binomial random variable because of its easy association with the number of successes in n trials.

A histogram of the binomial probability function with $n = 10$ and $p = .5$ is shown in Figure 4.6.

There are tables of values for the binomial probability function. For example, Table I in the Appendix contains values for the binomial probability function for $n = 1, 2, \ldots, 20$ and $p = .05, .10, .15, .20, .25, .30, .35, .40, .45,$ and $.50$. It should be noted that for values of $p > .5$ one can use the relationship

$$b(x; n, p) = b(n - x; n, 1 - p)$$

See Exercise 2 of Section 4.5 and Example 4.9.

EXAMPLE 4.7

(a)   Find the probability of obtaining exactly 4 heads in 10 flips of a coin.

(b)   Find the probability of obtaining exactly 12 sixes in 20 throws of a die.

   SOLUTION

(a)   Consider the event of heads in one flip of a coin as the event of success in a Bernoulli trial. We are then interested in finding the probability of exactly 4 successes in 10 Bernoulli trials with the probability of success being $p = .5$.

Thus

$$P(\text{exactly 4 heads in 10 flips}) = P(S_n = 4) = b(4; 10, .5)$$

$$= \binom{10}{4}(.5)^4 (.5)^6 = .2051$$

The value for $b(4; 10, .5)$ may be obtained from Table I.

(b) Consider the event of obtaining a six in a throw of a die as the event of success in a Bernoulli trial. We are then interested in finding the probability of exactly 12 successes in 20 Bernoulli trials with the probability of success being $p = 1/6$. Thus

$$P(\text{exactly 12 sixes in 20 throws}) = P(S_n = 12) = b(12; 20, 1/6)$$

$$= \binom{20}{12}\left(\frac{1}{6}\right)^{12}\left(\frac{5}{6}\right)^8 = .0000135$$

EXAMPLE 4.8

If the probability of recovery from a certain disease is .20 and 10 people came down with the disease, what is the probability that, at most, 3 of them will recover?

SOLUTION

Consider the event of recovery as the event of success in a Bernoulli trial. We are then interested in finding the probability of at most 3 successes in 10 Bernoulli trials, with the probability of success being $p = .20$. Thus

$P(\text{at most 3 out of 10 will recover})$

$= P(\text{at most 3 successes in 10 trials})$

$= P(\text{0 successes}) + P(\text{1 success}) + P(\text{2 successes}) + P(\text{3 successes})$

$= b(0; 10, .20) + b(1; 10, .2) + b(2; 10, .2) + b(3; 10, .2)$

$= .1074 + .2684 + .3020 + .2013 = .8791$

EXAMPLE 4.9

Typewriters manufactured in a certain plant are given a final inspection before they are shipped to retailers. If it is known from past experience that approximately 95% of the typewriters pass this inspection, what are the probabilities that, of 20 typewriters,
   (a) exactly 18 pass this inspection;
   (b) at most 18 pass this inspection;
   (c) at least 16 pass this inspection?

SOLUTION

Consider the event of a typewriter passing the inspection as the event of success in a Bernoulli trial.

In each of the following solutions we will use the relationship $b(x; n, p) = b(n - x; n, 1 - p)$.

(a)

$$P(\text{exactly 18 of 20 typewriters pass the inspection})$$
$$= P(\text{exactly 18 successes in 20 trials})$$
$$= b(18; 20, .95) = b(2; 20, .05) = .1887$$

(b)

$$P(\text{at most 18 of 20 typewriters pass the inspection})$$
$$= P(\text{at most 18 successes in 20 trials})$$
$$= \sum_{x=0}^{18} b(x; 20, .95) = \sum_{x=0}^{18} b(20 - x; 20, .05)$$
$$= \sum_{y=2}^{20} b(y; 20, .05) = .2641$$

(c)

$$P(\text{at least 16 of 20 typewriters pass the inspection})$$
$$= P(\text{at least 16 successes in 20 trials})$$
$$= \sum_{x=16}^{20} b(x; 20, .95) = \sum_{x=16}^{20} b(20 - x; 20, .05)$$
$$= \sum_{y=0}^{4} b(y; 20, .05) = .9975$$

EXAMPLE 4.10   SAMPLING WITH REPLACEMENT

The binomial distribution is often associated with sampling with replacement. The following problem is a typical case.

Note: we say that we have used random **sampling with replacement** to obtain a sample of size n from a set (or population) if the action of selecting an element at random from the set is repeated n times under the conditions that each element in the set has equal probability of being selected, and that after each selection, the element selected is replaced with another element of the same kind. The consequence of the replacement is that the composition of the set remains the same for each of the n selections. The n elements selected constitute the sample of size n.

A box contains 10 red and 30 black balls. Use random sampling with replacement to obtain a sample of size 5 from the set of 40 balls in the box (i.e., repeat the action of selecting a ball at random from the box 5 times under the condition that, after each selection, the ball selected is replaced with another ball of the same kind). Let X be the number of red balls in the sample (i.e., the number of times a red ball is selected). Find the probability function of X.

SOLUTION

Consider the event of selecting a red ball as the event of success in a Bernoulli trial. Then X is the number of successes in 5 Bernoulli trials. The probability p of success in each trial is the probability of selecting a red ball. Therefore, $p = 1/4$, and $P(X = x) = b(x; 5, 1/4)$ for $x = 0, 1, 2, 3, 4, 5$.

4.5.4   The Hypergeometric Distribution

Whereas sampling with replacement is associated with the binomial distribution (Example 4.10), sampling without replacement is associated with the **hypergeometric distribution**, which will be defined below. The following example provides a typical case of a random variable whose distribution is a hypergeometric distribution.

Note: we say that we have used random **sampling without replacement** to obtain a sample of size n from a set (or population) if a subset of size n is selected at random from the set of elements under the condition that every subset of size n from the set has an equal probability of being selected. The selected subset of size n is called the **sample** of size n.

EXAMPLE 4.11   SAMPLING WITHOUT REPLACEMENT

A box contains $N_1 + N_2$ items of which N_1 are of one type and N_2 are of a second type. If a sample of size n is selected from the box under sampling without replacement, and if X is the number of items of the first type in the sample, what is the probability function of X?

SOLUTION

To find $P(X = x)$, let us consider samples of size n from the box. Under sampling without replacement there are $\binom{N_1 + N_2}{n}$ ways of selecting n items from a box of $N_1 + N_2$ items. With respect to samples consisting of exactly x items of the first type, there are $\binom{N_1}{x}$ ways of selecting the x items of the first type, and for each of these ways, there are $\binom{N_2}{n - x}$ ways of selecting the remaining $n - x$ items that are of the second type. Hence there are $\binom{N_1}{x}\binom{N_2}{n - x}$ ways to select a sample of size n from the box so that there are exactly x items of the first type in the sample. Therefore

$$P(X = x) = \frac{\binom{N_1}{x}\binom{N_2}{n-x}}{\binom{N_1 + N_2}{n}} \qquad \text{for integers } x \text{ such that} \\ 0 \le x \le n \quad \text{and} \quad 0 \le x \le N_1$$

The probability function derived in the example above is called hypergeo-metric probability function, hypergeometric distribution function, or simply hypergeometric distribution. A formal definition follows.

Definition 4.6   Let n, N_1, and N_2 be positive integers with $n \le N_1 + N_2$. The function

$$h(x; n, N_1, N_2) = \frac{\binom{N_1}{x}\binom{N_2}{n-x}}{\binom{N_1 + N_2}{n}}$$

for integers x such that $0 \le x \le n$ and $0 \le x \le N_1$, is called the **hypergeo-metric probability function** with parameters n, N_1, and N_2.

EXAMPLE 4.12

A carton of 24 hand grenades contains 4 that are defective. If three hand grenades are randomly selected from this carton, what is the probability that exactly 2 of them are defective?

SOLUTION

Using Example 4.11 as a model, we would consider defective hand grenades as one type of item, and nondefective hand grenades as another type of item. Thus, $N_1 = 4$, $N_2 = 20$, and

P(exactly 2 defective hand grenades in a sample of 3 hand grenades)

$$= P(X = 2) = h(2; 3, 4, 20) = \frac{\binom{4}{2}\binom{20}{1}}{\binom{24}{3}} = .0593$$

EXAMPLE 4.13

From a group of 15 chess players, 8 are selected by lot to represent the group at a convention. What is the probability that the 8 selected include 3 of the 4 best players in the group?

SOLUTION

In this case, $n = 8$, $N_1 = 4$, and $N_2 = 11$. Therefore

$$P(3 \text{ of the 4 best players are in the sample of 8})$$

$$= P(X = 3) = h(3; 8, 4, 11) = \frac{\binom{4}{3}\binom{11}{8-3}}{\binom{4+11}{8}} = .287$$

When the sum of the parameters N_1 and N_2 is large compared to the sample size n, the hypergeometric probability $h(x; n, N_1, N_2)$ may be approximated by the binomial probability $b(x; n, p)$ with $p = N_1/(N_1 + N_2)$. The accepted rule-of-thumb is that if n is no more than 5% of $N_1 + N_2$, the approximation is good enough for most purposes. Figure 4.7 provides a comparison of these probabilities for the case of $n = 10$, $N_1 = 10$, and $N_2 = 90$.

x	$h(x; 10, 10, 90)$	$b(x; 10, .1)$
0	.3305	.3487
1	.4080	.3874
2	.2015	.1937
3	.0518	.0574
4	.0076	.0112
5	.0006	.0015
6	.0000	.0001
7	.0000	.0000
8	.0000	.0000
9	.0000	.0000
10	.0000	.0000

FIGURE 4.7
Binomial Approximation of the
Hypergeometric Distribution

$(n = 10, N_1 = 10, N_2 = 90,$
$p = N_1/(N_1 + N_2) = .1)$

Binomial approximation of hypergeometric probabilities

If $N_1 + N_2$ is large and n is small, then

$$h(x; n, N_1, N_2) \approx b(x; n, p)$$

where

$$p = N_1/(N_1 + N_2)$$

The formula for the approximation of hypergeometric probabilities with binomial probabilities can be justified intuitively by making the following observations. In reference to Example 4.11, suppose we think of random sampling without replacement from a set as n selections of elements from the set, provided that after each selection there is no replacement of the element selected. When the set is small, the proportions of type-1 and type-2 elements in the set are changed to a considerable extent after each selection. However, if the set is large (i.e., if $N_1 + N_2$ is large), then the proportions are not changed substantially. Consequently, sampling without replacement is similar to sampling with replacement when the number $N_1 + N_2$ of elements in the set is large. This in turn implies that hypergeometric probabilities are approximately binomial probabilities when $N_1 + N_2$ is large. (A direct verification of the approximation formula can be obtained by expressing both formulas for the hypergeometric probability and for the binomial probability in terms of fractions, and then observing that as $N_1 + N_2$ increases, the two expressions become numerical equivalences.)

One consequence of this approximation is that it allows a researcher to use binomial probabilities (which are easier to handle) even though the random sample may have been obtained through sampling without replacement. This of course is contingent on the size of the sample being relatively small compared to the size of the set from which the sample was obtained.

EXAMPLE 4.14

Consider a random sampling of 5 students from a sophomore class of 1,000 students. If 30% of the sophomores are pre-meds, what is the probability that 3 of the 5 students are pre-meds

(a)   if the sample was obtained under sampling without replacement;
(b)   if the sample was obtained under sampling with replacement.
(c)   Compare the answer to (a) with that to (b).

SOLUTION

(a)   Using the hypergeometric distribution, we obtain

$$P(3 \text{ of the 5 are pre-meds}) = h(3; 5, 300, 700) = .13211$$

(b)   Using the binomial distribution, we obtain

$$P(3 \text{ of the } 5 \text{ are pre-meds}) = b(3; 5, .30) = .1323$$

(c)   The probabilities are identical in the first three decimal places. Therefore, even if the sample was obtained under sampling without replacement, the desired probability could have been computed with considerable accuracy, assuming sampling with replacement.

4.5.5   The Poisson Distribution

Computation of binomial probabilities can be a tiresome task if the parameter n is large. Consider, for example, the calculations necessary to find

$$b(25; 100, .01) = \frac{100!}{(25!)(75!)} (.01)^{25}(.99)^{75}$$

Fortunately, there is a probability function, the Poisson probability function, that may be used to approximate the binomial probability function when n is large and p is suitably small. The formal definition of the Poisson probability function, a discussion of the use of Poisson probabilities to approximate binomial probabilities, and a discussion of random phenomena fitting a Poisson distribution are given below.

Definition 4.7   Let λ be a positive constant. The function

$$p(x; \lambda) = \frac{\lambda^x e^{-\lambda}}{x!} \qquad \text{for } x = 0, 1, 2, 3, \ldots$$

is called the **Poisson probability function** with parameter λ.

The Poisson probability function is also called the Poisson probability distribution, or simply, the Poisson distribution. A random variable is said to have a Poisson distribution, and is referred to as a Poisson random variable, if its probability function is the Poisson probability function.

4.5.5.1   Poisson approximation of binomial probabilities

When n is large and p is small, so that np is of a moderate size, then the Poisson probability $p(x; np)$ may be used to approximate the binomial probability $b(x; n, p)$. In particular, a good approximation may be obtained when $n \geq 20$ and $p \leq .05$. Excellent approximations can generally be obtained when

x	$p(x; 1)$	$b(x; 20, .05)$
0	.3679	.3585
1	.3679	.3774
2	.1839	.1887
3	.0613	.0596
4	.0153	.0133
5	.0031	.0022
6	.0005	.0003
7	.0001	.0000
8	.0000	.0000
9	.0000	.0000

FIGURE 4.8
Poisson Approximation of the
Binomial Distribution

$(n = 20, p = .05, \lambda = np = 1)$

$n \geq 100$ and $np \leq 10$. Figure 4.8 provides a comparison of the two distributions for the case of $n = 20$ and $p = .05$.

To understand why this approximation is possible, let us consider the
following. Let $\lambda = np$. Then

$$b(x; n, p) = \frac{n!}{x!(n - x)!} p^x (1 - p)^{n-x}$$

$$= \frac{n!}{x!(n - x)!} \left(\frac{\lambda}{n}\right)^x \left(1 - \frac{\lambda}{n}\right)^{n-x}$$

$$= \frac{n(n - 1)\cdots(n - x + 1)}{n^x} \frac{\lambda^x}{x!} \frac{\left(1 - \dfrac{\lambda}{n}\right)^n}{\left(1 - \dfrac{\lambda}{n}\right)^x}$$

Now, for n large and λ moderate

$$\left(1 - \frac{\lambda}{n}\right)^n \approx e^{-\lambda}$$

$$\frac{n(n - 1)\cdots(n - x + 1)}{n^x} \approx 1$$

and

$$\left(1 - \frac{\lambda}{n}\right)^x \approx 1$$

Hence, for n large and λ moderate

$$b(x; n, p) \approx \frac{\lambda^x e^{-\lambda}}{x!} = p(x; \lambda)$$

Summary

If n is large, and p is small, so that $\lambda = np$ is moderate, then

$$b(x; n, p) \approx p(x; \lambda)$$

EXAMPLE 4.15

If the probability is .0001 that any one passenger of a certain subway train will be robbed, what is the probability that ten of the next 50,000 passengers will be robbed? Use the Poisson approximation of the binomial distribution.

SOLUTION

Consider the event of being robbed as the event of success in a Bernoulli trial. We are interested in finding the probability of ten successes in 50,000 Bernoulli trials. Thus

$$P(10 \text{ out of } 50,000 \text{ will be robbed}) = b(10; 50,000, .0001)$$

Now, apply the Poisson approximation to obtain

$$b(10; 50,000, .0001) \approx p(10; 5) = \frac{5^{10} e^{-5}}{10!} = .0181$$

The final answer may be obtained by looking up $p(10; 5)$ in the table of Poisson probabilities.

For all practical purposes, the approximation formula given above provides the link between the binomial distribution and the Poisson distribution. Actually, a stronger statement can be made concerning the relationship between the two distributions. It can be shown that as n tends to infinity, the binomial probability $b(x; n, p)$ converges to the Poisson probability $p(x; \lambda)$ provided that the value of p is allowed to vary with n so that np remains equal to λ for all values of n. This is formally stated in the following theorem, which is known as **Poisson's limit law**.

Theorem 4.2   Poisson's Limit Law   *If p varies with n so that $np = \lambda$, where λ is a positive constant, then*

$$\lim_{n \to \infty} b(x; n, p) = p(x; \lambda)$$

Proof   Consider first of all the case where $x = 0$. Then we see that

$$\lim_{n \to \infty} b(0; n, p) = \lim_{n \to \infty} \binom{n}{0} p^0 (1 - p)^n = \lim_{n \to \infty} (1 - p)^n = \lim_{n \to \infty} \left(1 - \frac{\lambda}{n}\right)^n$$

$$= \lim_{n \to \infty} \left[\left(1 - \frac{\lambda}{n}\right)^{-(n/\lambda)}\right]^{-\lambda} = e^{-\lambda} = p(0; \lambda)$$

Note: we used the fact that

$$\lim_{t \to 0} (1 - t)^{-(1/t)} = e$$

Now, suppose that $x \neq 0$. Then

$$\lim_{n \to \infty} b(x; n, p) = \lim_{n \to \infty} \binom{n}{x} p^x (1 - p)^{n - x}$$

$$= \lim_{n \to \infty} \binom{n}{x} p^x (1 - p)^{-x} (1 - p)^n$$

$$= \lim_{n \to \infty} \binom{n}{x} p^x (1 - p)^{-x} b(0; n, p)$$

$$= \lim_{n \to \infty} \frac{n!}{x!(n - x)!} \left(\frac{\lambda}{n}\right)^x \left(1 - \frac{\lambda}{n}\right)^{-x} b(0; n, p)$$

$$= \lim_{n \to \infty} \frac{n(n - 1) \cdots (n - x + 1)}{n^x} \cdot \frac{\lambda^x}{x!} \cdot \frac{1}{\left(1 - \frac{\lambda}{n}\right)^x} \cdot b(0; n, p)$$

$$= \frac{\lambda^x}{x!} e^{-\lambda} = p(x; \lambda)$$

Note: we made use of the fact that

$$\lim_{n \to \infty} \frac{n(n - 1) \cdots (n - x + 1)}{n^x} = \lim_{n \to \infty} \left(\frac{n}{n}\right) \cdot \left(\frac{n - 1}{n}\right) \cdots \left(\frac{n - x + 1}{n}\right) = 1$$

$$\lim_{n \to \infty} \left(1 - \frac{\lambda}{n}\right)^x = 1$$

and

$$\lim_{n \to \infty} b(0; n, p) = e^{-\lambda}$$

4.5.5.2   Random phenomena with Poisson distributions

There are countless examples of random phenomena or random variables whose distributions are Poisson distributions. A number of examples of such random phenomena are listed below. The simplest of these are random phenomena whose distributions may be looked upon as binomial distributions (or approximately binomial distributions) with large parameters n. These distributions may then be considered approximately Poisson distributions because of the relationship between the binomial and the Poisson distributions. (See the approximation formula and the Poisson limit law of Section 4.5.5.1.) Derivation of the distributions for the more complex examples requires an understanding of the Poisson process, which is briefly discussed in Section 4.5.5.3 and again in Chapter 5 in conjunction with the discussion of the exponential distribution.

Examples of random phenomena with Poisson distributions

1. The number of incoming telephone calls within a fixed time interval at certain switchboards.

2. The number of misprints on a typical page of a book that is printed according to certain printing processes.

3. The number of flaws in certain bolts of cloth.

4. The number of raisins in a raisin cookie from a certain bakery.

5. The number of fish caught in a day in certain regions.

6. The number of job openings on a given day at the city hall of a certain community.

7. The number of letters lost in the mail on a given day in certain regions.

8. The number of robbers caught on a given day in certain cities.

9. The number of α-particles discharged in a fixed time interval from certain radioactive samples.

10. The number of hamburgers sold in a certain fast-food restaurant on a given day.

11. The number of emergency patients on a given day at a certain hospital.

12. The number of automobile accidents on a given day in certain counties.

13. The number of claims processed by a certain insurance company on a given day.

With very few additional assumptions, several of these random phenomena or random variables are essentially binomial random variables with large parameters n and small parameters p. Hence, they are essentially Poisson random variables. The following two examples explain why this is so for two of the random phenomena listed above. The reader may easily generalize from these examples to other random phenomena.

EXAMPLE 4.16

Let X be the number of misprints on a page selected at random from a certain large book. Explain why it may be reasonable to assume that X is a Poisson random variable.

SOLUTION

Suppose that each page of the book contains approximately the same number n of letters, numbers, and other printed symbols. If we consider the event of an error in the printing of a letter, number, or other symbol as the event of a success in a Bernoulli trial (so that these events are independent events with equal probabilities), then X is the number of successes in n Bernoulli trials, with the probability p of success being identical to the probability of an error in the printing of a letter, number, or other symbol. Thus

$$P(X = x) = b(x; n, p)$$

Now, it is reasonable to assume that n is large ($n > 100$) and p is small ($p < .1$) so we may be able to use the Poisson probability $p(x; np)$ to obtain an excellent approximation of the binomial probability $b(x; n, p)$. Thus, it is reasonable to assume that

$$P(X = x) = p(x; \lambda)$$

where $\lambda = np$. In other words, it is reasonable to assume that X is a Poisson random variable.

EXAMPLE 4.17

Let X be the number of raisins in a raisin cookie selected at random from a certain bakery. Explain why it may be reasonable to assume that X is a Poisson random variable.

SOLUTION

Suppose that both cookies and raisins are of uniform size so that we may assume that a cookie may be evenly divided into n regions with each region containing at most one raisin. If we consider the event of a raisin in a region as a success in a Bernoulli trial (so that these events are independent events with equal probabilities), then X is the number of successes in n Bernoulli trials with the probability p of success being identical to the probability of a raisin in the region. Now, suppose that the size of the cookies is large in relation to the size of the raisins, and that there are relatively few raisins per cookie; we may assume that n is large and p is small. Therefore, the binomial probabilities $P(X = x) = b(x; n, p)$ may be approximated by the Poisson probabilities $p(x; np)$, essentially making the random variable X a Poisson random variable.

EXAMPLE 4.18

Suppose that the number of emergency patients in a given day at a certain hospital is a Poisson random variable X with parameter $\lambda = 20$. What is the probability that in a given day there will be more than 30 emergency patients?

SOLUTION

P(more than 30 patients)

$= P$(exactly 31 patients) $+ P$(exactly 32 patients) $+ \cdots$

$= P(X = 31) + P(X = 32) + \cdots$

$= p(31; 20) + p(32; 20) + \cdots$

$= .0054 + .0034 + .0020 + .0012 + .0007 + .0004$

$\quad + .0002 + .0001 + .0001 + \cdots$

$= .0135$

(The Poisson probabilities $p(x; 20)$ may be obtained from Table II in the Appendix.)

4.5.5.3   Poisson processes

Suppose that a certain phenomenon is generated by a physical process and that within any fixed period of time, or within any fixed region of space, we can count the number of occurrences of this phenomenon. (For example, we may consider the phenomenon of an arrival of a telephone call at a certain switchboard within a fixed period of time, or the phenomenon of a penetration into a certain region in space of an atomic particle from a radioactive source.) If the physical process satisfies three given conditions, then it can be shown that the random variable X that represents the total number of occurrences of this phenomenon within the fixed period of time—or fixed region of space— is a Poisson random variable. An informal statement of these three conditions follows.

1. The number of occurrences of the phenomenon in any two or more disjoint intervals of time (or regions of space) must be independent of each other.

2. The probability of at least one occurrence of the phenomenon during any subinterval (or subregion) of length (or size) $1/n$ is the same as the probability of at least one occurrence of the phenomenon during any other subinterval (or subregion) of length (or size) $1/n$. Further, if p_n is the probability of at least one occurrence of the phenomenon during any given interval (or region) of length (or size) $1/n$, then there is a constant $\lambda > 0$ such that $\lim_{n \to \infty} (np_n) = \lambda$.

3. The probability of two of more occurrences in any given very short interval of time (or small region of space) must be smaller than the probability of just one occurrence. Furthermore, in any given sufficiently

short interval of time (or small region of space) the probability of two or more occurrences must be negligible in comparison with the probability of one occurrence.

A precise mathematical statement of these three conditions and a rigorous proof of the conclusion are beyond the scope of this book. The interested reader may find them in references 1, 3, and 4. However, a justification of the conclusion that X is a Poisson random variable is well within the technical level of this book if we modify condition 3 very slightly to include the statement that, in sufficiently short intervals of time (or small regions of space), the probability of more than one occurrence of the phenomenon is zero. It is easy to see that, for all practical purposes, such a modification does not alter the essential nature of the physical process. If we assume this modification of condition 3, then an argument to support the conclusion that X is a Poisson random variable may proceed in the manner that follows.

An Informal Proof That X Is a Poisson Random Variable   Without loss of generality, let us assume that the physical process is time- rather than space-related, and let us consider a unit interval of time. Divide the unit interval of time into n equal subintervals (each of length $1/n$). Because of condition 1, we can consider the event of at least one occurrence of the phenomenon within a subinterval of length $1/n$ as the event of success in a Bernoulli trial. Furthermore, condition 2 tells us that, with this interpretation of the event of success, the probability of success in each of the n Bernoulli trials is p_n. Then, because of condition 3 (with its modification), we can say that for all practical purposes, as n tends to infinity, the event of exactly x occurrences of the phenomenon within the unit interval is the same as the limit of the events of exactly x successes in the n Bernoulli trials. Therefore, it follows that

$$P(X = x) = P(\text{exactly } x \text{ occurrences of the phenomenon})$$

$$= \lim_{n \to \infty} P(\text{exactly } x \text{ successes in the } n \text{ Bernoulli trials})$$

$$= \lim_{n \to \infty} b(x; n, p_n) = \frac{\lambda^x e^{-\lambda}}{x!}$$

(The last step of the above statement is a result of the Poisson limit law and condition 2.) In other words, we have shown that the distribution of X is the Poisson distribution with parameter λ.

Because the exact distribution of the random variable that represents the total number of occurrences of the phenomenon is naturally dependent on the length of the interval (or the size of the region), the notation for the random variable usually contains some reference to the length of the interval (or the size of the region). If the physical process is time-related, we generally write X_t to denote the number of occurrences of the phenomenon from time zero to time t. The random variable X_t can then be looked upon as a function of t, and in this broader role it is called a Poisson process. It can be shown that if the three conditions listed above are satisfied, there is a constant $\lambda > 0$ such that for each given value of t the distribution of X_t is the Poisson distribution with parameter λt. That is

$$P(X_t = x) = \frac{(\lambda t)^x e^{-\lambda t}}{x!} \qquad \text{for} \quad x = 0, 1, 2, 3, \ldots$$

EXAMPLE 4.19

The number of telephone calls arriving at a certain switchboard within a time interval of length t (measured in minutes) is a Poisson process X_t with parameter $\lambda = 2$.

(a) Find the probability of no telephone calls arriving at this switchboard during a given 5-minute period.

(b) Find the probability of more than one telephone call arriving at this switchboard during a given $\frac{1}{2}$-minute period.

SOLUTION

(a) The distribution of X_t is the Poisson distribution given by

$$P(X_t = x) = \frac{(\lambda t)^x e^{-\lambda t}}{x!} = \frac{(2t)^x e^{-2t}}{x!}$$

Therefore

$$P(\text{no telephone calls in 5 minutes}) = P(X_5 = 0) = \frac{(2 \cdot 5)^0 e^{-2 \cdot 5}}{0!}$$

$$= e^{-10} = .000045$$

(b)

$$P(\text{more than one telephone call within } \tfrac{1}{2} \text{ minute})$$
$$= 1 - P(\text{none or one telephone call within } \tfrac{1}{2} \text{ minute})$$
$$= 1 - P(X_{1/2} = 0) - P(X_{1/2} = 1)$$
$$= 1 - \frac{[2(1/2)]^0 e^{-[2(1/2)]}}{0!} - \frac{[2(1/2)]^1 e^{-[2(1/2)]}}{1!}$$
$$= 1 - e^{-1} - e^{-1} = .26424$$

4.5.5.4   The parameter λ

The exact meaning of the parameter λ will be discussed in Chapter 6. In the meantime we can acquire an intuitive understanding of the parameter by reasoning as follows. It is intuitively obvious that we would expect, on the average, np successes in n Bernoulli trials if the probability of success in each Bernoulli trial is p. For example, in ten tosses of a coin we expect, on the average, five heads. (Note that in this case $np = 10 \times .5 = 5$.) This analysis gives us a clue to the meaning of λ. As we saw in Section 4.5.5.2, if X is a binomial random variable with parameters p and n, and if n is large, the distribution of X becomes essentially Poisson with parameter $\lambda = np$. Thus it seems reasonable to conclude that, if the numerical value of an experiment is a Poisson random variable X and if the experiment is repeated a number of times, we would expect the average value of X to be λ. For example, if the number of misprints on a page is a Poisson random variable X with parameter $\lambda = 3$, on the average 3 misprints per page can be expected.

Exercises

1. Show that the Bernoulli distribution is the binomial distribution with $n = 1$.

2. Prove that $b(x; n, p) = b(n - x; n, 1 - p)$.

✓ 3. Use Table 1 in the Appendix to find (a) $b(7; 10, .3)$; (b) $b(6; 16, .4)$; (c) $b(10; 15, .6)$; (d) $b(3; 7, .75)$. (*Hint:* refer to Exercise 2 for parts (c) and (d).)

✓ 4. Assume that 1% of model airplanes are sold with missing parts. If a child receives five model airplanes on her birthday, what is the probability that one of them has missing parts?

✓ 5. If the probability is .05 that a baseball game will go into an extra inning, find the probabilities that among 15 baseball games
   (a) exactly one will go into an extra inning;
   (b) at most 2 will go into extra innings;
   (c) at least 2 will go into extra innings.

✓ 6. If the probability that the Internal Revenue Service will audit an income tax return reporting gross income over \$100,000 is .65, find the probabilities that among 20 such returns
   (a) exactly 10 will be audited;
   (b) at least 10 will be audited;
   (c) between 8 and 12 (inclusive) will be audited.

7. If one-third of the employees of a certain company bring their lunches, find the probabilities that among 4 such employees
   (a) exactly two bring their lunches;
   (b) none of them bring their lunches;
   (c) more than one of them bring their lunches.

✓ 8. An editor wants to check whether her new proofreader meets the standards of missing typographical errors in no more than 1% of the galley proofs of a book. She decides to proofread 10 randomly selected galley proofs of a large book that has already been checked by her new proofreader. If all 10 pages are free of typographical errors, she will assume that the proofreader meets the standards. If more than 2 pages have uncorrected typographical errors, she will fire the proofreader. If one or two pages have typographical errors, she will read another 10 pages before passing judgment. Assuming that the editor is a flawless proofreader, what is the probability that after reading 10 pages she will

(a) fire the proofreader even though only 1% of the pages have typographical errors;

(b) assume that the proofreader meets the standards when 5% of the pages have typographical errors;

(c) burden herself with another 10 pages of proofreading when only 1% of the pages have typographical errors?

9. A quality-control inspector examines 10 out of every shipment of 1,000 packages of frozen chicken for spoilage. The shipment is approved if and only if all 10 packages pass inspection. Find the probability of a shipment being approved even though 10% of the packages are badly spoiled.

10. Ms. Jones drives into town once a week to see her psychiatrist. In the past she parked her car at a lot for $5, but she decided that in the next 5 weeks she will park at the fire hydrant and risk getting tickets with fines of $25 per offense. If the probability of getting a ticket is .1, what is the probability that Ms. Jones will pay more in fines in 5 weeks than she would pay in parking fees if she had opted not to break the law?

11. A quality-control inspector examines a random selection of 3 rifles from each case of 24 rifles that is ready for shipment to retailers. If such a case contains 5 rifles with imperfections, find the probabilities that the inspector's sample of 3 will

(a) contain none of the imperfect rifles;

(b) contain exactly one rifle with imperfections;

(c) contain at least one rifle with imperfections.

12. If 4 of 20 recently acquired works of art are frauds, what is the probability that an art inspector who randomly selects 5 of these works for inspection will have a sample that

(a) contains none of the frauds;

(b) contains all of the frauds;

(c) contains only one fraud.

13. An employer asks a random sampling of 3 of his 100 employees whether they would prefer a 9–5 or an 8–4 workday.

(a) If 50% of his employees prefer the 9–5 schedule, what is the probability that the sample will contain two who prefer the 9–5 schedule?

(b) Use the binomial distribution to approximate the answer to (a).

14. A random sampling of 4 members of a 150-member club has shown that 3 prefer no smoking in the clubhouse dining room. What is the probability that this will occur if in fact only 20% of members prefer no smoking in the dining room? Find this probability assuming that the sample was obtained under

(a) sampling without replacement, and

(b)   sampling with replacement.

(c)   Compare the two answers.

15.   Find the following binomial probabilities and compare with the Poisson approximations.

(a) $b(3; 20, .01)$ (b) $b(5, 100, .05)$

16.   Let X be the number of telephone calls arriving within a fixed time interval at a switchboard. Explain why it may be reasonable to assume that X is a Poisson random variable.

17.   Suppose that the number of letters lost in the mail in a certain city on a given day is a Poisson random variable X with parameter $\lambda = 5$. What is the probability that in a given day

(a)   exactly 5 letters will be lost in the mail;

(b)   at least one letter will be lost in the mail.

18.   Use the Poisson approximation of the binomial distribution to determine the probabilities that

(a)   there will be 15 heads in 30 tosses of a coin;

(b)   there will be fewer than four 1's in 60 throws of a die.

19.   A large shipment of disposable flashlights contains 1% that are defective. Use the Poisson approximation of the binomial distribution to find the probability that among 200 flashlights randomly selected from the shipment

(a)   exactly 3 will be defective;

(b)   at most 2 will be defective;

(c)   at least 3 will be defective.

20.   The number of public buses that need repair in any one day in a certain city is a Poisson random variable X with parameter $\lambda = 10$. What is the probability that on a given day more than 15 buses in this city will need repair?

21.   Show that the function $p(x; \lambda) = (\lambda^x e^{-\lambda})/x!$, for $x = 0, 1, 2, \ldots$, satisfies conditions (a) and (b) of Theorem 4.1 and is, therefore, truly a probability function. (*Hint:* $\sum_{x=0}^{\infty} (\lambda^x/x!) = e^{\lambda}$.)

22.   Show that the function $b(x; n, p) = \binom{n}{x} p^x (1 - p)^{n-x}$, for $x = 0, 1, \ldots, n$, satisfies conditions (a) and (b) of Theorem 4.1 and is, therefore, truly a probability function. (*Hint:* use the binomial expansion formula $(a + b)^n = \sum_{x=0}^{n} \binom{n}{x} a^x b^{n-x}$.)

23.   The number of cars passing through a certain toll gate at a certain bridge during any time interval of length t (measured in minutes) is a Poisson process with parameter $\lambda = .3$.

(a)   Find the probability of no car passing through this toll gate during a given 5-minute period.

(b)   Find the probability of more than one car passing through during a given 5-minute period.

(c)   Find the probability that between 15 and 25 cars pass through during a given one-hour period.

24. The number of customers arriving at a certain bank during any time interval of length t (measured in minutes) is a Poisson process with parameter $\lambda = 1$.

(a) Find the probability of 10 customers arriving at this bank during a given ten-minute period.

(b) Find the probability that customers will arrive at the rate of at least 1 per minute during a given period of 10 minutes.

4.6   Cumulative Distribution Functions of Discrete Random Variables

Often we are interested in estimating the chances that the values of a random phenomenon will remain at or below a certain fixed value. For example, what are the chances that the price of gold will remain at or below $425 per troy ounce? What are the chances that a certain candidate will get no more than 30% of the votes? What are the chances that 20 tosses of a coin will produce no more than 10 heads? In each case, we are concerned with the probability that a given random variable X will take on values that are less than or equal to some fixed value x. In other words, we wish to find $P(X \le x)$. This probability, considered as a function of x, is called the cumulative distribution function (cdf) or, more simply, the distribution function of the random variable X. A formal definition follows.

Definition 4.8   The function

$$F(x) = P(X \le x) \qquad \text{for} \quad -\infty < x < \infty$$

is called the **cumulative distribution function** or the distribution function of the random variable X.

EXAMPLE 4.20   FINDING THE CUMULATIVE DISTRIBUTION FUNCTION

Let X be a discrete random variable whose only possible values are -1, 2, and 5. Suppose that the probability function of X is

$$f(x) = \begin{cases} 1/4 & \text{for} \quad x = -1 \\ 1/2 & \qquad x = 2 \\ 1/4 & \qquad x = 5 \end{cases}$$

Find the cumulative distribution function of X.

SOLUTION

Step 1: $F(x)$ for $x < -1$. None of the possible values of X are less than -1. Hence, for all values of x that are less than -1, we have $P(X \leq x) = 0$. In other words, for values $x < -1$, $F(x) = P(X \leq x) = 0$.

Step 2: $F(x)$ for $-1 \leq x < 2$. Since -1 is the only possible value of the random variable X that is less than or equal to -1, it is clear that

$$F(-1) = P(X \leq -1) = f(-1) = 1/4$$

Now, pick a value, such as 1.5, that is between -1 and 2. We see that it is also true that -1 is the only possible value of the random variable X that is less than or equal to 1.5. Hence

$$F(1.5) = P(X \leq 1.5) = P(X = -1) = f(-1) = 1/4$$

Since this is true for any value between -1 and 2, it is clear that for all values of x such that $-1 \leq x < 2$,

$$F(x) = P(X \leq x) = P(X = -1) = f(-1) = 1/4$$

Step 3: $F(x)$ for $2 \leq x < 5$. Since -1 and 2 are the only possible values of the random variable X that are less than or equal to 2, it is clear that

$$\begin{aligned}
F(2) = P(X \leq 2) &= P[(X = -1) \quad \text{or} \quad (X = 2)] \\
&= P(X = -1) + P(X = 2) \\
&= f(-1) + f(2) \\
&= 1/4 + 1/2 \\
&= 3/4
\end{aligned}$$

Now, pick a value, such as 2.3, that is between 2 and 5. We see that it is also true that -1 and 2 are the only possible values of the random variable X that are less than or equal to 2.3. Hence

$$\begin{aligned}
F(2.3) = P(X \leq 2.3) &= P[(X = -1) \quad \text{or} \quad (X = 2)] \\
&= P(X = -1) + P(X = 2) = f(-1) + f(2) \\
&= 1/4 + 1/2 = 3/4
\end{aligned}$$

Since this is true for any value between 2 and 5, it is clear that for all values of x such that $2 \leq x < 5$

$$F(x) = P(X \leq x) = P(X = -1) + P(X = 2) = f(-1) + f(2) = 3/4$$

Step 4: $F(x)$ for $x \geq 5$. Since all of the possible values of X are less than or equal to 5, it is clear that if x is greater than or equal to 5, then

$$
\begin{aligned}
F(x) = P(X \leq x) &= P[(X = -1) \quad \text{or} \quad (X = 2) \quad \text{or} \quad (X = 5)] \\
&= P(X = -1) + P(X = 2) + P(X = 5) \\
&= f(-1) + f(2) + f(5) \\
&= 1/4 + 1/2 + 1/4 = 1
\end{aligned}
$$

Conclusion: $F(x)$ for $-\infty < x < \infty$. In summarizing the results of steps 1, 2, 3, and 4, we obtain

$$
F(x) = \begin{cases}
0 & \text{for} & x < -1 \\
1/4 & & -1 \leq x < 2 \\
3/4 & & 2 \leq x < 5 \\
1 & & 5 \leq x
\end{cases}
$$

The graph of the cumulative distribution function $F(x)$ is given in Figure 4.9.

In Example 4.20, we saw that the value of $F(x)$ is the sum of all the values of $f(t)$ for which t is less than or equal to x. In other words, $F(x) = \sum_{t \leq x} (f(t))$. This equality may be used as an alternate definition of the cumulative distribution function of a discrete random variable.

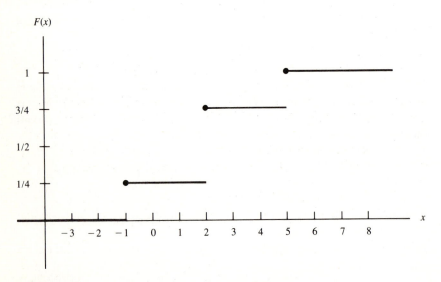

FIGURE 4.9
Graph of the Cumulative Distribution Function of Example 4.20

Alternate Definition 4.8a   The Cumulative Distribution Function of a Discrete Random Variable   Let f be the probability function of the discrete random variable X. The function

$$F(x) = \sum_{t \leq x} f(t) \qquad \text{for} \quad -\infty < x < \infty$$

is called the cumulative distribution function or the distribution function of the random variable X. (The summation is over all values of $t \leq x$ for which $f(t)$ is defined.)

EXAMPLE 4.21   FINDING PROBABILITIES WITH THE CUMULATIVE DISTRIBUTION FUNCTION

The following problem illustrates how we may use the cumulative distribution function to find probabilities and to find the probability function of a random variable.

Let X be a discrete random variable whose cumulative distribution function is

$$F(x) = \begin{cases} 0 & \text{for} & x < -3 \\ 1/6 & & -3 \leq x < 6 \\ 1/2 & & 6 \leq x < 10 \\ 1 & & 10 \leq x \end{cases}$$

(a) Find $P(X \leq 4)$, $P(-5 < X \leq 4)$, $P(X = -3)$, $P(X = 4)$.

(b) Find the probability function of X.

SOLUTION

(a) $P(X \leq 4) = F(4) = 1/6, P(-5 < X \leq 4) = P(X \leq 4) - P(X \leq -5) = F(4) - F(-5) = 1/6 - 0 = 1/6$, $P(X = -3) = 1/6$ because the magnitude of the jump in the value of $F(x)$ at $x = -3$ is $1/6$. (The alternate definition of the cumulative distribution function indicates that if $x_1, x_2, \ldots$ are the possible values of X with positive probabilities, then at each point $x_1, x_2, \ldots$ there is a jump in the value of $F(x)$. Furthermore, $f(x_i) = P(X = x_i)$ is the magnitude of the jump.) $P(X = 4) = 0$ because there is no jump in the value of $F(x)$ at $x = 4$.

(b) There are jumps in the value of $F(x)$ at $x = -3$, 6, and 10. The magnitudes of the jumps are $1/6$, $2/6$, and $3/6$, respectively. Therefore, the probability function of X is

$$f(x) = \begin{cases} 1/6 & \text{for} & x = -3 \\ 2/6 & & x = 6 \\ 3/6 & & x = 10 \end{cases}$$

Exercises

1. Let X be a discrete random variable whose only possible values are -5, -1, 0, and 7. Find and graph the cumulative distribution function of X if the probability function of X is defined by the values $f(-5) = .3$, $f(-1) = .1$, $f(0) = .2$, and $f(7) = .4$.

2. Consider a throw of two dice. Let X be the sum of the values on the two dice. Find and graph the cumulative distribution function of X.

3. Let X be a discrete random variable with cumulative distribution

$$F(x) = \begin{cases} 0 & \text{for} & x < 1 \\ .1 & & 1 \le x < 3 \\ .4 & & 3 \le x < 5 \\ .9 & & 5 \le x < 5.5 \\ 1.0 & & 5.5 \le x \end{cases}$$

   (a) Find $P(X \le 3)$, $P(X \le 4)$, $P(1.5 < X \le 5.2)$, $P(X = 3)$, $P(X = 3.5)$
   (b) Find the probability function of X.

4. Let $F(x)$ be the cumulative distribution function of a discrete random variable X. Explain why the following are true statements.
   (a) $\lim_{x \to -\infty} F(x) = 0$;
   (b) $\lim_{x \to \infty} F(x) = 1$;
   (c) If $a < b$ then $F(a) \le F(b)$.

Supplementary Exercises

1. How many times should you toss a coin if you want a probability of at least .9 of getting heads at least once?

2. If the probability that a certain book will get a favorable review from any reviewer is $1/3$, how many reviewers should you ask to review the book if you want a probability of at least .95 of getting at least one favorable review?

3. Suppose that your dream is to be a partner in a prestigious law firm. You feel that your chances of being accepted as a partner in any one of a number of such law firms is .1. How many prestigious law firms should there be to make your dream come true with a probability of at least .99?

4. Based on his past experience in marketing, a manufacturer of typewriters feels that when his two new models are introduced, 65% of consumers will prefer design A to design B. He takes a straw vote of 20 randomly chosen company employees and finds that 17 of them prefer design A. If the manufacturer's subjective probabilities were correct, what is the probability that so many (i.e., 17 or more) would prefer design A? Should he revise his subjective probabilities on the basis of these results?

5. To qualify for a certain job, an applicant must demonstrate a minimum proficiency in mathematics by obtaining a passing grade of at least 10 correct answers in any one of a number of multiple-choice tests consisting of 20 questions with 4 alternative answers each. Applicants may take as many tests as they wish. Suppose that John feels that he can always answer 5 of the 20 questions correctly and he can guess on the others. He decides to keep taking the tests until he passes one of them.

    (a) What is the probability that he will pass after taking fewer than 3 tests?

    (b) How many tests should he take to have a probability of at least .99 of passing?

    (c) Why should failing students press for makeup examinations, especially if the examinations are multiple-choice tests?

6. **The geometric distribution.** A random variable X has a *geometric distribution*, and is called a *geometric random variable*, if its probability function is

    $$f(x) = g(x; p) = q^{x-1}p \qquad \text{for} \quad x = 1, 2, 3, \ldots$$

    where p and q are constants such that $0 < p < 1$ and $p + q = 1$.

    (a) Consider the experiment of rolling a die until a six occurs. Show that if X is the required number of rolls then X is a geometric random variable with $p = 1/6$.

    (b) In the above experiment what is the probability that 4 throws of the die are required?

    (c) What is the probability that at least 3 throws are required?

7. Consider a sequence of Bernoulli trials where the probability of success in each trial is p. Show that if X is the required number of trials for the occurrence of a success, then X is a geometric random variable.

8. Show that the function $g(x; p) = q^{x-1}p$ for $x = 1, 2, 3, \ldots$ satisfies conditions (a) and (b) of Theorem 4.1 and is, therefore, truly a probability function. (*Hint:* use the geometric series $\sum_{n=0}^{\infty} t^n = 1/(1-t)$ for $0 < t < 1$.)

9. Suppose that at a certain industrial plant the probability of one or more cases of serious physical injury to any worker in any given working day is .0001. What is the probability of a whole year (260 working days) passing without a single case of a serious physical injury to a worker? (*Hint:* see Exercise 6.)

10. If .01% of the calculators manufactured by a certain company have defective parts, what is the probability that an inspector would have to inspect at least 2,000 calculators from this company before she inspects one with defective parts? (*Hint:* use the geometric probability function and the geometric series, $\sum_{n=0}^{\infty} t^n = 1/(1-t)$ for $|t| < 1$.)

11. **The negative binomial distribution.** A discrete random variable X is called a *negative binomial random variable*, and its probability function

$f(x)$ is called a *negative binomial probability function*, if there is a positive integer n, and constants p and q, with $0 < p < 1$ and $q = 1 - p$, such that

$$f(x) = f(x; n, p) = \binom{x-1}{n-1} p^n q^{x-n} \quad \text{for} \quad x = n, n+1, n+2, \ldots$$

Show that in a sequence of Bernoulli trials, if X is the number of the trial on which the nth success occurs, then X is a negative binomial random variable. (*Hint:* if the nth success occurs on the xth trial, then there are exactly $n - 1$ successes in the first $x - 1$ trials.)

References

1. Chung, Kai Lai. 1974. *Elementary probability theory with stochastic processes.* New York: Springer-Verlag.
2. Feller, William. 1950. *An introduction to probability theory and its applications,* Vol. I, 3d ed. New York: John Wiley & Sons, Inc.
3. Parzen, Emanuel. 1962. *Stochastic processes.* San Francisco: Holden-Day.
4. Ross, Sheldon. 1976. *A first course in probability.* New York: Macmillan Publishing Co., Inc.

See Reference 2 for an extensive treatment of discrete distributions. References 1, 3, and 4 contain discussions of the Poisson process at a more advanced level than that given in this book.

Chapter 5

CONTINUOUS RANDOM VARIABLES AND PROBABILITY DENSITY FUNCTIONS

5.1   Introduction

When the set of possible values of a random phenomenon is a continuous extent of real numbers (except in unusual cases, as in Example 5.5), the value of the random phenomenon is called a **continuous random variable**. We shall see, for example, that X is a continuous random variable

if X is the value of a number to be selected at random from the interval of real numbers between a and b;

if X is the distance from the center of a target to the hit position of an arrow that is to be shot at the target;

if X is the arrival time of the afternoon train from Chicago.

Continuous random variables differ from discrete random variables principally in the size of the sets of their possible values. If X is a discrete random variable, the set of possible values of X is either finite or can be placed in one-to-one correspondence with the set of positive integers. By contrast, if X is a continuous random variable, the set of possible values of X is either an entire interval of real numbers, a union of intervals of real numbers, or even the set of all real numbers. These differences mandate different techniques for the evaluation of probabilities for the two types of random variables.

5.2   Probability Density Functions

The probability function, $f(x) = P(X = x)$, plays a central role in the characterization of a discrete random variable X. Naturally, we seek the definition of a similar function that will characterize the continuous random variable. Due to a certain property of continuous random variables, however, a mere extension of the function f from the discrete to the continuous case is not possible. Consider, for example, the random variable X, which is the value of a number to be selected at random from the unit interval $(0, 1)$. We saw in Chapter 1, Section 1.4.5, that $P(X = 1/2) = 0$. Indeed, for any value x in $(0, 1)$, $P(X = x) = 0$. As explained, this does not mean that the event $X = 1/2$ cannot occur, but simply that its occurrence is so rare that any probability greater than zero is greater than the probability of its occurrence. We shall see in Section 5.4 that this property of $P(X = x) = 0$ is characteristic of all continuous random variables, because if X is a continuous random variable, then X has so many possible values that the occurrence of any particular fixed one of them is an extremely rare event. Thus, if X is a continuous random variable, it is meaningless to define a function f by the equation $f(x) = P(X = x)$.

Let us consider the problem of finding a function that will characterize the continuous random variable through another approach. Suppose that a certain expert archer is about to shoot an arrow at a target. Let X be the distance from the center of the target to the hit position of the arrow. The set of possible values of X is an entire interval of real numbers. Therefore, X cannot be a discrete random variable. However, if the value of X is to be rounded off to the nearest inch, X becomes a discrete random variable. There is a probability function and a probability histogram for this discrete random variable. Let us suppose that the probability histogram is the one given in Figure 5.1(a). Now, if the value of X is to be rounded off to the nearest half-inch, X becomes a somewhat different discrete random variable, as shown by Figure 5.1(b). In each of these cases, if the histogram is adjusted so that the area of each rectangle is the probability of the midpoint of the base of the

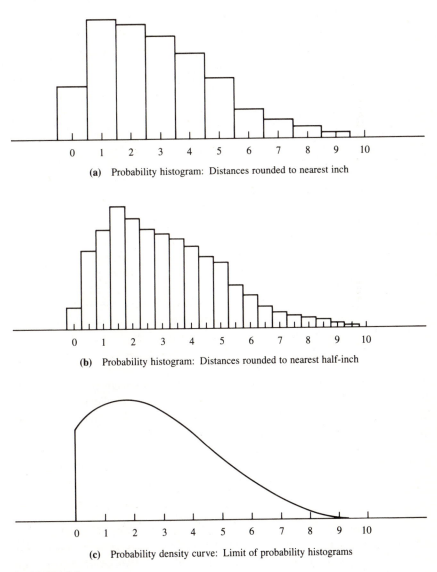

(a)   Probability histogram: Distances rounded to nearest inch

(b)   Probability histogram: Distances rounded to nearest half-inch

(c)   Probability density curve: Limit of probability histograms

FIGURE 5.1

rectangle, then the probability that X will take on a value between one midbase point and another is simply the sum of the areas of the rectangles from the one point to the other.

It is easy to see that if we continue to reduce the unit of measurement for rounding off the value of X, the histograms of the corresponding discrete random variables will converge to a smooth curve such as the one in Figure 5.1(c). This curve is called the **probability density curve** of the random variable X. The function $f(x)$ defined by this curve is called the **probability density function of** X. It is intuitively clear that the probability density curve, being a limit of histograms, must share the property of histograms that the area under this curve from point a to point b is equal to the probability that the random variable X will take on a value between a and b. In other words, $P(a \leq X \leq b) = \int_a^b f(x)\,dx$, and $P(X \leq x) = P(-\infty < X < x) = \int_{-\infty}^x f(t)\,dt$.

As we shall see in Section 5.3, the probability density function is used to characterize a continuous random variable.

EXAMPLE 5.1

Consider the experiment of selecting a number at random from the interval $(2, 4)$. Assume that the nature of the experiment ensures that if subintervals I_1 and I_2 of $(2, 4)$ are of equal lengths, the probability that the number selected will be in I_1 is equal to the probability that it will be in I_2. Let X be the value of the number selected. Use the method described above to find the probability density function of X.

SOLUTION

Suppose we round off the value of X to one decimal place. Then, the random variable X becomes a discrete random variable with possible values $2.0, 2.1, 2.2, 2.3, \ldots, 3.9$, and 4.0. The probability histogram of this discrete random variable is given in Figure 5.2(a). The histogram is constructed so that the area of each rectangle is the probability of the midpoint of the base of the rectangle. Thus, the height of each rectangle, except the first and last, is $1/2$. The heights of the first and last rectangles are $1/4$. The sum of the areas of the rectangles is 1.

Now, suppose we round off the value of X to two decimal places. Then, the random variable X becomes a discrete random variable with possible values $2.00, 2.01, 2.02, \ldots, 3.99$, and 4.00. It is clear that the probability histogram of this discrete random variable will be similar to the one in Figure 5.2(a) except that there will be more rectangles. However, the height of each rectangle except the first and last will remain $1/2$ and the heights of the first and last rectangles will remain $1/4$ so that the sum of the areas of the rectangles is 1.

We can see now that if we continue to increase the number of decimal places for rounding off the value of X, then the number of rectangles in the corresponding histogram increases but the height of each rectangle except the first and last will remain $1/2$. Consequently, the probability density function of X is defined by the probability density curve given in Figure 5.2(b). In other

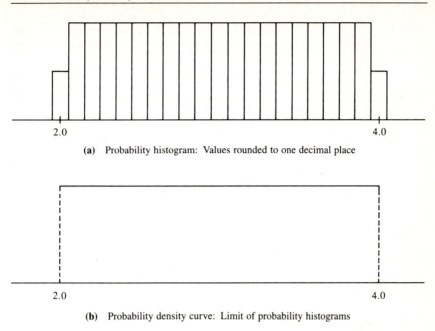

(a)   Probability histogram: Values rounded to one decimal place

2.0   4.0

(b)   Probability density curve: Limit of probability histograms

FIGURE 5.2

words, the probability density function of X is

$$f(x) = \begin{cases} 1/2 & \text{for} \quad 2 < x < 4 \\ 0 & \text{elsewhere} \end{cases}$$

Exercises

1. Which of the following statements are true?
    (a)   If the set of possible values of a random variable X is infinite, then X is a continuous random variable.
    (b)   The probability density function of a continuous random variable is the function $f(x) = P(X = x)$.
    (c)   If $P(X = x) = 0$ for all possible values x of a random variable, then X cannot be a discrete random variable.
    (d)   If $f(x)$ is the probability density function of X, then $P(a \leq X \leq b) = \int_a^b f(x)\,dx$.
    (e)   If X is a random variable, then there is at least one value x for which $P(X = x) > 0$.
    (f)   If the set of possible values of a random variable X is the interval $(0, 1)$, then X cannot be a discrete random variable.

2. Consider the experiment of selecting a number at random from the interval $(-1, 2)$. Assume that the nature of the experiment ensures that if

subintervals I_1 and I_2 of $(-1, 2)$ are of equal lengths, the probability that the number selected will be in I_1 is equal to the probability that it will be in I_2. Let X be the value of the number selected. Use the method of Example 5.1 to find the probability density function of X.

3. Consider the experiment of selecting a number at random from the interval (a, b), where $a \leq b$. Assume that the nature of the experiment ensures that if subintervals I_1 and I_2 of (a, b) are of equal lengths, the probability that the number selected will be in I_1 is equal to the probability that it will be in I_2. Let X be the value of the number selected. Use the method of Example 5.1 to find the probability density function of X.

4. Consider this experiment. Roll a die. If the outcome is a six, select a number at random from the interval $(0, 1)$. If the outcome is not a six, select a number at random from the interval $(1, 2)$. Let X be the value of the number selected.

    (a) Draw a histogram of this random variable, assuming the value of X is to be rounded off to one decimal place.

    (b) Find the probability density function of X.

5.3   Continuous Random Variables

The formal definition of a continuous random variable follows.

Definition 5.1   A real valued function X defined on a sample space S with probability set function P is called a **continuous random variable** provided that there is an integrable function f for which

$$P(X \leq x) = \int_{-\infty}^{x} f(t)\, dt \qquad \text{for} \quad -\infty < x < \infty$$

The function f is called a probability density function (abbreviated pdf) of the random variable X.

5.4   Finding Probabilities for Continuous Random Variables

The basic formula for finding probabilities for continuous random variables is the statement $P(X \leq x) = \int_{-\infty}^{x} f(t)\, dt$, given in Definition 5.1. It can be expanded to include the following rules.

Rules for Finding Probabilities for Continuous Random Variables

1. $P(X = x) = 0 \qquad \text{for} \quad -\infty < x < \infty$

2. $P(X < x) = P(X \leq x) = \int_{-\infty}^{x} f(t)\, dt \qquad \text{for} \quad -\infty < x < \infty$

3. $P(a \leq X \leq b) = P(a \leq X < b) = P(a < X < b) = P(a < X \leq b) = \int_{a}^{b} f(t)\, dt \qquad \text{for} \quad a < b$

We can justify these rules with the following observations. Observe that

$$P(a < X \le b) = P(X \le b) - P(X \le a)$$

$$= \int_{-\infty}^{b} f(t)\, dt - \int_{-\infty}^{a} f(t)\, dt$$

$$= \int_{a}^{b} f(t)\, dt$$

Therefore, the last equation of rule 3 is justified. Now, let us assume that

$$P(X = x) = \lim_{c \to 0} P(x - c < X \le x), \qquad \text{with} \quad c > 0$$

This assumption, which we feel to be reasonable, is a consequence of the Continuity Theorem of Probability (a theorem beyond the scope of this book). Hence, observe that

$$\lim_{c \to 0} P(x - c < X \le x) = \lim_{c \to 0} \int_{x-c}^{x} f(t)\, dt = 0$$

Therefore, $P(X = x) = 0$, and rule 1 is justified. The remaining rules are immediate consequences of rule 1 and of the last equation of rule 3. For instance, $P(a < X < b) = P(a < X \le b) - P(X = b) = P(a < X \le b) - 0 = P(a < X \le b)$.

EXAMPLE 5.2

Let X be a continuous random variable with probability density function

$$f(x) = \begin{cases} 2x & \text{for} \quad 0 < x < 1 \\ 0 & \text{elsewhere} \end{cases}$$

Find $P(X \le .4)$, $P(X > .2)$, $P(.2 < X < .4)$, $P(X = .5)$, $P(X < .4)$.

SOLUTION

$$P(X \le .4) = \int_{-\infty}^{.4} f(x)\, dx = \int_{-\infty}^{0} f(x)\, dx + \int_{0}^{.4} f(x)\, dx$$

$$= \int_{-\infty}^{0} 0\, dx + \int_{0}^{.4} 2x\, dx = .16$$

$$P(X > .2) = P(.2 < X < \infty) = \int_{.2}^{\infty} f(x)\, dx$$

$$= \int_{.2}^{1} 2x\, dx = .96$$

$$P(.2 < X < .4) = \int_{.2}^{.4} f(x)\, dx = \int_{.2}^{.4} 2x\, dx = .12$$

$$P(X = .5) = 0$$

$$P(X < .4) = P(X \le .4) - P(X = .4) = P(X \le .4) - 0 = P(X \le .4) = .16$$

EXAMPLE 5.3

Suppose that in a certain region the daily rainfall (in inches) is a continuous random variable X with probability density function $f(x)$ given by

$$f(x) = \begin{cases} (3/4)(2x - x^2) & \text{for} \quad 0 < x < 2 \\ 0 & \text{elsewhere} \end{cases}$$

Find the probability that on a given day in this region the rainfall is
  (a) not more than 1 inch;
  (b) greater than 1.5 inches;
  (c) between .5 and 1.5 inches;
  (d) equal to 1 inch;
  (e) less than 1 inch.

  SOLUTION

  (a)

$$P(\text{rainfall is not more than 1 inch}) = P(X \le 1) = \int_{-\infty}^{1} f(x)\, dx$$

$$= \int_{0}^{1} (3/4)(2x - x^2)\, dx = .5$$

  (b)

$$P(\text{rainfall is greater than 1.5 inches}) = P(X > 1.5)$$

$$= 1 - P(X \le 1.5) = 1 - \int_{-\infty}^{1.5} f(x)\, dx$$

$$= 1 - \int_{0}^{1.5} (3/4)(2x - x^2)\, dx$$

$$= 1 - .84375 = .15625$$

  (c)

$$P(\text{rainfall is between .5 and 1.5 inches}) = P(.5 < X < 1.5)$$

$$= \int_{.5}^{1.5} f(x)\, dx = \int_{.5}^{1.5} (3/4)(2x - x^2)\, dx = .34375$$

(d)

$$P(\text{rainfall is exactly 1 inch}) = P(X = 1) = 0$$

(e)

$$P(\text{rainfall is less than 1 inch}) = P(X < 1)$$
$$= P(X \le 1) - P(X = 1)$$
$$= P(X \le 1) = .5 \qquad (\text{see (a)})$$

EXAMPLE 5.4

The following problem gives an idea as to what you can and cannot find out about a continuous random variable simply by looking at a graph of the probability density function.

Suppose that Figure 5.3 is a graph of the probability density function of the continuous random variable X. Which of the following are true statements?

(a) $P(X = 1) > P(X = 2)$.
(b)   It is more likely that X will take on a value near 1 than near 2.
(c) $P(X \le 0) = 0$.
(d)   It is unlikely that X will take on values greater than 4.
(e)   The possible values of X between 2 and 3 are "uniformly distributed" in the sense that if a and b are values between 2 and 3, values of X are as likely to be in some small neighborhood around a as they are to be in some equivalently small neighborhood around b.

SOLUTION

(a)   False. Both $P(X = 1)$ and $P(X = 2)$ are equal to zero.
(b)   True, because $f(1) > f(2)$ and $f(x)$ is continuous at $x = 1$ and at $x = 2$. Therefore, if ε is a very small number, then

$$P(X \text{ will take on a value near 1})$$
$$= P(1 - \varepsilon < X < 1 + \varepsilon)$$
$$= \int_{1-\varepsilon}^{1+\varepsilon} f(x) \, dx > \int_{2-\varepsilon}^{2+\varepsilon} f(x) \, dx$$
$$= P(2 - \varepsilon < X < 2 + \varepsilon)$$
$$= P(X \text{ will take on a value near 2})$$

(c)   True, because $f(x) = 0$ for all values $x < 0$, so that $P(X \le 0) = \int_{-\infty}^{0} f(x) \, dx = 0$.
(d)   True. The area under the probability density curve from 4 to $+\infty$ is very small. Thus it is unlikely that X will take on values greater than 4.
(e)   True, because the values of $f(x)$ remain constant throughout the subinterval $(2, 3)$.

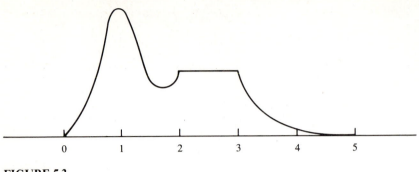

FIGURE 5.3

EXAMPLE 5.5   A THIRD TYPE OF RANDOM VARIABLE

The following is an example of a random variable that is neither discrete nor continuous.

Consider the following experiment. Toss a coin. If the outcome is heads, the player receives 10 ounces of beer. If the outcome is tails, the player presses a button and receives anywhere from 0 to 10 ounces of beer. Let X be the amount of beer received by the player in this experiment. Explain why X is neither a discrete nor a continuous random variable.

SOLUTION

The set of possible values of X is the entire interval of real numbers from 0 to 10. Therefore, X cannot be a discrete random variable.

The probability of heads on the toss of the coin is 1/2. Therefore, $P(X = 10) \geq 1/2$. Consequently X cannot be a continuous random variable because if X were continuous, then $P(X = 10) = 0$.

Exercises

1. Let X be a continuous random variable with probability density function

$$f(x) = \begin{cases} (2/3)x & \text{for} \quad 1 < x < 2 \\ 0 & \text{elsewhere} \end{cases}$$

Find $P(X \leq 1.2)$, $P(X > 1.2)$, $P(1.2 \leq X \leq 1.6)$, $P(X = 1.5)$, $P(X \leq .2)$, and $P(X > 3)$.

2. Let X be a continuous random variable with probability density function

$$f(x) = \begin{cases} -4x & \text{for} \quad -.5 < x < 0 \\ 4x & \quad\quad\quad 0 < x < .5 \\ 0 & \text{elsewhere} \end{cases}$$

(a)   Sketch the probability density curve.

(b)   Find $P(X \le -.3)$, $P(X \le .3)$, $P(-.2 \le X \le .2)$, $P(X > .2)$.

3.   Sixteen-ounce boxes of cereal packed automatically by a certain machine are sometimes overweight and sometimes underweight. The actual weight in ounces over or under 16 is a random variable X whose probability density function is

$$f(x) = \begin{cases} (3/4)(1 - x^2) & \text{for} \quad -1 < x < 1 \\ 0 & \text{elsewhere} \end{cases}$$

Negative values pertain to weight in ounces under the intended 16. Find the probability that a box of cereal packed by this machine will be

(a)   at least .5 ounces overweight;

(b)   neither underweight nor more than .5 ounces overweight;

(c)   neither overweight nor less than .3 ounces underweight;

(d)   between .3 and .5 ounces underweight;

(e)   exactly 16 ounces.

4.   The pressure (measured in pounds per cm^2) at a certain valve is a random variable X whose probability density function is

$$f(x) = \begin{cases} (6/27)(3x - x^2) & \text{for} \quad 0 < x < 3 \\ 0 & \text{elsewhere} \end{cases}$$

Find the probability that the pressure at this valve is

(a)   not more than 2 pounds per cm^2;

(b)   greater than 2 pounds per cm^2;

(c)   between 1.5 and 2.5 pounds per cm^2;

(d)   less than 1.5 pounds per cm^2.

5.   Suppose that the figure that follows is a graph of the probability density function of the continuous random variable X.

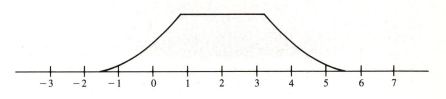

Which of the following are true statements?

(a) $P(X > 2) = P(X < 2)$.

(b) $P(X = 2) > P(X = 0)$.

(c)   It is more likely that X will take on a value near 2 than near 4.

(d) $P(X > 6) = 0$.

(e) $P(X < 7) = 1$.

(f)   It is unlikely that X will take on values greater than 5.

(g)   The possible values of X between 1 and 3 are "uniformly distributed" in the sense that if a and b are values between 1 and 3, then values of X are as likely to be in some small neighborhood around a as they are to be in some equivalently small neighborhood around b.

5.5 Cumulative Distribution Functions of Continuous Random Variables

As in the case of discrete random variables, we are often interested in the probability that the value of a particular continuous random variable will be no greater than a given number. Again, as in the case of discrete random variables, the mathematical function used to designate a probability of this type is called a cumulative distribution function. The formal definition follows.

Definition 5.2 Let X be a continuous random variable with probability density function f. The function

$$F(x) = P(X \le x) = \int_{-\infty}^{x} f(t)\, dt \qquad \text{for} \quad -\infty < x < \infty$$

is called the **cumulative distribution function** or the distribution function of the random variable X.

EXAMPLE 5.6

Suppose that the temperature of a certain fluid is never less than 50°F nor more than 51°F. Furthermore, if X is the temperature of the fluid above 50°F, the cumulative distribution function of X is given by

$$F(x) = \begin{cases} 0 & x < 0 \\ x & \text{for} \quad 0 \le x \le 1 \\ 1 & 1 < x \end{cases}$$

Find $P(X \le -1)$, $P(X \le .5)$, $P(X > .2)$, and $P(.2 < X \le .5)$.

SOLUTION

$$P(X \le -1) = F(-1) = 0$$
$$P(X \le .5) = F(.5) = .5$$
$$P(X > .2) = 1 - P(X \le .2) = 1 - F(.2) = .8$$
$$P(.2 < X \le .5) = P(X \le .5) - P(X \le .2) = F(.5) - F(.2)$$
$$= .5 - .2 = .3$$

EXAMPLE 5.7

Let X be a continuous random variable with probability density function

$$f(x) = \begin{cases} 1 & \text{for} \quad 0 < x < 1 \\ 0 & \text{elsewhere} \end{cases}$$

Find the cumulative distribution function of X. Sketch a graph of $F(x)$.

SOLUTION

Step 1:   If $x < 0$, then

$$F(x) = \int_{-\infty}^{x} f(t)\, dt = \int_{-\infty}^{x} 0\, dt = 0$$

Step 2:   If $0 \le x \le 1$, then

$$F(x) = \int_{-\infty}^{x} f(t)\, dt = \int_{-\infty}^{0} 0\, dt + \int_{0}^{x} 1\, dt$$

$$= 0 + x = x$$

Step 3:   If $1 < x$, then

$$F(x) = \int_{-\infty}^{x} f(t)\, dt = \int_{-\infty}^{0} 0\, dt + \int_{0}^{1} 1\, dt + \int_{1}^{x} 0\, dt$$

$$= 0 + 1 + 0 = 1$$

Conclusion: The cumulative distribution function of X is

$$F(x) = \begin{cases} 0 & \text{for} & x < 0 \\ x & & 0 \le x \le 1 \\ 1 & & 1 < x \end{cases}$$

A graph of $F(x)$ is given in Figure 5.4.

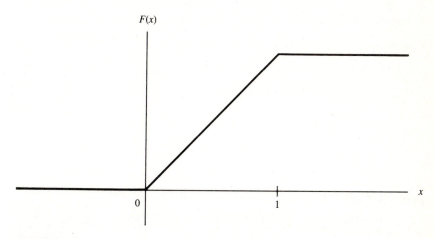

FIGURE 5.4

Exercises

1. Find the cumulative distribution function of the random variable of these exercises in Section 5.4:
   (a) Exercise 1;
   (b) Exercise 2;
   (c) Exercise 3.
   Sketch the graphs of the functions.

2. Let

$$F(x) = \begin{cases} 0 & \text{for} & x \le 0 \\ x^2 & & 0 < x < 1 \\ 1 & & x \ge 1 \end{cases}$$

   be the cumulative distribution function of the random variable X. Find $P(X \le -1)$, $P(X \le .5)$, $P(X \le 3)$, $P(X > .4)$, $P(.2 < X \le .5)$.

3. If a certain type of motor is in operating condition after 4 years of service, the total number of years it will be in operating condition is a random variable X with cumulative distribution function

$$F(x) = \begin{cases} 0 & \text{for} \quad x < 4 \\ 1 - 4/x^2 & \quad x \ge 4 \end{cases}$$

   Find the probability that a 4-year-old motor of this type will be in operating condition for
   (a) at least 5 years;
   (b) less than 10 years;
   (c) between 5 and 7 years.

4. The weekly profit (in thousands of dollars) from a certain concession is a random variable X whose cumulative distribution function is given by

$$F(x) = \begin{cases} 0 & \text{for} & x \le 0 \\ 3x - 3x^2 + x^3 & & 0 < x < 1 \\ 1 & & x \ge 1 \end{cases}$$

   (a) Find the probability of a weekly profit of at least $2,000.
   (b) Find the probability of a weekly profit of at least $500.

5.6   Properties of Probability Density and Cumulative Distribution Functions

Let X be a continuous random variable with probability density function $f(x)$ and cumulative distribution $F(x)$. The following properties of $f(x)$ and

$F(x)$ are immediate consequences of Definition 5.1 and Definition 5.2. (See Exercises 1–5.)

1. $\int_{-\infty}^{\infty} f(x)\,dx = 1$, and $f(x) \geq 0$ at all points x where $f(x)$ is continuous.
2. $\lim_{x \to -\infty} F(x) = 0$, $\lim_{x \to +\infty} F(x) = 1$, and $F(a) \leq F(b)$ for $a < b$.
3. $F(x)$ is continuous at all points x, $-\infty < x < \infty$.
4. $\dfrac{dF(x)}{dx}$ exists and is continuous at all points x where $f(x)$ is continuous.
5. $f(x) = \dfrac{dF(x)}{dx}$ at all points x where $\dfrac{dF(x)}{dx}$ exists.
6. The probability density function of a continuous random variable need not be unique.

   Property 1 is sometimes referred to as a definition of a probability density function because if f is a function with property 1, then there exists a continuous random variable whose probability density function is f.
   Because of property 6, it is preferable to write "a" probability density function of X rather than "the" probability density function of X. However, general practice permits the use of the definite article "the."

Exercises

1. Explain why property 1 must be true. (*Hint:* make use of the fact that $P(-\infty < X < \infty) = 1$, and consider what would happen if $f(x) < 0$ at some point x where $f(x)$ is continuous.)
2. Explain why property 2 must be true.
3. Explain why property 3 must be true. (*Hint:* make use of the fact that $\int_{-\infty}^{x} f(t)\,dt$ is a continuous function of x.)
4. Explain why property 4 must be true. (*Hint:* make use of the fact that $\int_{-\infty}^{x} f(t)\,dt$ is differentiable with respect to x at all points x where $f(x)$ is continuous.)
5. Explain why property 5 must be true.
6. Explain why property 6 must be true. (*Hint:* consider the function $g(x) = 1$ for $0 \leq x \leq 1$ and $g(x) = 0$ elsewhere. Can $g(x)$ be a probability density function for the random variable of Example 5.7?)
7. Which of the following functions are probability density functions? Explain. (*Hint:* see property 1)
    (a)

$$f(x) = \begin{cases} x & \text{for} \quad -.5 < x < .5 \\ 0 & \text{elsewhere} \end{cases}$$

(b)

$$g(x) = \begin{cases} 1/2 & \text{for } -1 < x < 1 \\ 0 & \text{elsewhere} \end{cases}$$

(c)

$$h(x) = \begin{cases} 10 & \text{for } x = -3 \\ 2x & 0 < x < 1 \\ 0 & \text{elsewhere} \end{cases}$$

(d)

$$q(x) = \begin{cases} 1/x & \text{for } 0 < x < 1 \\ 0 & \text{elsewhere} \end{cases}$$

(e)

$$m(x) = \begin{cases} 1/3 & \text{for } 0 < x < 1 \\ 2/3 & 2 < x < 3 \\ 0 & \text{elsewhere} \end{cases}$$

8. The probability density function of the continuous random variable X is

$$f(x) = \begin{cases} kx & \text{for } 1 < x < 5 \\ 0 & \text{elsewhere} \end{cases}$$

(a) Find the value of k.
(b) Sketch the graph of $f(x)$.
(c) Find the probability that X will take on a value that is less than 4; between 2 and 3; greater than 2.5; greater than 7.
(d) Find the cumulative distribution function of X.
(e) Use the cumulative distribution function to find the probabilities of (c).
(f) Sketch the graph of the cumulative distribution function.
(g) Is $F(x)$ continuous everywhere?
(h) Is $F(x)$ differentiable everywhere?

9. Which of the following are cumulative distribution functions of continuous random variables? Explain. (*Hint:* recall the properties discussed in Section 5.6.)

(a)

$$F(x) = \begin{cases} 0 & x < -1 \\ x^2 & -1 \le x < 1 \\ 1 & x \ge 1 \end{cases}$$

(b)

$$F(x) = \begin{cases} 0 & x < 0 \\ x^3 & 0 \le x < 1 \\ 1 & x \ge 1 \end{cases}$$

(c)

$$F(x) = \begin{cases} 0 & x < 0 \\ 1/2 & 0 \le x < 1 \\ 1 & 1 \le x \end{cases}$$

(d)

$$F(x) = \begin{cases} -1 & x < -1 \\ x & -1 \le x < 1 \\ 1 & x \ge 1 \end{cases}$$

10. The cumulative distribution function of the continuous random variable X is

$$F(x) = \begin{cases} 0 & \text{for} \quad x < 3 \\ 1 - k/x^2 & x \ge 3 \end{cases}$$

(a)   Find the value of k.
(b)   Sketch the graph of $F(x)$.
(c)   Find the probability that X will take on a value that is less than 4; between 4 and 5; greater than 6; less than 2.
(d)   Find a probability density function of X.
(e)   Sketch the graph of the probability density function.

5.7   Several Important Continuous Distributions

In this section three of the most important continuous random variables and their distributions will be discussed: the continuous uniform distribution, the exponential distribution, and the normal distribution.

5.7.1   The Continuous Uniform Distribution

A continuous random variable X has a uniform distribution, is said to be uniformly distributed, and is called a continuous uniform random variable if its probability density function can be defined as follows:

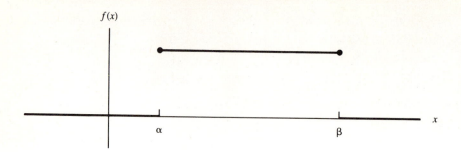

FIGURE 5.5
The Uniform Probability Density Function

Definition 5.3 Let α and β be constants with $\alpha < \beta$. The function

$$f(x) = \begin{cases} \dfrac{1}{\beta - \alpha} & \text{for} \quad \alpha < x < \beta \\ 0 & \text{elsewhere} \end{cases}$$

is called the **uniform probability density function**. (See Figure 5.5.)

The characteristics of a continuous uniform random variable X are:

1. The possible values of the random variable are restricted to some interval (α, β) of real numbers.
2. Within this interval of possible values, any value is as likely to occur as any other value. By this we mean that if I_1 and I_2 are two subintervals of (α, β) and are of equal lengths, the probability that X will take on a value in I_1 is equal to the probability that it will take on a value in I_2.

The following is an example of a random phenomenon whose numerical values are the values of a continuous uniform random variable. The example also shows that if X is a random variable with characteristics 1 and 2, then X is a continuous random variable.

EXAMPLE 5.8

Let X be the amount of coffee in ounces that will be dispensed when a certain automatic coffee dispensing machine is activated. Suppose that the machine never dispenses less than 6 or more than 8 ounces. Within this range, any quantity is as likely to be dispensed as any other (in the sense of characteristic 2 above). Show that X is a continuous uniform random variable.

SOLUTION

In this problem we are given a random variable X (amount of coffee dispensed) that possesses the two characteristics of a continuous uniform random variable. Our task is to show that such a random variable is, in fact, a continuous uniform variable.

Characteristic 2 implies that if I is a subinterval of $(6, 8)$ then the probability that X will take on a value in I is equal to the length of I divided by the length of the interval $(6, 8)$. Therefore, in this particular example the cumulative distribution function of X is

$$F(x) = P(X \leq x) = \begin{cases} 0 & \text{for} & x \leq 6 \\ \dfrac{x - 6}{8 - 6} & & 6 < x < 8 \\ 1 & & x \geq 8 \end{cases}$$

We can now make use of the fact that if $f(x)$ is the probability density function of X, then $f(x) = [dF(x)]/dx$ at all points x where $[dF(x)]/dx$ exists (property 5, Section 5.6). Therefore, the probability density function of X is

$$f(x) = \begin{cases} 1/2 & \text{for} \quad 6 < x < 8 \\ 0 & \text{elsewhere} \end{cases}$$

This probability density function is the uniform probability density function with $\alpha = 6$ and $\beta = 8$. Therefore, the random variable X is a continuous uniform random variable.

EXAMPLE 5.9

The label on a certain brand of syrup states that it is 99% pure maple syrup. If, in fact, the percentage of pure maple syrup in any given bottle of this brand is a uniform random variable X with values ranging from 98.5 to 99.5, what is the probability that a bottle of this brand is

(a) less than 99% pure maple syrup;
(b) more than 99.4% pure maple syrup;
(c) exactly 99% pure maple syrup?

SOLUTION

Since X is a uniform random variable with values ranging from 98.5 to 99.5, the probability density function of X is

$$f(x) = \begin{cases} 1 & \text{for} \quad 98.5 < x < 99.5 \\ 0 & \text{elsewhere} \end{cases}$$

(a)

$$P(\text{less than 99\% pure maple syrup}) = P(X < 99)$$

$$= \int_{-\infty}^{99} f(x)\, dx = \int_{98.5}^{99} 1\, dx = .5$$

(b)

$$P(\text{more than 99.4\% pure maple syrup}) = P(X > 99.4)$$

$$= \int_{99.4}^{\infty} f(x)\, dx = \int_{99.4}^{99.5} 1\, dx = .1$$

(c)

$$P(\text{exactly 99\% pure maple syrup}) = P(X = 99) = 0$$

5.7.2 The Exponential Distribution

A continuous random variable X has an exponential distribution, is said to be exponentially distributed, and is called an exponential random variable if its probability density function can be defined as follows:

Definition 5.4 Let $\theta > 0$ be a constant. The function

$$f(x) = \begin{cases} \dfrac{1}{\theta} e^{-x/\theta} & \text{for} \quad x > 0 \\ 0 & \text{elsewhere} \end{cases}$$

is called the **exponential probability density function** with parameter θ. (See Figure 5.6.)

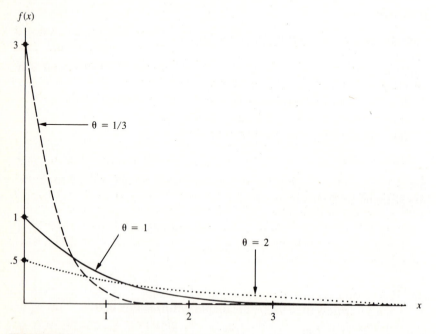

FIGURE 5.6
The Exponential Probability Density Function

EXAMPLE 5.10

The time interval (measured in minutes) between telephone calls arriving at a certain switchboard is an exponential random variable with parameter $\theta = .5$. Find the probability that

(a) there will be a wait of at least 1 minute between the first and the second telephone calls arriving at this switchboard on a given day;

(b) there will be a wait of at least 1 minute between each of the first 5 telephone calls arriving at this switchboard on a given day.

SOLUTION

(a) Let X be the time interval in minutes. Then, the probability density function of X is

$$f(x) = \begin{cases} (1/.5)e^{-x/.5} & \text{for} \quad x > 0 \\ 0 & \text{elsewhere} \end{cases}$$

Therefore

$$P(\text{wait of at least 1 minute}) = P(X > 1) = \int_1^\infty f(x)\, dx$$

$$= \int_1^\infty (1/.5)e^{-x/.5}\, dx = e^{-2} = .1353$$

(b) Consider the event of a time interval of at least 1 minute between telephone calls as a success in a Bernoulli trial. Then, the probability of an interval of at least 1 minute between each of the first 5 telephone calls is the probability of 5 successes in 5 Bernoulli trials, with the probability of success in each of them being equal to .1353. Therefore, the answer to (b) is

$$(.1353)^5 = .000045$$

Exponential random variables are often associated with a "waiting time" that precedes the occurrence of certain specific events. For instance, the time that precedes a breakdown in certain systems, the interval between telephone calls at certain switchboards, and the time before an earthquake strikes in certain regions are all random variables that may reasonably be assumed to be exponential. In particular, random variables that satisfy either of the following two conditions are exponential random variables.

1. If a certain phenomenon is generated by a physical process, and if the random variable X_t that represents the total number of occurrences of the phenomenon from time zero to any given time t is a Poisson random variable (also called a Poisson process) with parameter λt (i.e., with $P(X_t = x) = p(x; \lambda t) = (\lambda t e^{-\lambda t})/x!$ for $x = 0, 1, 2, 3, \ldots$), then the waiting

time until the first occurrence of the phenomenon (or the waiting time between occurrences of the phenomenon) is an exponential random variable with parameter $\theta = 1/\lambda$. (A justification of this statement follows condition 2.) For example, if the number of automobile accidents from time zero to time t is a Poisson random variable with parameter $\lambda t = (1/15)t$, then the waiting time until the first automobile accident (or the waiting time between automobile accidents) is an exponential random variable with parameter $\theta = 1/\lambda = 15$. (See Example 5.11.)

2. If T is a nonnegative random variable that is "memoryless" in the sense that

$$P(T > s + t \mid T > t) = P(T > s) \qquad \text{for all} \quad s, t \geq 0$$

then T is an exponential random variable. (A justification of this statement follows.) For example, suppose that a certain electrical system is constantly maintained to run at an operationally perfect level. It is reasonable to assume that the chances that the system will not fail for at least $s + t$ hours—given that it had not failed for at least t hours—is the same as the chances that it will not fail for at least s hours. In other words, if T is the time until failure occurs in the system, then

$$P(T > s + t \mid T > t) = P(T > s) \qquad \text{for all} \quad s, t \geq 0$$

and we can assume that T is an exponential random variable.

Justification of Statement 1   We will show that if X_t is a Poisson random variable with parameter λt, and if, furthermore, X_t represents the total number of occurrences of a certain phenomenon from time zero to time t, then the waiting time until the first occurrence of the phenomenon is an exponential random variable T with parameter $\theta = 1/\lambda$.

First of all, let us consider the event $T > t$. That is, let us consider the event that from time zero to time t there are no occurrences of the phenomenon. By applying the properties of Poisson processes we see that the probability of this event is

$$P(T > t) = P(X_t = 0) = \frac{(\lambda t)^0 e^{-t}}{0!} = e^{-\lambda t}$$

Next, let $F(t)$ be the cumulative distribution function of T. It is clear that for values of $t \leq 0$

$$F(t) = P(T \leq t) = 0$$

whereas, for values of $t \geq 0$

$$F(t) = P(T \leq t) = 1 - P(T > t) = 1 - e^{-\lambda t}$$

Therefore, the probability density function of T is

$$f(t) = \frac{dF(t)}{dt} = \begin{cases} \lambda e^{-\lambda t} & \text{for} \quad t > 0 \\ 0 & \text{elsewhere} \end{cases}$$

In other words, T is an exponential random variable with parameter $\theta = 1/\lambda$.

Justification of Statement 2   We will show that if T is a nonnegative random variable that is "memoryless" in the sense that

$$P(T > s + t \mid T > t) = P(T > s) \qquad \text{for all} \quad s, t \geq 0$$

then T is an exponential random variable.

First, we will prove the following lemma.

Lemma   *If the function ϕ defined on $[0, \infty]$ is nonincreasing, $\phi(0) \neq 0$, and satisfies the equation*

$$\phi(s + t) = \phi(s)\phi(t) \qquad \text{for} \quad s, t \geq 0$$

then $\phi(t) = e^{-\lambda t}$ for some $\lambda \geq 0$.

Proof of Lemma

Step 1: $\phi(0) = \phi(0 + 0) = \phi(0)\phi(0)$. Therefore, $\phi(0) = 1$.

Step 2:   Let $\alpha = \phi(1)$. Then $\alpha = \phi(1) = \phi[1/n + 1/n + \cdots + 1/n] = [\phi(1/n)]^n$. Therefore, $\phi(1/n) = \alpha^{1/n}$, and $\phi(m/n) = [\phi(1/n)]^m = \alpha^{m/n}$, for integers m and n.

Step 3:   Let t be a fixed number in $[0, \infty]$, and let m and n be integers such that

$$m/n \leq t \leq (m + 1)/n$$

Then, because ϕ is nonincreasing, we have

$$\phi[(m + 1)/n] \leq \phi(t) \leq \phi(m/n)$$

and because of Step 2, we have

$$\alpha^{(m + 1)/n} \leq \phi(t) \leq \alpha^{m/n}$$

Step 4:   From Step 3 it follows that

$$\lim_{n \to \infty} \alpha^{(m + 1)/n} \leq \lim_{n \to \infty} \phi(t) \leq \lim_{n \to \infty} \alpha^{m/n}$$

Now, observe that

$$\lim_{n \to \infty} \alpha^{(m+1)/n} = \alpha^t$$

$$\lim_{n \to \infty} \phi(t) = \phi(t)$$

and

$$\lim_{n \to \infty} \alpha^{m/n} = \alpha^t$$

Therefore

$$\phi(t) = \alpha^t$$

Step 5: Since $\phi(t)$ is nonincreasing (hypothesis of the lemma), and $\phi(t) = \alpha^t$ (Step 4), we conclude that α must be a number such that

$$0 < \alpha \le 1$$

Step 6: Let $\lambda = -\ln \alpha$. Then λ is a number such that $\lambda \ge 0$ and

$$\alpha = e^{-\lambda}$$

Hence, it follows from Step 4 that

$$\phi(t) = \alpha^t = e^{-\lambda t}$$

and we have completed the proof of the lemma.

Let us return to a proof of the statement that if T is a nonnegative random variable that is "memoryless" in the sense that

$$P(T > s + t \,|\, T > t) = P(T > s) \qquad \text{for all} \quad s, t \ge 0$$

then T is an exponential random variable. Observe that if T has the "memoryless" property, then

$$\frac{P(T > s + t)}{P(T > s)} = \frac{P(T > s + t \quad \text{and} \quad T > s)}{P(T > s)}$$

$$= P(T > s + t \,|\, T > s) = P(T > t)$$

Therefore

$$P(T > s + t) = P(T > s)P(T > t) \qquad \text{for all} \quad s, t \ge 0$$

Let us set

$$\phi(t) = P(T > t) \qquad \text{for} \quad t \geq 0$$

Then we see that all of the conditions of the lemma are satisfied for $\phi(t)$. Hence

$$P(T > t) = \phi(t) = e^{-\lambda t} \qquad \text{for some constant} \quad \lambda > 0$$

Note: since T is a random variable, it does not make sense to have $P(T > t) = 1$ for all $t \geq 0$. Therefore, we may eliminate the possibility of $\lambda = 0$ in the above equation.

The reader may now complete the proof that T is an exponential random variable by referring to the last few lines of the justification of Statement 1.

EXAMPLE 5.11

Suppose that the number of automobile accidents from any given time zero to any given time t (measured in hours) along a certain segment of Highway Z is a Poisson random variable with parameter $\lambda t = (1/15)t$. Furthermore, suppose that on a certain day there is an automobile accident along this segment of the highway at 10 A.M.

(a)   Find the probability that there will not be another automobile accident along this segment of the highway for at least 24 hours.

(b)   Find the probability that there will be another automobile accident along this segment of the highway within the same day.

   SOLUTION

Let T be the time (measured in hours) until the next automobile accident occurs along this segment of the highway. Then T is an exponential random variable with parameter $\theta = 15$.

(a)

P(there will not be another automobile accident for at least 24 hours)

$$= P(X > 24) = \int_{24}^{\infty} (1/15)e^{-x/15}\, dx = e^{-24/15} = .2019$$

(b)

P(there will be another automobile accident within the same day)

$$= P(X \leq 14) = \int_{0}^{14} (1/15)e^{-x/15} = 1 - e^{-14/15} = .6068$$

EXAMPLE 5.12

Suppose that a certain electrical system is constantly maintained to run at an operationally perfect level so that it is reasonable to assume that the chances that the system will not fail for at least $s + t$ hours—given that it had not failed for at least t hours—is the same as the chances that it will not fail for at least s hours. Furthermore, suppose that the probability that it will not fail for at least 10 hours is .99. Find the probability that it will not fail for at least 100 hours.

SOLUTION

Let T be the time (measured in hours) until there is a failure in the system. Then, according to the statement above, it is reasonable to assume that

$$P(T > s + t \mid T > t) = P(T > s) \qquad \text{for all} \quad s, t \geq 0$$

By Statement 2 on page 152 this means that we may assume T to be an exponential random variable. To find the parameter θ for the distribution of T, observe that

$$.99 = P(T \geq 10) = \int_{10}^{\infty} (1/\theta)e^{-x/\theta} \, dx = e^{-10/\theta}$$

Therefore

$$\theta = \frac{-10}{\ln .99} = 994.99$$

This means that

P(system will not fail for at least 100 hours)

$$= P(T \geq 100) = \int_{100}^{\infty} (1/994.99)e^{-x/994.99} \, dx = e^{-100/994.99} = .904$$

Before we end this discussion on the exponential random variable we should mention the following additional important relationship between the exponential random variable and the Poisson random variable: if the waiting time between events of a specific kind is an exponential random variable with parameter θ, the number of events of this kind from any time zero to any time t is a Poisson random variable with parameter $\lambda t = (1/\theta)t$. The proof of this statement is beyond the scope of this book (see references 1 or 2).

EXAMPLE 5.13

With reference to the switchboard of Example 5.10, what is the probability that more than 3 telephone calls will arrive within a 5-minute period?

SOLUTION

The time interval (measured in minutes) between telephone calls arriving at this switchboard is an exponential random variable with parameter $\theta = .5$.

Therefore, the number of telephone calls arriving at this switchboard from any time zero to any time t is a Poisson random variable with parameter $\lambda t = (1/.5)t$. Let X be the number of telephone calls arriving at this switchboard within a 5-minute period. Then X is a Poisson random variable with parameter $\lambda = (1/.5) \cdot (5) = 10$. Therefore

P(more than 3 telephone calls within a 5-minute period)

$= P(X > 3)$

$= 1 - P(X \leq 3) = 1 - p(0; 10) - p(1; 10) - p(2; 10) - p(3; 10)$

$= 1 - .0099 = .99$

A discussion of the exact meaning of the parameter θ is given in Chapter 6. In the meantime, we can acquire an intuitive understanding of the parameter by reasoning in the following manner. In Chapter 4 we saw that if X is a Poisson random variable with parameter λ, on the average the observed value of X will be λ. Condition 1 establishes a relationship between Poisson random variables and exponential random variables. In particular, if the number of events of a specific type within a unit time interval is a Poisson random variable with parameter λ, the waiting time between events of this specific type is an exponential random variable with parameter $\theta = 1/\lambda$. This means that if on the average there are λ events of a specific type within a unit-time interval, then on the average the time between events of this specific type is $\theta = 1/\lambda$. In other words, if X is an exponential random variable with parameter θ, then on the average the observed value of X will be θ. For example, if the time in minutes until the occurrence of the first automobile accident after 4:30 P.M. is an exponential random variable with parameter $\theta = 15$, then on the average, that time is 15 minutes.

5.7.3   The Normal Distribution

A continuous random variable X has a normal distribution, is said to be normally distributed, and is called a normal random variable if its probability density function can be defined as follows:

Definition 5.5   Let μ and σ be constants with $-\infty < \mu < \infty$ and $\sigma > 0$. The function

$$f(x) = \frac{1}{\sqrt{2\pi}\,\sigma}\, e^{-(x-\mu)^2/(2\sigma^2)} \qquad \text{for} \quad -\infty < x < \infty$$

is called the **normal probability density function** with parameters μ and σ. (See Figure 5.7.)

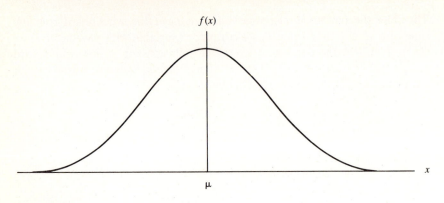

FIGURE 5.7
The Normal Probability Density Function

The normal probability density function is without question the most important of probability density functions. This is a consequence of the design of nature as well as humanity's ingenuity in analyzing that design. In any case, empirical data point to the existence of innumerable random phenomena whose distributions are normal, or approximately normal. The distribution of IQ's, SAT scores, heights of 5-year-old children, displacements of certain gas molecules in time, and errors made in measuring certain physical quantities are just a few among the countless measurements whose distributions are normal. In particular, measurements that are the result of innumerable independent components tend to be normally distributed. A partial explanation of this property of the normal distribution is provided in Chapter 8, where the implications of the central limit theorem will be discussed.

5.7.3.1 Finding probabilities for normal random variables

Unlike the uniform or exponential probability density functions, the normal probability density function does not have an antiderivative that can be expressed in terms of a simple function. Consequently, finding probabilities for normal random variables would be a horrendous task if it were not for the fact that extensive tables of values have been tabulated for such a purpose. Table III in Appendix C contains values of $P(0 < X < z)$ for normal random variables X with parameters $\mu = 0$ and $\sigma = 1$. (Such random variables are called **standard normal** random variables.) Values of $P(-z < X < 0)$ may be obtained from the relationship

$$P(-z < X < 0) = P(0 < X < z)$$

(This relationship is valid because the probability density function of the standard normal random variable is symmetric about the vertical axis at $x = 0$.)

EXAMPLE 5.14

Let X be a standard normal random variable. Use Table III to find
(a) $P(0 < X < 2)$,
(b) $P(-2 < X < 0)$,
(c) $P(-1.5 < X < 2)$,
(c) $P(X > 1.14)$.

SOLUTION

(a) Using Table III we obtain $P(0 < X < 2) = .4772$.
(b) The standard normal probability density function is symmetric about the line $x = 0$. Therefore

$$P(-2 < X < 0) = P(0 < X < 2) = .4772$$

(c)

$$P(-1.5 < X < 2) = P(-1.5 < X < 0) + P(0 < X < 2)$$
$$= P(0 < X < 1.5) + P(0 < X < 2)$$
$$= .4332 + .4772 = .9104$$

(d) The standard normal probability density function is symmetric about the line $x = 0$. Therefore, $P(-\infty < X < 0) = P(0 < X < \infty) = .5$. It follows that

$$P(X > 1.14) = P(0 < X < \infty) - P(0 < X < 1.14)$$
$$= .5 - .3729 = .1271$$

Finding probabilities for nonstandard normal random variables is discussed formally in Chapter 6 where we will also discuss the meaning of the parameters μ and σ. However, we shall give a brief outline here of the method for finding such probabilities. It will be shown in Chapter 6 that if X is a normal random variable with parameters μ and σ, then $Z = (X - \mu)/\sigma$ is a standard normal random variable. Thus, to find $P(a < X < b)$, we would proceed as follows

$$P(a < X < b) = P\left(\frac{a - \mu}{\sigma} < \frac{X - \mu}{\sigma} < \frac{b - \mu}{\sigma}\right)$$

$$= P\left(\frac{a - \mu}{\sigma} < Z < \frac{b - \mu}{\sigma}\right)$$

We could then use the method outlined in Example 5.14 to obtain the value of the last term above.

EXAMPLE 5.15

Let X be a normal random variable with parameters $\mu = 2$ and $\sigma = .5$. Find $P(1 < X < 2.1)$.

SOLUTION

$$P(1 < X < 2.1) = P[(1 - 2)/.5 < (X - \mu)/\sigma < (2.1 - 2)/.5]$$
$$= P(-2 < Z < .2)$$
$$= P(-2 < Z < 0) + P(0 < Z < .2)$$
$$= P(0 < Z < 2) + P(0 < Z < .2)$$
$$= .4772 + .0793$$
$$= .5565$$

EXAMPLE 5.16

Suppose that any given day's gross receipts (in thousands of dollars) from a certain sales unit may be treated as a normal random variable X with parameters $\mu = 20$ and $\sigma = 2$. Find the probability that
  (a)   gross receipts on a certain day are between $15,000 and $25,000;
  (b)   gross receipts on a certain day are more than $24,000.

  SOLUTION

(a)

P(gross receipts are between 15 and 25 (thousand dollars))

$$= P(15 < X < 25) = P\left(\frac{15 - 20}{2} < \frac{X - \mu}{\sigma} < \frac{25 - 20}{2}\right)$$
$$= P(-2.5 < Z < 2.5) = P(-2.5 < Z < 0) + P(0 < Z < 2.5)$$
$$= P(0 < Z < 2.5) + P(0 < Z < 2.5) = (2)(.4938) = .9876$$

(b)

P(gross receipts are more than 24 (thousand dollars))

$$= P(X > 24) = P\left(\frac{X - \mu}{\sigma} > \frac{24 - 20}{2}\right) = P(Z > 2)$$
$$= .5 - P(0 < Z \le 2) = .5 - .4772 = .0228$$

EXAMPLE 5.17

Suppose that when the thermostat is set at d degrees Celsius, the actual temperature of a certain container of liquid is a normal random variable with parameters $\mu = d$ and $\sigma = .5$.
  (a)   If the thermostat is set at 90°C, what is the probability that the actual temperature of the container of liquid is less than 89°C?
  (b)   What is the lowest setting of the thermostat that will maintain a temperature of at least 80°C with a probability of .99?

SOLUTION

Let X be the temperature of the liquid. Let Z be the standard normal random variable.

(a)

$$P(\text{temperature is less than } 89°C) = P(X < 89)$$

$$= P\left(\frac{X - \mu}{\sigma} < \frac{89 - 90}{.5}\right) = P(Z < -2) = .5 - P(0 < Z \le 2)$$

$$= .5 - .4772 = .0228$$

(b) Let d be the lowest setting of the thermostat to maintain a temperature of at least 80°C with a probability of .99. Then

$$.99 = P(\text{temperature of at least } 80°C) = P(X > 80)$$

$$= P\left(\frac{X - \mu}{\sigma} > \frac{80 - d}{.5}\right) = P\left(Z > \frac{80 - d}{.5}\right)$$

This means that

$$\frac{80 - d}{.5} = 2.327$$

and

$$d = 78.8365°C$$

Exercises

1. The chairman of a certain board of directors never arrives for a meeting more than 5 minutes late or 5 minutes early. Within that range of time, however, he is as likely to arrive at one time as another, so that if X is the number of minutes that he will be early or late, the probability density function of X is

$$f(x) = \begin{cases} 1/10 & \text{for } -5 \le x \le 5 \\ 0 & \text{elsewhere} \end{cases}$$

Negative values pertain to early arrivals and positive values pertain to late arrivals. Find the probability that the chairman will be
   (a) at least 3 minutes late;
   (b) at least 2 minutes early;
   (c) right on time;
   (d) no more than 1 minute either early or late.

2. Suppose that the annual rainfall in a certain region is a uniform random variable with values ranging from 12 to 15 inches.

   (a) Find the probability that in a given year the region's rainfall will be less than 13 inches.

   (b) Find the probability that the region will have rainfalls of less than 13 inches 3 out of 4 years.

3. Suppose that the operating lifetime of a certain type of battery is an exponential random variable with parameter $\theta = 2$ (measured in years). Find the probability that

   (a) a battery of this type will have an operating lifetime of over 4 years;

   (b) a battery of this type will have an operating lifetime of between 1 and 3 years;

   (c) at least one out of 5 batteries of this type will have operating lifetimes of over 4 years.

4. Suppose that the number of emergency calls arriving at a certain police headquarters from any given time zero to any given time t (measured in hours) is a Poisson random variable X_t with parameter $\lambda t = (1/2)t$.

   (a) Find the probability that on a certain day no emergency calls will arrive from 12 noon to 3 P.M.

   (b) Find the probability that there will be at least one emergency call between 12 noon and 5 P.M.

5. Suppose that miracles do happen, but that there is no telling when they will happen because the time of the next occurrence of a miracle is independent of the time of the last occurrence of a miracle. If the probability of a miracle within the next 1,000 years is .99, what is the probability that there will be a miracle within the next 200 years?

6. Suppose that a certain computerized accounting system is nearly flawless. The designer of the system claims that the probability of an error in any one day is .00001. If we assume that the designer's claim is correct, and if we assume that the error-making capacity of the system is "memoryless" in the sense of Statement 2 of Section 5.7.2, what is the probability that the system will be error-free for a period of (a) 1 year; (b) 10 years?

7. Let X be a standard normal random variable. Find the probability that X will take on a value that is

   (a) less than 1.5;

   (b) greater than 2.14;

   (c) between 1.5 and 2.14;

   (d) less than -1.32;

   (e) greater than -2.52;

   (f) between -2.52 and 1.64;

   (g) between -2.52 and -1.5;

   (h) exactly 2.

8. Let X be a standard normal random variable. Find z if the probability that X will take on a value

   (a) less than z is .9842;

   (b) greater than z is .3085;

   (c) greater than z is .5675;

    (d)   less than z is .1292;

    (e)   between $-z$ and z is .8904.

9.   The temperature of a certain liquid in a research laboratory, measured in degrees Celsius, is a standard normal random variable. Find the probability that the temperature of the liquid will be

    (a)   warmer than $2°C$;

    (b)   cooler than $1°C$;

    (c)   between $-.5°C$ and $+.5°C$;

    (d)   cooler than $-1°C$.

10.   A certain type of packaged length of string is labeled as being 12 feet long. The deviation from 12 feet, measured in inches, is a standard normal random variable. Negative values pertain to lengths in inches under 12 feet, and positive values pertain to lengths in inches over 12 feet. Find the probability that one of these packages of string will be

    (a)   at least 2 inches longer than 12 feet;

    (b)   no more than 1.5 inches short of 12 feet;

    (c)   more than 2 inches short of 12 feet;

    (d)   within 1.5 inches of 12 feet;

    (e)   exactly 12 feet long.

11.   Suppose that the average temperature in July in a certain region is a normal random variable with parameters $\mu = 90$ (measured in degrees Fahrenheit) and $\sigma = 5$. Find the probability that in a given year the average temperature in July in this region will be

    (a)   above $100°F$;

    (b)   below $95°F$;

    (c)   between $85°F$ and $95°F$;

    (d)   below $88°F$.

12.   Suppose that the growth in inches during the tenth year of life of a North American boy is a normal random variable with parameters $\mu = 2$ and $\sigma = 1$. Find the probability that a randomly selected North American boy will grow

    (a)   at least 1 inch in his tenth year;

    (b)   less than $\frac{1}{2}$ inch in his tenth year;

    (c)   more than 3 inches in his tenth year;

    (d)   between 1 and 2 inches in his tenth year.

13.   Prove that if T is an exponential random variable, it has the "memoryless" property of $P(T > s + t \mid T > t) = P(T > s)$ for all s, $t \geq 0$. (Combined with Statement 2 of Section 5.7.2 this means that a nonnegative random variable T is an exponential random variable if and only if $P(T > s + t \mid T > t) = P(T > s)$ for all s, $t \geq 0$.)

14.   Show that for all positive values of t and small values of Δt

$$P(T \leq t + \Delta t \mid T > t) \approx \frac{\Delta t}{\theta}$$

for exponential random variables T with parameter θ. (*Hint:* use the "memoryless" property of T.)

15. **The Gamma Distribution.** A random variable X has a *gamma distribution*, and is called a *gamma random variable*, if its probability density function is given by

$$f(x) = \begin{cases} \dfrac{1}{\beta^\alpha \Gamma(\alpha)} x^{\alpha-1} e^{-x/\beta} & \text{for} \quad x > 0 \\ 0 & \text{elsewhere} \end{cases}$$

where $\alpha > 0$ and $\beta > 0$. (Γ is the gamma function defined by $\Gamma(\alpha) = \int_0^\infty y^{\alpha-1} e^{-y} \, dy$ for $\alpha > 0$. When α is a positive integer $\Gamma(\alpha) = (\alpha - 1)!$) Show that the exponential distribution is a gamma distribution with $\alpha = 1$ and $\beta = \theta$.

16. In a certain city, the daily consumption of water (measured in units of 100,000 gallons) is a gamma random variable with $\alpha = 2$ and $\beta = 2$. What is the probability that this city will use more than 600,000 gallons of water on a randomly selected day?

17. **The Beta Distribution.** A random variable X has a *beta distribution*, and is called a *beta random variable*, if its probability density function is given by

$$f(x) = \begin{cases} \dfrac{\Gamma(\alpha + \beta)}{\Gamma(\alpha)\Gamma(\beta)} x^{\alpha-1}(1 - x)^{\beta-1} & \text{for} \quad 0 < x < 1 \\ 0 & \text{elsewhere} \end{cases}$$

where $\alpha > 0$ and $\beta > 0$. Show that if $\alpha = 1$ and $\beta = 1$, the beta probability density function is the uniform probability density function $f(x) = 1$ for $0 < x < 1$ and $f(x) = 0$ elsewhere.

18. In a certain city, the proportion of voters who will vote for the Republican candidate can be treated as a beta random variable with $\alpha = 2$ and $\beta = 1$. If an election is won with a majority (more than 50%) of the votes, then what is a Republican candidate's probability of winning an election in that city?

19. Let X be the time elapsed until a certain specific event occurs. Suppose that the following assumptions can be made concerning X.
    (a)   There is a constant $\theta > 0$ such that for any sufficiently short time interval $\varDelta_x$ of length $\varDelta_x$

$$P(\text{the event occurs during the time interval } \varDelta_x) = \varDelta_x/\theta$$

    (b)   If $\varDelta_1, \varDelta_2, \ldots, \varDelta_n$ are consecutive, nonoverlapping, and sufficiently short time intervals of lengths $\varDelta_1, \varDelta_2, \ldots, \varDelta_x$, respectively, then

$$P(\text{the event does not occur during } \varDelta_1, \varDelta_2, \ldots, \text{ and } \varDelta_n)$$
$$= P(\text{the event does not occur during } \varDelta_1) \cdot P(\text{the event does not occur during } \varDelta_2) \cdots P(\text{the event does not occur during } \varDelta_n)$$

Show that X is an exponential random variable with parameter θ. (*Hint:* divide the time interval from 0 to x into n equal subintervals $\Delta_1, \Delta_2, \ldots,$ Δ_n of lengths $\Delta_x = x/n$. Then, for n sufficiently large

$$
\begin{aligned}
P(X > x) &= P(\text{the event does not occur from time 0 to time } x) \\
&= P(\text{the event does not occur during } \Delta_1, \Delta_2, \ldots, \Delta_n) \\
&= P(\text{the event does not occur during } \Delta_1) \cdots P(\text{the} \\
&\quad\ \text{event does not occur during } \Delta_n) = [1 - (\Delta_x/\theta)]^n
\end{aligned}
$$

Consider this probability as n tends to infinity.)

References

1. Chung, Kai Lai. 1974. *Elementary probability theory with stochastic processes.* New York: Springer-Verlag.
2. Ross, Sheldon. 1976. *A first course in probability.* New York: Macmillan Publishing Co., Inc.
3. Hoel, P. G., S. C. Port, C. J. Stone. 1971. *Introduction to probability theory.* New York: Houghton Mifflin Co.

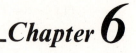

Chapter 6

MATHEMATICAL EXPECTATION

Standardized Random Variables
   Related theorems
   Computing probabilities for the normal random variable
More Applications
   Estimating the means of distributions
   Estimating the probability of events
Derivation of the Means and Variances of
   Special Probability Distributions

6.1 The Mathematical Expectation of a Random Variable

6.1.1 Introduction

With few exceptions, the single most important number associated with a random variable is the mathematical expectation (also called expectation, expected value, or mean). We shall see that ingenious uses of this number may reap vast fortunes for the cunning gambler, gain insights into complex problems and solve integrals and differential equations for the informed mathematician, engineer, and scientist, and play an important role in the decision-making processes of government and industry. To acquire a working knowledge of this most important entity let us begin with the definition and proceed to derive useful theorems and techniques.

6.1.2 The Mathematical Expectation $E(X)$

Definition 6.1 The **mathematical expectation of a discrete random variable** X with probability function $f(x)$ is the quantity

$$E(X) = \sum_x xf(x)$$

provided that $\sum_x |x| f(x) < \infty$ (i.e., the series converges absolutely). The notation $\sum_x$ denotes summation over all values x that are possible values of the random variable X.

Note: $E(X)$ is often denoted by the Greek letter μ.

EXAMPLE 6.1

Let X be the numerical outcome of one roll of a die. Find $E(X)$.

   SOLUTION

The possible values of X are the numbers $1, 2, 3, 4, 5$, and 6. The probability function of X is $f(x) = 1/6$ for $x = 1, 2, 3, 4, 5$, and 6. Therefore

$$E(X) = \sum_x xf(x) = 1 \cdot \frac{1}{6} + 2 \cdot \frac{1}{6} + 3 \cdot \frac{1}{6} + 4 \cdot \frac{1}{6} + 5 \cdot \frac{1}{6} + 6 \cdot \frac{1}{6}$$

$$= 3.5$$

EXAMPLE 6.2

Let X be a discrete random variable with probability function

$$f(x) = \begin{cases} 1/3 & \text{for} \quad x = 3 \\ 2/3 & \phantom{\text{for} \quad} x = 9 \end{cases}$$

Find $E(X)$.

SOLUTION

$$E(X) = \sum_x f(x) = 3 \cdot \frac{1}{3} + 9 \cdot \frac{2}{3} = 7$$

Definition 6.2   The **mathematical expectation of a continuous random variable** X with probability density function $f(x)$ is the quantity

$$E(X) = \int_{-\infty}^{\infty} xf(x)\, dx$$

provided that the integral exists.

EXAMPLE 6.3

Let X be a continuous random variable with probability density function

$$f(x) = \begin{cases} 81x^2 & \text{for} \quad 0 < x < 1/3 \\ 0 & \text{elsewhere} \end{cases}$$

Find $E(X)$.

SOLUTION

$$E(X) = \int_{-\infty}^{\infty} xf(x)\, dx = \int_0^{1/3} x \cdot 81x^2\, dx = \frac{1}{4}$$

EXAMPLE 6.4

Let X be the numerical outcome of selecting a number at random from the interval $(0, 1)$. Find $E(X)$.

SOLUTION

The probability density function of X is the uniform probability density function $f(x) = 1$ for $0 < x < 1$ and $f(x) = 0$ for all other values of x. Therefore

$$E(X) = \int_{-\infty}^{\infty} xf(x)\, dx = \int_0^1 x \cdot 1\, dx = \frac{1}{2}$$

6.1.3   Why $E(X)$ is Called the Mean of X

In mathematics we refer to the average of a set of values as the mean of the set. We see in Example 6.1 that $E(X) = 3.5$, and that the average of the possible values in rolling a die is also 3.5. However, with respect to the random variable X in Example 6.2 we see that $E(X) = 7$, whereas the average of the possible values of X is 6. In other words, $E(X)$ is not always the same as the mean of the set of all possible values of X. One wonders, then, why $E(X)$ is called the mean of X. Let us consider the following two examples.

EXAMPLE 6.5

A class of N students is given a test. The possible grades are A, B, C, D, and F with the numerical values of $4, 3, 2, 1$, and 0, respectively. Suppose we select a student at random from the class and let X be the numerical grade obtained by the student. What is $E(X)$?

SOLUTION

Let N_x denote the number of students receiving the grade x. Then

$$f(x) = \frac{N_x}{N} \qquad \text{for} \quad x = 0, 1, 2, 3, 4$$

and

$$E(X) = \sum_x xf(x) = \sum_x x \cdot \frac{N_x}{N} = \frac{\sum_x x \cdot N_x}{N}$$

$$= \frac{\text{(sum of all the grades obtained by the students in the class)}}{\text{(total number of students in the class)}}$$

$$= \text{average grade}$$

We see in Example 6.5 that if X is the numerical grade of a student selected at random from a class, then $E(X)$ is the average grade, or the mean of the set of all grades obtained by students in the class. It is in this sense that $E(X)$ is called the mean of X. In other words, the mean of X is not to be interpreted as the average of the possible values of X, but rather as the average of the values weighted by their frequencies. If the value x_1 occurs with a frequency that is k times that of the value x_2, then x_1 is weighted k times as heavily as x_2 in the averaging process. In the averaging process used in Example 6.5 to obtain $E(X)$, we see that each value x is weighted (i.e., multiplied) by its frequency N_x. Referring to Example 6.2, we note that corresponding to the fact that the frequency of 9 is twice that of 3, the value of 9 is weighted twice as heavily as the value 3 in the computation of $E(X)$.

EXAMPLE 6.6

Show that if X is the numerical value of an element selected at random from a finite set A of numbers, then $E(X)$ is the mean of the set A of numbers.

SOLUTION

Suppose that $A = \{a_1, a_2, \ldots, a_n\}$ where each element a_i is a number. (*Note:* the elements $a_1, a_2, \ldots, a_n$ need not be distinct numbers.) By definition, the mean of the set A of numbers is the quantity

$$\frac{a_1 + a_2 + \cdots + a_n}{n}$$

Now, if X is the numerical value of an element selected at random from the set A of numbers, then the probability function of X is given by

$$f(x_i) = P(X = x_i) = k_i/n \qquad \text{for} \quad i = 1, 2, 3, \ldots, m$$

where $x_1, x_2, \ldots, x_m$ are the distinct numbers in the set A and where k_i is the number of elements in A that have the numerical value x_i. Therefore

$$E(X) = \sum_{i=1}^{m} x_i f(x_i) = \sum_{i=1}^{m} x_i(k_i/n) = \left(\sum_{i=1}^{m} x_i k_i \right) \Big/ n$$

$$= \frac{a_1 + a_2 + \cdots + a_n}{n}$$

$$= \text{the mean of the set } A \text{ of numbers}$$

It is easy to see that the result of Example 6.5 is compatible (in a sense, identical) with the result of Example 6.6. If A is the set of all grades obtained by the students of Example 6.5 and X is the numerical value of an element selected at random from A, then X is also the numerical grade of a student selected at random from the class and the mean of the set A of numbers is also the average grade for the class.

6.1.4   The Strong Law of Large Numbers and Why $E(X)$ Is Called the Mathematical Expectation of X

In Example 6.1, we saw that if X is the numerical outcome of one roll of a die, then $E(X) = 3.5$. Now, when we roll a die we certainly never expect to obtain the value 3.5, so why do we call 3.5 the expected value of X? To understand this we must look to the strong law of large numbers. A statement of the strong law of large numbers for repeated observations of a random variable is given below. A more general statement of the theorem may be found in most advanced texts in probability.

Theorem 6.1   The Strong Law of Large Numbers   Let $\mathscr{E}$ be an experiment with sample space S. Let X be a random variable defined on S with $E(X) = \mu$. If $\mathscr{E}$ is repeated n times under identical conditions and X_i is the random variable X on the ith performance of $\mathscr{E}$, then

$$P\left(\lim_{n \to \infty} \frac{X_1 + X_2 + \cdots + X_n}{n} = \mu \right) = 1$$

The statement of the theorem implies that if n is large, there is a great probability that the average of the observed values of the random variables $X_1, X_2, \ldots, X_n$ will be approximately equal to the mathematical expectation μ. That is, for n large

$$\frac{X_1 + X_2 + \cdots + X_n}{n} \approx \mu$$

with great probability. In other words, if X is a random variable associated with an experiment, and the experiment is repeated many times, it is very likely that the average of the many observed values of X will be approximately equal to $E(X)$. Now, if something is very likely to happen, then we often say that we expect it to happen. Thus, another way of making the above statement is to say that we expect the average of many observed values of X to be approximately equal to $E(X)$. It is in that sense that we call $E(X)$ the mathematical expectation, or the expected value, of the random variable X. Thus, when we say that if X is the numerical outcome of one roll of a die, the expected value of X is 3.5, we do not mean that we expect to obtain the value 3.5 when we roll a die. We mean, rather, that the average value of many rolls of a die is expected to be approximately 3.5.

EXAMPLE 6.7

Suppose it is known that if X is the length of life, measured in years, of a certain type of battery, then

$$f(x) = \begin{cases} (3/4)(2x - x^2) & \text{for} \quad 0 < x < 2 \\ 0 & \text{elsewhere} \end{cases}$$

(a) Show that $E(X) = 1$.

(b) Use the strong law of large numbers to show that if we bought 2,000 of these batteries, we should expect the average length of life of the 2,000 batteries to be approximately 1 year.

SOLUTION

(a)

$$E(X) = \int_{-\infty}^{\infty} xf(x)\, dx = \int_{0}^{2} x(3/4)(2x - x^2)\, dx = 1$$

(b) Consider the experiment of selecting at random a battery of the type described above. Let X be the length of life, measured in years, of the battery selected. Repeat the experiment n times (with $n = 2000$) and let X_i be the random variable X on the ith performance of the experiment. That is, X_i is the length of life of the ith battery, and $(X_1 + X_2 + \cdots + X_n)/n$ is the average length of life of the 2,000 batteries. Now, since n is large, the strong law of large numbers tells us that

$$\frac{X_1 + X_2 + \cdots + X_n}{n} \approx 1$$

with great probability. That is, an approximate 1-year average length of life for the 2,000 batteries is very likely and to be expected.

Exercises

1. X is a discrete random variable with probability function

$$f(x) = \begin{cases} 1/9 & \text{for} \quad x = -3 \\ 2/9 & \qquad x = \quad 0 \\ 6/9 & \qquad x = \quad 6 \end{cases}$$

Find $E(X)$.

2. Find $E(X)$ if X is a continuous random variable with probability density function given by

(a) $$f(x) = \begin{cases} 2(1 - x) & \text{for} \quad 0 < x < 1 \\ 0 & \text{elsewhere} \end{cases}$$

(b) $$f(x) = \begin{cases} 1/2 & \text{for} \quad 0 < x < 1 \\ 1/2 & \qquad 2 < x < 3 \\ 0 & \text{elsewhere} \end{cases}$$

3. Find $E(X)$ if X is
    (a)   the sum of the numerical outcomes of two rolls of a die
    (b)   the maximum of the numerical outcomes of two rolls of a die

4. If a set of 3 balls is selected at random from a box containing 10 balls, of which 4 are white and 6 are black, what is the expected number of black balls in the set of 3?

5. If X is the weight in ounces of an orange selected at random from a certain box of 100 oranges, then X is a random variable with probability function

$$f(x) = \begin{cases} 1/6 & \text{for} \quad x = 4 \\ 3/6 & \qquad x = 5 \\ 2/6 & \qquad x = 6 \end{cases}$$

What is the average weight of the oranges in the box?

6. The average annual income of families in a certain city is \$15,000. If X is the income of a family selected at random from the city, what is $E(X)$?

7. The number of employees who are absent on any working day at a certain small factory is a random variable X whose probability function

is given by

$$f(x) = \begin{cases} 6/10 & \text{for} \quad x = 0 \\ 1/10 & \quad\quad\; x = 1, 2, 3, 4 \end{cases}$$

(a)   Find the expected number of employees who will be absent on any one working day.

(b)   Thinking in terms of the years to come, about how many employees will be absent per day, on the average?

(c)   What theorem can you cite to justify your answer to (b)?

8.   The actual amount of jelly (in ounces) in a jar filled by a certain machine is a random variable X having the probability density function

$$f(x) = \begin{cases} 0 & \text{for} \quad\quad\quad x < 4.9 \\ 5 & \quad\quad 4.9 \le x \le 5.1 \\ 0 & \quad\quad\quad\;\; x > 5.1 \end{cases}$$

(a)   Find the expected amount of jelly in a jar filled by this machine.

(b)   In a shipment of thousands of jars of jelly filled by this machine, would you be surprised if the average amount of jelly per jar were 5.1 ounces? Why?

9.   Find the expected number of aces in a poker hand of 5 cards.

10.   Find the expected number of trump cards in a bridge hand of 13 cards. (A trump is a card of a predetermined suit.)

11.   Suppose that the pressure in pounds per square inch at a certain valve is a random variable X whose pdf is

$$f(x) = \begin{cases} (3/4)(2x - x^2) & \text{for} \quad 0 < x < 2 \\ 0 & \text{elsewhere} \end{cases}$$

What is the expected amount of pressure at the valve?

12.   Suppose that the profit that a certain contractor will make on any one job, in thousands of dollars, is a random variable X with probability density function given by

$$f(x) = \begin{cases} (1/4)(4x - x^3) & \text{for} \quad 0 < x < 2 \\ 0 & \text{elsewhere} \end{cases}$$

(a)   Find $E(X)$.

(b)   On the average, how much profit can this contractor be expected to make per job?

(c)   Should you be surprised if he makes a profit of less than $400 on his next job? Why?

(d)   Should you be surprised if on the next 2,000 jobs the contractor makes a profit of less than $400 on each of them? Why?

6.2 Applications

The examples in this section are, by necessity, of an elementary nature. The purpose here is to outline the fundamental ideas involved. In-depth studies of each area of application are available in more advanced texts.

6.2.1 Fair Games

In Example 6.1, we saw that if X is the numerical outcome of one roll of a die, then $E(X) = 3.5$. Let us put the implications of this statement into monetary terms. Let us suppose that a gambling house offers the following game. For the price of $4 a player rolls a balanced die and collects in dollars the numerical outcome of the die. Is this a profitable game for the house? Anyone who understands the strong law of large numbers would say yes, because the expected cost to the house is $3.50 per game and it collects $4 per game. This means that if the game is played over and over again, then the house is almost certain to make, on the average, a profit of 50¢ per game.

In the language of games, a fair game is one in which the expected net gain of each player is zero.

EXAMPLE 6.8

This is a version of the popular gambling game called chuck-a-luck. A player pays a fee of $1 and then chooses one of the numbers 1 through 6. Three dice are then rolled. If the number chosen by the player appears 1, 2, or 3 times, the player receives a $2, $3, or $4, respectively, "payoff" from the house. Is this a fair game? If not, what should the fee be to make this a fair game?

SOLUTION

Let X be the net gain of the player in units of $1. If the fee is $1, the possible values of X are $-1, 1, 2,$ or 3.

To find the probability function of X, let us consider each roll of the die as a Bernoulli trial and the appearance of the chosen number as a success in that trial. Then, the probability of success is 1/6 and the probabilities for X are

$$P(X = -1) = b(0; 3, 1/6) = 125/216$$
$$P(X = 1) = b(1; 3, 1/6) = 75/216$$
$$P(X = 2) = b(2; 3, 1/6) = 15/216$$
$$P(X = 3) = b(3; 3, 1/6) = 1/216$$

The expected value of X is

$$E(X) = (-1)(125/216) + (1)(75/216) + (2)(15/216) + (3)(1/216) = -.0787$$

Therefore, the game is not a fair game. On the average, a player who plays the game many times is expected to lose approximately 8¢ per game.

To determine the fee that will result in a fair game, let K denote the desired amount. Then, find the value K for which $E(X) = 0$. Now, if K is the fee, then

$$E(X) = (0 - K)(125/216) + (2 - K)(75/216) + (3 - K)(15/216) \\ + (4 - K)(1/216)$$

Therefore, if $E(X) = 0$ we must have

$$K = 199/216 \approx .92$$

6.2.2   Decision Problems

Another area where the mathematical expectation of a random variable plays an important role is in decision theory. Government, management in industry, or individuals may be faced with a set of alternatives. The problem is to choose the alternative that maximizes gains or minimizes losses. We will illustrate with two simple examples.

EXAMPLE 6.9

A newly formed company plans to send a number of employees each year to an annual one-day conference held in an out-of-town hotel. The hotel offers a bargain rate of \$30 per day per person if reservations are received 30 or more days in advance. However, there is a cancellation fee of \$15 per person. The regular rate is \$40 per day per person. The company cannot be certain, 30 days in advance of the conference, of the exact number of employees it will be sending. However, management feels that if X is the number it will send, then X is a random variable with probability function

$$f(x) = 1/4 \qquad \text{for} \quad x = 4, 5, 6, 7$$

Should the company make advance reservations? If so, how many? Give the answer that will minimize costs in the long run.

SOLUTION

Let Y_i be the cost in dollars if the company makes reservations for i people. Since there will be at least 4 and at most 7 attenders, we are interested in $E(Y_i)$ for $i = 4, 5, 6,$ or 7.

The possible values of Y_4 are 120 (if 4 attend), 160 (if 5 attend), 200 (if 6 attend) and 240 (if 7 attend). Each of these values are equally likely. Therefore

$$E(Y_4) = (120)(1/4) + (160)(1/4) + (200)(1/4) + (240)(1/4) = 180$$

The equally likely possible values of Y_5 are 135, 150, 190 and 230. Therefore

$$E(Y_5) = (135)(1/4) + (150)(1/4) + (190)(1/4) + (230)(1/4) = 176.25$$

In a similar manner we find

$$E(Y_6) = (150)(1/4) + (165)(1/4) + (180)(1/4) + (220)(1/4) = 178.75$$
$$E(Y_7) = (165)(1/4) + (180)(1/4) + (195)(1/4) + (210)(1/4) = 187.50$$

Now, observe that $E(Y_5) = 176.25$ is the smallest of the values $E(Y_4)$, $E(Y_5)$, $E(Y_6)$, and $E(Y_7)$. This means that if the company makes advance reservations for 5 people per year for many years, on the average the costs per year can be expected to be approximately \$176.25. If, on the other hand, the company makes advance reservations for 4, 6, or 7 people for many years, on the average the costs per year can be expected to be more than \$176.25. Therefore, the company should make advance reservations for 5 people; in the long run, this is the alternative that can be expected to minimize costs.

EXAMPLE 6.10   AN INVENTORY PROBLEM

The owner of a newsstand feels that in any one week the number of requests for a certain magazine is a random variable X with probability function

$$f(x) = \begin{cases} 5/12 & \text{for} \quad x = 3 \\ 4/12 & x = 4 \\ 2/12 & x = 5 \\ 1/12 & x = 6 \end{cases}$$

The magazine sells for \$1, and the cost to the owner is 50¢ (unsold copies cannot be returned to the publisher). How many magazines should he order from the publisher per week if he wishes to maximize expected profit?

SOLUTION

Let Y_i be the profit in dollars if the owner of the newsstand orders i magazines from the publisher. We wish to find $E(Y_i)$ for $i = 3, 4, 5,$ or 6. Observe that $Y_3 = 1.50$, regardless of the value of X. Therefore

$$E(Y_3) = 1.50 \cdot P(Y_3 = 1.50) = 1.50$$

The possible values of Y_4 are 1 if $x = 3$ and 2 if $x = 4, 5$ or 6. Therefore

$$\begin{aligned} E(Y_4) &= 1 \cdot P(Y_4 = 1) + 2 \cdot P(Y_4 = 2) \\ &= 1 \cdot P(X = 3) + 2[P(X = 4) + P(X = 5) + P(X = 6)] \\ &= (1)(5/12) + (2)[(4/12) + (2/12) + (1/12)] \\ &= 19/12 = 1.58 \end{aligned}$$

In a similar manner we obtain

$$E(Y_5) = (.5)P(Y_5 = .5) + (1.5)P(Y_5 = 1.5) + (2.5)P(Y_5 = 2.5)$$
$$= (.5)P(X = 3) + (1.5)P(X = 4) + (2.5)[P(X = 5) + P(X = 6)]$$
$$= (.5)(5/12) + (1.5)(4/12) + (2.5)[(2/12) + (1/12)]$$
$$= 16/12 = 1.33$$
$$E(Y_6) = 0 \cdot P(Y_6 = 0) + 1 \cdot P(Y_6 = 1) + 2 \cdot P(Y_6 = 2) + 3 \cdot P(Y_6 = 3)$$
$$= 0 \cdot P(X = 3) + 1 \cdot P(X = 4) + 2 \cdot P(X = 5) + 3 \cdot P(X = 6)$$
$$= (0)(5/12) + (1)(4/12) + (2)(2/12) + (3)(1/12)$$
$$= 11/12 = .92$$

If the owner's speculation concerning the random variable X is correct, he should order 4 magazines per week if he wishes to maximize expected profit.

6.2.3   Monte Carlo Methods

We have seen in Sections 6.2.1 and 6.2.2 that the importance of the strong law of large numbers goes beyond the justification of calling $E(X)$ the expected value or mathematical expectation of X. Indeed, it might be more correct to say that $E(X)$ is important because of the implications of the strong law of large numbers. Let us examine again the implications of this law. It tells us that there is a great probability that the average of many observations of a random variable X will be approximately equal to $E(X)$. This means that if we can define the solution to a problem—any problem—in terms of the mathematical expectation of some random variable X, then, with great probability, we can obtain a good approximation of that solution by taking the average of many observations of X. This method of solving problems is sometimes called a Monte Carlo method because its use of random numbers reminds one of casino games.

EXAMPLE 6.11   MONTE CARLO METHOD FOR
EVALUATING DEFINITE INTEGRALS

Let us consider a simple case where $\phi(u)$ is continuous on $a \le u \le b$ and is bounded below and above by $0 \le \phi(u) \le M$. A sketch of $\phi(u)$ in the region of the rectangle $a \le u \le b, 0 \le v \le M$ may look like the graph in Figure 6.1.

We will outline a Monte Carlo method for evaluating $\int_a^b \phi(u)\,du$.

Consider the following experiment $\mathscr{E}$. Select at random a point (u, v) from the rectangle $a \le u \le b, 0 \le v \le M$. Let X be the random variable defined as follows.

$$X = \begin{cases} 1 & \text{if the point selected is below the curve of } \phi(u) \\ 0 & \text{otherwise} \end{cases}$$

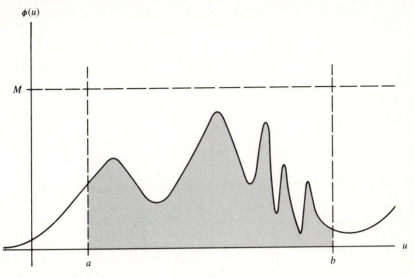

FIGURE 6.1

Observe that

$$P(X = 1) = \frac{\text{area under the curve}}{\text{area of the rectangle}}$$

$$= \frac{\displaystyle\int_a^b \phi(u)\,du}{M(b-a)}$$

$$P(X = 0) = 1 - P(X = 1)$$

Therefore

$$E(X) = 1 \cdot P(X = 1) + 0 \cdot P(X = 0) = \frac{\displaystyle\int_a^b \phi(u)\,du}{M(b-a)}$$

and

$$\int_a^b \phi(u)\,du = M(b-a)E(X)$$

Now, suppose we perform the experiment $\mathscr{E}$ n times and n is a large number. Let X_i be the value of X on the ith performance of $\mathscr{E}$. By the strong law of large numbers

$$\frac{\displaystyle\sum_{i=1}^{n} X_i}{n} \approx E(X)$$

with great probability. Therefore

$$\int_a^b \phi(u) \, du \approx M(b - a) \cdot \frac{\sum_{i=1}^n X_i}{n}$$

with great probability.

Example 6.11 outlines a general Monte Carlo method for finding definite integrals. Let us consider a specific case.

EXAMPLE 6.12   MONTE CARLO ESTIMATE OF $\int_0^1 e^{-(u^2/2)} \, du$

Applying the method outlined in Example 6.11, we have $\phi(u) = e^{-(u^2/2)}$, $M = 1, a = 0, b = 1$ so that

$$\int_0^1 e^{-(u^2/2)} \, du = E(X)$$

and

$$\int_0^1 e^{-(u^2/2)} \, du \approx \frac{\sum_{i=1}^n X_i}{n}$$

To perform the experiment $\mathscr{E}$ outlined in Example 6.11 we have to select a point (u, v) at random from the unit square $0 \leq u \leq 1, 0 \leq v \leq 1$. One method is to refer to a table of random numbers such as Table IV in Appendix C. Turning to the table, we see that the first number listed is 04839. Place a decimal point in front of the 5 digits and convert it to .04839. The second number on the list is 68086. Convert it to .68086. The pair of numbers (.04839, .68086) is then a point in the unit square. Since we obtained these numbers from a table of random numbers the point (.04839, .68086) may be considered as randomly selected from the unit square.

To determine the value of the random variable X we need to check if (.04839, .68086) lies below the curve of $\phi(u) = e^{-(u^2/2)}$. In this case

$$\phi(.04839) = e^{-(.04839)^2/2} = .99883$$

Since $.68086 < \phi(.04839)$, we conclude that the point (.04839, .68086) lies below the curve of $\phi(u)$, and therefore $X = 1$.

To repeat the experiment $\mathscr{E}$ we would continue down the list of random numbers. Using the first column of the table of random numbers we would obtain $n = 17$ points (u, v) from the unit square. Table 6.1 shows the result.

Using these results, we obtain the Monte Carlo estimate

$$\int_0^1 e^{-(u^2/2)} \, du \approx \frac{\sum_{i=1}^{17} X_i}{17} = \frac{15}{17} = .8824$$

TABLE 6.1

u	v	$\phi(u) = e^{-(u^2/2)}$	X
.04839	.68086	.99883	1
.39064	.25669	.92654	1
.64117	.87917	.81419	0
.62797	.95876	.82105	0
.29888	.73577	.95632	1
.27958	.90999	.96167	1
.18845	.94824	.98240	1
.35605	.33362	.93858	1
.88720	.39475	.67465	1
.06990	.40980	.99756	1
.83974	.33339	.70287	1
.31662	.93526	.95111	1
.20492	.04153	.97922	1
.05520	.47498	.99848	1
.23167	.23792	.97352	1
.85900	.42559	.69147	1
.14349	.17403	.98976	1

Let us compare this with the value we would obtain from the table of values for the normal distributions.

From Table III, Appendix C

$$\int_0^1 \frac{1}{\sqrt{2\pi}} e^{-(u^2/2)} \, du = .3413$$

Therefore

$$\int_0^1 e^{-(u^2/2)} \, du = \sqrt{2\pi} \int_0^1 \frac{1}{\sqrt{2\pi}} e^{-(u^2/2)} \, du = .8555$$

Thus, the Monte Carlo estimate is .8824 and the value, using the table, is .8555.

Note: A clever way of selecting numbers at random from (0, 1) is to flip coins. *n* flips of a fair coin yield one number in (0, 1) as follows. Let

$$x_i = \begin{cases} 1 & \text{if } i\text{th flip is heads} \\ 0 & \text{if } i\text{th flip is tails} \end{cases}$$

The number $\sum_{i=1}^{n} (x_i/2^i)$ is a number in (0, 1).

We should mention that the example we have chosen here is an integral whose value may be obtained only through approximation techniques because the function $\phi(u) = e^{-(u^2/2)}$ does not have an antiderivative in terms of elementary functions. When faced with integrals of this sort, sometimes it is more efficient to use a Monte Carlo method rather than the traditional methods of estimation.

The problems that can be solved by Monte Carlo methods are limitless. We include here a novel example.

EXAMPLE 6.13   MONTE CARLO METHOD FOR OBTAINING THE COMPLETE WORKS OF SHAKESPEARE

Suppose we wish to obtain the complete works of Shakespeare—in chronological order and without error. One Monte Carlo method would require the following items: an immortal monkey, a typewriter that will never break down, and an endless roll of typing paper. The monkey should be capable of typing away in a completely random fashion forever.

Let us use the following notations.

$T =$ number of symbols available on the typewriter, including the empty space as one of them

$M =$ number of typewriter spaces necessary to reproduce the complete works of Shakespeare

We place the monkey in front of the typewriter and it begins to type away forever. Let us assume that at the end of each line the typing advances to the next line automatically.

Now, let $\mathscr{E}$ be the experiment consisting of M consecutive strikes on the typewriter by the monkey. Let X be the following random variable.

$$X = \begin{cases} 1 & \text{if the } M \text{ consecutive strikes produce the} \\ & \text{complete works of Shakespeare—chronologically} \\ & \text{and without error} \\ 0 & \text{otherwise} \end{cases}$$

Then

$$P(X = 1) = P(\text{the } M \text{ strikes produce the complete works of Shakespeare})$$
$$= P(\text{first strike is correct})$$
$$\cdot P(\text{second strike is correct})$$
$$\vdots$$
$$\cdot P(M\text{th strike is correct})$$
$$= (1/T)^M$$
$$P(X = 0) = 1 - P(X = 1)$$
$$E(X) = 1 \cdot P(X = 1) + 0 \cdot P(X = 0)$$
$$= P(X = 1) = (1/T)^M$$

The monkey types on forever. Therefore, the experiment $\mathscr{E}$ is repeated infinitely often and, letting X_i be the random variable X on the ith performance of $\mathscr{E}$, we have

$$X_i = \begin{cases} 1 & \text{if the } i\text{th repetition of } \mathscr{E} \text{ produces the complete} \\ & \text{works of Shakespeare} \\ 0 & \text{otherwise} \end{cases}$$

By the strong law of large numbers

$$P\left[\lim_{n \to \infty} \frac{\sum\limits_{i=1}^{n} X_i}{n} = \left(\frac{1}{T}\right)^M\right] = 1 \tag{1}$$

Now, if

$$\lim_{n \to \infty} \sum_{i=1}^{n} X_i < \infty, \quad \text{then} \quad \lim_{n \to \infty} \frac{\sum\limits_{i=1}^{n} X_i}{n} = 0$$

Therefore, if

$$\lim_{n \to \infty} \frac{\sum\limits_{i=1}^{n} X_i}{n} = \left(\frac{1}{T}\right)^M > 0$$

it follows that

$$\lim_{n \to \infty} \sum_{i=1}^{n} X_i = \infty$$

Therefore, Statement (1) implies that

$$P\left(\lim_{n \to \infty} \sum_{i=1}^{n} X_i = \infty\right) = 1$$

This means that with probability 1 the complete works of Shakespeare will be reproduced, not once, but infinitely often. (But—how long would we have to wait for the first appearance on the typewriter of the complete works of Shakespeare? See Exercise 11 on p. 218.

Another interesting use of Monte Carlo methods is to solve differential equations. Appendix B provides an example for differential equations of a particular type.

Exercises

1. If player A pays player B $1 when the outcome of one roll of a die is 1, 2, 3, 4, or 5, then, in order to make this a fair game, what should player B pay player A?

2. A balanced die is rolled until the same result appears twice. If this takes place in less than 6 rolls of the die, player A pays player B $1 per roll. Otherwise, player B pays player A $25. Is this a fair game? Would you rather be player A or player B? What is player A's expected net gain per game?

3. Describe a Monte Carlo method for estimating π. (*Hint:* $\int_0^1 (1 - x^2)^{1/2} dx = \pi/4$.) Use the table of random numbers (Table IV in Appendix C) to select at random 20 points from the unit square to obtain an actual estimate of π. If you have a calculator with a random number generator, you might want to use that function instead of a table of random numbers.

4. Describe a Monte Carlo method for estimating ln 2. (*Hint:* $\int_1^2 (1/x) \, dx = \ln 2$.) How might we use Table IV, or a random number generator on a calculator, to select points at random from the rectangle $1 \le u \le 2$, $0 \le v \le 1$? Use 20 points to estimate ln 2.

5. Ms. Jones works in midtown and must take her car to work each day. She has the choice of parking her car at a parking lot at the rate of $5 per day or parking her car illegally on the street at the risk of paying a $25 fine if caught. If it is known that the probability of being caught is .10, is it less expensive in the long run for Ms. Jones to park her car in the parking lot? Give reasons for your answer.

6. The manager of a fish market speculates that the number of requests for salmon on any given day is a random variable X with the probability function

$$f(x) = \begin{cases} 1/4 & \text{for} \quad x = 0 \\ 1/2 & x = 1 \\ 1/4 & x = 2 \end{cases}$$

There is a profit of $2 on each salmon he sells and a loss (due to spoilage) of $1 on each salmon he does not sell. Assuming that each salmon can be sold only on the day it is up for sale, that each request is for a single salmon, and that the manager's speculation is correct, find the number of salmons the market should have per day to maximize profit.

7. Answer the questions of Example 6.9 if $f(4) = .2, f(5) = .3, f(6) = .3$, and $f(7) = .2$.

8. Answer the questions of Example 6.10 if $f(3) = .1, f(4) = .3, f(5) = .4$, and $f(6) = .2$.

9. A wealthy real estate speculator is faced with the following decision problem. A parcel of land is up for sale for $1 million. There is a proposal to build an airport adjacent to the parcel of land. A firm decision is to be

made one year hence. If the proposal to build the airport is adopted, the land may be resold at \$4 million. However, if the proposal is not adopted, the value of the parcel will drop to \$.5 million. There is also the problem of raising the money to buy the parcel. Although the speculator has vast resources, he would have to divert an investment of \$1 million to purchase the parcel; this means a loss of \$250,000 for the one-year time that will elapse until the decision is made about the proposal. If his expert advisors estimate the probability of adoption of the proposal to build the airport to be .6, what should the real estate speculator do to maximize his expected profit?

6.3   The Mathematical Expectation of a Function of a Random Variable

6.3.1   Function of a Random Variable

We are often interested not only in the expected value of a random variable but also in the expected value of a function of a random variable.

Definition 6.3   Let X be a random variable defined on sample space S. Let g be a real valued function defined for values x that are possible values of X (or for $-\infty < x < \infty$). The random variable $g(X)$ is defined on S as follows. For each s in S,

$$g(X)(s) = g[X(s)]$$

EXAMPLE 6.14

Let $\mathcal{E}$ be the experiment of rolling two dice, and let $S = \{(i, j)|i = 1, 2, \ldots, 6;$ $j = 1, 2, \ldots, 6\}$ be the sample space for $\mathcal{E}$. Let X be the sum of the values of the two dice so that

$$X(i, j) = i + j \qquad \text{for} \quad (i, j) \text{ in } S$$

Let g be the function

$$g(x) = x^2 \qquad \text{for} \quad -\infty < x < \infty$$

What is $g(X)$?

   SOLUTION

Following Definition 6.3

$$g(X)(i, j) = g[X(i, j)] = g(i + j) = (i + j)^2 \qquad \text{for} \quad (i, j) \text{ in } S$$

In other words, $g(X)$ is the random variable that represents the square of the sum of the values of the two dice.

When, as in the above example, a function g is defined by $g(x) = x^2$, we often write X^2 when we are referring to the random variable $g(X)$. This scheme extends to all functions g, so that, for example, if $g(x) = a_0 + a_1 x + \cdots + a_n x^n$, then $g(X)$ is written as $a_0 + a_1 X + a_2 X^2 + \cdots + a_n X^n$, and if $g(x) = e^x$, then $g(X)$ is written as e^X. In this manner we may also define the random variable which is a constant. Let c be a constant, S a sample space, X a random variable defined on S, and g the function $g(x) = c$ for $-\infty < x < \infty$. Then $g(X)$ is a random variable, and according to the scheme outlined above we would write c when we are referring to $g(X)$. We note that for any s in S, $c(s) = c$.

6.3.2 The Mathematical Expectation of a Function of a Random Variable

Definition 6.4 Let X be a random variable with probability (or probability density) function $f(x)$. Let g be a real valued function defined for values x that are possible values of X (or for $-\infty < x < \infty$). The mathematical expectation of the random variable $g(X)$ is the quantity

$$E[g(X)] = \begin{cases} \sum_x g(x)f(x) & \text{if } X \text{ is discrete} \\ \int_{-\infty}^{\infty} g(x)f(x)\,dx & \text{if } X \text{ is continuous} \end{cases}$$

provided that the series converges absolutely (if X is discrete) and that the integral is defined (if X is continuous).

EXAMPLE 6.15

Let $g(X)$ be the random variable of Example 6.14. Find $E[g(X)]$.

SOLUTION

$$E[g(X)] = \sum_x g(x)f(x) = \sum_{x=2}^{12} x^2 f(x)$$

$$= (2)^2(1/36) + (3)^2(2/36) + (4)^2(3/36)$$

$$+ (5)^2(4/36) + (6)^2(5/36)$$

$$+ (7)^2(6/36) + (8)^2(5/36) + (9)^2(4/36)$$

$$+ (10)^2(3/36) + (11)^2(2/36) + (12)^2(1/36)$$

$$= 1974/36 = 54.83$$

EXAMPLE 6.16

The length of life measured in hours of a certain rare type of insect is a random variable X with probability density function

$$f(x) = \begin{cases} (3/4)(2x - x^2) & \text{for } 0 < x < 2 \\ 0 & \text{elsewhere} \end{cases}$$

If the amount of food measured in milligrams consumed in a lifetime by such an insect is defined by the function $g(x) = x^2$, where x is the length of life measured in hours, find the expected amount of food that will be consumed by an insect of this type.

SOLUTION

$$\text{expected amount of food} = E[g(X)] = \int_{-\infty}^{\infty} g(x)f(x)\, dx$$

$$= \int_{0}^{2} x^2[(3/4)(2x - x^2)]\, dx$$

$$= 1.2\, \text{mg}$$

EXAMPLE 6.17

The surface area, measured in square feet, of a flat metal disk manufactured by a certain process is a random variable X with probability density function

$$f(x) = \begin{cases} 6(x - x^2) & \text{for} \quad 0 < x < 1 \\ 0 & \text{elsewhere} \end{cases}$$

Find the expected radius, measured in feet, of a flat metal disk manufactured by this process.

SOLUTION

Let X and R be the respective area and radius of the disk. Then, $X = \pi R^2$ and $R = (X/\pi)^{1/2}$. In other words, if $g(x) = (x/\pi)^{1/2}$, then $R = g(X)$. Therefore, the expected radius is

$$E(R) = E(\sqrt{X/\pi}) = E[g(X)] = \int_{-\infty}^{\infty} g(x)f(x)\, dx$$

$$= \int_{0}^{1} (\sqrt{x/\pi}) \cdot 6(x - x^2)\, dx$$

$$= .3869\, \text{feet}$$

Since every random variable X is also the random variable $g(X)$ if g is the identity function $g(x) = x$, we need to verify that, where this is the case, $E(X) = E[g(X)]$. Indeed, this is so because if $g(x) = x$, then

$$E[g(X)] = \left\{ \begin{array}{c} \sum_{x} g(x)f(x) = \sum_{x} xf(x) \\ \int_{-\infty}^{\infty} g(x)f(x)\, dx = \int_{-\infty}^{\infty} xf(x)\, dx \end{array} \right\} = E(X)$$

Exercises

1. Let X be the random variable and S be the sample space of Example 6.14. Let g be the function defined by $g(x) = x^{1/2}$. What is $g(X)(i, j)$ for (i, j) in S? Find $E[g(X)]$.

2. Let X be a random variable with probability function given by

$$f(x) = \begin{cases} .1 & \text{for} \quad x = 1 \\ .4 & x = 4 \\ .5 & x = 9 \end{cases}$$

Find $E(X^2)$, $E(X^{1/2})$, and $E(2X^{1/2})$.

3. Let X be a random variable with probability density function given by

$$f(x) = \begin{cases} (3/4)(1 - x^2) & \text{for} \quad -1 < x < 1 \\ 0 & \text{elsewhere} \end{cases}$$

Find $E(X^2)$, $E(X - 1)$, and $E(|X|)$.

4. The diameters (measured in inches) of circular disks cut by a certain machine may be looked upon as values of a random variable X with probability density function given by

$$f(x) = \begin{cases} (1/9)(4x - x^2) & \text{for} \quad 1 < x < 4 \\ 0 & \text{elsewhere} \end{cases}$$

Find the expected area of a circular disk cut by this machine.

5. In a certain business, the gross receipts in hundreds of dollars for any one day may be looked upon as a random variable X with probability density function given by

$$f(x) = \begin{cases} (6/181)(11x - x^2) & \text{for} \quad 5 < x < 6 \\ 0 & \text{elsewhere} \end{cases}$$

The profit that the business will make on a day when the gross receipts is x is given by the function

$$g(x) = .5x - 1$$

Find the expected daily profit for this business.

6. Let X and Y_i, $i = 4, 5, 6, 7$, be the random variables of Example 6.9. Define g_i so that $g_i(X)$ is the random variable Y_i. What is $g_i(x)$ for $i = 4, 5, 6, 7$, and $x = 4, 5, 6, 7$? Find $E[g_i(X)]$ for $i = 4, 5, 6, 7$, using Definition 6.4.

7. Let X and Y_i, $i = 3, 4, 5, 6$, be the random variables of Example 6.10. Define g_i so that $g_i(X)$ is the random variable Y_i. Find $E[g_i(X)]$ for $i = 3, 4, 5, 6$, using Definition 6.4.

8. Let X be the random variable of Exercise 6 of Section 6.2. Let $g_i(x)$ be the profit in dollars if there are x requests for salmon on a day when the fish

market has i salmons for sale. Find $g_i(x)$ for $i = 1, 2$, and $x = 0, 1, 2$. Find $E[g_i(X)]$ for $i = 1, 2$.

9. *The rth moment of a random variable.* The rth moment of a random variable X is defined to be the quantity $E(X^r)$. Find the second and third moments of each of the random variables of Exercises 1, 2, and 3 of Section 6.1.

6.4   Theorems on Mathematical Expectation

Some immediate consequences of the definitions on mathematical expectation are the following theorems.

Theorem 6.2   *If c is a constant, then $E(c) = c$.*

Proof   There is only one possible value for the random variable c, namely c itself. Therefore, $E(c) = c \cdot P(c = c) = c \cdot 1 = c$.

Theorem 6.3   *Let $a_0, a_1, \ldots, a_n$ be constants. If $E(X), E(X^2), \ldots$, and $E(X^n)$ exist, then*

$$E(a_0 + a_1 X + a_2 X^2 + \cdots + a_n X^n) = a_0 + a_1 E(X^2) + \cdots + a_n E(X^n)$$

Corollary   *If a and b are constants, then $E(aX + b) = aE(X) + b$.*

Note: Theorem 6.3 is a special case of a more general theorem which is considered in Exercise 4.

Proof   We shall consider the discrete case here. For a proof for the continuous case, see Exercise 1 of the Supplementary Exercises.

Let g be the function $g(x) = a_0 + a_1 x + a_2 x^2 + \cdots + a_n x^n$. Then

$$a_0 + a_1 X + a_2 X^2 + \cdots + a_n X^n = g(X)$$

and

$$E(a_0 + a_1 X + a_2 X^2 + \cdots + a_n X^n) = E[g(X)] = \sum_x g(x) f(x)$$

$$= \sum_x (a_0 + a_1 x + \cdots + a_n x^n) f(x)$$

$$= a_0 \sum_x f(x) + a_1 \sum_x x f(x) + \cdots + a_n \sum_x x^n f(x)$$

$$= a_0 + a_1 E(X) + \cdots + a_n E(X^n)$$

The corollary is an immediate consequence of Theorem 6.3.

EXAMPLE 6.18

Let X be the random variable of Example 6.2. Find $E(3 - 2X)$ and $E(X + 4X^2)$.

SOLUTION

$$E(X) = 7$$

$$E(X^2) = \sum_x x^2 f(x) = (3)^2(1/3) + (9)^2(2/3) = 57$$

$$E(3 - 2X) = 3 - 2E(X) = 3 - (2)(7) = -11$$

$$E(X + 4X^2) = E(X) + 4E(X^2) = 7 + (4)(57) = 235$$

Exercises

1. Let X be the random variable of Example 6.2. Find $E(X^2 - 2X + 1)$ by first finding $E(X)$ and $E(X^2)$ and then applying Theorem 6.3.

2. Let X be the random variable of Example 6.3. Find $E(X^3 - X + 7)$ by first finding $E(X)$ and $E(X^3)$ and then applying Theorem 6.3.

3. Let X be a random variable whose first, second, and third moments are -1, 3, and 5, respectively. Find $E(4 - 2X + 3X^2 + 6X^3)$.

    Note: the rth moment of a random variable is defined in Section 6.3, Exercise 9.

4. Prove the following theorem: if X is a random variable, $g_1, g_2, \ldots, g_k$ are functions, and if $E[g_1(X)]$, $E[g_2(X)], \ldots, E[g_k(X)]$ exist, then

    $$\begin{aligned} E[g_1(X) &+ g_2(X) + \cdots + g_k(X)] \\ &= E[g_1(X)] + E[g_2(X)] + \cdots + E[g_k(X)] \end{aligned}$$

    (*Hint:* let $g(X) = g_1(X) + g_2(X) + \cdots + g_k(X)$, then apply the definition of $E[g(X)]$.)

5. Let X be a random variable whose probability density function is given by

    $$f(x) = \begin{cases} 2x & \text{for } 0 < x < 1 \\ 0 & \text{elsewhere} \end{cases}$$

    If $g_1(x) = x^2, g_2(x) = 1 - x$, and $g_3(x) = x^{1/2}$, what is $E[g_1(X) + g_2(X) + g_3(X)]$? (*Hint:* see Exercise 4.)

6. At a certain printing company the cost of labor, in dollars, for any one job contracted by the company may be looked upon as a random variable X with mean $\mu = 2,000$. If the profit Y that the company will make on the job is given by the formula $Y = .2X + 100$, what is the expected profit?

7. Use the theorems of this section to solve the problem in Exercise 5 of Section 6.3.

6.5   The Variance and Standard Deviation of a Random Variable

6.5.1   Definition of Variance and Standard Deviation

Definition 6.5   The **variance** of a random variable X with mathematical expectation μ is the quantity

$$\text{Var}(X) = E[(X - \mu)^2]$$

provided that the expectation $E[(X - \mu)^2]$ exists.

Definition 6.6   The positive square root of $\text{Var}(X)$ is called the **standard deviation of** X.

Note: the standard deviation of a random variable X is often denoted by the Greek letter σ. $\text{Var}(X)$ is denoted by the symbol σ^2.

EXAMPLE 6.19

The variance of the random variable of Example 6.1 is

$$\text{Var}(X) = E[(X - \mu)^2] = E[(X - 3.5)^2]$$

$$= \sum_{x=1}^{6} (x - 3.5)^2 f(x)$$

$$= (1 - 3.5)^2 \cdot \frac{1}{6} + (2 - 3.5)^2 \cdot \frac{1}{6} + (3 - 3.5)^2 \cdot \frac{1}{6}$$

$$+ (4 - 3.5)^2 \cdot \frac{1}{6} + (5 - 3.5)^2 \cdot \frac{1}{6} + (6 - 3.5)^2 \cdot \frac{1}{6}$$

$$= \frac{175}{60} = 2.917$$

EXAMPLE 6.20

The variance of the random variable X in Example 6.4 is the quantity

$$\text{Var}(X) = E[(X - \mu)^2] = E\left[\left(X - \frac{1}{2}\right)^2\right]$$

$$= \int_{-\infty}^{\infty} \left(x - \frac{1}{2}\right)^2 f(x)\, dx$$

$$= \int_{0}^{1} \left(x - \frac{1}{2}\right)^2 \cdot 1\, dx$$

$$= \frac{1}{12}$$

EXAMPLE 6.21

The variance of the random variable of Example 6.3 is the quantity

$$\text{Var}(X) = E[(X - \mu)^2] = E\left[\left(X - \frac{1}{4}\right)^2\right]$$

$$= \int_{-\infty}^{\infty} \left(x - \frac{1}{4}\right)^2 f(x)\, dx$$

$$= \int_0^{1/3} \left(x - \frac{1}{4}\right)^2 \cdot 81x^2\, dx$$

$$= .00417$$

6.5.2   The Variance as a Measure of Dispersion

What does the quantity $\text{Var}(X)$ tell us about the random variable X? By definition, $\text{Var}(X)$ is the mathematical expectation of the random variable $(X - \mu)^2$. Therefore, it is the mean or expectation of the square of the deviation of X from the mean of X. For example, if we were to roll a die many, many times, we would expect the average of the outcomes to be 3.5 since the mean is 3.5 and we would expect the average of the square of the deviations from 3.5 to be 2.9 since the variance is 2.9. Thus, the variance may be looked upon as a measure of dispersion.

EXAMPLE 6.22

Let X be a discrete random variable with mean μ and variance σ^2. Show that if the variance is small, observed values of the random variable will tend to cluster (in a small neighborhood) about the mean whereas if the variance is large, it is likely that observed values of the random variable will be spread out over a large region.

SOLUTION

By definition

$$0 \le \sigma^2 = \text{Var}(X) = E[(X - \mu)^2] = \sum_x (x - \mu)^2 f(x)$$

Therefore, if σ^2 is a small quantity, then for each possible value x of the random variable X, $(x - \mu)^2 f(x)$ must also be small. This means that if $(x - \mu)^2$ is large, then $f(x)$, which is $P(X = x)$, must be very small. Conversely, if $f(x)$ is large, then $(x - \mu)^2$ must be very small. Thus, if σ^2 is small, it is very unlikely that X will take on values very far away from μ. In other words, a small variance implies that observed values of the random variable will tend to be clustered about the mean whereas, if the variance is large, it is likely that observed values of the random variable will be spread out over a large region.

EXAMPLE 6.23

Suppose it is known that if the temperature of a liquid is $t°C$, the measurement of the temperature with instrument A is a random variable X with mean equal to t and variance equal to 9, while the measurement of the temperature with instrument B is a random variable Y with mean equal to t and variance equal to .01. If you plan to measure the temperature of a liquid by taking one reading from either instrument A or instrument B, which instrument is preferable (if the criterion is accuracy of measurement)?

SOLUTION

Instrument B is preferable. Since the variance for instrument B is considerably smaller than the variance for instrument A, the readings from B will tend to be closer to the true temperature of the liquid.

To illustrate the role of the variance as a measure of dispersion we have drawn in Figure 6.2 the probability density functions of three normal random variables with respective variances of .06, .25, and 1. The mean of each of the random variables is zero.

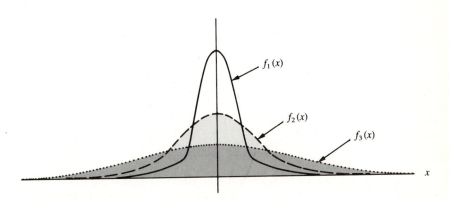

FIGURE 6.2

f_1 is normal density function with $\mu = 0$ and $\sigma^2 = .06$
f_2 is normal density function with $\mu = 0$ and $\sigma^2 = .25$
f_3 is normal density function with $\mu = 0$ and $\sigma^2 = 1.00$

Note: in Section 6.8 we shall see that the mean and variance of the normal random variable X with probability density function

$$f(x) = \frac{1}{\sqrt{2\pi}\sigma} e^{-(x-\mu)^2/(2\sigma^2)}$$

are the parameters μ and σ^2, respectively.

Observe in the figure that, as the variance increases, the density function becomes less peaked about the mean, thus indicating less concentration of values there.

The standard deviation plays a distinctively important role in the normal random variable. It can be shown that approximately 68.3% of the area under the curve of the probability density function lies within 1 standard deviation from the mean; 95.4% is within 2 standard deviations, and 99.7% is within 3 standard deviations from the mean (Figure 6.3).

We should mention that although the variance of X tells us something about what we should expect concerning the deviation of X from the mean of X, it is not equivalent to the expected deviation of X from the mean of X. Technically, the expected deviation of X from the mean of X is the quantity $E(X - \mu)$, and it is zero. (*Note:* $E(X - \mu) = E(X) - \mu = 0$.) Nor is it the expected magnitude of the deviation of X from the mean. To obtain that we would need to compute $E(|X - \mu|)$. One wonders, then, why there is so much more interest in $E[(X - \mu)^2]$, which is $\mathrm{Var}(X)$, than in $E(|X - \mu|)$. The reason is purely one of simplicity and convenience! It so happens that for many, and certainly for the most important, random variables the quantity $E[(X - \mu)^2]$ is easily obtained and is a simple expression of parameters of the distribution; therefore it lends itself readily to further use in considerations of a theoretical nature. By contrast, the quantity $E(|X - \mu|)$ is often complex and difficult to obtain. For example, consider the most important case of the normal random

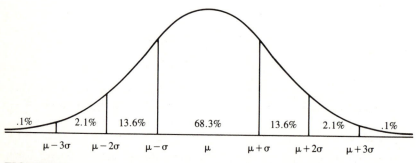

FIGURE 6.3

Percentage of Area Under the Standard Normal Density Curve

variable X with probability density function

$$f(x) = \frac{1}{\sqrt{2\pi}\,\sigma}\, e^{-(x-\mu)^2/(2\sigma^2)}$$

We shall see that $\mathrm{Var}(X) = E[(X - \mu)^2] = \sigma^2$. Nice! Now, try finding $E(|X - \mu|)$!

6.5.3 Relating the Variance of a Random Variable to the Variance of a Set of Numbers

How do we relate the variance of a random variable with the variance of a set of numbers? By definition, the variance of a set $A = \{a_1, a_2, \ldots, a_N\}$ of not necessarily distinct numbers is the quantity

$$\frac{\sum\limits_{i=1}^{N} (a_i - \bar{a})^2}{N}$$

where

$$\bar{a} = \frac{\sum\limits_{i=1}^{N} a_i}{N} = \text{the mean (i.e., average) of the set } \{a_1, \ldots, a_N\}$$

A little arithmetic shows that the variance of the set A can also be expressed as the quantity

$$\frac{\sum\limits_{i=1}^{N} a_i^2}{N} - (\bar{a})^2$$

The arithmetic goes like this:

$$\frac{\sum\limits_{i=1}^{N} (a_i - \bar{a})^2}{N} = \frac{\sum\limits_{i=1}^{N} (a_i^2 - 2a_i\bar{a} + (\bar{a})^2)}{N} = \frac{\sum\limits_{i=1}^{N} a_i^2}{N} - \frac{2\bar{a}\sum\limits_{i=1}^{N} a_i}{N} + \frac{N(\bar{a})^2}{N}$$

$$= \frac{\sum\limits_{i=1}^{N} a_i^2}{N} - 2\bar{a}\cdot\bar{a} + (\bar{a})^2 = \frac{\sum\limits_{i=1}^{N} a_i^2}{N} - (\bar{a})^2$$

Suppose we now consider the following experiment. Select an element at random from the set A in such a manner that each element has an equal probability of being selected. Let X be the numerical value of the element selected.

What are the possible values of X? Clearly, the possible values of X are the distinct numbers in the set A. Denote these distinct numbers by the symbols $x_1, x_2, \ldots, x_n$.

What is the probability function of X? Clearly

$$f(x_i) = P(X = x_i) = k_i/N$$

where k_i is the number of elements in A that have the numerical value x_i.
What is the variance of X? We have

$$\sigma^2 = \mathrm{Var}(X) = E(X^2) - E^2(X) = \sum_{i=1}^{n} x_i^2 f(x_i) - \left[\sum_{i=1}^{n} x_i f(x_i) \right]^2$$

$$= \sum_{i=1}^{n} x_i^2 (k_i/N) - \left[\sum_{i=1}^{n} x_i (k_i/N) \right]^2$$

$$= \frac{\sum_{i=1}^{N} a_i^2}{N} - \left(\frac{\sum_{i=1}^{n} a_i}{N} \right)^2$$

$$= \frac{\sum_{i=1}^{N} (a_i - \bar{a})^2}{N}$$

$$= \text{variance of the set } A$$

In other words, if X is the numerical value of the element selected at random from the set A of numbers, then the variance of X is the variance of the set A of numbers.

Exercises

1. Find the variance and the standard deviation of the random variables defined in Exercises 1, 2, 3, 4, 5, and 8 of Section 6.1.

2. Suppose that the weight in ounces of a hen selected at random from group I is a normal random variable X_1 with mean $\mu_1 = 48$ and standard deviation $\sigma_1 = 3$, whereas the weight in ounces of a hen from group II is a random variable X_2 with mean $\mu_2 = 48$ and $\sigma_2 = 6$. What is the average weight of hens in group I? What is the average weight of hens in group II? In which group are you more likely to find hens over 60 ounces? What is the probability of doing so in group I? In group II? (*Hint:* refer to Figure 6.3.)

3. Ms. Smith is considering the purchase of one of two businesses. The first business under consideration has a daily expected profit (in dollars) of $\mu = 100$ and a standard deviation of $\sigma = 28.87$ (dollars). The second business under consideration has a daily expected profit in dollars of $\mu = 100$ and a standard deviation of $\sigma = 51.96$ (dollars). If Ms. Smith is interested in a business as a steady source of income, which business should she purchase?

4. Two methods may be used to measure the temperature of a material. If t is the actual temperature in degrees Celsius of the material, then the measurement of the temperature with the first method is a random variable X_1 with mean $\mu_1 = t$ and standard deviation $\sigma_1 = .5$. The measurement with the second method is a random variable X_2 with mean $\mu_2 = t$ and standard deviation $\sigma_2 = 2$. In terms of accuracy of measurement, which of these two methods would you say is the superior one? Why?

5. The heights of the 9 children on a certain baseball team are $4'2''$, $4'3''$, $4'4''$, $4'4''$, $4'5''$, $4'5''$, $4'5''$, $4'6''$, and $4'8''$, respectively. Find the mean and variance of this set of heights. If a child is selected at random from this team and X is the height of the child selected, what is $E(X)$ and what is $\text{Var}(X)$?

6.6 Theorems on Variance

Theorem 6.4

$$\text{Var}(X) = E(X^2) - E^2(X)$$

Notation: $E^2(X)$ is the notation for $[E(X)]^2$.

Proof Using Theorems 6.2 and 6.3, we have

$$\begin{aligned}
\text{Var}(X) &= E[(X - \mu)^2] = E(X^2 - 2\mu X + \mu^2) \\
&= E(X^2) + E(-2\mu X) + E(\mu^2) \\
&= E(X^2) - 2\mu E(X) + \mu^2 \\
&= E(X^2) - E^2(X)
\end{aligned}$$

Theorem 6.5

$$\text{Var}(aX + b) = a^2\,\text{Var}(X)$$

Proof Using the definition of variance and the corollary to Theorem 6.3

$$\begin{aligned}
\text{Var}(aX + b) &= E[\{(aX + b) - E(aX + b)\}^2] \\
&= E[\{aX + b - aE(X) - b\}^2] \\
&= E[\{aX - aE(X)\}^2] \\
&= E[a^2\{X - E(X)\}^2] \\
&= a^2 E[\{X - E(X)\}^2] \\
&= a^2\,\text{Var}(X)
\end{aligned}$$

Exercises

1. Let X be a random variable with $E(X) = 1$ and $E[X(X - 1)] = 4$. Find $\text{Var}(X)$ and $\text{Var}(2 - 3X)$.

2. Consider the random variable X in each of the Exercises 1, 2, and 3 of Section 6.1. Use Theorem 6.3 to find the variance of X by first finding $E(X)$ and $E(X^2)$ in each of these cases.

3. Find the variance of X and the variance of $g(X)$ in Exercise 5 of Section 6.3.

4. If the standard deviation of X in Exercise 6 of Section 6.4 is $\sigma_x = 500$, what is the standard deviation of Y?

6.7   Chebyshev's Inequality

The following theorem of P. L. Chebyshev (a nineteenth-century Russian mathematician) is generally called Chebyshev's inequality. The theorem has important implications of a theoretical nature. In addition, it provides a rough measure of the dispersion of the values of a random variable in terms of the variance (or standard deviation) of the random variable.

Theorem 6.6   Chebyshev's Inequality   *Let X be a random variable. If $E(X^2) < \infty$, then for any constant $c > 0$ we have*

$$P(|X| \geq c) \leq \frac{E(X^2)}{c^2}$$

Corollary   *Let X be a random variable with finite mean μ and finite variance σ^2. For constants $d > 0$ and $k > 0$ we have*

(a)

$$P(|X - \mu| \geq d) \leq \frac{\sigma^2}{d^2}$$

and

(b)

$$P(|X - \mu| < k\sigma) \geq 1 - \frac{1}{k^2}$$

Note: the corollary is also referred to as Chebyshev's inequality.

Proof   We will prove the theorem for the case of discrete random variables. The proof for continuous random variables and the proof of the corollary are considered in Exercises 1 and 2.

Observe that

$$
\underset{\substack{\text{(see Statement 1)} \\ \downarrow}}{} \quad \underset{\substack{\text{(see Statement 2)} \\ \downarrow}}{}
$$

$$
E(X^2) = \sum_x x^2 f(x) \geq \sum_{\substack{x \\ |x| \geq c}} x^2 f(x) \geq \sum_{\substack{x \\ |x| \geq c}} c^2 f(x)
$$

$$
\underset{\substack{\text{(see Statement 3)} \\ \downarrow}}{}
$$

$$
\geq c^2 \sum_{\substack{x \\ |x| \geq c}} f(x) \geq c^2 P(|X| \geq c)
$$

Therefore

$$
P(|X| \geq c) \leq \frac{E(X^2)}{c^2}
$$

Statement 1:   The notation

$$
\sum_{\substack{x \\ |x| \geq c}} x^2 f(x)
$$

is used to denote the summation of $x^2 f(x)$ for all values of x for which $|x| \geq c$. Since each term in the sum $\sum_x x^2 f(x)$ is positive, the given inequality follows.

Statement 2:   Since $|x| \geq c$ implies that $x^2 \geq c^2$, the given inequality follows.

Statement 3:   Since X is discrete

$$
P(|X| \geq c) = \sum_{\substack{x \\ |x| \geq c}} P(X = x) = \sum_{\substack{x \\ |x| \geq c}} f(x)
$$

EXAMPLE 6.24

When a certain speed-control device is set at m miles per hour, the actual speed in miles per hour of an automobile equipped with this device is a random variable X with mean $\mu = m$ and standard deviation $\sigma = 3$. If the device is set at 45 miles per hour in a zone with a 50-mile-per-hour speed limit, what is the Chebyshev estimate of the probability that the speed of the automobile will exceed the speed limit?

SOLUTION

Assuming that X is a continuous random variable, we have

$$
P(\text{actual speed exceeds speed limit})
$$

$$
= P(X > 50) = P(X \geq 50) \leq P(|X - 45| \geq 5)
$$

$$
= P(|X - \mu| \geq 5) \leq \frac{\sigma^2}{5^2} = \frac{9}{25} = .36
$$

$$
\uparrow
$$

(corollary to Chebyshev's inequality)

Exercises

1. Prove Chebyshev's inequality for the case of continuous random variables. (*Hint:* follow the proof for the discrete case but replace sums with integrals.)

2. Prove the corollary to Theorem 6.6. (*Hint:* consider the random variable $Y = X - \mu$. Apply Chebyshev's inequality to Y.)

3. Use Chebyshev's inequality to estimate the probability that the first method of Exercise 4 of Section 6.5 will provide a measurement that is within 1 degree of the actual temperature. Estimate the probability of the same event for the second method.

4. Use Chebyshev's inequality to estimate the probability that in any one day the first business of Exercise 3 of Section 6.5 will make either less than $60 or more than $140 in profit. Estimate the probability of the same event for the second business.

5. Chebyshev's inequality should be used to estimate probabilities only when the distribution is unknown, because the estimated probabilities tend to be far from the exact probabilities. For example, suppose that the daily profits for the two businesses of Exercise 3 of Section 6.5 are uniform random variables with parameters $\alpha = 50$ and $\beta = 150$ for the first business, and parameters $\alpha = 10$ and $\beta = 190$ for the second business. Find the exact probabilities of the events of Exercise 4 and compare these probabilities with the Chebyshev estimates.

   Note: the means and variances of the random variables defined here are the same as those of Exercise 3 of Section 6.5.

6. *The Law of Large Numbers* (for Bernoulli trials). Let S_n be the number of successes in the first n of a sequence of Bernoulli trials where the probability of success in each trial is p. The law of large numbers states that for any constant $c > 0$

$$\lim_{n \to \infty} P\left(\left| \frac{S_n}{n} - p \right| < c \right) = 1$$

   Use Chebyshev's inequality to prove this statement.

   Note: since this statement is weaker than the strong law of large numbers it is sometimes referred to as the weak law of large numbers.

6.8   The Means and Variances of Special Probability Distributions

Since the mean, variance, and standard deviation of a random variable are completely determined by the distribution of the random variable, we also refer to them as the mean, variance, and standard deviation of the distribution.

Table 6.2 gives the means and variances of the distributions studied in Chapters 4 and 5. The derivation of these means and variances is provided in Section 6.11. An ability to derive the means and variances of the distributions is, of course, recommended for all serious students who intend to pursue theoretical studies in probability and statistics. It is not, however, necessary for a meaningful comprehension of applied aspects of the field.

EXAMPLE 6.25

If exposed to a certain contagious disease, the probability that a person will become infected is .05. If 1,000 people are exposed, what is the expected number that will become infected?

SOLUTION

Let X be the number of people who will become infected. Clearly, X is a binomial random variable with $p = P(\text{become infected}) = .05$ and $n = 1,000$ people. Therefore, $E(X) = np = 1000 \cdot .05 = 50$, the expected number of people who will become infected.

EXAMPLE 6.26

If the distribution of raisins per cookie in a certain type of raisin cookies is Poisson with $\lambda = 4$, what is the expected number of raisins per cookie?

SOLUTION

Let X be the number of raisins per cookie. X is a Poisson random variable with parameter $\lambda = 4$. Therefore, $E(X) = \lambda = 4$, the expected number of raisins per cookie.

EXAMPLE 6.27

If X is the amount of soda in ounces per cup dispensed by a certain soda dispensing machine, then

$$f(x) = \begin{cases} 1 & 6.5 < x < 7.5 \\ 0 & \text{elsewhere} \end{cases}$$

Find the expected amount of soda per cup.

SOLUTION

The probability density function $f(x)$ indicates that X is a uniform random variable with parameters $\alpha = 6.5$ and $\beta = 7.5$. Therefore, $E(X) = (\alpha + \beta)/2 = (6.5 + 7.5)/2 = 7$, the expected number of ounces per cup.

TABLE 6.2
Means and Variances of Special Probability Distributions

Distribution	pdf or pf	Mean	Variance
Binomial	$b(x; n, p) = \binom{n}{x} p^x q^{n-x}$ $x = 0, 1, 2, \ldots, n$ $0 < p < 1, \quad q = 1 - p$	np	npq
Hyper-geometric	$h(x; n, N_1, N_2) = \dfrac{\binom{N_1}{x}\binom{N_2}{n-x}}{\binom{N_1 + N_2}{n}}$ positive integers x, $0 \leq x \leq n$ and $0 \leq x \leq N_1$	$\dfrac{nN_1}{N_1 + N_2}$	$\dfrac{nN_1 N_2(N_1 + N_2 - n)}{(N_1 + N_2)^2(N_1 + N_2 - 1)}$
Poisson	$p(x; \lambda) = \dfrac{\lambda^x e^{-\lambda}}{x!}$ $x = 0, 1, 2, \ldots; \quad \lambda > 0$	λ	λ
Geometric	$g(x; p) = q^{x-1} p$ $x = 1, 2, 3, \ldots;$ $0 < p < 1; \quad q = 1 - p$	$1/p$	q/p^2
Negative Binomial	$f(x; n, p) = \binom{x-1}{n-1} p^n q^{x-n}$ $x = n, n+1, n+2, \ldots;$ $0 < p < 1; \quad q = 1 - p$	n/p	nq/p^2
Uniform	$f(x) = \begin{cases} \dfrac{1}{\beta - \alpha} & \text{for } \alpha < x < \beta \\ \\ 0 & \text{elsewhere} \end{cases}$	$\dfrac{\alpha + \beta}{2}$	$\dfrac{(\beta - \alpha)^2}{12}$
Normal	$f(x) = \dfrac{1}{\sqrt{2\pi}\,\sigma} e^{-(x-\mu)^2/(2\sigma^2)}$ $-\infty < x < \infty$	μ	σ^2
Exponential	$f(x) = \begin{cases} \dfrac{1}{\theta} e^{-x/\theta} & x > 0 \\ \\ 0 \end{cases}$	θ	θ^2
Gamma	$f(x) = \begin{cases} \dfrac{1}{\beta^\alpha \Gamma(\alpha)} x^{\alpha-1} e^{-x/\beta} & x > 0 \\ \\ x \leq 0 \quad x \leq 0 \end{cases}$	$\alpha\beta$	$\alpha\beta^2$
Beta	$f(x) = \begin{cases} \dfrac{\Gamma(\alpha + \beta)}{\Gamma(\alpha)\Gamma(\beta)} x^{\alpha-1}(1 - x)^{\beta-1} \\ \text{for } 0 < x < 1 \\ \\ 0 \quad \text{elsewhere} \end{cases}$	$\dfrac{\alpha}{\alpha + \beta}$	$\dfrac{\alpha\beta}{(\alpha + \beta)^2(\alpha + \beta + 1)}$

EXAMPLE 6.28

If the length of phone calls in minutes arriving at a certain switchboard is a random variable X with probability density function

$$f(x) = \begin{cases} \frac{1}{5}e^{-(x/5)} & x > 0 \\ 0 & x \le 0 \end{cases}$$

what is the expected length per call?

SOLUTION

X is exponentially distributed with parameter $\theta = 5$. Therefore, $E(X) = \theta = 5$, the expected length of individual phone calls.

EXAMPLE 6.29

If X is the income in thousands of dollars of a family selected at random from Woodland County, then X can be treated as a random variable with pdf

$$f(x) = \frac{1}{\sqrt{2\pi}\,5}e^{-(x-25)^2/(2 \cdot 25)} \qquad -\infty < x < \infty$$

whereas, if Y is the income in thousands of dollars of a family selected at random from High Park County, then Y can be treated as a random variable with pdf

$$g(y) = \frac{1}{\sqrt{2\pi}\,10}e^{-(x-30)^2/(2 \cdot 100)} \qquad -\infty < x < \infty$$

(a) Which county has the higher average family income?
(b) Which county has the higher percentage of families with incomes over $50,000?
(c) Which county has the higher percentage of families with incomes below $10,000?
(d) In terms of family income, which is the more homogeneous community?

SOLUTION

(a) The pdf's of X and Y indicate that they are normal random variables with means of 25 and 30, respectively, and standard deviations of 5 and 10, respectively. Therefore, the average family income in Woodland County is $25,000, and the average family income in High Park County is $30,000. High Park County has the higher average family income.

(b) In Woodland County, an income of $50,000 is 5 standard deviations away from the mean. Therefore, the percentage of families with incomes over that figure is approximately zero (see Figure 6.3 on p. 194). In High Park County, an income of $50,000 is only 2 standard deviations away from the mean. Therefore, approximately 2.2% of families there have incomes over $50,000. (Again, see Figure 6.3.) High Park County has the higher percentage of families with incomes over $50,000.

(c)   In Woodland County, an income of $10,000 is 3 standard deviations away from the mean. In High Park County, an income of $10,000 is 2 standard deviations away from the mean. High Park County has the higher percentage of families with incomes below $10,000.

(d)   Answers to (b) and (c) imply that Woodland County is the more homogeneous community because the standard deviation is smaller there, and therefore, there is less variability in size of income.

Exercises

1.   What is the expected number of sixes in 30 rolls of a balanced die? What is the standard deviation?

2.   Suppose that the probability that a certain football team wins a game is $p = .7$. If the team plays 15 games, what is the expected wins?

3.   Of 100 jars filled by the machine in Exercise 8 of Section 6.1, how many are expected to contain more than five ounces of jelly?

4.   What is the expected amount of jelly in jars filled by the machine of Exercise 8 of Section 6.1? What is the variance?

5.   If a family is selected at random from a certain densely populated area, its annual income in thousands of dollars can be treated as a random variable X with pdf

$$f(x) = \frac{1}{\sqrt{2\pi}\,2}\,e^{-(x-10)^2/8} \qquad \text{for} \quad -\infty < x < \infty$$

What is the average income of families in this area?
What is the standard deviation?

6.   The time in minutes required to perform a certain job is a random variable X with probability density function

$$f(x) = \frac{1}{\sqrt{2\pi}\,2}\,e^{-(x-60)^2/8} \qquad \text{for} \quad -\infty < x < \infty$$

What is the expected amount of time required to perform this job? What is the variance?

7. Each year, a certain company uses thousands of light bulbs which burn continuously day and night. Suppose that under these conditions, the life of such a light bulb may be regarded as a random variable that is normally distributed with a mean of 50 days and a standard deviation of 19 days. On January 1 of a certain year, the company put 5,000 new light bulbs into service. How many could be expected to last till February 1 of the same year?

8. Chocolate chips are distributed in a certain type of cookie according to a Poisson distribution with parameter $\lambda = 5$. What is the expected number of chocolate chips per cookie? In a bag of 1,000 such cookies, what number can be expected to have exactly five chocolate chips?

9. If the number of emergency patients treated at a certain hospital in any one day can be looked upon as a random variable X with probability density function given by

$$f(x) = \frac{9^x e^{-9}}{x!} \qquad \text{for} \quad x = 0, 1, 2, \ldots$$

   what number of emergency patients can this hospital expect to treat per day?

10. The time in hours required to repair a certain piece of equipment is a random variable X with probability density function

$$f(x) = \begin{cases} \dfrac{1}{3} e^{-(x/3)} & \text{for} \quad x > 0 \\ 0 & x \le 0 \end{cases}$$

   What is the expected amount of time required to repair this piece of equipment? What is the standard deviation?

11. If the number of automobile accidents from time zero to time t is a Poisson random variable with parameter $\lambda t = (1/15)t$, what is the time that can be expected to elapse until the first automobile accident occurs? (*Hint:* see Chapter 5, Section 5.7.2, Statement 1.)

12. If the number of telephone calls arriving at a certain switchboard between any time zero to any time t (measured in minutes) is a random variable X_t with probability function given by

$$f(x) = P(X_t = x) = (t^x e^{-t})/x! \qquad \text{for} \quad x = 0, 1, 2, \ldots$$

   what is the expected waiting time between telephone calls at this switchboard?

13. In reference to Exercise 12, how many telephone calls can be expected at this switchboard per hour?

6.9   Standardized Random Variables

6.9.1   Related Theorems

The following results, which relate a random variable to its mean and standard deviation, are necessary for the definition of standardized random variables.

Theorem 6.7   *Let X be a random variable with mean μ and standard deviation σ. If*

$$Y = \frac{X - \mu}{\sigma}$$

then

$$E(Y) = 0 \quad and \quad \text{Var}(Y) = 1$$

Proof

$$E(Y) = E\left(\frac{x - \mu}{\sigma}\right) = \frac{1}{\sigma} E(X - \mu)$$

$$= \frac{1}{\sigma}[E(X) - \mu] = 0$$

$$\text{Var}(Y) = \text{Var}\left(\frac{X - \mu}{\sigma}\right) = \frac{1}{\sigma^2} \text{Var}(X - \mu)$$

$$= \frac{1}{\sigma^2} \text{Var}(X) = \frac{\sigma^2}{\sigma^2} = 1$$

A random variable having mean 0 and variance 1 is said to be in standard form. The random variable $(X - \mu)/\sigma$ therefore is in standard form; it is called the **standardized random variable** corresponding to the random variable X.

As applied to the normal random variable, the following theorem provides a more powerful result than Theorem 6.7. The standardized normal random variable is called a **standard normal random variable**.

Theorem 6.8   *Let X be a random variable whose distribution is normal with mean μ and variance σ². Let*

$$Z = \frac{X - \mu}{\sigma}$$

Z is standard normal.

Proof Let $g(z)$ and $G(z)$ be the probability density function and cumulative distribution function of Z. We wish to show that

$$g(z) = \frac{1}{\sqrt{2\pi}} e^{-(z^2/2)} \qquad -\infty < z < \infty$$

Observe that

$$G(z) = P(Z \le z) = P\left(\frac{X - \mu}{\sigma} \le z\right) = P(X \le \sigma z + \mu)$$

$$= \int_{-\infty}^{\sigma z + \mu} \frac{1}{\sqrt{2\pi}\,\sigma} e^{-(x-\mu)^2/(2\sigma^2)}\, dx$$

$$g(z) = \frac{d\,G(z)}{dz} = \frac{d \displaystyle\int_{-\infty}^{\sigma z + \mu} \frac{1}{\sqrt{2\pi}\,\sigma} e^{-(x-\mu)^2/(2\sigma^2)}\, dx}{dz}$$

$$= \frac{d \displaystyle\int_{-\infty}^{\sigma z + \mu} \frac{1}{\sqrt{2\pi}\,\sigma} e^{-(x-\mu)^2/(2\sigma^2)}\, dx}{d(\sigma z + \mu)} \cdot \frac{d(\sigma z + \mu)}{dz}$$

$$= \left[\frac{1}{\sqrt{2\pi}\,\sigma} e^{-(\sigma z + \mu - \mu)^2/(2\sigma^2)}\right] \cdot \sigma$$

$$= \frac{1}{\sqrt{2\pi}} e^{-(z^2/2)}$$

We performed the above differentiation by using the rule from calculus that

$$\frac{d}{dx}\int_a^x f(u)\, du = f(x)$$

6.9.2   Computing Probabilities for the Normal Random Variable

Theorem 6.8 enables us to use Table III in Appendix C to compute probabilities for normal random variables. Let X be a normal random variable with mean μ and variance σ^2. To find $P(a < X < b)$ use Theorem 6.8 to write

$$P(a < X < b) = P\left(\frac{a - \mu}{\sigma} < \frac{X - \mu}{\sigma} < \frac{b - \mu}{\sigma}\right)$$

$$= P\left(\frac{a - \mu}{\sigma} < Z < \frac{b - \mu}{\sigma}\right)$$

Z is standard normal so we may use Table III to find the desired probability.

EXAMPLE 6.30

Let X be normal with mean 1 and variance 4. Find $P(2 < X < 3)$.

SOLUTION

$$P(2 < X < 3) = P\left(\frac{2 - \mu}{\sigma} < \frac{X - \mu}{\sigma} < \frac{3 - \mu}{\sigma}\right)$$

$$= P\left(\frac{2 - 1}{2} < \frac{X - \mu}{\sigma} < \frac{3 - 1}{2}\right) = P(.5 < Z < 1) = .1498$$

Exercises

1. If X is a random variable having a normal distribution with mean $\mu = 2$ and standard deviation $\sigma = .5$, find the probability that
    (a) X takes on a value greater than 3;
    (b) X takes on a value less than 1.5;
    (c) X takes on a value greater than 2.5;
    (d) X takes on a value less than 1;
    (e) X takes on a value between -1 and 1.

2. The lifetime of a certain kind of electronic unit is a normally distributed random variable with mean $\mu = 1.1$ years and standard deviation $\sigma = .1$ year. Find the probability that such a unit will last
    (a) less than 1 year;
    (b) at least 1.25 years;
    (c) anywhere from .9 to 1.2 years.

3. If families are selected at random from a certain densely populated area, their annual income can be treated as a normally distributed random variable with mean $\mu = 10,000$ (dollars) and standard deviation $\sigma = 2,000$ (dollars).
    (a) What is the probability that a family selected at random from this community has an income greater than $12,000?
    (b) Find the income below which we will find the poorest 10% of families in this area.

4. X is normal with mean -1 and variance 4. Find the value x for which the probability is .2676 that X will take on a value less than x.

5. If X is normally distributed with $\sigma = 10$, and $P(X > 16.34) = .1212$, what is the mean (expected value) of X?

6. The time required to complete an inspection of a certain machine is a random variable having a normal distribution with a mean of 60 minutes and a standard deviation of 10 minutes.
    (a) What is the probability that an inspection will take anywhere from 40 to 50 minutes?

(b) Find the value t for which it can be said that 90% of the time an inspection is complete in less than t minutes.

7. The lifetime of a certain kind of TV tube is a normally distributed random variable with a mean of 390 hours. Find the standard deviation of this distribution if the probability is .33 that any one of these TV tubes will last less than 380 hours.

8. X is normally distributed with mean K and variance K^2. If $P(X < 8) =$ .8413, what is K?

6.10   More Applications

6.10.1   Estimating the Means of Distributions

One method of estimating the mean μ of a distribution is to apply the strong law of large numbers. This method works if we are able to obtain repeated observed values of the random variable associated with the distribution. If X is the random variable and X_i the ith observed value of X, then the strong law of large numbers tells us that

$$\frac{\sum_{i=1}^{n} X_i}{n} \approx \mu$$

with great probability if n is large. This method of estimating the mean of a distribution is discussed in great detail in Chapter 11.

EXAMPLE 6.31

Table IV in Appendix C is a list of random numbers. By placing a decimal point in front of each number we obtain a table of random numbers from the interval $(0, 1)$. If these numbers are truly randomly selected from $(0, 1)$ and if X is a random variable that is uniform on $(0, 1)$, we may consider the numbers in the table as successive observed values of X. In that case if X_i denotes the ith observed value, and n is large, then with great probability,

$$\frac{\sum_{i=1}^{n} X_i}{n} \approx .500$$

Find the average of the first 70 numbers from the table of random numbers and confirm that it is indeed close to .500.

SOLUTION

The average of the first 70 numbers from the table of random numbers is .510, which is indeed close to .500. This is an indication that the numbers are truly random.

EXAMPLE 6.32

Suppose that in Example 6.31 we use a list of numbers from a telephone book instead of the numbers from Table IV. Table 6.3 is a list of the last 3 digits of the first 70 numbers from column 4, page 1534, of the 1978–79 Manhattan, New York, phone book. We could speculate that by placing a decimal point in front of each number we would obtain a list of 70 numbers selected at random from $(0, 1)$. Let x_i denote the ith number on the list. We find that

$$\frac{\sum_{i=1}^{70} x_i}{70} = \frac{.770 + .184 + \cdots + .361}{70} = .513$$

We see that 5.13 is also very close to the expected value of .500. Does this mean that the numbers from the phone book can be considered as randomly selected from $(0, 1)$? Such questions are considered in Chapter 12, in the section on the testing of statistical hypotheses.

TABLE 6.3

The last three digits of telephone numbers from a page of the 1978–79 Manhattan, New York, telephone book.

1. 770	15. 882	29. 227	43. 493	57. 521
2. 184	16. 287	30. 393	44. 376	58. 294
3. 721	17. 572	31. 249	45. 812	59. 148
4. 616	18. 798	32. 713	46. 925	60. 788
5. 421	19. 625	33. 846	47. 876	61. 077
6. 116	20. 421	34. 939	48. 280	62. 161
7. 716	21. 810	35. 569	49. 111	63. 400
8. 142	22. 707	36. 421	50. 658	64. 989
9. 989	23. 660	37. 410	51. 964	65. 588
10. 816	24. 453	38. 203	52. 212	66. 555
11. 587	25. 310	39. 615	53. 271	67. 611
12. 042	26. 117	40. 638	54. 025	68. 886
13. 406	27. 809	41. 654	55. 623	69. 773
14. 093	28. 299	42. 084	56. 800	70. 361

sum = 35.908, mean = .51297, variance = .07597, standard deviation = .2756

6.10.2   Estimating the Probability of Events

Let A be an event of the experiment $\mathscr{E}$. How might we estimate p, where $p = P(A)$?

Suppose we perform the experiment $\mathscr{E}$ and define X to be the following random variable.

$$X = \begin{cases} 1 & \text{if event } A \text{ occurs} \\ 0 & \text{if event } A \text{ does not occur} \end{cases}$$

Then

$$E(X) = 1 \cdot P(X = 1) + 0 \cdot P(X = 0) = 1 \cdot P(A)$$

Now, suppose we repeat $\mathscr{E}$ under identical conditions n times and let X_i be the value of X on the ith performance of $\mathscr{E}$. The strong law of large numbers tells us that

$$P\left(\lim_{n \to \infty} \frac{\sum_{i=1}^{n} X_i}{n} = p\right) = 1 \tag{2}$$

and therefore

$$\frac{\sum_{i=1}^{n} X_i}{n} \approx p \tag{3}$$

with great probability if n is large.

Since $\sum_{i=1}^{n} X_i$ provides a count of the number of times we observe the occurrence of the event A in the n performances of the experiment $\mathscr{E}$, the ratio $\sum_{i=1}^{n} X_i/n$ is the relative frequency ratio of the number of occurrences of the event A relative to the number of performances of the experiment $\mathscr{E}$. In terms of the notations used in Chapter 1, Equations 2 and 3 tell us that if n denotes the number of repetitions of $\mathscr{E}$, n_A denotes the number of times A occurs, and n is large, then the ratio n_A/n should provide a good estimate of p most of the time. In other words, if we wish to estimate p, where $p = P(A)$, then we should use the method of empirical probabilities and set

$$p = \frac{n_A}{n}$$

Exercises

1. How might you estimate the average income of residents of a very large city without finding the income of every resident? What theorem may be used to justify this method?

2.   A certain die is suspected of being loaded. How might you check out this die? Justify your method.

6.11   Derivation of the Means and Variances of Special Probability Distributions

Theorem 6.9   ***Mean and Variance of the Binomial Distribution***   *Let X be a random variable with probability function*

$$b(x; n, p) = \binom{n}{x} p^x q^{n-x} \quad \text{for} \quad x = 0, 1, 2, \ldots, n, \quad 0 < p < 1, \quad q = 1 - p$$

The mean and variance of X are the quantities

$$E(X) = np, \qquad \text{Var}(X) = npq$$

Proof

$$E(X) = \sum_n x f(x) = \sum_{x=0}^{n} x b(x; n, p)$$

$$= \sum_{x=0}^{n} x \binom{n}{x} p^x q^{n-x}$$

$$= \sum_{x=1}^{n} x \binom{n}{x} p^x q^{n-x}$$

$$= \sum_{x=1}^{n} x \frac{n!}{x!(n-x)!} p^x q^{n-x}$$

$$= \sum_{x=1}^{n} \frac{n!}{(x-1)!(n-x)!} p^x q^{n-x}$$

$$= np \sum_{x=1}^{n} \frac{(n-1)!}{(x-1)!(n-x)!} p^{x-1} q^{n-x}$$

$$= np \sum_{r=0}^{m} \frac{m!}{r!(m-r)!} p^r q^{m-r}$$

$$= np$$

The last two inequalities are attained when we substitute $r = x - 1$, $m = n - 1$ and apply the binomial theorem $(a + b)^m = \sum_{r=0}^{m} \binom{m}{r} a^r b^{m-r}$ with $a = p$, $b = q$.

$$\text{Var}(X) = E(X^2) - E^2(X)$$
$$E(X^2) = E[X(X-1)] + E(X)$$

In finding $E[X(X-1)]$ we will use the substitution $r = x - 2$ and $m = n - 2$ in the binomial theorem stated above.

$$E[X(X-1)] = \sum_{x=0}^{n} x(x-1)b(x; n, p)$$

$$= \sum_{x=2}^{n} x(x-1) \frac{n!}{x!(n-x)!} p^x q^{n-x}$$

$$= \sum_{x=2}^{n} \frac{n!}{(x-2)!(n-x)!} p^x q^{n-x}$$

$$= n(n-1)p^2 \sum_{x=2}^{n} \frac{(n-2)!}{(x-2)!(n-x)!} p^{x-2} q^{n-x}$$

$$= n(n-1)p^2 \sum_{r=0}^{m} \frac{m!}{r!(m-r)!} p^r q^{m-r}$$

$$= n(n-1)p^2$$

Therefore

$$\text{Var}(X) = E(X^2) - E^2(X) = E[X(X-1)] + E(X) - E^2(X)$$
$$= n(n-1)p^2 + np - n^2 p^2$$
$$= npq$$

Theorem 6.10   Mean and Variance of the Poisson Distribution   *Let X be a random variable with probability function*

$$p(x; \lambda) = \frac{\lambda^x e^{-\lambda}}{x!} \qquad \text{for} \quad x = 0, 1, 2, \ldots, \quad \lambda > 0$$

The mean and variance of X are the quantities

$$E(X) = \lambda, \qquad \text{Var}(X) = \lambda$$

Proof

$$E(X) = \sum_{x=0}^{\infty} xp(x; \lambda) = \sum_{x=0}^{\infty} x \cdot \frac{\lambda^x e^{-\lambda}}{x!}$$

$$= \sum_{x=1}^{\infty} x \cdot \frac{\lambda^x e^{-\lambda}}{x!} = \sum_{x=1}^{\infty} \frac{\lambda^x e^{-\lambda}}{(x-1)!}$$

$$= \lambda e^{-\lambda} \sum_{x=1}^{\infty} \frac{\lambda^{x-1}}{(x-1)!} = \lambda$$

The last equality is attained when we observe that $e^\lambda = \sum_{r=0}^{\infty}(\lambda^r/r!)$.

By changing the variable x to the variable $r = x - 2$ and using the fact that $e^\lambda = \sum_{r=0}^{\infty}(\lambda^r/r!)$, we obtain

$$E[X(X-1)] = \sum_{x=0}^{\infty} x(x-1)\frac{\lambda^x e^{-\lambda}}{x!}$$

$$= \sum_{x=2}^{\infty} x(x-1)\frac{\lambda^x e^{-\lambda}}{x!} = \lambda^2 e^{-\lambda}\sum_{x=2}^{\infty}\frac{\lambda^{x-2}}{(x-2)!}$$

$$= \lambda^2 e^{-\lambda}\sum_{r=0}^{\infty}\frac{\lambda^r}{r!} = \lambda^2 e^{-\lambda}e^\lambda = \lambda^2$$

$$\text{Var}(X) = E(X^2) - E^2(X) = E[X(X-1)] + E(X) - E^2(X)$$
$$= \lambda^2 + \lambda - \lambda^2 = \lambda$$

Theorem 6.11   Mean and Variance of the Uniform Distribution   *Let X be a random variable with probability density function*

$$f(x) = \begin{cases} \dfrac{1}{\beta - \alpha} & \alpha < x < \beta \\ 0 & \text{elsewhere} \end{cases}$$

The mean and variance of X are the quantities

$$E(X) = \frac{\alpha + \beta}{2}, \qquad \text{Var}(X) = \frac{(\beta - \alpha)^2}{12}$$

Proof

$$E(X) = \int_{-\infty}^{\infty} xf(x)\,dx = \int_{\alpha}^{\beta} x \cdot \frac{1}{(\beta - \alpha)}\,dx = \frac{\alpha + \beta}{2}$$

$$E(X^2) = \int_{\alpha}^{\beta} x^2 \cdot \frac{1}{\beta - \alpha}\,dx = \frac{\beta^2 + \alpha\beta + \alpha^2}{3}$$

$$\text{Var}(X) = E(X^2) - E^2(X)$$

$$= \frac{\beta^2 + \alpha\beta + \alpha^2}{3} - \frac{(\alpha + \beta)^2}{4}$$

$$= \frac{(\beta - \alpha)^2}{12}$$

Theorem 6.12   Mean and Variance of the Exponential Distribution   *Let X be a random variable with probability density function*

$$f(x) = \begin{cases} \dfrac{1}{\theta} e^{(x/\theta)} & x > 0 \\ 0 & x \leq 0 \end{cases}$$

The mean and variance of X are the quantities

$$E(X) = \theta, \qquad \text{Var}(X) = \theta^2$$

Proof   Employ in this proof the method of integration by parts by using the formula $\int u\, dv = uv - \int v\, du$ and make the substitution $v = -e^{-(x/\theta)}$, $dv = (1/\theta) e^{-(x/\theta)}\, dx$, $u = x$, $du = dx$, then

$$E(X) = \int_{-\infty}^{\infty} xf(x)\, dx = \int_{0}^{\infty} x(1/\theta)\, e^{-(x/\theta)}\, dx$$

$$= -xe^{-(x/\theta)} \Big|_{0}^{\infty} + \int_{0}^{\infty} e^{-(x/\theta)}\, dx$$

$$= -xe^{-(x/\theta)} \Big|_{0}^{\infty} + \theta \int_{0}^{\infty} \frac{1}{\theta} e^{-(x/\theta)}\, dx$$

$$= \theta$$

Using two successive integrations by parts, we obtain

$$E(X^2) = \int_{-\infty}^{\infty} x^2 f(x)\, dx$$

$$= \int_{0}^{\infty} x^2 (1/\theta) e^{-(x/\theta)}\, dx = 2\theta^2$$

$$\text{Var}(X) = E(X^2) - E^2(X) = 2\theta^2 - \theta^2 = \theta^2$$

Theorem 6.13   Mean and Variance of the Normal Distribution   *Let X be a random variable with probability density function*

$$f(x) = \frac{1}{\sqrt{2\pi}\, \sigma} e^{-(x-\mu)^2/(2\sigma^2)} \qquad -\infty < x < \infty$$

The mean and variance of X are the quantities

$$E(X) = \mu, \qquad \text{Var}(X) = \sigma^2$$

Proof   Use the substitution $w = (x - \mu)/\sigma$, then

$$E(X) = \int_{-\infty}^{\infty} xf(x)\, dx = \int_{-\infty}^{\infty} x \cdot \frac{1}{\sqrt{2\pi}\,\sigma}\, e^{-(x-\mu)^2/(2\sigma^2)}\, dx$$

$$= \int_{-\infty}^{\infty} \left[(\sigma w + \mu)\frac{1}{\sqrt{2\pi}\,\sigma}\, e^{-(w^2/2)}\right] \cdot \sigma\, dw$$

$$= \frac{\sigma}{\sqrt{2\pi}} \int_{-\infty}^{\infty} w\, e^{-(w^2/2)}\, dw + \mu \int_{-\infty}^{\infty} \frac{1}{\sqrt{2\pi}}\, e^{-(w^2/2)}\, dw$$

Now, observe that if $h(w) = w\, e^{-(w^2/2)}$, then $h(-w) = -h(w)$. Therefore

$$\int_{-\infty}^{\infty} w\, e^{-(w^2/2)}\, dw = \int_{-\infty}^{0} w\, e^{-(w^2/2)}\, dw + \int_{0}^{\infty} w\, e^{-(w^2/2)}\, dw$$

$$= -\int_{0}^{\infty} w\, e^{-(w^2/2)}\, dw + \int_{0}^{\infty} w\, e^{-(w^2/2)}\, dw = 0$$

Also, observe that

$$\int_{-\infty}^{\infty} \frac{1}{\sqrt{2\pi}}\, e^{-(w^2/2)}\, dw = 1$$

since the integrand is the standard normal density function.
    Therefore

$$E(X) = \mu$$

Let $w = (x - \mu)/\sigma$, then

$$\text{Var}(X) = E[(X - \mu)^2] = \int_{-\infty}^{\infty} (x - \mu)^2 f(x)\, dx$$

$$= \int_{-\infty}^{\infty} (x - \mu)^2 \frac{1}{\sqrt{2\pi}\,\sigma}\, e^{-(x-\mu)^2/(2\sigma^2)}\, dx$$

$$= \int_{-\infty}^{\infty} \left[\sigma^2 w^2 \frac{1}{\sqrt{2\pi}\,\sigma}\, e^{-(w^2/2)}\right] \cdot \sigma\, dw$$

$$= \sigma^2 \int_{-\infty}^{\infty} \frac{1}{\sqrt{2\pi}}\, w^2\, e^{-(w^2/2)}\, dw = \sigma^2$$

We used the fact that

$$\int_{-\infty}^{\infty} \frac{1}{\sqrt{2\pi}}\, w^2\, e^{-(w^2/2)}\, dw = 1$$

This result may be obtained from a standard table of integrals or through integration by parts.

Supplementary Exercises

1. Prove Theorem 6.3 for continuous random variables. (*Hint:* refer to the proof for the discrete case and replace summations with integrations.)

2. Keno is a popular game in many gambling casinos. In one version of this game the casino selects 20 numbers at random from the set of numbers 1 through 80. A player selects 10 numbers. A win occurs if at least one of the player's chosen numbers match any of the 20 numbers selected by the casino. The payoffs are as follows:

Keno Payoffs

Number of matches	Dollars won for each $1 bet
0–4	0
5	1
6	17
7	179
8	1,299
9	2,599
10	24,999

Find the expected net gain for the player. Is this a fair game?

3. In certain games of dice, sums of 7 or 11 in a throw of a pair of dice are considered lucky. If a pair of dice is thrown 100 times, what is the expected number of 7's? What is the expected number of 11's?

4. How many times should you throw a pair of dice if you want a probability of at least .95 of getting a sum of 7 at least once?

5. Players A, B, C take turns at a fair game involving two players. To begin, A plays B while C is out. In the next game, C plays the winner while the loser is out. The players continue in this manner until one of them wins a second game. The first player to win 2 games is declared the winner of the series and collects the single prize of $100. Is this a fair game? If not, why not?

6. Can the following function be a probability density function of a random variable?

$$f(x) = \begin{cases} 1/x^2 & \text{for} \quad x > 1 \\ 0 & \text{elsewhere} \end{cases}$$

If the answer is yes, is $E(X)$ defined?

7. Suppose that the random variable of Example 6.24 is normally distributed. Find the exact probability that the speed of the automobile will exceed the speed limit. Compare this probability with the Chebyshev estimate computed in that example.

8. Suppose that the number of flounders that a certain fisherman will catch in any one day may be looked upon as a Poisson random variable with parameter $\lambda = 50$. Furthermore, suppose that the profit in dollars that the fisherman will make on the flounders on a day when he catches x flounders is given by the function

$$g(x) = 3x - 45$$

What is the daily expected profit from flounders for this fisherman? What is the standard deviation?

9. Suppose that the average daytime temperature in degrees F on any one day during the month of August in a certain city is a random variable whose distribution is normal with mean $\mu = 90$ and standard deviation $\sigma = 5$. Furthermore, suppose that on a day when the average daytime temperature is x degrees the cost (in dollars) of operating the air conditioning system for a certain company in this city is given by the function

$$g(x) = 100x + 50$$

In the month of August, what is the expected daily cost of operating the air conditioning system for this company? What is the standard deviation?

10. Suppose that the age in years of any one of a certain type of motor to be reconditioned at a certain shop may be looked upon as a random variable whose distribution is exponential with parameter $\theta = 5$. Furthermore, suppose that the time in hours required to recondition a motor of this type that is x years old is given by the function

$$g(x) = 2x + 10$$

Find the expected time required to recondition motors of this type at this shop. What is the standard deviation?

11. The number of typewritten spaces necessary to reproduce the complete works of Shakespeare is approximately 10 million. If the monkey of Example 6.13 types at the rate of 250 strokes per minute, then what is an estimate of the expected waiting time until the first appearance on the typewritten page of the complete works of Shakespeare? (*Hint:* use the geometric distribution with $p = (1/T)^M$ where $T = 50$ and $M = 10$ million.)

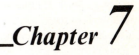

Chapter 7

MULTIVARIATE DISTRIBUTIONS

7.1   Introduction

Up to this point we have dealt with random variables as isolated entities, such as the numerical outcome of one roll of a die, or the income of a family selected at random from a certain community. In more complex situations and especially in prediction problems, problems of correlation, and theoretical aspects of statistical inference, it is necessary to deal with more than one random variable simultaneously. For example, a manufacturer may be interested in predicting the volume of sales of a product (random variable X) based on the projected volume of sales of a foreign competing product (random variable Y), or an educator may be interested in relating family income (random variable X) with academic achievement (random variable Y), or an engineer may be interested in the combined output of generator A (random variable X_1), generator B (random variable X_2), and generator C (random variable X_3).

     When several random variables are considered simultaneously, the underlying probability structure is defined in terms of multivariate distributions. This chapter is concerned with a discussion of the basic elements of multivariate distributions, the use of multivariate distributions to define the concept of independent random variables, and the use of multivariate distributions to find probabilities for jointly distributed random variables.[1]

7.2   Multivariate Distributions

7.2.1   Multivariate Probability Functions

     Definition 7.1   The **multivariate probability function** of the discrete random variables $X_1, X_2, \ldots, X_n$ is the function f defined by

$$f(x_1, x_2, \ldots, x_n) = P(X_1 = x_1, X_2 = x_2, \ldots, X_n = x_n)$$

for all possible values $x_1, x_2, \ldots, x_n$ of the random variables $X_1, X_2, \ldots, X_n$, respectively.

     A multivariate probability function is also called a multivariate probability distribution. Furthermore, if only two discrete random variables are involved, the multivariate probability function is usually called the joint (or bivariate) probability function or probability distribution of the two discrete random variables.

---

[1] Readers who are anxious to get to the topic of statistical inference may omit this chapter if they are willing to accept a few theorems without proofs in the following chapters.

EXAMPLE 7.1

Consider the experiment of rolling a die and tossing a coin. Let X_1 be the value on the die, and let X_2 be the number of heads. Find the joint probability function of X_1 and X_2.

SOLUTION

Since X_1 is the value on the die, and X_2 is the number of heads in 1 toss of a coin, the events $X_1 = x_1$ and $X_2 = x_2$ (for all relevant values of x_1 and x_2) are clearly independent events. Therefore, the joint probability function for X_1 and X_2 is

$$f(x_1, x_2) = P(X_1 = x_1, X_2 = x_2) = P(X_1 = x_1)P(X_2 = x_2)$$

$$= \frac{1}{6} \cdot \frac{1}{2} = \frac{1}{12}$$

for $x_1 = 1, 2, 3, 4, 5, 6$ and $x_2 = 0, 1$.

EXAMPLE 7.2

Consider two throws of a balanced die. Let X be the value on the first throw, and Y the absolute value of the difference of the values in the two throws. Find the joint probability function of X, Y.

SOLUTION

The joint probability function of two discrete random variables are often given in the form of a table of values such as Table 7.1. In this example the possible values of X are the integers 1, 2, 3, 4, 5, and 6, which appear along the top of the table. The possible values of Y are the integers 0, 1, 2, 3, 4, and 5, which appear along the left edge of the table. The corresponding values of $f(x, y)$ are shown within the table boxes where the respective values of X and Y intersect. Table 7.2 gives the values of the joint probability function of X, Y according to the scheme described here.

TABLE 7.1

	x					
y	1	2	3	4	5	6
0	$f(1,0)$	$f(2,0)$	$f(3,0)$	$f(4,0)$	$f(5,0)$	$f(6,0)$
1	$f(1,1)$	$f(2,1)$	$f(3,1)$	$f(4,1)$	$f(5,1)$	$f(6,1)$
2	$f(1,2)$	$f(2,2)$	$f(3,2)$	$f(4,2)$	$f(5,2)$	$f(6,2)$
3	$f(1,3)$	$f(2,3)$	$f(3,3)$	$f(4,3)$	$f(5,3)$	$f(6,3)$
4	$f(1,4)$	$f(2,4)$	$f(3,4)$	$f(4,4)$	$f(5,4)$	$f(6,4)$
5	$f(1,5)$	$f(2,5)$	$f(3,5)$	$f(4,5)$	$f(5,5)$	$f(6,5)$

TABLE 7.2

	x					
	1	**2**	**3**	**4**	**5**	**6**
0	$\frac{1}{36}$	$\frac{1}{36}$	$\frac{1}{36}$	$\frac{1}{36}$	$\frac{1}{36}$	$\frac{1}{36}$
1	$\frac{1}{36}$	$\frac{2}{36}$	$\frac{2}{36}$	$\frac{2}{36}$	$\frac{2}{36}$	$\frac{1}{36}$
2	$\frac{1}{36}$	$\frac{1}{36}$	$\frac{2}{36}$	$\frac{2}{36}$	$\frac{1}{36}$	$\frac{1}{36}$
3	$\frac{1}{36}$	$\frac{1}{36}$	$\frac{1}{36}$	$\frac{1}{36}$	$\frac{1}{36}$	$\frac{1}{36}$
4	$\frac{1}{36}$	$\frac{1}{36}$	0	0	$\frac{1}{36}$	$\frac{1}{36}$
5	$\frac{1}{36}$	0	0	0	0	$\frac{1}{36}$

y labels the rows.

As an illustration of how we might find the values of $f(x, y)$, let us consider the case where $x = 3$ and $y = 1$. Then

$$f(3, 1) = P(X = 3, Y = 1)$$

= P(a value of 3 on the first throw, and an absolute
value of 1 on the difference of the 2 throws)

= P(a value of 3 on the first throw and a value of
2 on the second throw)

+ P(a value of 3 on the first throw and a value of
4 on the second throw)

$$= \frac{1}{36} + \frac{1}{36} = \frac{2}{36}$$

EXAMPLE 7.3

Suppose that if X is the number of outgoing local telephone calls and if Y is the number of outgoing long distances telephone calls from a certain switch-board during a 5-minute period, the joint probability function of X and Y is given by the following table of values.

		x			
	0	**1**	**2**	**3**	**4**
0	.23	.23	.11	.04	.01
1	.11	.11	.05	.02	.01
2	.03	.03	.01	.01	.00

y labels the rows.

$$f(x, y)$$

(a)   Find the probability of exactly 2 outgoing local telephone calls and exactly 2 outgoing long distance telephone calls from this switchboard during a 5-minute period.

(b)   Find the probability of more than 3 outgoing telephone calls from this switchboard during a 5-minute period.

(c)   Find the probability of no outgoing local telephone calls from this switchboard during a 5-minute period.

SOLUTION

(a)

$$P(\text{exactly 2 local and 2 long distance calls}) = P(X = 2, Y = 2)$$
$$= f(2, 2) = .01$$

(b)

$$P(\text{more than 3 calls})$$
$$= P(X = 2, Y = 2) + P(X = 3, Y = 1) + P(X = 4, Y = 0)$$
$$+ P(X = 3, Y = 2) + P(X = 4, Y = 1) + P(X = 4, Y = 2)$$
$$= f(2, 2) + f(3, 1) + f(4, 0) + f(3, 2) + f(4, 1) + f(4, 2)$$
$$= .01 + .02 + .01 + .01 + .01 + .00 = .06$$

(c)

$$P(\text{no outgoing local telephone calls})$$
$$= P(X = 0) = P(X = 0, Y = 0) + P(X = 0, Y = 1)$$
$$+ P(X = 0, Y = 2)$$
$$= f(0, 0) + f(0, 1) + f(0, 2) = .23 + .11 + .03 = .37$$

7.2.2   Multivariate Cumulative Distribution Functions

Definition 7.2   The **multivariate cumulative distribution function** of the random variables $X_1, X_2, \ldots, X_n$ is the function F defined by

$$F(x_1, x_2, \ldots, x_n) = P(X_1 \le x_1, X_2 \le x_2, \ldots, X_n \le x_n)$$

for values $-\infty < x_i < \infty$, $i = 1, 2, \ldots, n$.

A multivariate cumulative distribution function is also called, more simply, a multivariate distribution function. Furthermore, if only two random variables are involved, the multivariate cumulative distribution function is usually called the joint (or bivariate) cumulative distribution function or distribution function of the two random variables. It should also be noted here that the definition of the multivariate cumulative distribution function is identical in format for both discrete and continuous random variables.

EXAMPLE 7.4

Consider the random variables X and Y of Example 7.2. Find $F(2, 1)$, $F(6, 0)$, $F(7, 8)$, $F(-2, 1)$, and $F(-3, -10)$.

SOLUTION

$$F(2, 1) = P(X \leq 2, Y \leq 1)$$
$$= P(X = 1, Y = 0) + P(X = 1, Y = 1)$$
$$+ P(X = 2, Y = 0) + P(X = 2, Y = 1)$$
$$= f(1, 0) + f(1, 1) + f(2, 0) + f(2, 1)$$
$$= \frac{1}{36} + \frac{1}{36} + \frac{1}{36} + \frac{2}{36} = \frac{5}{36}$$

$$F(6, 0) = P(X \leq 6, Y \leq 0)$$
$$= P(X = 1, Y = 0) + P(X = 2, Y = 0)$$
$$+ P(X = 3, Y = 0) + P(X = 4, Y = 0)$$
$$+ P(X = 5, Y = 0) + P(X = 6, Y = 0)$$
$$= 6/36$$

$$F(7, 8) = P(X \leq 7, Y \leq 8)$$
$$= 1, \text{ since } X \text{ is always } \leq 7 \text{ and } Y \leq 8$$

$$F(-2, 1) = 0, \text{ since } X \text{ is never } \leq -2$$
$$F(-3, -10) = 0, \text{ since } X \text{ is never } \leq -3 \text{ and } Y \text{ is never } \leq -10$$

EXAMPLE 7.5

Suppose that if X and Y represent the operating lives measured in years, of components A and B of a certain system, then their joint cumulative distributive function is given by

$$F(x, y) = \begin{cases} (1 - e^{-x})(1 - e^{-y}) & \text{for } x \geq 0, \quad y \geq 0 \\ 0 & \text{elsewhere} \end{cases}$$

Find the probabilities of the following events:
   (a)   Both components have operating lives shorter than or equal to 1 year.
   (b)   Component A has an operating life no greater than 1 year, and component B has an operating life no greater than 2 years.
   (c)   Both components have operating lives greater than 2 years.

SOLUTION

(a)

P(both have lives shorter than or equal to 1)
$$= P(X \leq 1, Y \leq 1) = F(1, 1) = (1 - e^{-1})(1 - e^{-1}) = .3996$$

(b)

P(component A has life no greater than 1, and
component B has life no greater than 2)

$$= P(X \leq 1, Y \leq 2) = F(1, 2) = (1 - e^{-1})(1 - e^{-2}) = .5466$$

(c)

P(both have operating lives greater than 2)

$$= P(X > 2, Y > 2) = 1 - P(X \leq 2 \text{ or } Y \leq 2)$$
$$= 1 - [P(X \leq 2) + P(Y \leq 2) - P(X \leq 2, Y \leq 2)]$$
$$= 1 - P(X \leq 2, Y < \infty) - P(X < \infty, Y \leq 2)$$
$$+ P(X \leq 2, Y \leq 2)$$
$$= 1 - F(2, \infty) - F(\infty, 2) + F(2, 2)$$
$$= 1 - (1 - e^{-2}) - (1 - e^{-2}) + (1 - e^{-2})(1 - e^{-2}) = .0183$$

7.2.3   Multivariate Probability Density Functions

Definition 7.3   The **multivariate probability density function** of the continuous random variables $X_1, X_2, \ldots, X_n$ with multivariate cumulative distribution function $F(x_1, x_2, \ldots, x_n)$ is the function f defined by

$$f(x_1, x_2, \ldots, x_n) = \frac{\partial^n F(x_1, x_2, \ldots, x_n)}{\partial x_1 \partial x_2 \cdots \partial x_n}$$

at all values $x_1, x_2, \ldots, x_n$ where the partial derivative is defined. (At values $x_1, x_2, \ldots, x_n$ where the partial derivatives do not exist, values for f may be arbitrarily assigned.)

When only two continuous random variables are involved, the multivariate probability density function is usually called the joint (or bivariate) probability density function of the two continuous random variables.

EXAMPLE 7.6

Find the joint probability density function of the random variables of Example 7.5.

SOLUTION

$$f(x, y) = \frac{\partial^2 F(x, y)}{\partial x \partial y} = \begin{cases} e^{-x} e^{-y} & \text{for } x > 0, \quad y > 0 \\ 0 & \text{elsewhere} \end{cases}$$

An immediate consequence of Definition 7.3 is that if X and Y are continuous random variables with joint probability density function $f(x, y)$, then

1. The joint cumulative distribution function of X and Y is

$$F(x, y) = \int_{-\infty}^{y} \int_{-\infty}^{x} f(s, t)\, ds\, dt \qquad \text{for} \quad -\infty < x < \infty, \quad -\infty < y < \infty$$

2. For values $a_1 < x < b_1, a_2 < y < b_2$

$$P(a_1 < X < b_1, a_2 < Y < b_2) = \int_{a_2}^{b_2} \int_{a_1}^{b_1} f(x, y)\, dx\, dy$$

and

3. If C is a region of points (x, y) in the plane $-\infty < x < \infty, -\infty < y < \infty$, and $P[(X, Y) \in C]$ denotes the probability that the random variables X and Y will take on respective values x and y so that (x, y) is a point in C, then

$$P[(X, Y) \in C] = \iint_{C} f(x, y)\, dx\, dy$$

The integration is over the region C, and the probability is defined only if the integral exists.

These three statements can, of course, be extended to statements concerning more than two continuous random variables.

EXAMPLE 7.7

Suppose that X and Y are continuous random variables with joint probability density function

$$f(x, y) = \begin{cases} (1/2)xe^{-y} & \text{for} \quad 0 < x < 2, \quad y > 0 \\ 0 & \text{elsewhere} \end{cases}$$

(a) Find the joint cumulative distribution function of X, Y.
(b) Find $P(X \le 1, Y \le 2)$ and $P(1 \le X \le 2, 3 \le Y \le 4)$.

SOLUTION

(a)

Step 1:   For $x \le 0, -\infty < y < \infty$

$$F(x, y) = \int_{-\infty}^{y} \int_{-\infty}^{x} f(s, t)\, ds\, dt = \int_{-\infty}^{y} \int_{-\infty}^{x} 0\, ds\, dt = 0$$

Step 2:   For $-\infty < x < \infty$, $y \le 0$

$$F(x, y) = \int_{-\infty}^{y} \int_{-\infty}^{x} f(s,t)\, ds\, dt = \int_{-\infty}^{y} \int_{-\infty}^{x} 0\, ds\, dt = 0$$

Step 3:   For $0 < x < 2$, $y > 0$

$$F(x, y) = \int_{-\infty}^{y} \int_{-\infty}^{x} f(s,t)\, ds\, dt = \int_{0}^{y} \int_{0}^{x} (1/2)se^{-t}\, ds\, dt$$

$$= \left(\int_{0}^{y} e^{-t}\, dt \right) \left(\int_{0}^{x} (1/2)s\, ds \right) = (1/4)x^2(1 - e^{-y})$$

Step 4:   For $x \ge 2$, $y > 0$

$$F(x, y) = \int_{-\infty}^{y} \int_{-\infty}^{x} f(s,t)\, ds\, dt = \int_{0}^{y} \int_{0}^{2} (1/2)se^{-t}\, ds\, dt$$

$$= \left(\int_{0}^{y} e^{-t}\, dt \right) \left(\int_{0}^{2} (1/2)s\, ds \right) = 1 - e^{-y}$$

Conclusion: Combining the results of Steps 1, 2, 3, and 4, we obtain

$$F(x, y) = \begin{cases} 0 & \text{for} \quad x \le 0, \quad -\infty < y < \infty \\ 0 & \quad\quad\quad -\infty < x < \infty, \quad y \le 0 \\ (1/4)x^2(1 - e^{-y}) & \quad\quad\quad\;\; 0 < x < 2, \quad y > 0 \\ 1 - e^{-y} & \quad\quad\quad\;\; x \ge 2, \quad y > 0 \end{cases}$$

(b)

$$P(X \le 1, Y \le 2) = F(1, 2) = (1/4)(1 - e^{-2})$$

$$P(1 \le X \le 2,\ 3 \le Y \le 4) = \int_{3}^{4} \int_{1}^{2} f(x, y)\, dx\, dy$$

$$= \int_{3}^{4} \int_{1}^{2} (1/2)xe^{-y}\, dx\, dy$$

$$= (3/4)(e^{-3} - e^{-4})$$

EXAMPLE 7.8

Let X and Y be continuous random variables with joint probability density function

$$f(x, y) = \begin{cases} 2xy + (3/2)y^2 & \text{for} \quad 0 < x < 1, \quad 0 < y < 1 \\ 0 & \text{elsewhere} \end{cases}$$

Let C be the region of points (x, y) in the plane $-\infty < x < \infty$, $-\infty < y < \infty$ such that $x + y < 1$. Find $P[(X, Y) \in C]$.

SOLUTION

$$P[(X, Y) \in C] = \iint_C f(x, y)\, dx\, dy = \int_0^1 \int_0^{1-y} [2xy + (3/2)y^2]\, dx\, dy$$

$$= \int_0^1 [(1 - y)^2 y + (3/2)(1 - y)y^2]\, dy = 5/24$$

7.2.4   Marginal Distributions

Let us introduce the concept of a marginal distribution through an example.

EXAMPLE 7.9

Let X and Y be discrete random variables with joint probability function $f(x, y)$. Suppose that the values of $f(x, y)$ are given by the table of values in Table 7.2.

(a)   Find $P(X = 1)$, $P(X = 2)$, $P(Y = 0)$, $P(Y = 1)$.

(b)   Derive a general formula for finding $P(X = x)$ for $x = 1, 2, 3, 4, 5$, and 6.

(c)   Derive a general formula for finding $P(Y = y)$ for $y = 0, 1, 2, 3, 4$, and 5.

(d)   Show that if X and Y are the random variables of Example 7.2, the probabilities derived in (a), (b), and (c) are identical to the probabilities derived in the usual manner for these random variables.

SOLUTION

(a)   The possible values of Y are $0, 1, 2, 3, 4$, and 5. Therefore,

$P(X = 1)$

$= P\left(\begin{array}{l}(X = 1 \text{ and } Y = 0) \text{ or } (X = 1 \text{ and } Y = 1) \text{ or } (X = 1 \text{ and } Y = 2) \\ \text{or } (X = 1 \text{ and } Y = 3) \text{ or } (X = 1 \text{ and } Y = 4) \text{ or } (X = 1 \text{ and } Y = 5)\end{array}\right)$

$= P(X = 1, Y = 0) + P(X = 1, Y = 1) + P(X = 1, Y = 2)$

$\quad + P(X = 1, Y = 3) + P(X = 1, Y = 4) + P(X = 1, Y = 5)$

$= f(1, 0) + f(1, 1) + f(1, 2) + f(1, 3) + f(1, 4) + f(1, 5)$

$= \dfrac{1}{36} + \dfrac{1}{36} + \dfrac{1}{36} + \dfrac{1}{36} + \dfrac{1}{36} + \dfrac{1}{36} = \dfrac{1}{6}$

Following the method for finding $P(X = 1)$, we have

$P(X = 2)$

$$= P\begin{pmatrix} (X = 2 \text{ and } Y = 0) \text{ or } (X = 2 \text{ and } Y = 1) \text{ or } (X = 2 \text{ and } Y = 2) \\ \text{or } (X = 2 \text{ and } Y = 3) \text{ or } (X = 2 \text{ and } Y = 4) \text{ or } (X = 2 \text{ and } Y = 5) \end{pmatrix}$$

$$= P(x = 2, Y = 0) + P(X = 2, Y = 1) + P(X = 2, Y = 2)$$

$$+ P(X = 2, Y = 3) + P(X = 2, Y = 4) + P(X = 2, Y = 5)$$

$$= f(2,0) + f(2,1) + f(2,2) + f(2,3) + f(2,4) + f(2,5)$$

$$= \frac{1}{36} + \frac{2}{36} + \frac{1}{36} + \frac{1}{36} + \frac{1}{36} + \frac{0}{36} = \frac{1}{6}$$

It is now evident, from the calculations above, that since the possible values of X are $1, 2, 3, 4, 5,$ and 6, it follows that

$$P(Y = 0) = P(X = 1, Y = 0) + P(X = 2, Y = 0) + P(X = 3, Y = 0)$$

$$+ P(X = 4, Y = 0) + P(X = 5, Y = 0) + P(X = 6, Y = 0)$$

$$= f(1,0) + f(2,0) + f(3,0) + f(4,0) + f(5,0) + f(6,0)$$

$$= \frac{6}{36} = \frac{1}{6}$$

and

$$P(Y = 1) = f(1,1) + f(2,1) + f(3,1) + f(4,1) + f(5,1) + f(6,1)$$

$$= \frac{10}{36}$$

(b)   It is clear, from the calculations of (a) that a general formula for finding $P(X = x)$ is

$$P(X = x) = \sum_y f(x, y) \qquad \text{for} \quad x = 1, 2, 3, 4, 5, \text{ and } 6$$

The summation is over all possible values y of the random variable Y.

(c)

$$P(Y = y) = \sum_x f(x, y) \qquad \text{for} \quad y = 0, 1, 2, 3, 4, \text{ and } 5$$

The summation is over all possible values x of the random variable X.

(d)   This is left as an exercise for the reader.

The results of (b) and (c) of Example 7.9 provide the formula for the marginal distribution of a random variable, formally stated in the definition that follows.

Definition 7.4

(a)   Let X and Y be discrete random variables with joint probability function $f(x, y)$. The marginal probability functions of X and Y, respectively, are given by

$$f_X(x) = \sum_y f(x, y) \quad \text{and} \quad f_Y(y) = \sum_x f(x, y)$$

(b)   Let X and Y be continuous random variables with joint probability density function $f(x, y)$. The marginal probability density functions of X and Y, respectively, are given by

$$f_X(x) = \int_{-\infty}^{\infty} f(x, y)\, dy \quad \text{and} \quad f_Y(y) = \int_{-\infty}^{\infty} f(x, y)\, dx$$

EXAMPLE 7.10

(a)   Show that if X and Y are discrete random variables, the marginal probability function of X is identical to the probability function of X, and the marginal probability function of Y is identical to the probability function of Y.

(b)   Show that if X and Y are continuous random variables, the marginal probability density function and the probability density function may be used interchangeably in the computation of probabilities for the random variables. That is, show that

$$P(a < X < b) = \int_a^b f(x)\, dx = \int_a^b f_X(x)\, dx$$

SOLUTION

(a)   Let $f(x)$ and $f_X(x)$ be, respectively, the probability function and the marginal probability function of X. Suppose that the possible values of Y are the numbers $y_1, y_2, y_3, \ldots$. Then,

$$f(x) = P(X = x) = P[(X = x) \text{ and } (Y = y_1 \text{ or } y_2 \text{ or } y_3 \text{ or} \ldots)]$$
$$= P(X = x, Y = y_1) + P(X = x, Y = y_2) + P(X = x, Y = y_3) + \cdots$$
$$= f(x, y_1) + f(x, y_2) + f(x, y_3) + \cdots = \sum_y f(x, y) = f_X(x)$$

A similar argument shows that the marginal probability function of Y is identical to the probability function of Y.

(b)   Let $f(x)$ and $f_X(x)$ be, respectively, the probability density function and the marginal probability density function of X. Then,

$$P(a < X < b) = P(a < X < b, -\infty < Y < \infty) = \int_{-\infty}^{\infty} \int_a^b f(x, y)\, dx\, dy$$
$$= \int_a^b \left(\int_{-\infty}^{\infty} f(x, y)\, dy \right) dx = \int_a^b f_X(x)\, dx$$

EXAMPLE 7.11

Consider the continuous random variables X and Y of Example 7.7. Find the marginal probability density functions of X and Y, respectively.

SOLUTION

$$f_X(x) = \int_{-\infty}^{\infty} f(x, y)\, dy = \begin{cases} \int_0^{\infty} (1/2)xe^{-y}\, dy = (1/2)x & \text{for } 0 < x < 2 \\ 0 & \text{elsewhere} \end{cases}$$

$$f_Y(y) = \int_{-\infty}^{\infty} f(x, y)\, dx = \begin{cases} \int_0^2 (1/2)xe^{-y}\, dx = e^{-y} & \text{for } y > 0 \\ 0 & \text{elsewhere} \end{cases}$$

EXAMPLE 7.12

Find the marginal probability density function of the random variable X of Example 7.5. Use this marginal probability density function to find the probability that component A will have an operating life greater than 2 years. (*Note:* the joint probability density function was obtained in Example 7.6)

SOLUTION

The joint probability density function of X and Y is

$$f(x, y) = \begin{cases} e^{-x}e^{-y} & \text{for } x > 0, \ y > 0 \\ 0 & \text{elsewhere} \end{cases}$$

Therefore, the marginal probability density function of X is

$$f_X(x) = \int_{-\infty}^{\infty} f(x, y)\, dy = \begin{cases} \int_0^{\infty} e^{-x}e^{-y}\, dy = e^{-x} & \text{for } x > 0 \\ 0 & \text{elsewhere} \end{cases}$$

and

P(component A will have an operating life greater than 2 years)

$$= P(X > 2) = \int_2^{\infty} f(x)\, dx = \int_2^{\infty} f_X(x)\, dx = \int_2^{\infty} e^{-x}\, dx = e^{-2} = .1353$$

The extension of marginal distributions to more than two random variables is considered in Exercises 18, 19, 20, and 21.

7.2.5   Conditional Distributions

The concept of the conditional distribution is modeled after the concept of conditional probabilities. Suppose that X and Y are discrete random variables with marginal probability functions $f_X(x)$ and $f_Y(y)$, respectively, and joint probability function $f(x, y)$. Let x_0 and y_0 be fixed values, and consider the events $X = x_0$ and $Y = y_0$. Then,

$$P(X = x_0 \mid Y = y_0) = \frac{P[(X = x_0) \cap (Y = y_0)]}{P(Y = y_0)} = \frac{f(x_0, y_0)}{f_Y(y_0)}$$

provided that $P(Y = y_0) > 0$. Now, suppose we let $f_X(x \mid y_0)$ be the function in x defined by

$$f_X(x \mid y_0) = P(X = x \mid Y = y_0) = \frac{f(x, y_0)}{f_Y(y_0)}$$

The function $f_X(x \mid y_0)$ is called the conditional probability function of X, given that the random variable Y is equal to y_0, or simply the conditional probability function of X given $Y = y_0$.

Definition 7.5

(a)   Let X and Y be discrete random variables with marginal probability functions $f_X(x)$ and $f_Y(y)$, respectively, and joint probability function $f(x, y)$. The conditional probability function of X given $Y = y$ and the conditional probability function of Y given $X = x$, respectively, are given by

$$f_X(x \mid y) = \frac{f(x, y)}{f_Y(y)} \quad \text{and} \quad f_Y(y \mid x) = \frac{f(x, y)}{f_X(x)}$$

The function $f_X(x \mid y)$, for fixed value y at which $f_Y(y) > 0$, is defined for all possible values x of the random variable X. Similarly, the function $f_Y(y \mid x)$, for fixed value x at which $f_X(x) > 0$, is defined for all possible values y of the random variable Y.

(b)   Let X and Y be continuous random variables with marginal probability density functions $f_X(x)$ and $f_Y(y)$, respectively, and joint probability density function $f(x, y)$. The conditional probability density function of X given $Y = y$ and the conditional probability density function of Y given $X = x$, respectively, are given by

$$f_X(x \mid y) = \frac{f(x, y)}{f_Y(y)} \quad \text{and} \quad f_Y(y \mid x) = \frac{f(x, y)}{f_X(x)}$$

The function $f_X(x \mid y)$, for fixed value y at which $f_Y(y) > 0$, is defined for $-\infty < x < \infty$. Similarly, the function $f_Y(y \mid x)$, for fixed value x at which $f_X(x) > 0$, is defined for $-\infty\ y < \infty$.

It should be noted here that the conditional probability function, or the conditional probability density function, of a random variable also satisfies the conditions of a probability function or probability density function, respectively. In other words, if $f_X(x \mid y)$ is the conditional probability function of a discrete random variable X given $Y = y$, then $0 \le f_X(x|y) \le 1$ for each possible value x of the random variable X, and $\sum_x f_X(x|y) = 1$; if $f_X(x|y)$ is the conditional probability density function of a continuous random variable X given $Y = y$, then $f_X(x|y) \ge 0$ for $-\infty < x < \infty$, and $\int_{-\infty}^{\infty} f_X(x|y)\, dx = 1$ (see Exercises 22 and 23). The function $f_X(x|y)$, with y as a fixed value, is in fact the probability function, or probability density function, of the random variable X operating under the condition that $Y = y$.

EXAMPLE 7.13

Let X and Y be the continuous random variables of Example 7.8.
   (a)   Find the marginal probability density functions of X and Y, respectively.
   (b)   Find the conditional probability density functions of X and Y, respectively.
   (c)   Find $P(0 \le X \le .5)$ and $P(0 \le Y \le .4)$.
   (d)   If $Y = .3$, what is the probability that X will take on a value that is between 0 and .5?
   (e)   If $X = .2$, what is the probability that Y will take on a value that is between 0 and .4?

   SOLUTION

   (a)

$$f_X(x) = \int_{-\infty}^{\infty} f(x, y)\, dy$$

$$= \begin{cases} \int_0^1 [2xy + (3/2)y^2]\, dy = x + (1/2) & \text{for } 0 < x < 1 \\ \\ 0 & \text{elsewhere} \end{cases}$$

$$f_Y(y) = \int_{-\infty}^{\infty} f(x, y)\, dx$$

$$= \begin{cases} \int_0^1 [2xy + (3/2)y^2]\, dx = y + (3/2)\, y^2 & \text{for } 0 < y < 1 \\ \\ 0 & \text{elsewhere} \end{cases}$$

   (b)   For $0 < y < 1$

$$f_X(x|y) = \frac{f(x, y)}{f_Y(y)}$$

$$= \begin{cases} \dfrac{2xy + (3/2)y^2}{y + (3/2)y^2} = \dfrac{2x + (3/2)y}{1 + (3/2)y} & \text{for } 0 < x < 1 \\ \\ 0 & \text{elsewhere} \end{cases}$$

For $0 < x < 1$

$$f_Y(x \mid y) = \frac{f(x, y)}{f_X(x)} = \begin{cases} \dfrac{2xy + (3/2)y^2}{x + (1/2)} & \text{for} \quad 0 < y < 1 \\ 0 & \text{elsewhere} \end{cases}$$

(c)

$$P(0 \le X \le .5) = \int_0^{.5} f_X(x) \, dx = \int_0^{.5} [x + (1/2)] \, dx = .375$$

$$P(0 \le Y \le .4) = \int_0^{.4} f_Y(y) \, dy = \int_0^{.4} [y + (3/2)y^2] \, dy = .112$$

(d)

$$P(0 \le X \le .5 \mid Y = .3) = \int_0^{.5} f_X(x \mid .3) \, dx = \int_0^{.5} \frac{2x + (3/2)(.3)}{1 + (3/2)(.3)} \, dx = .3276$$

(e)

$$P(0 \le Y \le .4 \mid X = .2) = \int_0^{.4} f_Y(y \mid .2) \, dy = \int_0^{.4} \frac{2(.2) y + (3/2)y^2}{.2 + (1/2)} \, dy = .0914$$

7.2.6   The Multinomial Distribution

The multinomial distribution, which takes its name from the multinomial coefficient $n!/(x_1! x_2! \cdots x_k!)$, is a multivariate probability function of considerable importance. The definition of the distribution and examples illustrating the use of the distribution are given below.

Definition 7.6   Let n and k be positive integers. Let $p_1, p_2, \ldots, p_k$ be numbers such that $\sum_{i=1}^{k} p_i = 1$ and $0 \le p_i \le 1$ for $i = 1, 2, \ldots, k$. The multinomial probability function is the function f defined by

$$f(x_1, x_2, \ldots, x_n) = \frac{n!}{x_1! \, x_2! \cdots x_k!} \, p_1^{x_1} p_2^{x_2} \cdots p_k^{x_k}$$

for nonnegative integers $x_1, x_2, \ldots, x_k$ such that $\sum_{i=1}^{k} x_i = n$.

A frequent use of the multinomial distribution occurs in problems of the following sort. Suppose that $A_1, A_2, \ldots, A_k$, are the k distinct possible outcomes of an experiment, and that $p_1, p_2, \ldots, p_k$ are their respective probabilities. Furthermore, suppose that the experiment is to be performed n times, and X_1 is the number of times the outcome A_1 occurs; X_2 is the number of times the outcome A_2 occurs, ..., and X_k is the number of times the outcome A_k occurs. If we wish to find the probability of exactly x_1 occurrences of the outcome A_1, exactly x_2 occurrences of the outcome A_2, ..., and exactly x_k occurrences of the outcome A_k, then we should use the formula

$$P(\text{exactly } x_i \text{ occurrences of the outcome } A_i \text{ for } i = 1, 2, \ldots, k)$$

$$= P(X_1 = x_1, X_2 = x_2, \ldots, X_k = x_k)$$

$$= \frac{n!}{x_1! x_2! \cdots x_k!} p_1^{x_1} p_2^{x_2} \cdots p_k^{x_k}$$

In other words, we can show that the multivariate probability function of the random variables $X_1, X_2, \ldots$, and X_k, is the multinomial probability function of Definition 7.6.

To show that this formula is correct, let E be the event that each of the first x_1 performances of the experiment results in the outcome A_1, that each of the next x_2 performances of the experiment results in the outcome A_2, ..., and that each of the final x_k performances of the experiment results in the outcome A_k. Because of the property of independence (i.e., the occurrence or nonoccurrence of A_i on any performance of the experiment is independent of the occurrence or nonoccurrence of A_j on any other performance of the experiment—see Section 3.3.2 of Chapter 3), the probability of the event E is clearly $p_1^{x_1} p_2^{x_2} \cdots p_k^{x_k}$. Observe that the event E is only one of a number of ways in which the n performances of the experiment can result in exactly x_1 outcomes of type A_1, exactly x_2 outcomes of type A_2, ..., and exactly x_k outcomes of type A_k. In fact, the distinct number of ways in which this can happen is $n!/(x_1! x_2! \cdots x_k!)$ (i.e., the multinomial coefficient—see Section 2.4 of Chapter 2). Furthermore, each of these distinct ways is an event with probability $p_1^{x_1} p_2^{x_2} \cdots p_k^{x_k}$. Therefore

$$P(X_1 = x_1, X_2 = x_2, \ldots, X_k = x_k)$$

$$= P(\text{exactly } x_i \text{ occurrences of the outcome } A_i \text{ for } i = 1, 2, \ldots, k)$$

$$= \frac{n!}{x_1! x_2! \cdots x_k!} p_1^{x_1} p_2^{x_2} \cdots p_k^{x_k}$$

EXAMPLE 7.14

Past records at a certain clinic showed that $\frac{1}{2}$ of all patients under treatment A benefit from the treatment, $\frac{1}{3}$ of them are not affected by the treatment, and $\frac{1}{6}$ of them encounter adverse side effects. If each of 4 new patients were to be given treatment A, what is the probability that exactly 2 of them will benefit from the treatment, 1 will not be affected by the treatment, and 1 will encounter adverse side effects?

SOLUTION

Consider the experiment of giving treatment A to a patient. Let A_1 be the event that the patient will benefit from the treatment, let A_2 be the event that the patient will not be affected by the treatment, and let A_3 be the event that the patient will encounter adverse side effects. Then, $p_1 = P(A_1) = 1/2$, $p_2 = P(A_2) = 1/3$, and $p_3 = P(A_3) = 1/6$. Now, consider 4 performances of this experiment (i.e., give treatment A to 4 patients), and let X_1 be the number of times A_1 occurs, X_2 be the number of times A_2 occurs, and X_3 be the number of times A_3 occurs. According to the formula given above, the solution to our problem is

$$P(X_1 = 2, X_2 = 1, X_3 = 1) = \frac{4!}{2!\,1!\,1!}(1/2)^2(1/3)^1(1/6)^1 = 1/6$$

7.2.7   The Bivariate Normal Distribution

The bivariate normal distribution is a multivariate distribution of major importance (see, for example, Chapter 14 on regression and correlation).

Definition 7.7   Let $\mu_1, \mu_2, \sigma_1, \sigma_2$, and ρ be constants with $\sigma_1 > 0, \sigma_2 > 0$, and $-1 < \rho < 1$. The function

$$f(x, y) = \frac{e^{-\frac{1}{2(1-\rho^2)}\left[\left(\frac{x-\mu_1}{\sigma_1}\right)^2 - 2\rho\left(\frac{x-\mu_1}{\sigma_1}\right)\left(\frac{y-\mu_2}{\sigma_2}\right) + \left(\frac{y-\mu_2}{\sigma_2}\right)^2\right]}}{2\pi\sigma_1\sigma_2\sqrt{1-\rho^2}} \qquad \begin{array}{l} \text{for } -\infty < x < \infty, \\ -\infty < y < \infty \end{array}$$

is called the **bivariate normal probability density function.**

Random variables X and Y are said to be jointly normally distributed random variables and their joint distribution is said to be bivariate normal if their joint probability density function $f(x, y)$ is a bivariate normal probability density function. It can be shown (see ref. 1) that if the joint probability density function of X and Y is the bivariate normal probability density function defined above, then X and Y are normal random variables with means equal to μ_1 and μ_2, respectively, and with standard deviations equal to σ_1 and σ_2, respectively. The constant ρ is the correlation coefficient of X and Y. (See Chapter 8 for the definition of correlation coefficients.) Furthermore, it can be shown that if X and Y are bivariate normal random variables, they are independent if and only if $\rho = 0$. (See Section 7.3 for the definition of independent random variables.)

Exercises

1. Consider the experiment of a single roll of a die and a single toss of 3 coins. Let X_1 be the value on the die, and let X_2 be the number of heads. Find the joint probability function of X_1 and X_2.

2. Consider two throws of a balanced die. Let X be the value on the second throw, and let Y be the value on the first throw minus the value on the second throw. Find the joint probability function of X, Y.

3. Among 4 identical pieces of luggage delivered to a certain airport, there are 2 that contain a time bomb. The pieces of luggage are to be searched one at a time until the bombs are located. Let X be the number of pieces of luggage searched until the first bomb is found, and let Y be the number of additional pieces searched until the second bomb is found. Find the joint probability function of X and Y.

4. Consider the random variables of Exercise 2. Find $F(1, 3)$, $F(2, 4)$, $F(5, 3)$, $F(-1, -1)$, $F(3, 0)$, $F(-3, 2)$, $F(5, -4)$.

5. Let X and Y be continuous random variables with joint cumulative distribution

$$F(x, y) = \begin{cases} 0 & \text{for} \quad x \le 0, \quad -\infty < y < \infty \\ 0 & -\infty < x < \infty, \quad y \le 0 \\ x^2 y^2 & 0 < x < 1, \quad 0 < y < 1 \\ x^2 & 0 < x < 1, \quad y \ge 1 \\ y^2 & x \ge 1, \quad 0 < y < 1 \\ 1 & x \ge 1, \quad y \ge 1 \end{cases}$$

Find the probability of each of the following events:
(a) Both random variables take on values that are less than or equal to .5.
(b) X takes on a value that is less than or equal to .5, and Y takes on a value that is not greater than .75.
(c) Both random variables take on values that are greater then .75.
(d) Both random variables take on values that are greater than 2.

6. Find the joint probability density function of the random variables X and Y of Exercise 5.

7. Consider the random variables X and Y of Exercise 6.
(a) Find the probability that X will take on a value that is between .25 and .5 and that Y will take on a value that is between 0 and .5.
(b) Find the probability that the sum of the values of X and Y will be less than 1.
(c) Find the probability that the sum of the values of X and Y will be between .5 and 1.

8. Suppose that X and Y are continuous random variables with joint probability density function

$$f(x, y) = \begin{cases} (4/9)xy & \text{for} \quad 1 < x < 2, \quad 1 < y < 2 \\ 0 & \text{elsewhere} \end{cases}$$

(a) Find the joint cumulative distribution function of X and Y.

(b) Find $P(X \le 1, \quad Y \le .5)$, $P(1.5 \le X \le 1.75, \quad 1.5 \le Y \le 1.75)$, $P(X > 1.2, Y > 1.4)$.
(c) Find the probability that the value of X will be greater than the value of Y.

9. Let X and Y be discrete random variables with joint probability function f defined by the following values: $f(-1,0) = .2$, $f(-1,1) = .1$, $f(-1,2) = .05$, $f(0,0) = .05$, $f(0,1) = .2$, $f(0,2) = .3$, $f(3,0) = .02$, $f(3,1) = .04$, and $f(3,2) = .04$.
   (a) Find the marginal probability function of X.
   (b) Find the marginal probability function of Y.
   (c) Find $P(-1 \le X \le 2)$, $P(-1 \le Y \le 2)$.

10. Let X and Y be the continuous random variables of Exercise 8.
    (a) Find the marginal probability density function of X.
    (b) Find the marginal probability density function of Y.
    (c) Find $P(-1 \le X \le 1.5)$, $P(1.5 \le Y \le 1.75)$.

11. Let X and Y be the discrete random variables of Exercise 9.
    (a) Find the conditional probability function of X given $Y = 1$.
    (b) Find the conditional probability function of Y given $X = 3$.
    (c) If $Y = 1$, what is the probability that X will take on a value that is less than 2?
    (d) If $X = 0$, what is the conditional probability that Y will take on a value that is greater than 0?

12. Let X and Y be the continuous random variables of Exercise 8.
    (a) Find the conditional probability density function of X given $Y = 1.5$.
    (b) Find the conditional probability density function of Y given $X = 1.4$.
    (c) Find the conditional probability of X taking on a value that is between -1 and 1.5 if $Y = 1.5$.
    (d) Find the conditional probability of Y taking on a value that is between 1.5 and 1.75 if $X = 1.4$.

13. Suppose that the probabilities are, respectively, .60, .35, and .05 that a call received at a certain telephone of a department store is to place an order, to obtain information, and is a wrong number. Find the probability that among 10 such calls, 5 will be to place an order, 3 will be for information, and 2 will be wrong numbers.

14. If 45%, 55%, and 5% of voters will vote for candidates A, B, and C, respectively, then what is the probability that of 6 randomly selected voters, 4 will vote for candidate A, 1 will vote for candidate B, and 1 will vote for candidate C?

15. In 6 throws of a die, what is the probability of obtaining each value exactly once?

16. In 7 throws of a die, what is the probability of obtaining each value at least once?

17. If 35%, 35%, and 30% of a certain shipment of tulip bulbs will produce, respectively, red, yellow, and white flowers, what is the probability that 3 randomly selected bulbs will produce flowers of all available colors?

18. (If X_1, X_2, and X_3 are discrete random variables with multivariate probability function $f(x_1, x_2, x_3)$, their respective marginal probability functions are

$$f_{X_1}(x_1) = \sum_{x_3} \sum_{x_2} f(x_1, x_2, x_3),$$

$$f_{X_2}(x_2) = \sum_{x_3} \sum_{x_1} f(x_1, x_2, x_3)$$

and

$$f_{X_3}(x_3) = \sum_{x_2} \sum_{x_1} f(x_1, x_2, x_3)$$

The respective joint marginal probability functions of X_1 and X_2, of X_1 and X_3, and of X_2 and X_3 are

$$f_{X_1, X_2}(x_1, x_2) = \sum_{x_3} f(x_1, x_2, x_3),$$

$$f_{X_1, X_3}(x_1, x_3) = \sum_{x_2} f(x_1, x_2, x_3)$$

and

$$f_{X_2, X_3}(x_2, x_3) = \sum_{x_1} f(x_1, x_2, x_3)$$

(This definition is extended to any finite number of discrete random variables.)) Suppose that X_1, X_2, and X_3 are discrete random variables with multivariate probability function given by

$$f(x_1, x_2, x_3) = \frac{x_1 x_2 x_3}{162}$$

for $x_1 = 1, 2, 3;$ $x_2 = 1, 2;$ $x_3 = 4, 5$

(a) Find the marginal probability function of X_1.
(b) Find the joint marginal probability function of X_2 and X_3.

19. Suppose that X_1, X_2, and X_3 are discrete random variables with multivariate probability function given by

$$f(x_1, x_2, x_3) = \frac{x_1 + x_2 + x_3}{78}$$

for $x_1 = 1, 2, 3;$ $x_2 = 1, 2;$ $x_3 = 2, 4$

(a) Find the marginal probability function of X_2.
(b) Find the joint marginal probability function of X_1 and X_3.

20. If X_1, X_2, and X_3 are continuous random variables with multivariate probability density function $f(x_1, x_2, x_3)$, their respective marginal

probability functions are

$$f_{X_1}(x_1) = \int_{-\infty}^{\infty} \int_{-\infty}^{\infty} f(x_1, x_2, x_3) \, dx_2 \, dx_3,$$

$$f_{X_2}(x_2) = \int_{-\infty}^{\infty} \int_{-\infty}^{\infty} f(x_1, x_2, x_3) \, dx_1 \, dx_3$$

and

$$f_{X_3}(x_3) = \int_{-\infty}^{\infty} \int_{-\infty}^{\infty} f(x_1, x_2, x_3) \, dx_1 \, dx_2$$

The respective joint marginal probability density functions of X_1 and X_2, of X_1 and X_3, and of X_2 and X_3 are

$$f_{X_1, X_2}(x_1, x_2) = \int_{-\infty}^{\infty} f(x_1, x_2, x_3) \, dx_3,$$

$$f_{X_1, X_3}(x_1, x_3) = \int_{-\infty}^{\infty} f(x_1, x_2, x_3) \, dx_2$$

and

$$f_{X_2, X_3}(x_2, x_3) = \int_{-\infty}^{\infty} f(x_1, x_2, x_3) \, dx_1$$

(This definition is extended to any finite number of continuous random variables.) Suppose that X_1, X_2, and X_3 are continuous random variables with multivariate probability density function given by

$$f(x_1, x_2, x_3) = \begin{cases} 8x_1 x_2 x_3 & \text{for } 0 < x_1 < 1, \\ & \quad\ 0 < x_2 < 1, \\ & \quad\ 0 < x_3 < 1 \\ 0 & \text{elsewhere} \end{cases}$$

(a) Find the marginal probability density function of X_2.
(b) Find the joint marginal probability density function of X_1 and X_3.

21. Suppose that X_1, X_2, and X_3 are continuous random variables with multivariate probability density function given by

$$f(x_1, x_2, x_3) = \begin{cases} \dfrac{2(x_1 + x_2 + x_3)}{3} & \text{for } 0 < x_1 < 1, \\ & \quad\ 0 < x_2 < 1, \\ & \quad\ 0 < x_3 < 1 \\ 0 & \text{elsewhere} \end{cases}$$

(a) Find the marginal probability density function of X_3.
(b) Find the joint marginal probability density function of X_1 and X_2.

22.   Let $f_X(x \mid y)$ be the conditional probability function of a discrete random variable X given $Y = y$. Show that the following statements are true.

(a)

$$0 \le f_X(x \mid y) \le 1 \qquad \text{for each possible value } x \text{ of } X$$

(b)

$$\sum_x f_X(x \mid y) = 1$$

(Hint: $f_X(x \mid y) = [f(x, y)]/f_Y(y)$.)

23.   Let $f_X(x \mid y)$ be the conditional probability density function of a continuous random variable X given $Y = y$. Show that the following statements are true.

(a)

$$f_X(x \mid y) \ge 0 \qquad \text{for} \quad -\infty < x < \infty$$

(b)

$$\int_{-\infty}^{\infty} f_X(x \mid y) \, dx = 1$$

(Hint: $f_X(x \mid y) = [f(x, y)]/f_Y(y)$.)

24.   Let X and Y be continuous random variables with joint probability density function given by

$$f(x, y) = \begin{cases} e^{-y} & \text{for} \quad 0 < x < 1, \quad y > 0 \\ 0 & \text{elsewhere} \end{cases}$$

(a)   Find the conditional probability density functions of X and Y, respectively.
(b)   If $Y = 3$, what is the probability that X will take on a value that is between 0 and .5?
(c)   If $X = .5$, what is the probability that Y will take on a value that is between 1 and 4?

7.3   Independent Random Variables

As we shall see in Chapter 8, some of the most important theorems in probability and statistics are concerned with independent random variables. Informally speaking, two random variables are independent if the values of either random variable do not affect in any way the values of the other random variable. For example, if X_1 is the value of the first throw of a die and X_2 is the value of the second throw of the die, then X_1 and X_2 are independent random variables. The formal definition of independent random variables is stated in terms of the distributions of the random variables. We shall see in the sections to follow that the formal definition and the informal definition are in agreement.

7.3.1   Definition of Independent Random Variables

Definition 7.8   Let $X_1, X_2, \ldots, X_n$ be random variables (discrete or continuous) with probability functions or probability density functions $f_1(x_1), f_2(x_2), \ldots, f_n(x_n)$, respectively, and multivariate probability function or probability density function $f(x_1, x_2, \ldots, x_n)$. The n random variables are said to be **independent** if

$$f(x_1, x_2, \ldots, x_n) = f_1(x_1)f_2(x_2)\cdots f_n(x_n)$$

for all possible values $x_1, x_2, \ldots, x_n$ of the random variables $X_1, X_2, \ldots, X_n$.

EXAMPLE 7.15

Use Definition 7.8 to show that the random variables X_1 and X_2 of Example 7.1 are independent.

SOLUTION

The respective probability functions of X_1 and X_2 are

$$f_1(x_1) = P(X_1 = x_1) = 1/6 \quad \text{for} \quad x_1 = 1, 2, 3, 4, 5, 6$$

and

$$f_2(x_2) = P(X_2 = x_2) = 1/2 \quad \text{for} \quad x_2 = 0, 1$$

From Example 7.1, we see that the joint probability function is

$$f(x_1, x_2) = 1/12 \quad \text{for} \quad x_1 = 1, 2, 3, 4, 5, 6 \quad \text{and} \quad x_2 = 0, 1$$

Therefore

$$f(x_1, x_2) = 1/12 = (1/6)(1/2) = f_1(x_1)f_2(x_2)$$

for all possible values of x_1 and x_2. The last statement is the condition for independence given in Definition 7.8.

EXAMPLE 7.16

Are the random variables X and Y of Example 7.13 independent?

SOLUTION

The answer is no. In this example

$$f_X(x) = \begin{cases} x + (1/2) & \text{for } 0 < x < 1 \\ 0 & \text{elsewhere} \end{cases}$$

$$f_Y(y) = \begin{cases} y + (3/2)y^2 & \text{for } 0 < y < 1 \\ 0 & \text{elsewhere} \end{cases}$$

$$f(x, y) = \begin{cases} 2xy + (3/2)y^2 & \text{for } 0 < x < 1, \quad 0 < y < 1 \\ 0 & \text{elsewhere} \end{cases}$$

Therefore, $f(x, y) \neq f_X(x)f_Y(y)$ and the conditions of independence are not met.

7.3.2 The Meaning of Independence

As it stands, the formal definition of independence sheds little light on the meaning of independence of random variables. A closer look, however, shows that it is in agreement with the informal definition given at the beginning of Section 7.3.1.

Let us consider the case of two independent discrete random variables X_1 and X_2. Suppose that $f_1(x_1)$ and $f_2(x_2)$ are their respective probability functions and that $f(x_1, x_2)$ is their joint probability function. The condition of independence is that

$$f(x_1, x_2) = f_1(x_1)f_2(x_2)$$

for all possible values of x_1 and x_2. Now, since we are considering discrete random variables, by definition we have

$$f_1(x_1) = P(X_1 = x_1), \qquad f_2(x_2) = P(X_2 = x_2)$$

and

$$f(x_1, x_2) = P(X_1 = x_1, X_2 = x_2)$$

for all possible values of x_1 and x_2.

Let A be the event that $X_1 = x_1$, and let B be the event that $X_2 = x_2$. Then

$$P(A) = P(X_1 = x_1) = f_1(x_1), \qquad P(B) = P(X_2 = x_2) = f_2(x_2)$$

and

$$P(A \cap B) = P(X_1 = x_1, X_2 = x_2) = f(x_1, x_2)$$

We see, then, that the condition of independence for X_1 and X_2 is equivalent to the condition that

$$P(A \cap B) = P(A)P(B)$$

This last equation is the condition of independence of events A and B. Thus, we see that the random variables X_1 and X_2 are independent if and only if, for all possible values of x_1 and x_2, the events $(X_1 = x_1)$ and $(X_2 = x_2)$ are independent events. This phenomenon relates independence of random variables with independence of events. This is also the meaning we have in mind when we say that two random variables are independent if the values of either random variable do not affect in any way the values of the other random variable.

7.3.3   Independent Random Variables Derived from Independent Experiments

Although Definition 7.8 is the formal definition of independent random variables, in practical applications we seldom refer to that definition when we are checking for independence of random variables. More likely, we will encounter random variables $X_1, X_2, \ldots, X_n$ that satisfy either condition (a) or (b) that follow. When such is the case, the random variables can be assumed to be independent random variables.

(a) $\mathscr{E}$ is an experiment. S is a sample space for $\mathscr{E}$. X is a random variable defined on S. $\mathscr{E}$ is repeated n times under identical conditions. The random variables $X_1, X_2, \ldots, X_n$ are independent random variables if X_1 is the random variable X on the first performance of $\mathscr{E}$, X_2 is the random variable X on the second performance of $\mathscr{E}, \ldots$, and X_n is the random variable X on the nth performance of $\mathscr{E}$.

(b) $\mathscr{E}_1, \mathscr{E}_2, \ldots, \mathscr{E}_n$ are separate and distinct experiments. The outcome of $\mathscr{E}_i$ is in no way dependent on the outcome of $\mathscr{E}_j$, for $i \neq j$. $S_1, S_2, \ldots, S_n$ are the respective sample spaces for $\mathscr{E}_1, \mathscr{E}_2, \ldots, \mathscr{E}_n$. If $X_1, X_2, \ldots, X_n$ are random variables defined on $S_1, S_2, \ldots,$ and S_n, respectively, then they are independent random variables.

It can be shown that if $X_1, X_2, \ldots, X_n$ are random variables satisfying either condition (a) or (b), then they are independent random variables satisfying the conditions of Definition 7.8. (See Ref. 3.)

EXAMPLE 7.17

Let $\mathscr{E}$ be the experiment of selecting at random a student from Queens College, and let X be the grade point average of the student selected. Suppose that the experiment is to be repeated n times under identical conditions, and X_i is the grade point average of the student selected on the ith performance of $\mathscr{E}$. (Note that this implies that the same student may be selected more than once.) Show that the random variables $X_1, X_2, \ldots, X_n$ are independent random variables.

SOLUTION

The random variables satisfy condition (a) above and are therefore independent random variables.

EXAMPLE 7.18

Consider the experiment of tossing two coins and throwing one die. Let X_1 be the number of heads and let X_2 be the value on the die. Show that X_1 and X_2 are independent random variables.

SOLUTION

We can consider the tossing of two coins as one experiment, and the throwing of the die as a second separate and distinct experiment. It is clear that outcomes of either experiment do not affect in any way the outcome of the other experiment. X_1 is a random variable related to the first experiment alone, and X_2 is a random variable related to the second experiment alone. Therefore, according to condition (b) above the random variables are independent.

Exercises

1. Are the random variables of Exercise 9 of Section 7.2 independent random variables? Explain.

2. Are the random variables of Exercise 8 of Section 7.2 independent random variables? Explain.

3. Check the random variables of Section 7.2 Exercises 18, 19, 20, and 21. Which of these sets of random variables are independent random variables? Explain.

4. Consider 3 throws of a die. Let X_1, X_2, and X_3 be the values of the first, second and third throws, respectively. Are the random variables X_1, X_2, and X_3 independent random variables? Explain.

5. Let X_1 be the annual income of a randomly selected family from Phoenix, Arizona; let X_2 be the IQ of a University of California student selected at random; and let X_3 be the number of heads in 10 tosses of a coin. Are the random variables X_1, X_2, and X_3 independent random variables? Explain.

6. Suppose that if X_1 is the time in hours until a failure occurs in Computer System I, then the distribution of X_1 is exponential, with parameter $\theta_1 = 35$; suppose also that if X_2 is the time in hours until a failure occurs in Computer System II, then the distribution of X_2 is exponential, with parameter $\theta_2 = 24$. If X_1 and X_2 are independent random variables, what is the joint probability density function of X_1 and X_2?

7. With reference to Exercise 6, what is the probability that neither Computer System I nor Computer System II will fail within a period of 2 days?

References

Reference 1 contains a comprehensive discussion of the bivariate normal distribution. Reference 2 contains some excellent exercises on jointly distributed discrete random variables. Reference 3 contains some excellent exercises on jointly distributed discrete as well as continuous random variables.

1. De Groot, M. H. 1975. *Probability and statistics*. Menlo Park, CA: Addison-Wesley Publishing Co.
2. Feller, William. 1968. *An introduction to probability theory and its applications*, Vol. I, 3d ed. New York: John Wiley & Sons, Inc.
3. Ross, Sheldon. 1976. *A first course in probability*. New York: Macmillan Publishing Co., Inc.

Chapter *8*

SUMS OF RANDOM VARIABLES, THE CENTRAL LIMIT THEOREM

8.1   Introduction

Applied problems in probability and statistical inference often involve sums (and averages) of several random variables. Relevant statements about them are generally the result of a number of theorems specifically concerned with sums of random variables. Of these theorems, the two most important ones are the strong law of large numbers—discussed in Chapter 6—and the central limit theorem. This chapter concerns sums (and averages) of random variables, distributions of sums of random variables, and a discussion of the central limit theorem.

8.2   Sums of Random Variables

Definition 8.1   Let $X_1, X_2, \ldots, X_n$ be random variables defined on a sample space S. For each point s in S define

$$Y(s) = X_1(s) + X_2(s) + \cdots + X_n(s)$$

The random variable Y is called the **sum of the random variables** X_1, $X_2, \ldots, X_n$, and we write

$$Y = X_1 + X_2 + \cdots + X_n$$

Remark: it should be mentioned that a set of random variables X_1, $X_2, \ldots, X_n$ can always be interpreted as random variables defined on the same sample space S. (See Exercise 1.)

EXAMPLE 8.1

The following are examples of random variables that are sums of random variables.

(a)   Consider n Bernoulli trials. Let $X_1, X_2, \ldots, X_n$ be random variables defined by

$$X_i = \begin{cases} 1 & \text{if the } i\text{th trial is a success} \\ 0 & \text{if the } i\text{th trial is a failure} \end{cases}$$

Let $S_n = X_1 + X_2 + \cdots + X_n$. The random variable S_n is a sum of random variables and is the number of successes in n Bernoulli trials.

(b)   A certain insurance company handles life, fire, and hospital insurance policies. The number of claims arriving at the company office in any one day in each of the categories are $X_1, X_2,$ and X_3, respectively. Let $Y = X_1 + X_2 + X_3$. The random variable Y is a sum of random variables and is the total number of claims arriving at the office in any one day.

Exercises

1. Let X_1 and X_2 be discrete random variables defined on sample spaces S_1 and S_2, respectively. Show that there exist random variables X_1^* and X_2^*, and a sample space S such that (a) S is a sample space for both X_1^* and X_2^*, and (b) the distributions of X_1^* and X_2^* are identical to the distributions of X_1 and X_2, respectively. As a consequence, X_1 and X_2 may be interpreted as random variables defined on the same sample space. (*Hint:* let S be the set of all elements (s_1, s_2) such that s_1 is an element of S_1 and s_2 is an element of S_2. Let $X_1^*(s_1, s_2) = X_1(s_1)$ and $X_2^*(s_1, s_2) = X_2(s_2)$.)

2. Let X_1 be the numerical outcome of one roll of a die. Let X_2 be the number of heads in 1 toss of a coin. Find a sample space S such that both X_1 and X_2 are random variables defined on S.

3. Give three examples of random variables that are sums of random variables.

8.3   Means and Variances of Sums of Random Variables

Since a sum of random variables is itself a random variable, it is meaningful to speak of the mean and variance of a sum of random variables. In other words, if $Y = X_1 + X_2 + \cdots + X_n$, then it is meaningful to speak of $E(Y)$ and $\text{Var}(Y)$. Depending on the known properties of Y, there are a number of ways to find $E(Y)$ and $\text{Var}(Y)$. If the multivariate distribution of $X_1, X_2, \ldots,$ and X_n is given, one possible method is to apply the following definition.

Definition 8.2   Let $X_1, X_2, \ldots, X_n$ be random variables with multivariate probability, or probability density function $f(x_1, x_2, \ldots, x_n)$.

(a)   If the random variables are discrete random variables, then

$$E(X_1 + X_2 + \cdots + X_n)$$

$$= \sum_{x_n} \cdots \sum_{x_2} \sum_{x_1} (x_1 + x_2 + \cdots + x_n) f(x_1, x_2, \ldots, x_n)$$

provided that $\sum_{x_n} \cdots \sum_{x_2} \sum_{x_1} (|x_1| + |x_2| + \cdots + |x_n|) f(x_1, x_2, \ldots, x_n) < \infty$, and

$$\text{Var}(X_1 + X_2 + \cdots + X_n)$$

$$= \sum_{x_n} \cdots \sum_{x_2} \sum_{x_1} (x_1 + x_2 + \cdots + x_n - \mu)^2 f(x_1, x_2, \ldots, x_n)$$

provided that the sum exists and $\mu = E(X_1 + X_2 + \cdots + X_n)$ exists. The summation is over all values x_1 that are possible values of X_1, all values x_2 that are possible values of $X_2, \ldots,$ and all values x_n that are possible values of X_n.

(b)   If the random variables are continuous random variables, then

$$E(X_1 + X_2 + \cdots + X_n)$$

$$= \int_{-\infty}^{\infty} \cdots \int_{-\infty}^{\infty} (x_1 + x_2 + \cdots + x_n) f(x_1, x_2, \ldots, x_n)\, dx_1 \cdots dx_n$$

provided that the integral exists, and

$$\mathrm{Var}(X_1 + X_2 + \cdots + X_n)$$

$$= \int_{-\infty}^{\infty} \cdots \int_{-\infty}^{\infty} (x_1 + x_2 + \cdots + x_n - \mu)^2 f(x_1, x_2, \ldots, x_n)\, dx_1 \cdots dx_n$$

provided that the integral exists and $\mu = E(X_1 + X_2 + \cdots + X_n)$ exists.

EXAMPLE 8.2

Let X_1 and X_2 be discrete random variables with probability function f defined by the following values: $f(-1,0) = .5$, $f(-1,1) = .2$, $f(2,0) = .1$, and $f(2,1) = .1$. Find the mean and variance of $X_1 + X_2$.

SOLUTION

$$E(X_1 + X_2)$$
$$= (-1+0)f(-1,0) + (-1+1)f(-1,1)$$
$$\quad + (2+0)f(2,0) + (2+1)f(2,1)$$
$$= 0$$
$$\mathrm{Var}(X_1 + X_2)$$
$$= (-1+0-0)^2 f(-1,0) + (-1+1-0)^2 f(-1,1)$$
$$\quad + (2+0-0)^2 f(2,0) + (2+1-0)^2 f(2,1)$$
$$= 2$$

In terms of methods for finding the mean and variance of a sum of random variables, the following two theorems are perhaps more useful than the definition.

Theorem 8.1
(a)   *Let $X_1, X_2, \ldots, X_n$ be random variables with finite expectations. Then*

$$E(X_1 + X_2 + \cdots + X_n) = E(X_1) + E(X_2) + \cdots + E(X_n)$$

(b)   *More generally*

$$E(a_1 X_1 + a_2 X_2 + \cdots + a_n X_n + a_{n+1})$$
$$= a_1 E(X_1) + a_2 E(X_2) + \cdots + a_n E(X_n) + a_{n+1}$$

for constants $a_1, a_2, \ldots, a_{n+1}$.

Proof   Here we shall consider only the case of two discrete random variables. The proof of the theorem for more than two discrete random variables is simply a generalization of this case. The proof for continuous random variables is considered in Exercise 8.

By definition

$$E(X_1 + X_2) = \sum_{x_2} \sum_{x_1} (x_1 + x_2) f(x_1, x_2)$$

$$= \sum_{x_2} \sum_{x_1} x_1 f(x_1, x_2) + \sum_{x_2} \sum_{x_1} x_2 f(x_1, x_2)$$

Now observe that

$$\sum_{x_2} \sum_{x_1} x_2 f(x_1, x_2) = \sum_{x_2} x_2 \sum_{x_1} f(x_1, x_2) = \sum_{x_2} x_2 f_{X_2}(x_2) = E(X_2)$$

By symmetry

$$\sum_{x_2} \sum_{x_1} x_1 f(x_1, x_2) = E(X_1)$$

Therefore

$$E(X_1 + X_2) = E(X_1) + E(X_2)$$

The proof of (b) is considered in Exercise 9.

Theorem 8.2   *Let* $X_1, X_2, \ldots, X_n$ *be independent random variables with finite means and variances. Then,*

(a)

$$\mathrm{Var}(X_1 + X_2 + \cdots + X_n) = \mathrm{Var}(X_1) + \mathrm{Var}(X_2) + \cdots + \mathrm{Var}(X_n)$$

and

(b)

$$\mathrm{Var}(a_1 X_1 + a_2 X_2 + \cdots + a_n X_n + a_{n+1})$$
$$= a_1^2 \mathrm{Var}(X_1) + a_2^2 \mathrm{Var}(X_2) + \cdots + a_n^2 \mathrm{Var}(X_n)$$

for constants $a_1, a_2, \ldots, a_{n+1}$.

Proof   Here we shall consider only the case of two continuous random variables. See Exercise 10 for an application to discrete random variables.

Let $\mu_1 = E(X_1)$, $\mu_2 = E(X_2)$, and $\mu = E(X_1 + X_2) = \mu_1 + \mu_2$. Then

$$\text{Var}(X_1 + X_2) = \int_{-\infty}^{\infty} \int_{-\infty}^{\infty} (x_1 + x_2 - \mu)^2 f(x_1, x_2) \, dx_1 \, dx_2$$

$$= \int_{-\infty}^{\infty} \int_{-\infty}^{\infty} (x_1 + x_2 - \mu_1 - \mu_2)^2 f(x_1, x_2) \, dx_1 \, dx_2$$

$$= \int_{-\infty}^{\infty} \int_{-\infty}^{\infty} (x_1 - \mu_1)^2 f(x_1, x_2) \, dx_1 \, dx_2$$

$$+ \int_{-\infty}^{\infty} \int_{-\infty}^{\infty} (x_2 - \mu_2)^2 f(x_1, x_2) \, dx_1 \, dx,$$

$$+ \int_{-\infty}^{\infty} \int_{-\infty}^{\infty} 2(x_1 - \mu_1)(x_2 - \mu_2) f(x_1, x_2) \, dx_1 \, dx_2$$

Now, observe that by independence of X_1 and X_2, we have

$$\int_{-\infty}^{\infty} \int_{-\infty}^{\infty} (x_2 - \mu_2)^2 f(x_1, x_2) \, dx_1 \, dx_2$$

$$= \int_{-\infty}^{\infty} \int_{-\infty}^{\infty} (x_2 - \mu_2)^2 f_{X_1}(x_1) f_{X_2}(x_2) \, dx_1 \, dx_2$$

$$= \int_{-\infty}^{\infty} (x_2 - \mu_2)^2 f_{X_2}(x_2) \, dx_2 \int_{-\infty}^{\infty} f_{X_1}(x_1) \, dx_1$$

$$= \int_{-\infty}^{\infty} (x_2 - \mu_2)^2 f_{X_2}(x_2) \, dx_2$$

$$= \text{Var}(X_2)$$

By symmetry, we also have $\int_{-\infty}^{\infty} \int_{-\infty}^{\infty} (x_1 - \mu_1)^2 f(x_1, x_2) \, dx_1 \, dx_2 = \text{Var}(X_1)$. It can easily be shown (see Exercise 11) that

$$\int_{-\infty}^{\infty} \int_{-\infty}^{\infty} 2(x_1 - \mu_1)(x_2 - \mu_2) f(x_1, x_2) \, dx_1 \, dx_2 = 0$$

Therefore

$$\text{Var}(X_1 + X_2) = \text{Var}(X_1) + \text{Var}(X_2)$$

The proof of (b) is considered in Exercise 12.

EXAMPLE 8.3

The average weight of a certain type of apple is 3.5 ounces, and the standard deviation is .5 ounces. If these apples are packaged in gift boxes of 10 apples per box, and Y is the weight of apples in one of these boxes, what is $E(Y)$ and $\text{Var}(Y)$?

SOLUTION

Let $X_1, X_2, \ldots, X_{10}$ be the weights of 10 randomly selected apples. Then, $X_1, X_2, \ldots, X_{10}$ are independent randoms. All of these random variables have means equal to 3.5 and standard deviations equal to .5. Therefore, by Theorem 8.1 and Theorem 8.2 we have

$$E(Y) = E(X_1 + X_2 + \cdots + X_{10}) = E(X_1) + E(X_2) + \cdots + E(X_{10})$$
$$= 35 \text{ ounces}$$
$$\text{Var}(Y) = \text{Var}(X_1 + X_2 + \cdots + X_{10}) = \text{Var}(X_1) + \cdots + \text{Var}(X_{10})$$
$$= 2.5 \text{ ounces}$$

EXAMPLE 8.4

A certain restaurant is open daily for lunch and dinner. On any one day, the net take from lunches is a random variable L with a mean of $200 and a standard deviation of $50. The net take from dinners is a random variable D with a mean of $500 and a standard deviation of $100. If the owner's profit is 20% on lunches and 30% on dinners, what is the average combined daily profit from both lunches and dinners? What is the standard deviation? (Assume that L and D are independent random variables.)

SOLUTION

Let Y be the daily profit. Then, $Y = .2L + .3D$. From Theorem 8.1 and Theorem 8.2 we have

$$E(Y) = .2E(L) + .3E(D) = \$190$$

and

$$\sqrt{\text{Var}(Y)} = \sqrt{(.2)^2 \text{Var}(L) + (.3)^2 \text{Var}(D)}$$
$$= \$31.62$$

Therefore, on the average, the daily combined profit from both lunches and dinners is $190, and the standard deviation is $31.62.

EXAMPLE 8.5

The following seemingly difficult problem becomes easy when we apply Theorem 8.1.

Assume that birthdays of people in general are equally distributed among the 12 months. In a class of n students what is

(a)   the expected number of months that have birthdays of exactly k students;

(b)   the expected number of months that have birthdays of more than 1 student.

SOLUTION

(a) Let N_k = the number of months that have birthdays of exactly k students. Let

$$X_i = \begin{cases} 1 & \text{if the } i\text{th month has birthdays of exactly } k \text{ students} \\ 0 & \text{otherwise} \end{cases}$$

Then, $N_k = X_1 + X_2 + \cdots + X_{12}$.

By Theorem 8.1, we have

$$E(N_k) = E(X_1) + E(X_2) + \cdots + E(X_{12})$$

Computing for the mean of X_i, we have

$$\begin{aligned} E(X_i) &= 1 \cdot P(X_i = 1) + 0 \cdot P(X_i = 0) \\ &= 1 \cdot P(i\text{th month has birthdays of exactly } k \text{ students}) \\ &= 1 \cdot b(k; n, 1/12) \end{aligned}$$

Therefore, the solution to (a) is

$$E(N_k) = 12 \cdot b(k; n, 1/12)$$

(b) Let N = the number of months that have birthdays of more than 1 student. Clearly, $N = N_2 + N_3 + \cdots + N_{12} = 12 - N_0 - N_1$. By applying Theorem 8.1, we see that the solution to (b) is

$$\begin{aligned} E(N) &= 12 - E(N_0) - E(N_1) \\ &= 12 - 12 \cdot b(0; n, 1/12) - 12 \cdot b(1; n, 1/12) \end{aligned}$$

Exercises

1. Let X_1 and X_2 be discrete random variables with joint probability function f defined by the following values: $f(-2, -1) = .3$, $f(-2, 0) = .1$, $f(-2, 1) = .05$, $f(1, -1) = .05$, $f(1, 0) = .4$, and $f(1, 1) = .1$. Find the mean and variance of $X_1 + X_2$.

2. Let X_1 and X_2 be continuous random variables with joint probability density function given by

$$f(x_1, x_2) = \begin{cases} (4/3)x_1 x_2 & \text{for } 1 < x_1 < 2, \ 1 < x_2 < 2 \\ 0 & \text{elsewhere} \end{cases}$$

Find the mean and variance of $X_1 + X_2$.

3.   Let X_1 and X_2 be continuous random variables with joint probability density function given by

$$f(x_1, x_2) = \begin{cases} 2x_1x_2 + (3/2)x_2^2 & \text{for } 0 < x_1 < 1, \quad 0 < x_2 < 1 \\ 0 & \text{elsewhere} \end{cases}$$

Find the mean and variance of $X_1 + X_2$.

4.   Let $X_1, X_2, \ldots, X_n$ be independent and identically distributed random variables with means equal to μ and variances equal to σ^2. Let $\bar{X} = (X_1 + X_2 + \cdots + X_n)/n$. Show that $E(\bar{X}) = \mu$ and $\text{Var}(\bar{X}) = \sigma^2/n$.

5.   Let X_1, X_2, X_3 be independent random variables with means $-1, 2, 3$ and variances $4, 1, 9$, respectively. Find (a) $E(X_1 + X_2 + X_3)$, (b) $\text{Var}(X_1 + X_2 + X_3)$, (c) $E(2X_1 - 3X_2 + X_3 - 2)$, (d) $\text{Var}(2X_1 - 3X_2 + X_3 - 2)$.

6.   During the busiest time of the day the elevator in a certain office building operates at maximum capacity of 15 passengers. If the weight of any one passenger is a random variable with a mean of 130 pounds and a standard deviation of 15 pounds, what is the expected total weight of passengers in the elevator during the busy period? What is the standard deviation?

7.   With reference to Example 8.5, list the birthdays (by month) of all students in your class. Compare the expected number of months that have birthdays for more than 1 student with the actual figure for your class.

8.   Prove Theorem 8.1(a) for continuous random variables. Let $n = 2$. (*Hint:* follow the proof for discrete random variables. Replace summations with integrals.)

9.   Prove Theorem 8.1(b) for both discrete and continuous random variables. Let $n = 2$.

10.   Prove Theorem 8.2(a) for discrete random variables. Let $n = 2$. (*Hint:* follow the proof for continuous random variables. Replace integrals with summations.)

11.   Show that if X_1 and X_2 are independent and continuous random variables with respective means μ_1 and μ_2, then $\int_{-\infty}^{\infty} \int_{-\infty}^{\infty} (x_1 - \mu_1) (x_2 - \mu_2) f(x_1, x_2) \, dx_1 \, dx_2 = 0$. (*Hint:* independence of X_1 and X_2 implies that $f(x_1, x_2) = f_1(x_1)f_2(x_2)$.)

12.   Prove Theorem 8.2(b) for both discrete and continuous random variables. Let $n = 2$.

13.   Let $X_1, X_2, \ldots, X_n$ be independent and identically distributed random variables with means equal to μ and variances equal to σ^2. Let $\bar{X} = (X_1 + X_2 + \cdots + X_n)/n$, and let $S^2 = [1/(n-1)] \sum_{i=1}^{n} (X_i - \bar{X})^2$. There is a theorem to the effect that $E(S^2) = \sigma^2$. Verify that this is so for $n = 2$ and for $n = 3$.

14. Let $X_1^{(1)}, X_2^{(2)}, \ldots, X_n^{(1)}, X_1^{(2)}, X_2^{(2)}, \ldots, X_n^{(2)}$ be independent random variables with $E(X_1^{(1)}) = E(X_2^{(1)}) = \cdots = E(X_n^{(1)}) = \mu_1$ and $E(X_1^{(2)}) = E(X_2^{(2)}) = \cdots = E(X_n^{(2)}) = \mu_2$. Show that if $\bar{X}_1 = (X_1^{(1)} + X_2^{(1)} + \cdots + X_n^{(1)})/n$ and $\bar{X}_2 = (X_1^{(2)} + X_2^{(2)} + \cdots + X_n^{(2)})/n$ then $E(\bar{X}_1 - \bar{X}_2) = \mu_1 - \mu_2$.

8.4   Distributions of Sums of Independent Random Variables

A wonderful thing happens when we look for the distributions of sums of independent and identically distributed random variables. Very often we find, to our delight, that the distributions are the well-known ones! This section is a discussion of the distributions of sums of independent and identically distributed Bernoulli, Poisson, geometric, exponential, and normal random variables. A discussion of the distributions of sums of independent and identically distributed random variables with unknown distributions is contained in the next section on the central limit theorem.

Let us begin our discussion with a review of what we mean when we say that the random variables $X_1, X_2, \ldots, X_n$ are independent and identically distributed. Chapter 7 gives a formal definition of independent random variables. An informal interpretation of the definition would state that two random variables are independent if the values of either random variable do not affect in any way the values of the other. More generally, $X_1, X_2, \ldots, X_n$ are independent random variables if the values of each of the random variables do not affect in any way the values of the other random variables. For example, if $X_1, X_2, \ldots, X_n$ are the respective numerical results of n repetitions of an experiment performed under identical conditions, then $X_1, X_2, \ldots, X_n$ are independent random variables. A collection of random variables is said to be identically distributed if their distributions are identical. Thus, the respective numerical results $X_1, X_2, \ldots, X_n$ of n repetitions of an experiment performed under identical conditions are not only independent but are also identically distributed random variables.

The distributions of sums of independent and identically distributed Bernoulli, Poisson, geometric, exponential, and normal random variables are given below.

1. Let $X_1, X_2, \ldots, X_n$ be independent and identically distributed Bernoulli random variables with parameter p. The distribution of $Y = X_1 + X_2 + \cdots + X_n$ is the binomial distribution with parameters p and n. That is, $P(Y = y) = b(y; n, p) = \binom{n}{y} p^y (1 - p)^{n-y}$.

2. Let $X_1, X_2, \ldots, X_n$ be independent Poisson random variables with parameters $\lambda_1, \lambda_2, \ldots, \lambda_n$, respectively. The distribution of $Y = X_1 + X_2 + \cdots + X_n$ is the Poisson distribution with parameter $\lambda = \lambda_1 + \lambda_2 + \cdots + \lambda_n$.

That is

$$P(Y = y) = p(y; \lambda) = \frac{\lambda^y e^{-\lambda}}{y!}$$

(Note that the random variables $X_1, X_2, \ldots, X_n$ need not be identically distributed.)

3. Let $X_1, X_2, \ldots, X_n$ be independent geometric random variables with $P(X_i = k) = q^{k-1}p$ for $k = 1, 2, \ldots$, $p > 0$, $q = 1 - p$, and $i = 1, 2, \ldots, n$. The distribution of $Y = X_1 + X_2 + \cdots + X_n$ is the negative binomial distribution with parameters p and n. That is, $P(Y = r) = f(r; n, p) = \binom{r-1}{n-1}p^n q^{r-n}$ for $r = n, n + 1, n + 2, \ldots$.

4. Let $X_1, X_2, \ldots, X_n$ be independent and identically distributed exponential random variables with means equal to θ. The distribution of $Y = X_1 + X_2 + \cdots + X_n$ is the gamma distribution with mean equal to $n\theta$. That is, the probability density function of Y is

$$f(y) = \begin{cases} \dfrac{1}{\theta^n(n-1)!} \, y^{n-1}e^{-y/\theta} & \text{for} \quad y > 0 \\ 0 & \text{elsewhere} \end{cases}$$

5. Let $X_1, X_2, \ldots, X_n$ be independent and identically distributed normal random variables with means equal to μ and variances equal to σ^2. The distribution of $Y = X_1 + X_2 + \cdots + X_n$ is the normal distribution with mean equal to $n\mu$ and variance equal to $n\sigma^2$.

6. Let $X_1, X_2, \ldots, X_n$ be independent normal random variables with means equal to $\mu_1, \mu_2, \ldots, \mu_n$, respectively, and variances equal to $\sigma_1^2, \sigma_2^2, \ldots, \sigma_n^2$, respectively. The distribution of $Y = a_1 X_1 + a_2 X_2 + \cdots + a_n X_n$ is the normal distribution with mean $\mu = a_1 \mu_1 + a_2 \mu_2 + \cdots + a_n \mu_n$ and variance $\sigma^2 = a_1^2 \sigma_1^2 + a_2^2 \sigma_2^2 + \cdots + a_n^2 \sigma_n^2$.

Proofs of these statements may be obtained through the use of moment generating functions and are given in Appendix A. In this chapter, we are primarily concerned with the use of these statements to solve problems.

EXAMPLE 8.6

A certain insurance company offers life, fire, and automobile coverage. The number of claims arriving on any day on these three types of policy are independent Poisson random variables with means equal to 30, 20, and 50, respectively. What is the probability, on any given day, the company will receive claims on more than 120 policies of all three types?

SOLUTION

Let $X_1, X_2,$ and X_3, be the number of life, fire, and automobile claims, respectively. According to Statement 2, the distribution of $Y = X_1 + X_2 + X_3$

is the Poisson distribution with mean $\lambda = 30 + 20 + 50 = 100$. Therefore, the solution to the problem is

$$P(Y > 120) = \sum_{k=121}^{\infty} p(k; 100) = \sum_{k=121}^{\infty} \frac{100^k e^{-100}}{k!} = 1 - \sum_{k=0}^{120} \frac{100^k e^{-100}}{k!}$$

The very industrious reader may be able to obtain the exact answer to the above expression. We shall see in Example 8.18, however, that an estimate of the above expression is easily obtained through the use of a normal approximation of the Poisson distribution.

EXAMPLE 8.7

A party of 25 wealthy friends decide that all will play at slot machines at a casino until each of them makes a win. If these slot machines use $1 coins and the probability of a win is 1/200, what is the probability that, as a group, the friends will spend a total of over $7,000 to satisfy their whim?

SOLUTION

Let $p = 1/200$ be the probability of a win, and $q = 199/200$ be the probability of a loss. Let X_i be the required number of games for player i. Then

$$P(X_i = k) = q^{k-1} p \qquad \text{for} \quad k = 1, 2, \ldots$$

Thus, $X_1, X_2, \ldots, X_{25}$ are independent and identically distributed geometric random variables. According to Statement 3, the distribution of $Y = X_1 + X_2 + \cdots + X_{25}$ is the negative binomial distribution with parameters $p = 1/200$ and $n = 25$. Since Y is the total number of games as well as the total amount in dollars that will be spent by the party of friends, the solution to the problem is

$$P(Y > 7,000) = 1 - P(Y \leq 7,000)$$

$$= 1 - \sum_{r=25}^{7,000} f(r; 25, 1/200)$$

To sum 6,976 terms is a tiresome task. But good news is ahead. A simple normal approximation of the negative binomial distribution (see Section 8.6) provides an easy solution. This example is continued in Example 8.19.

EXAMPLE 8.8

A certain electronic system has an automatic back-up device so that, should component A fail, it is instantly replaced, and the system continues to operate without a break. If this replacement feature can operate at most 2 times, and if the length of life of the component is an exponential random variable with mean $\theta = 100$ hours, what is the probability that the system fails, due to a failure in component A, in less than 200 hours of operation?

SOLUTION

Let X_1 be the length of life of the initial component. Let X_2, X_3 be the length of life of the two available replacements. Let $Y = X_1 + X_2 + X_3$. The solution to the problem is the probability $P(Y < 200)$.

Statement 4 tells us that the probability density function of Y is the gamma density function $g(y)$ with $\theta = 100$ and $n = 3$.

$$P(Y < 200) = \int_0^{200} g(y)\, dy = \int_0^{200} \frac{1}{(100^3) \cdot 2!}\, y^2 e^{-y/100}\, dy$$

$$= \frac{1}{2 \cdot 100^3} \int_0^{200} y^2 e^{-y/100}\, dy$$

$$= \frac{e^{-x/100}}{2 \cdot 100^3} \left[-100x^2 - 2 \cdot 100^2 x - 2 \cdot 100^3 \right] \Big|_0^{200}$$

$$= \frac{e^{-2}}{2 \cdot 100^3} \left[-10^2 \cdot (2 \cdot 10^2)^2 - 2 \cdot 10^4 \cdot 2 \cdot 10^2 - 2 \cdot 10^6 \right]$$

$$- \frac{1}{2 \cdot 100^3} \cdot (-2 \cdot 100^3)$$

$$= 1 - .677 = .323$$

The integration was accomplished through repeated use of the formula

$$\int x^m e^{ax}\, dx = \left[(x^m e^{ax})/a \right] - \left[(m/a) \int x^{m-1} e^{ax}\, dx \right]$$

EXAMPLE 8.9

Let X_1, X_2, X_3 be independent normal random variables with means $3, 1, -2$ and variances $11, 4, 9$, respectively. Find $P(5 < X_1 - 2X_2 + X_3 < 11)$.

SOLUTION

Let $Y = X_1 - 2X_2 + X_3$. From Theorems 8.1 and 8.2, the mean and variance of Y are

$$\mu = E(Y) = E(X_1 - 2X_2 + X_3)$$
$$= E(X_1) - 2E(X_2) + E(X_3)$$
$$= -1$$

and

$$\sigma^2 = \text{Var}(Y) = \text{Var}(X_1 - 2X_2 + X_3)$$
$$= \text{Var}(X_1) + 4\text{Var}(X_2) + \text{Var}(X_3)$$
$$= 36$$

By Statement 6 the distribution of Y is normal with mean $\mu = -1$ and variance $\sigma^2 = 36$. Thus, the standardized random variable $Z = (Y - \mu)/\sigma$ is standard normal. Therefore, the answer to the problem is

$$P(5 < X_1 - 2X_2 + X_3 < 11) = P(5 < Y < 11)$$

$$= P\left(\frac{5 - \mu}{\sigma} < \frac{Y - \mu}{\sigma} < \frac{11 - \mu}{\sigma}\right)$$

$$= P(1 < Z < 2) = .1359$$

EXAMPLE 8.10

Suppose that if X is the IQ of a college student selected at random from the United States student population, then X is a normal variable with mean 120 and standard deviation 15. What is the probability that the average IQ of 9 college students selected at random from the U.S. student population will be less than 115?

SOLUTION

Let X_i be the IQ of the ith student selected. The random variables $X_1, X_2, \ldots, X_9$, then, are independent and identically distributed normal random variables with means equal to 120 and standard deviations equal to 15. By Statement 5, the distribution of the random variable $Y = X_1 + X_2 + \cdots + X_9$ is normal with mean $\mu = 9(120) = 1080$ and variance $\sigma^2 = 9(15)^2 = 2025$. Therefore, the answer to the problem is

$$P(\text{average IQ of the 9 students} < 115) = P\left(\frac{Y}{9} < 115\right) = P(Y < 1035)$$

$$= P\left(\frac{Y - \mu}{\sigma} < \frac{1035 - 1080}{45}\right)$$

$$= P\left(\frac{Y - \mu}{\sigma} < -1\right) = .1587$$

EXAMPLE 8.11

Every workday Mr. Jones and Ms. Cohen arrive at the bus stop to take the 8 A.M. express to Manhattan. Mr. Jones' time of arrival at the bus stop can be thought of as a random variable that is normally distributed with mean at 7:50 A.M. and standard deviation of 8 minutes. Ms. Cohen's time of arrival at the bus stop is a normal random variable with mean at 7:55 A.M. and standard deviation of 4 minutes. What is the probability that, on a given day, Ms. Cohen will be at the bus stop before Mr. Jones?

SOLUTION

Let X_1 be the time measured in minutes after 7:50 A.M. of Mr. Jones' arrival at the bus stop. Let X_2 be the time measured in minutes after 7:50 A.M. of Ms.

Cohen's arrival at the bus stop. The means and variances of X_1 and X_2 are

$$E(X_1) = 0, \quad E(X_2) = 5$$
$$\text{Var}(X_1) = 64, \quad \text{Var}(X_2) = 16$$

The solution to the problem is the probability

$$P(X_2 < X_1) = P(X_2 - X_1 < 0)$$

Let $Y = X_2 - X_1$. Then $\mu = E(Y) = E(X_2) - E(X_1) = 5 - 0 = 5$ and $\sigma^2 = \text{Var}(Y) = \text{Var}(X_2) + \text{Var}(X_1) = 64 + 16 = 80$. Statement 6 tells us that Y is normal with mean $\mu = 5$ and variance $\sigma^2 = 80$. Therefore

$$P(X_2 < X_1) = P(X_2 - X_1 < 0) = P(Y < 0)$$
$$= P\left(\frac{Y - \mu}{\sigma} < \frac{0 - \mu}{\sigma}\right) = P(Z < -.559) = .2877$$

Exercises

1. With reference to Example 8.6, what is the daily number of claims that the company can expect to receive? What is the standard deviation?

2. Let X_1, X_2, and X_3 be independent Poisson random variables with means 1, 3, and 5, respectively. Find $P(7 < X_1 + X_2 + X_3 < 9)$.

3. Let X_1, X_2, and X_3 be independent and identically distributed geometric random variables with $P(X_i = k) = q^{k-1}p$ for $k = 1, 2, \ldots,$ and with $p = 1/3, q = 2/3$. Find $P(X_1 + X_2 + X_3 = 4)$.

4. Suppose that incomes of families in Newport Harbor are normally distributed with a mean of $750,000 and a standard deviation of $250,000.
   (a) If 4 families are selected at random from this community, what is the probability that their combined incomes exceed $4 million?
   (b) What is the probability that the average income of these 4 families exceeds $800,000?

5. Robert's friends Judy and Alice have similar-sounding voices. Both of them call him daily during his lunch hour from 12 noon to 1 P.M. If Judy's time of call is a normal random variable with a mean at 12:15 P.M. and a standard deviation of 5 minutes, whereas Alice's time of call is a normal random variable with a mean at 12:30 P.M. and a standard deviation of 10 minutes, how safe is it for Robert to greet the first caller by the name Judy? That is, what are his chances of making a mistake?

6. Show that if $X_1, X_2, \ldots, X_n$ are independent and identically distributed normal random variables with means equal to μ and variances equal to

σ^2, and if $\bar{X} = (X_1 + X_2 + \cdots + X_n)/n$, then the random variable $\bar{X}$ is normal with mean μ and variance σ^2/n. (*Hint:* see Statement 6 of Section 8.4 and observe that $\bar{X} = (1/n)X_1 + (1/n)X_2 + \cdots + (1/n)X_n$.)

7. With reference to Exercise 6, what is the distribution of $(\bar{X} - \mu)/(\sigma/n^{1/2})$?

8. The collection, measured in dollars, from any one parking meter on Main Street on any given day is a normal random variable with mean $\mu = 2$ and standard deviation $\sigma = 1$. If there are 100 parking meters on Main Street, what is
   (a)   the expected amount that will be collected in one day;
   (b)   the probability that in any one day the total collection from all parking meters will exceed $250.00? Assume that the collection from any one parking meter does not affect in any way the collection from any of the other parking meters.

9. Suppose that in any given day, the amount of oil measured in barrels from wells A, B, and C are independent normal random variables with means equal to 100, 150, and 200, respectively, and standard deviations equal to 10, 5, and 15, respectively. What is the probability that on a given day the amount of oil from all three wells combined will fall below 400 barrels?

10. Suppose that the length of life, measured in hours, of any one of a certain type of fuse is an exponential random variable X with mean $\theta = 100$. What is the probability that the combined length of life of two of these fuses will be between 180 and 220 hours? (*Hint:* see Statement 4, Section 8.4.)

8.5   The Central Limit Theorem

Section 8.4 gave us the distributions of sums of independent and identically distributed Bernoulli, Poisson, geometric, exponential, and normal random variables. In this section we will discuss a most remarkable theorem. It is the theorem that tells us that even if the distributions of the random variables are unknown, the distribution of sums of independent and identically distributed random variables is approximately normal provided only that there be enough random variables in the sum. In other words, regardless of the specific types of distributions, if $X_1, X_2, \ldots, X_n$ are independent and identically distributed random variables with means equal to μ and variances equal to σ^2, the distribution of $Y = X_1 + X_2 + \cdots + X_n$ is approximately normal with mean $n\mu$ and variance $n\sigma^2$ provided that n is sufficiently large. A formal statement of the central limit theorem follows.

Theorem 8.3   The Central Limit Theorem   *Let* $X_1, X_2, \ldots, X_n$ *be independent and identically distributed random variables with means equal to* μ *and variances* *equal to σ^2. Let $\bar{X}$ be the random variable defined by $\bar{X} = (X_1 + X_2 + \cdots + X_n)/n$. The distribution of*

$$\frac{\bar{X} - \mu}{\sigma/\sqrt{n}}$$

tends to the standard normal distribution as $n \to \infty$.

The proof of the central limit theorem as it is stated above is beyond the scope of this text. However, if the random variables under consideration possess moment generating functions, the proof is relatively simple; it is given in Appendix A. Justification of the following three consequences is considered in Exercise 6.

Consequences of the central limit theorem

1. For n large, the distribution of $(\bar{X} - \mu)/(\sigma/n^{1/2})$ is approximately standard normal.

2. For n large, the distribution of $\bar{X}$ is approximately normal with mean μ and variance σ^2/n.

3. For n large, the distribution of $Y = X_1 + X_2 + \cdots + X_n$ is approximately normal with mean $n\mu$ and variance $n\sigma^2$.

In each of the above consequences there is a condition that the value n should be large. An immediate question would be "how large is large enough?" An exact answer to this question can be obtained only if the desired degree of precision of the approximation is given and if the exact form of the distribution that is common to all the random variables is known. There is general agreement among statisticians, however, that if the value of n is at least 50, then the approximation is good enough for most purposes.

As stated in the leading paragraph, a remarkable feature of the central limit theorem is that no restrictions are made on the type of distribution that is common to all of the random variables $X_1, X_2, \ldots, X_n$. The common distribution of the random variables may be binomial, exponential, or any arbitrary type, yet the distribution of the sum will be approximately normal, provided that all the conditions of the theorem are satisfied.

Another version of the central limit theorem, developed by Liapounoff, removes the restriction of identical distributions but adds a requirement of bounded third moments about the mean for each of the random variables. With the central limit theorem stated in this general form, one can understand the frequent occurrence of random variables whose distributions are normal. Many measurements, for example IQ's, are the result of innumerable independent components. The central limit theorem tells us that measurements of this sort will be approximately normally distributed.

EXAMPLE 8.12

What is the probability that the average of 100 numbers selected at random from the unit interval $(0, 1)$ will be within .01 of the midpoint of the interval?

SOLUTION

Let X_i be the value of the ith number selected at random from $(0, 1)$. The random variables $X_1, X_2, \ldots, X_{100}$ are independent and identically distributed random variables. The distribution that is common to all the random variables is the uniform distribution with mean .5 and variance $1/12$ (see Chapter 6 on the mean and variance of the uniform distribution). Therefore, by the central limit theorem, $(\bar{X} - .5)/(.289/10)$ is approximately standard normal, and

$$P[|(\text{average of 100 numbers}) - .5| \le .01] = P(|\bar{X} - .5| \le .01)$$

$$= P\left(\left|\frac{\bar{X} - .5}{.289/10}\right| \le \frac{.1}{.289/10}\right) \approx P(|Z| \le .3460) = .2706$$

where Z is the standard normal random variable.

EXAMPLE 8.13

The burning time of a certain type of lamp is an exponential random variable with mean $\theta = 30$ hours. What is the probability that 144 of these lamps will provide a total of more than 4,500 hours of burning time?

SOLUTION

Let X_i be the burning time of the ith lamp. The random variables $X_1, X_2, \ldots, X_{144}$ are independent and identically distributed random variables with means equal to 30 and variances equal to 30^2. By the central limit theorem the distribution of $(\bar{X} - 30)/(30/12)$ is approximately standard normal. Therefore

$$P(\text{a total of more than 4,500 hours}) = P(X_1 + X_2 + \cdots + X_{144} > 4,500)$$

$$= P\left(\frac{\bar{X} - 30}{30/12} > \frac{(4,500/144) - 30}{30/12}\right) \approx P(Z > .5) = .3085$$

(We use the letter Z to denote the standard normal random variable.)

EXAMPLE 8.14

A newly employed switchboard operator was told by her employer that, on the average, there would be 1 incoming call per minute. During her first 4 hours on duty she handled 280 incoming calls. Does she have reason to suspect that her employer lied to her concerning the workload? Assume that the number of incoming calls in any fixed interval of time may be considered a Poisson random variable.

SOLUTION

Let X_i be the number of calls during the ith minute. Let $N = 60 \cdot 4 = 240$ minutes. If the employer's claim is correct, then $E(X_i) = \lambda = 1$ and $\text{Var}(X_i) = \lambda = 1$ for $i = 1, 2, \ldots, 240$. Therefore, if the employer's claim is correct, the distribution of $(\bar{X} - 1)/(1/240^{1/2})$ is approximately standard normal, and

$$P(\text{a total of 280 or more calls in 4 hours})$$

$$= P(X_1 + X_2 + \cdots + X_{240} \geq 280) = P\left(\bar{X} \geq \frac{280}{240}\right)$$

$$= P\left(\frac{\bar{X} - 1}{1/\sqrt{240}} \geq \frac{(280/240) - 1}{1/\sqrt{240}}\right) \approx P(Z \geq 2.582) = .0049$$

(Again, Z denotes the standard normal random variable.)

The calculation tells us that if the employer's claim is correct, it is very unlikely (probability of .0049) that there would be 280 or more incoming calls in 4 hours. The switchboard operator has good reason to suspect that the information from her employer is unreliable.

Problems of this sort are also treated in the chapter on the testing of statistical hypotheses.

Exercises

Use the central limit theorem whenever applicable.

1. What is the probability that the average of the outcomes of 100 rolls of a die will be less than 3.4?

2. What is the probability that there will be more than 40 heads in 64 flips of a fair coin? (*Hint:* let X_i be the number of heads on the ith flip.)

3. If the burning time of a certain type of lamp is an exponential random variable with mean $\theta = 30$ hours, what is the probability that the average burning time of 100 of these lamps will be within 6 minutes of 30 hours?

4. Let us assume that sons of alumni of a certain prestigious men's college will eventually apply for admission to that college. If the probability that an alumnus will have 0, 1, 2, 3, or 4 sons are 1/10, 4/10, 3/10, 1/10, and 1/10, respectively, what is the probability that the 1960 graduating class of 1,000 will produce over 1,600 applicants to the college?

5. The organizers of a campaign to solicit contributions from alumnae of a prestigious college estimated that the average contribution per alumna

should be $1,000 with a standard deviation of $300. After 100 alumnae were contacted, the total contribution was less than $80,000. Was the initial estimate overly optimistic?

6. Show that Statements 1, 2, and 3 of Section 8.5 are consequences of the central limit theorem. (*Hint:* if Y is a random variable with mean μ and standard deviation σ, and if $(Y - \mu)/\sigma$ is standard normal, then Y is normal.)

8.6   Normal Approximations of the Binomial, Poisson, and Other Distributions

8.6.1   The Approximation Formulas

In Section 8.4 we saw that sums of independent and identically distributed Bernoulli, Poisson, geometric, and exponential random variables are, respectively, binomial, Poisson, negative binomial, and gamma random variables. It can also be shown (Exercise 18) that every binomial, Poisson, negative binomial, and gamma distribution (of the type given in Statement 4 of Section 8.4) is, respectively, the distribution of a sum of independent and identically distributed Bernoulli, Poisson, geometric, and exponential random variables. Therefore, on application of the central limit theorem, it follows that the binomial, Poisson, and negative binomial distributions, and gamma distributions of a certain type may be approximated by normal distributions. (The approximation also holds for gamma distributions of the general type, but the general case is not a simple consequence of the central limit theorem.) The following statements provide the formulas for these approximations (see Exercise 19 for a justification of these formulas.)

1. Let S_n be a binomial random variable with $P(S_n = k) = b(k; n, p)$ for $k = 0, 1, 2, \ldots, n$, $0 \le p \le 1$, and $q = 1 - p$. For large values of n, the distribution of $(S_n - np)/(npq)^{1/2}$ is approximately standard normal. (The approximation is improved to a considerable extent if it is done with a correction for continuity as described in Section 8.6.2. In that case a good approximation is obtain when both np and nq are greater than 5.)

2. Let X be a Poisson random variable with $P(X = k) = p(k; \lambda) = (\lambda^k e^{-\lambda})/k!$ for $k = 0, 1, 2, \ldots$ and $\lambda > 0$. For large values of λ, the distribution of $(X - \lambda)/\lambda^{1/2}$ is approximately standard normal. (As in the case of the binomial distribution, the approximation is greatly improved if it is done with a correction for continuity as described in Section 8.6.2. A good approximation may then be obtained if λ is greater than 10.)

3. Let X be a negative binomial random variable with probability function $P(X = r) = f(r; n, p) = \binom{r-1}{n-1} p^n q^{r-n}$ for $r = n, n + 1, n + 2, \ldots, 0 \le p \le 1$,

$q = 1 - p$, and positive integer n. For large values of n, the distribution of

$$\frac{X - \dfrac{n}{p}}{\sqrt{nq/p^2}}$$

is approximately standard normal.

4. Let X be a gamma random variable with probability density function

$$f(x) = \begin{cases} \dfrac{1}{\beta^\alpha \Gamma(\alpha)} x^{\alpha-1} e^{-x/\beta} & \text{for} \quad x > 0 \\[2mm] 0 & \text{elsewhere} \end{cases}$$

where $\alpha > 0$ and $\beta > 0$. For large values of α, the distribution of $(X - \alpha\beta)/(\alpha\beta^2)^{1/2}$ is approximately standard normal. (The approximation is greatly improved if it is done with a correction for skewness as described in Section 8.6.3.)

As illustrations of the approximation of the binomial, Poisson, and gamma distributions by normal distributions, Figure 8.1, Figure 8.2, and Figure 8.3 show that as n, λ, and α increase, the respective contours of the

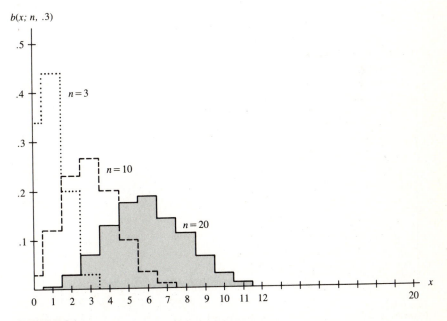

FIGURE 8.1
Normal Approximation of the Binomial Distribution
The binomial probability function $b(x; n, .3)$ for $n = 3$, 10, and 20.

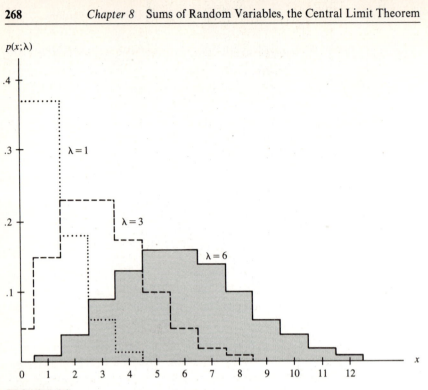

FIGURE 8.2
Normal Approximation of the Poisson Distribution

The Poisson probability function $p(x; \lambda)$ for $\lambda = 1, 3$, and 6.

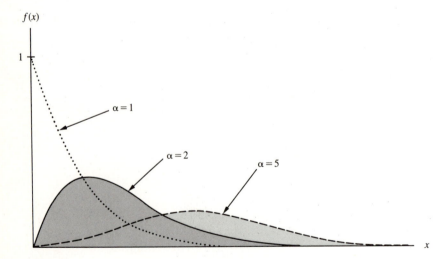

FIGURE 8.3
Normal Approximation of the Gamma Distribution

The gamma probability density function $f(x)$ for $\beta = 1$, and $\alpha = 1, 2$, and 5.

histogram of the binomial probability function, the histogram of the Poisson probability function, and the curve of the gamma probability density function, approximate more closely curves of normal probability density functions.

EXAMPLE 8.15

Use the normal approximation of the binomial distribution as given in Statement 1 above to estimate the probability of getting anywhere from 45 to 55 heads in 100 flips of a fair coin.

   SOLUTION

Let $n = 100$, $p = P(\text{heads}) = 1/2$, S_n = the number of heads in 100 flips of a fair coin, and Z = the standard normal random variable. Then

$$P(\text{from 45 to 55 heads}) = P(45 \leq S_n \leq 55)$$

$$= P\left(\frac{45 - np}{\sqrt{npq}} \leq \frac{S_n - np}{\sqrt{npq}} \leq \frac{55 - np}{\sqrt{npq}}\right)$$

$$\approx P\left(\frac{45 - 50}{5} \leq Z \leq \frac{55 - 50}{5}\right)$$

$$= P(-1 \leq Z \leq +1) = .6826$$

EXAMPLE 8.16

In order to be elected to a certain office in the coming election, candidate A, in running against candidate B, must obtain more than 50% of the votes cast in a certain large metropolitan area. To assess her popularity, her campaign organization has polled 100 registered voters selected at random from the area and has found that only 48 of them prefer candidate A. Should candidate A feel that defeat on election day is inevitable?

   SOLUTION

Suppose that, in spite of the unfavorable poll, candidate A will get 51% of the votes on election day. Let $n = 100$ = the number of voters selected, $p = P(\text{a vote for candidate A}) = .51$, S_n = the number of votes for candidate A from the 100 voters, and Z = the standard normal random variable. Then, by Statement 1

$$P(\text{at most 48 votes for candidate A}) = P(S_n \leq 48)$$

$$= P\left(\frac{S_n - np}{\sqrt{npq}} \leq \frac{48 - 51}{4.999}\right) \approx P(Z \leq -.6) = .2743$$

   The calculations show that even if candidate A were to win the election with 51% of the votes, the chances are considerable (probability of .2743) that in a random sample of 100 voters at most 48 would have voted for her. She should not feel she is doomed to be defeated on election day.

EXAMPLE 8.17

A certain television network wishes to estimate the popularity of program A, which is shown on Monday evenings at 7 P.M. in a very large metropolitan area. During the program broadcast calls will be made to residents selected at random from a telephone book. The resident will be asked if he or she is watching television, and if so, if it is program A. The ratio of the number of respondents watching program A to the total number of respondents watching television will give an estimate of the actual proportion of viewers watching program A. If n is the number of respondents watching television, how large should n be for us to be 95% sure that our method of estimating the actual proportion of viewers watching A will be good to one decimal place?

   SOLUTION

Let n be the number of respondents who are watching television, let S_n be the number of respondents who are watching program A, and let p be the actual proportion of viewers (in the entire area) who are watching program A. We wish to find n so that

$$P(|(S_n/n) - p| < 1/10) = .95$$

   Applying Statement 1 and letting Z be the standard normal random variable, we see that n is the number such that

$$.95 = P\left(\left|\frac{S_n}{n} - p\right| < 1/10\right) = P\left(\left|\frac{S_n - np}{n}\right| < 1/10\right)$$

$$= P\left(\left|\frac{S_n - np}{\sqrt{npq}}\right| < \frac{\sqrt{n}}{10\sqrt{pq}}\right)$$

$$\approx P\left(|Z| < \frac{\sqrt{n}}{10\sqrt{pq}}\right)$$

Therefore

$$\frac{\sqrt{n}}{10\sqrt{pq}} > 1.96$$

or

$$n > (19.6)^2\, pq$$

   Now, the problem is that we don't know the value of p or q. After all, p is what we wish to estimate. However, suppose we define the function $h(p)$ to be

$$h(p) = p(1 - p) = pq$$

Then

$$h'(p) = 1 - 2p$$

and

$$h'(p) = 0 \text{ at } p = 1/2$$

In other words the function $h(p)$ has a maximum at $p = q = 1/2$. This means that

$$96.04 = (19.6)^2(1/2)(1/2) \geq (19.6)^2 \, pq$$

so that if

$$n > 96.04$$

then n is sufficiently large for the equation

$$P\left(\left|\frac{S_n}{n} - p\right| < 1/10\right) = .95$$

to hold.

Our answer, then, is that if $n = 97$, we could be 95% sure that our method of estimating p will be good to one decimal place. The number 97 may nevertheless be larger than necessary.

The chapter on confidence interval estimation will also deal with problems of this sort.

EXAMPLE 8.18

Use the normal approximation of the Poisson distribution (Statement 2) to solve the problem of Example 8.6.

SOLUTION

The random variable Y of Example 8.6 is a Poisson random variable with mean $\lambda = 100$ and standard deviation $\lambda^{1/2} = 10$. By Statement 2, the distribution of $(Y - \lambda)/\lambda^{1/2}$ is approximately standard normal. Therefore

$$P(Y > 120) = P\left(\frac{Y - \lambda}{\sqrt{\lambda}} > \frac{120 - \lambda}{\sqrt{\lambda}}\right) \approx P\left(Z > \frac{120 - 100}{10}\right)$$

$$= P(Z > 2) = .0228$$

where Z denotes the standard normal random variable.

EXAMPLE 8.19

Use the normal approximation of the negative binomial distribution (Statement 3) to solve the problem of Example 8.7.

SOLUTION

In Example 8.7, Y is the total number of games as well as the total amount in dollars that will be spent by the group. Furthermore, Y is a negative binomial random variable with $P(Y = r) = f(r; 25, 1/200)$. By Statement 3, the distribution of $(Y - 5000)/995000^{1/2}$ is approximately standard normal. Therefore

$$P(\text{more than \$7,000 will be spent by the group}) = P(Y > 7000)$$

$$= P\left(\frac{Y - 5000}{\sqrt{995000}} > \frac{7000 - 5000}{\sqrt{995000}}\right) \approx P(Z > 2.005) = .02$$

where Z denotes the standard normal random variable.

EXAMPLE 8.20

Use the normal approximation of the gamma distribution (Statement 4) to solve the problem of Example 8.8.

SOLUTION

In Example 8.8, Y is the length of time measured in hours before the system fails. It is shown that Y is a gamma random variable with parameters $\alpha = 3$ and $\beta = 100$. By Statement 4, the distribution $(Y - 300)/(100 \cdot 3^{1/2})$ is approximately standard normal. Therefore, the solution to the problem is

$$P(Y < 200) = P\left(\frac{Y - 300}{100\sqrt{3}} < \frac{200 - 300}{100\sqrt{3}}\right) \approx P(Z < -.57735) = .28$$

8.6.2   Correction for Continuity

In approximating the binomial distribution with the normal distribution, in effect we took the discrete random variable S_n, and approximated it with a continuous random variable X. The possible values of S_n are the distinct integers $0, 1, 2, \ldots, n$, whereas the possible values of X are all of the numbers on the line of real numbers. Using the standard method of rounding off values at the midpoint of unit intervals, it would seem reasonable to assume that a good approximation of the event $S_n = k$ would be the event $k - 1/2 \le X < k + 1/2$. Extending this scheme to more than one value for S_n, a good approximation for $a \le S_n \le b$ would be the event $a - 1/2 \le X < b + 1/2$. Example 8.21 illustrates the use of this scheme to obtain a normal approximation of a binomial probability. The example also shows that this method of correction for continuity improves the approximation. If we are approximating the binomial distribution with the normal distribution, and if we use this method of correction for continuity, then a good approximation may be obtained if both np and $n(1 - p)$ are greater than 5.

EXAMPLE 8.21

Use the normal approximation of the binomial distribution, without correction for continuity and then with correction for continuity, to estimate the probability of getting anywhere from 8 to 12 heads in 20 flips of a fair coin. Compare these estimates with the exact calculation using the table of binomial probabilities.

SOLUTION

Let $n = 20$, $p = P(\text{heads}) = 1/2$, S_n = the number of heads in 20 flips of a fair coin, and Z = the standard normal random variable.

(a)   *Normal approximation without correction for continuity:*

$$P(\text{from 8 to 12 heads}) = P(8 \leq S_n \leq 12)$$

$$= P\left(\frac{8 - np}{\sqrt{npq}} \leq \frac{S_n - np}{\sqrt{npq}} \leq \frac{12 - np}{\sqrt{npq}}\right)$$

$$\approx P\left(\frac{8 - 10}{\sqrt{5}} \leq Z \leq \frac{12 - 10}{\sqrt{5}}\right)$$

$$= P(-.8944 \leq Z \leq .8944) = .6266$$

(b)   *Normal approximation with correction for continuity:* Let X be a normal random variable with mean np and variance npq so that $(X - np)/(npq)^{1/2}$ is the standard normal random variable Z.

$$P(\text{from 8 to 12 heads}) = P(8 \leq S_n \leq 12) \approx P(7.5 \leq X < 12.5)$$

$$= P\left(\frac{7.5 - np}{\sqrt{npq}} \leq Z < \frac{12.5 - np}{\sqrt{npq}}\right) = P(-1.118 \leq Z < 1.118)$$

$$= .7364$$

(c)   *Exact calculation using the table of binomial probabilities:*

$$P(\text{from 8 to 12 heads}) = P(8 \leq S_n \leq 12)$$

$$= \sum_{x=8}^{12} b(x; 20, .5) = .7368$$

The method of correction for continuity outlined above for the binomial random variable is also appropriate for the Poisson, negative binomial, and all other discrete random variables.

EXAMPLE 8.22

Let Y be a Poisson random variable with mean $\lambda = 9$. Use the normal approximation with correction for continuity to find $P(Y = 15)$. Compare the answer with the exact probability.

SOLUTION

(a)　Let X be a normal random variable with the same mean and variance as Y. That is, $E(X) = E(Y) = \lambda = 9$ and $\text{Var}(X) = \text{Var}(Y) = \lambda = 9$. Then

$$P(Y = 15) \approx P(14.5 \leq X < 15.5)$$

$$= P\left(\frac{14.5 - \lambda}{\sqrt{\lambda}} \leq \frac{X - \lambda}{\sqrt{\lambda}} < \frac{15.5 - \lambda}{\sqrt{\lambda}}\right)$$

$$= P(1.83 \leq Z < 2.17) = .0186$$

where Z denotes the standard normal random variable.

(b)　The exact probability is

$$P(Y = 15) = p(15; 9) = \frac{9^{15}e^{-9}}{15!} = .0194$$

EXAMPLE 8.23

Use the correction for continuity in Example 8.18. Compare it with the answer obtained without correction for continuity. Observe that when we are dealing with large numbers, such as $\lambda = 100$ and $Y > 120$, the correction for continuity is unimportant.

SOLUTION

(a)　*With correction for continuity*

$$P(Y > 120) \approx P(X \geq 120.5) = P\left(\frac{X - \lambda}{\sqrt{\lambda}} \geq \frac{120.5 - 100}{10}\right)$$

$$= P(Z \geq 2.05) = .0202$$

(b)　*Without correction for continuity*　(obtained in Example 8.18)

$$P(Y > 120) \approx .0228$$

8.6.3　Correction for Skewness

Just as the normal approximation of the binomial, Poisson, and other discrete distributions may be improved with a correction for continuity, the normal approximation of the gamma distribution may be improved with a correction for skewness. R. A. Fisher showed that if X is a gamma random variable with parameters α and β, and if we set

$$Y = 2\sqrt{X/\beta}$$

then the distribution of Y, for large values of α, becomes approximately normal with mean $\mu = (4\alpha - 1)^{1/2}$ and standard deviation $\sigma = 1$.

A better, but more complex, correction for skewness is offered by Wilson and Hilferty. They suggested setting

$$W = (X/\beta\alpha)^{1/3}$$

For large values of α, the distribution of W is approximately normal with mean $\mu = (9\alpha - 1)/(9\alpha)$ and standard deviation $\sigma = 1/(3\alpha^{1/2})$

EXAMPLE 8.24

(a) Use the Fisher correction for skewness to solve the problem of Example 8.20.

(b) Use the Wilson and Hilferty correction for skewness to solve the problem of Example 8.20.

(c) Compare the answers with the solution obtained without correction for skewness in Example 8.20, and with the exact calculation obtained in Example 8.8.

SOLUTION

In Examples 8.8 and 8.20, Y is a gamma random variable with parameters $\alpha = 3$ and $\beta = 100$. The problem is to find $P(Y < 200)$.

(a) *Normal approximation with Fisher's correction*   Let $F = 2(Y/\beta)^{1/2}$, $\mu = (4\alpha - 1)^{1/2}$, and $\sigma = 1$. Then the standardized random variable $Z = (F - \mu)/\sigma$ is approximately standard normal. Therefore

$$P(Y < 200) = P\left(\frac{Y}{\beta} < \frac{200}{\beta}\right)$$
$$= P(2\sqrt{Y/\beta} < 2\sqrt{200/\beta})$$
$$= P(F < 2\sqrt{200/\beta})$$
$$= P\left(\frac{F - \mu}{\sigma} < 2\sqrt{200/\beta} - \sqrt{4\alpha - 1}\right)$$
$$= P(Z < -.488) \approx .312$$

(b) *Normal approximation with the Wilson and Hilferty correction*   Let $W = (Y/\beta\alpha)^{1/3}$, $\mu = (9\alpha - 1)/9\alpha$, and $\sigma = 1/(3\alpha^{1/2})$. Then, the standardized random variable $Z = (W - \mu)/\sigma$ is approximately standard normal. Therefore

$$P(Y < 200) = P[(Y/\beta\alpha) < (200/\beta\alpha)]$$
$$= P[(Y/\beta\alpha)^{1/3} < (200/\beta\alpha)^{1/3}] = P[W < (2/3)^{1/3}]$$
$$= P\left(\frac{W - \mu}{\sigma} < -.4644\right) = P(Z < -.4644) \approx .323$$

(c)   *Comparison*

Approximation without correction: $P(Y < 200) \approx .281$
Approximation with Fisher's correction: $P(Y < 200) \approx .312$
Approximation with Wilson–Hilferty correction: $P(Y < 200) \approx .323$
Exact calculation: $P(Y < 200) = .323$

Exercises

Use a normal approximation whenever applicable. Include a correction for continuity or for skewness if the relevant parameter is less than 50.

1. Find the probability of getting anywhere from 25 to 35 sixes in 180 throws of a die.

2. What is the probability of obtaining more than 475 heads in 900 tosses of a fair coin?

3. One hundred cans of contaminated tuna were inadvertently distributed randomly among 10,000 cans of tuna. If 1,000 cans were sold to the public before a recall was issued by the tuna cannery, what are the chances that at least 3 cans of contaminated tuna were sold to the public?

4. The manufacturer of a certain product would like to keep its production of defective items below the 5% level. As an initial step, the quality control department decides to examine a random selection of n of these items. If more than 4% of the items among the selection of n items are defective, they will recommend a more extensive examination of the production process. How large should n be so that the probability of recommending the extensive examination is .99 if the process produces defective items 5% of the time?

5. Suppose that on a given summer evening, the number of mosquitoes on Mr. Brown's back porch is a Poisson random variable with mean $\lambda = 64$. Find the probability that there will be at least 45 mosquitoes on Mr. Brown's back porch the evening of next July 10.

6. Suppose that the number of fawns that will be born in any one season in a certain mountain region is a Poisson random variable with mean $\lambda = 100$. Find the probability that there will be fewer than 75 newborn fawns in this mountain region next season.

7. Suppose that a certain company wishes to hire 100 trainees for a certain type of work and will interview applicants until the 100th qualified person is found. If the required number of interviews is a negative binomial random variable with $p = 2/3$, what is the probability that there will be at least 50 interviews of unqualified applicants before the search for trainees is completed?

8. Suppose that the operating time, measured in hours, before a complete breakdown of a certain system is a gamma random variable with

parameters $\alpha = 100$ and $\beta = 3$. Find the probability that a complete breakdown will occur before 250 hours.

9. In 36 throws of a die, what is the probability of getting (a) exactly six 1's; (b) fewer than five 1's; (c) from five to seven 1's?

10. In reference to Exercise 4 of Section 8.4, what is the probability that a random selection of 35 families from Newport Harbor will include at least two families each of whose income exceeds \$1 million?

11. Suppose it is known that 85% of all patients with a certain disease can be cured with a certain drug. What is the probability that among 40 patients with this disease at most 3 of them cannot be cured with the drug?

12. Consider the fuses of Exercise 10 of Section 8.4. What is the probability that, of 20 such fuses, more than 10 of them will last more than 110 hours?

13. Suppose that the number of rainy days in a certain city in the month of October is a Poisson random variable with mean $\lambda = 9$. What is the probability of more than 10 rainy days in this city next October?

14. Suppose that on any given day the number of reported burglaries in a certain city is a Poisson random variable with mean $\lambda = 20$. Find the probability of exactly 20 reported burglaries in that city on a given day. Compare the normal approximation with the exact probability.

15. Use the normal approximation of the negative binomial distribution to find the probability of getting at least 10 tails before 5 heads in a sequence of tosses of a fair coin. Compare the normal approximation with the exact probability.

16. With reference to Examples 8.7 and 8.19, what is the probability that the 25 friends will collectively spend less than \$5,000 to reach their goal?

17. With reference to Example 8.8, if the replacement feature can operate up to 10 times, what is the probability that there will not be a failure in the system due to a failure in component A for over 1,000 hours?

18. (a) Let $b(x; n, p)$ be a binomial probability function. Show that it is the probability function of a sum of independent and identically distributed Bernoulli random variables. (*Hint:* follow the definition of X_i in Example 8.1(a).
    (b) Let $p(x; \lambda)$ be a Poisson probability function. Show that it is the probability function of a sum of n independent and identically distributed Poisson random variables. (*Hint:* let $X_1, X_2, \ldots, X_n$ be independent and identically distributed Poisson random variables with means equal to λ/n.)
    (c) Show that every negative binomial probability function is the probability function of a sum of independent and identically distributed geometric random variables.
    (d) Show that every gamma probability density function of the type given in Statement 4 of Section 8.4 is the probability density function of a sum of independent and identically distributed exponential random variables.

19. Use the central limit theorem to justify Statements 1, 2, 3, and 4 of Section 8.6. (*Hint:* apply the central limit theorem to the conclusions of Exercise 18.)

8.7   Products of Random Variables

Products of random variables are defined in a manner that is similar to the definition of sums of random variables. For that reason we are including a discussion of products of random variables in this chapter. The formal definition of products of random variables follows.

Definition 8.3   Let $X_1, X_2, \ldots, X_n$ be random variables defined on a sample space S. For each point s in S define

$$Y(s) = X_1(s)X_2(s)\cdots X_n(s)$$

The random variable Y is called the product of the random variables $X_1, X_2, \ldots, X_n$, and we write

$$Y = X_1 X_2 \cdots X_n$$

Note: the definition of sums of random variables and of products of random variables is extended to the definition of a function of random variables as follows. Let $X_1, X_2, \ldots, X_n$ be random variables defined on a sample space S. Let $h(x_1, x_2, \ldots, x_n)$ be a function of n variables. For each point s in S define $h(X_1, X_2, \ldots, X_n)(s) = h[X_1(s), X_2(s), \ldots, X_n(s)]$.

EXAMPLE 8.25

The following are examples of random variables that are products of random variables:

   (a) If the length and width of a sheet of aluminum cut by a certain process are random variables X and Y, respectively, then the area of such a sheet of aluminum is the random variable $Z = XY$.

   (b) If the length, width, and height of a paper carton manufactured by a certain process are random variables X, Y, and Z, respectively, then the volume of such a carton is the random variable $V = XYZ$.

EXAMPLE 8.26

Suppose that the length and width (measured in feet) of a sheet of aluminum cut by a certain process are random variables X and Y, respectively, with joint probability density function given by

$$f(x, y) = \begin{cases} 25 & \text{for } .95 < x < 1.15, \quad .95 < y < 1.15 \\ 0 & \text{elsewhere} \end{cases}$$

Find the probability that a sheet of aluminum cut by this process will have an area that is less that one square foot.

SOLUTION

P(aluminum sheet has an area that is less than one square foot)

$$= P(XY < 1) = \iint_C f(x, y) \, dx \, dy$$

where the region C is the region of all points (x, y) such that $xy < 1$. Since $f(x, y) = 0$ except where $.95 < x < 1.15$ and $.95 < y < 1.15$, the integration can be further restricted to the region of all points (x, y) such that $xy < 1$, $.95 < x < 1.15$, and $.95 < y < 1.15$. In other words, the integration is over the region of all points that are within the rectangle $.95 < x < 1.15$ and $.95 < y < 1.15$, and that are below the curve defined by the equation $xy = 1$ (i.e., the curve of the function $y = 1/x$). (See the illustration that follows.) Therefore, the probability we seek is

$$P(XY < 1) = \int_{.95}^{1.05} \int_{.95}^{1/y} 25 \, dx \, dy = 25 \int_{.95}^{1.05} \left(\frac{1}{y} - .95 \right) dy$$

$$= 25(1n \ y - .95y) \Big|_{.95}^{1.05} = .127$$

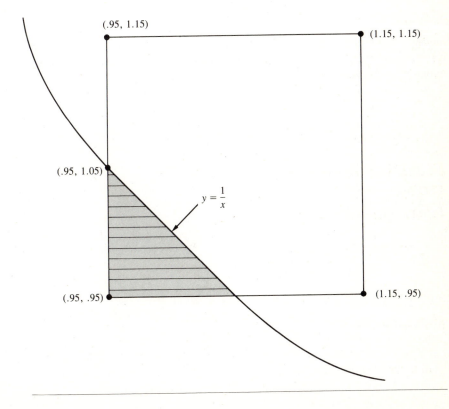

The mathematical expectation of a product of random variables is defined as follows.

Definition 8.4

(a) Let $X_1, X_2, \ldots, X_n$ be discrete random variables with multivariate probability function $f(x_1, x_2, \ldots, x_n)$. The mathematical expectation of $X_1 X_2 \cdots X_n$ is the quantity

$$E(X_1 X_2 \cdots X_n) = \sum_{x_n} \cdots \sum_{x_2} \sum_{x_1} x_1 x_2 \cdots x_n f(x_1, x_2, \ldots, x_n)$$

provided that the series converges absolutely.

(b) Let $X_1, X_2, \ldots, X_n$ be continuous random variables with multivariate probability density function $f(x_1, x_2, \ldots, x_n)$. The mathematical expectation of $X_1 X_2 \cdots X_n$ is the quantity

$$E(X_1 X_2 \cdots X_n)$$
$$= \int_{-\infty}^{\infty} \cdots \int_{-\infty}^{\infty} \int_{-\infty}^{\infty} x_1 x_2 \cdots x_n f(x_1, x_2, \ldots, x_n) \, dx_1 \, dx_2 \cdots dx_n$$

provided that the integral exists.

Note: the definition of the mathematical expectation of a product of random variables is extended to the definition of the mathematical expectation of a function of random variables as follows. If the random variables are discrete, then

$$E[h(X_1, X_2, \ldots, X_n)] = \sum_{x_n} \cdots \sum_{x_2} \sum_{x_1} h(x_1, x_2, \ldots, x_n) f(x_1, x_2, \ldots, x_n)$$

provided that the series converges absolutely. If the random variables are continuous, then

$$E[h(X_1, X_2, \ldots, X_n)]$$
$$= \int_{-\infty}^{\infty} \cdots \int_{-\infty}^{\infty} \int_{-\infty}^{\infty} h(x_1, x_2, \ldots, x_n) f(x_1, x_2, \ldots, x_n) \, dx_1 \, dx_2 \cdots dx_n$$

provided that the integral exists.

EXAMPLE 8.27

Let Z be the area of a sheet of aluminum cut by the process of Example 8.25. Find the mean and variance of Z.

SOLUTION

$$E(Z) = E(XY) = \int_{.95}^{1.15} \int_{.95}^{1.15} xy(25) \, dx \, dy = 1.1025$$

$$Var(Z) = E(Z^2) - E^2(Z) = E(X^2 Y^2) - 1.215$$

$$= \int_{.95}^{1.15} \int_{.95}^{1.15} x^2 y^2(25) \, dx \, dy - 1.215 = 1.1105$$

The following theorem relates expectation with independence.

Theorem 8.4   *If X and Y are independent random variables, then*

$$E(XY) = E(X)E(Y)$$

Proof   If X and Y are discrete and independent random variables, then

$$E(XY) = \sum_y \sum_x xyf(x, y) = \sum_y \sum_x xyf_X(x)f_Y(y)$$

$$= \sum_x xf_X(x) \sum_y yf_Y(y) = E(X)E(Y)$$

The proof for continuous random variables is left as an exercise for the reader. (*Hint:* replace summations with integrals.)

Two quantities that are of considerable importance in the study of random variables are the covariance and the correlation coefficient of random variables. The definition of these quantities is in terms of the mathematical expectation of the product of two random variables and is given below. A discussion of the use of these quantities in analyzing the relationship between random variables is given in Chapter 14 under the topic of correlation analysis.

Definition 8.5   Let X and Y be random variables with means equal to μ_x and μ_y, respectively, and variances equal to σ_x^2 and σ_y^2, respectively. The covariance of X and Y is the quantity

$$Cov(X, Y) = E[(X - \mu_x)(Y - \mu_y)]$$

The correlation coefficient of X and Y is the quantity

$$\rho(X, Y) = E\left(\frac{X - \mu_x}{\sigma_x} \cdot \frac{Y - \mu_y}{\sigma_y}\right)$$

The following theorem offers an alternate expression for the covariance and for the correlation coefficient of random variables X and Y.

Theorem 8.5   *Let X and Y be random variables with means equal to μ_x and μ_y, respectively, and variances equal to σ_x^2 and σ_y^2, respectively. Then*

$$\text{Cov}(X, Y) = E(XY) - \mu_x\mu_y \quad \text{and} \quad \rho(X, Y) = \frac{\text{Cov}(X, Y)}{\sigma_x\sigma_y} = \frac{E(XY) - \mu_x\mu_y}{\sigma_x\sigma_y}$$

Proof

$$\text{Cov}(X, Y) = E[(X - \mu_x)(Y - \mu_y)] = E[XY - \mu_xY - \mu_yX + \mu_x\mu_y]$$
$$= E(XY) - \mu_xE(Y) - \mu_yE(X) + \mu_x\mu_y$$
$$= E(XY) - \mu_x\mu_y - \mu_x\mu_y + \mu_x\mu_y = E(XY) - \mu_x\mu_y$$
$$\rho(X, Y) = E\left(\frac{X - \mu_x}{\sigma_x} \cdot \frac{Y - \mu_y}{\sigma_y}\right) = (1/\sigma_x\sigma_y)E[(X - \mu_x)(Y - \mu_y)]$$
$$= (1/\sigma_x\sigma_y)\text{Cov}(X, Y) = [E(XY) - \mu_x\mu_y]/(\sigma_x\sigma_y)$$

EXAMPLE 8.28

(a)   Let X and Y be discrete random variables with means equal to $46/21$ and $33/21$, respectively, with variances equal to $278/441$ and $108/441$, respectively, and with joint probability function given by

$$f(x, y) = (x + y)/21 \quad \text{for} \quad x = 1, 2, 3; \quad y = 1, 2.$$

Find the covariance and correlation coefficient of X and Y.

(b)   Let X and Y be continuous random variables with means equal to $7/12$ and $7/12$, respectively, with variances equal to $11/144$ and $11/144$, respectively, and with joint probability density function given by

$$f(x, y) = \begin{cases} x + y & \text{for} \quad 0 < x < 1, \quad 0 < y < 1 \\ 0 & \text{elsewhere} \end{cases}$$

Find the covariance and correlation coefficient of X and Y.

SOLUTION

(a)

$$E(XY) = \sum_y \sum_x xyf(x, y)$$
$$= (1)(1)(2/21) + (2)(1)(3/21) + (3)(1)(4/21)$$
$$+ (1)(2)(3/21) + (2)(2)(4/21) + (3)(2)(5/21)$$
$$= 72/21$$

Therefore

$$Cov(X, Y) = E(XY) - \mu_x\mu_y = 72/21 - (46/21)(33/21)$$
$$= -2/147 = -.0136$$

and

$$\rho(X, Y) = \frac{Cov(X, Y)}{\sigma_x\sigma_y} = \frac{-.0136}{\sigma_x\sigma_y} = -.0346$$

(b)

$$E(XY) = \int_{-\infty}^{\infty} \int_{-\infty}^{\infty} xyf(x, y) \, dx \, dy = \int_{0}^{1} \int_{0}^{1} xy(x + y) \, dx \, dy$$
$$= 1/3$$

Therefore

$$Cov(X, Y) = E(XY) - \mu_x\mu_y = (1/3) - (7/12)(7/12) = -.0069$$

and

$$\rho(X, Y) = \frac{Cov(X, Y)}{\sigma_x\sigma_y} = \frac{-.0069}{\sigma_x\sigma_y} = -.0903$$

Exercises

1. Find the probability that a sheet of aluminum cut by the process of Example 8.26 will have an area that is less than 1.21 square feet.

2. Suppose that the mass (measured in grams) and the acceleration (measured in centimeters per second square) of a certain moving object may be looked upon as random variables X and Y, respectively, with joint probability density function given by

$$f(x, y) = \begin{cases} 1/10 & \text{for } 1 < x < 2, \quad 10 < y < 15 \\ 0 & \text{elsewhere} \end{cases}$$

    (a)   Find the probability that the object will exert a force that is less than 15 dynes.
    (b)   Find the probability that the object will exert a force that is more than 25 dynes. (*Note:* the force (in dynes) exerted by an object is equal to the product of the mass (in grams) of the object and the acceleration (in centimeters per second square) of the object.)

3. Let Z be the force exerted by the object of Exercise 2. Find the mean and variance of Z.

4. Consider two rolls of a die. Let X and Y be the values on the first and second rolls, respectively. Find (a) $E(XY)$, (b) $\text{Var}(XY)$, (c) $\text{Cov}(X, Y)$, and (d) $\rho(X, Y)$.

5. Show that if X and Y are independent random variables, then (a) $\text{Cov}(X, Y) = 0$, and (b) $\rho(X, Y) = 0$.

6. The joint probability function of the discrete random variables X and Y is given by the values $f(-1, 2) = .3$, $f(3, 2) = .1$, $f(4, 2) = .2$, $f(-1, 5) = .2$, $f(3, 5) = .1$, and $f(4, 5) = .1$. Find (a) $E(XY)$, (b) $\text{Var}(XY)$, (c) $\text{Cov}(X, Y)$, and (d) $\rho(X, Y)$. Are the random variables independent?

7. The joint probability density function of the continuous random variables X and Y is given by

$$f(x, y) = \begin{cases} (2/3)(2x + y) & \text{for } 0 < x < 1, \ 0 < y < 1 \\ 0 & \text{elsewhere} \end{cases}$$

Find (a) $E(XY)$, (b) $\text{Var}(XY)$, (c) $\text{Cov}(X, Y)$, and (d) $\rho(X, Y)$.

Part II

STATISTICAL INFERENCE

Chapter *9*

DESCRIPTIVE STATISTICS

9.1   Introduction

Descriptive statistics is concerned with the organization and reduction of data into meaningful summaries and graphs. Although the topic technically is not a part of statistical inference, we have included it here because methods of descriptive statistics are sometimes applied to random samples (discussed in Chapter 10), and random samples are fundamental to the subject of statistical inference. (This chapter may, nevertheless, be omitted if the reader is interested primarily in those topics of statistical inference that are covered in the remaining chapters of this book, because the methods of descriptive statistics that are necessary for an understanding of the later chapters are briefly summarized in Chapter 10.) Surveys (such as the U.S. census) and experiments (such as a national pilot program in remedial mathematics) often result in the accumulation of an enormous quantity of data. Raw data, however, are generally useless unless they are organized or reduced to more suitable forms. In this chapter we shall discuss some of the more important standard techniques that can be used to accomplish that task.

9.2   Frequency Distributions

Very often, the first step in data analysis is its arrangement into a frequency distribution. A frequency distribution is an organization of the data into separate classes with a count of the number of elements in each class. The organized data are often presented in the form of a table of values called a **frequency table**. If desired, the information in a frequency table can then be used to construct a relative frequency table and/or a percentage table for the data. The format of such tables varies according to the nature of the data and the type of information desired. The following examples illustrate the construction of a variety of frequency tables, relative frequency tables, and percentage tables.

EXAMPLE 9.1

In a survey of the size of families in a certain neighborhood the following set of data of the number of persons in each family was obtained:

$$2, 2, 5, 6, 3, 3, 7, 4, 7, 5, 2, 2, 2, 4, 3, 5, 9,$$
$$1, 1, 2, 5, 4, 3, 2, 6, 4, 4, 3, 2, 1, 1, 2, 4, 1,$$
$$2, 2, 6, 6, 3, 4, 3, 1, 6, 2, 5, 4, 2, 5, 4, 2, 6,$$
$$3, 3, 1, 5, 3, 4, 7, 2, 5, 5, 3, 4, 2, 3, 4, 4, 5,$$
$$1, 1, 3, 2, 4, 3, 5, 5, 4, 2, 4, 3, 4, 4, 7, 5, 5.$$

A table of values that gives the frequency and relative frequency of each family size may be constructed by reading through the list of family sizes, keeping a running tally of the different family sizes, and then calculating the frequency and relative frequency of each family size. See Table 9.1. The markings for the running tally are generally omitted from the final presentation of the data.

TABLE 9.1
Frequency and Relative Frequency Distributions of Family Size
(survey of 85 families)

(1) Family size	(2) Tally	(3) Frequency (f)	(4) Relative frequency (f/n)
1	~~1111~~ 1111	9	.11
2	~~1111~~ ~~1111~~ ~~1111~~ 111	18	.21
3	~~1111~~ ~~1111~~ ~~1111~~	15	.18
4	~~1111~~ ~~1111~~ ~~1111~~ 111	18	.21
5	~~1111~~ ~~1111~~ 1111	14	.16
6	~~1111~~ 1	6	.07
7	1111	4	.05
8		0	.00
9	1	1	.01

The data in the table show at a glance that the number of families in the survey with 1, 2, 3, 4, 5, 6, 7, 8, and 9 members are, respectively, 9, 18, 15, 18, 14, 6, 4, 0, and 1. Furthermore, the proportion (i.e., relative frequency) of families in the survey with 1, 2, 3, 4, 5, 6, 7, 8, and 9 members are, respectively .11, .21, .18, .21, .16, .07, .05, .00, and .01. The information in columns (1) and (3) constitutes a frequency table for the data. The information in columns (1) and (4) constitutes a relative frequency table for the data. The information in columns (3) and (4) is called, respectively, a **frequency distribution** and a **relative frequency distribution** of the data.

EXAMPLE 9.2

In a survey of the weekly income of taxicab drivers in a certain city, the following set of incomes, rounded off to the nearest dollar, was obtained.

225, 195, 145, 200, 215, 227, 231, 262, 295, 333, 245,
213, 187, 197, 340, 233, 256, 296, 288, 167, 245, 303,
175, 270, 188, 267, 275, 285, 305, 312, 299, 266, 214,
155, 168, 257, 288, 232, 250, 256, 288, 267, 256, 255,
177, 245, 232, 213, 286, 257, 313, 317, 178, 199, 299,
210, 213, 285, 290, 250, 214, 241, 231, 250, 194, 182.

A scan of the data shows that the smallest and the largest weekly incomes are $145 and $340, respectively. Furthermore, we can see that there are many distinct values. Consequently, instead of counting the number of elements belonging to each distinct value, it seems reasonable to group the data into subgroups or classes and then count the number of elements in each class. A frequency distribution and a relative frequency distribution of these data, with weekly incomes grouped into classes, are given in Table 9.2.

TABLE 9.2
Frequency and Relative Frequency Distributions of Weekly Incomes
(survey of 66 weekly incomes)

Weekly income	Frequency	Relative frequency
125–149	1	.02
150–174	3	.05
175–199	10	.15
200–224	8	.12
225–249	11	.17
250–274	14	.21
275–299	12	.18
300–324	5	.08
325–349	2	.03

The information in the table provides the following summary of the survey: the number of taxicab drivers with weekly incomes in the classes (range of values) of 125–249, 150–174, 175–199, 200–224, 225–249, 250–274, 275–299, 300–324, and 325–349, are, respectively, 1, 3, 10, 8, 11, 14, 12, 5, and 2. Furthermore, the proportion (i.e., relative frequency) of taxicab drivers in each of these respective classes are .02, .05, .15, .12, .17, .21, .18, .08, and .03.

The numbers 125 and 149 are called the **class limits** of the first class, the numbers 150 and 174 are called the class limits of the second class,..., and the numbers 325 and 349 are called the class limits of the last class in the frequency table. Since the weekly incomes were rounded off to the nearest dollar, every weekly income in the survey belongs to one of the classes in this frequency table (Table 9.2.) However, if the data had not been rounded off to the nearest dollar, a weekly income such as $149.50 would not belong to any of the classes in the given frequency table. To take care of problems of this sort (and also to accommodate data obtained from a continuous distribution of values), we could use **class boundaries** instead of class limits to define the classes in a frequency table. A frequency table of the data of Example 9.2 with classes defined by class boundaries instead of class limits is given in Table 9.3. The class boundaries of the first class in this table are the numbers 124.50 and 149.50 (the class consists of all numbers from 124.50 up to, but not including, 149.50), the class boundaries of the second class in this table are the numbers 149.50 and 174.50 (the class consists of all numbers from 149.50 up to, but not including, 174.50),..., and the class boundaries of the last class are the numbers 324.50 and 349.50. The midpoint of each class is called the **class mark** of the class. For example, the class mark of the first class in Table 9.2—as well as Table 9.3—is 137. The class mark of the second class in each of these frequency tables is 162. When all the classes in a frequency table are of the same length, as they are in both of these frequency tables, the classes are called **class intervals**, and the length of the interval between any two successive class marks is called the **length of the class interval**. In both Tables 9.2 and 9.3 the length of the class interval is 25.

TABLE 9.3
Frequency Distribution of Weekly Incomes
(survey of 66 weekly incomes)

Weekly income	Frequency
124.50–under 149.50	1
149.50–under 174.50	3
174.50–under 199.50	10
199.50–under 224.50	8
224.50–under 249.50	11
249.50–under 274.50	14
274.50–under 299.50	12
299.50–under 324.50	5
324.50–under 349.50	2

A different choice of class size would result in both a different frequency distribution and a different relative frequency distribution. For example, another frequency distribution and relative frequency distribution for the same set of data are given in Table 9.4.

TABLE 9.4
Frequency and Relative Frequency Distributions of Weekly Incomes
(survey of 66 weekly incomes)

Weekly income	Frequency	Relative frequency
100–149	1	.02
150–199	13	.20
200–249	19	.29
250–299	26	.39
300–349	7	.11

In each instance of data reduction, the choice of class size and of the number of classes depends to a large extent on the number of elements in the set of data to be reduced and on the range of values of the data (i.e., the difference between the largest number and the smallest number in the set of data). However, it is seldom meaningful to use fewer than 5 classes. On the other hand, the use of more than 15 classes tends to defeat the purpose of a substantial reduction of the data. Generally speaking, it is also advisable to use classes of equal size in any one frequency table. However, there are occasions, such as in the two examples that follow, when we may wish to use classes of different sizes.

EXAMPLE 9.3

The frequency and percentage distributions of SAT scores of Roosevelt High School's class of 1983 are given in Table 9.5. The table of values shows that only 2% of the class obtained SAT scores lower than or equal to 300, whereas 8% of the class scored higher than 700.

TABLE 9.5
Frequency and Percentage Distributions of SAT Scores
(class of 1982; 122 SAT scores)

SAT score	Frequency	Percentage
Lower than 301	2	2
301–400	5	4
401–500	30	25
501–600	50	41
601–700	25	20
701–800	10	8

EXAMPLE 9.4

The frequency and percentage distributions of IQ scores of students enrolled in the Emerson School for Gifted Children are given in Table 9.6. The table of values shows that all the students have IQ scores of at least 139. Furthermore, 6% of the students have IQ scores of 160 or higher.

The class of all SAT scores that are lower than 301 and the class of all IQ scores that are 160 or higher in Tables 9.5 and 9.6, respectively, are called **open classes** because they are classes lacking either a lower limit (or lower boundary) or an upper limit (or upper boundary). Open classes are particularly useful when very few elements of the data are in either the lower or upper ends of the data range, as in the case of the number of SAT scores lower than 301 or as in the case of the number of IQ scores greater than 160.

With the exception of Table 9.3, in each of these examples we have given either the relative frequency distribution or the percentage distribution along with the frequency distribution of the data. In many instances, such as in Table 9.3, only one of the distributions will be given. However, if the total number of items in the set of data is known, any one of these distributions may be converted to any of the other of these distributions.

In addition to the frequency distribution, it is sometimes useful to reduce a set of numerical data into a format that provides at a quick glance the respective number (frequency) and proportion (relative frequency) of elements that are less than or equal to each of a number of successively larger values.

TABLE 9.6
Frequency and Percentage Distributions of IQ Scores
(Emerson School for Gifted Children; 348 students)

IQ	Frequency	Percentage
139–144	28	8
145–149	87	25
150–154	115	33
155–159	97	28
160 or higher	21	6

When this is done, we have what is called a **cumulative frequency distribution**. A cumulative frequency distribution of the data of Example 9.1 is given in Table 9.7. At one glance, the table provides the information that there are 9, 27, 42, 60, 74, 80, 84, 84, and 85 families with no more than 1, 2, 3, 4, 5, 6, 7, 8, and 9 members, respectively. In addition, the table shows that the proportion of families with no more than 1, 2, 3, 4, 5, 6, 7, 8, and 9 members are, respectively, .11, .32, .50, .71, .87, .94, .99, .99, and 1.00. A cumulative frequency distribution of the data of Example 9.2 is given in Table 9.8; it indicates that, among the taxicab drivers in the survey, 1 has a weekly income no greater than $149, 4 have weekly incomes no greater than $174, 14 have weekly incomes no greater than $199,..., and 66 (i.e., all of them) have weekly incomes no greater than $349. The Cumulative Relative Frequency column shows that 2% of the cabdrivers in the survey have weekly incomes no greater than $149, 7% of them have weekly incomes no greater than $174,..., and 100% of them have weekly incomes no greater than $349.

TABLE 9.7
Cumulative Frequency and Relative Frequency Distributions of Family Size
(survey of 85 families)

Family size	Cumulative frequency	Cumulative relative frequency
Less than or equal to 1	9	.11
Less than or equal to 2	27	.32
Less than or equal to 3	42	.50
Less than or equal to 4	60	.71
Less than or equal to 5	74	.87
Less than or equal to 6	80	.94
Less than or equal to 7	84	.99
Less than or equal to 8	84	.99
Less than or equal to 9	85	1.00

TABLE 9.8
Cumulative Frequency and Relative Frequency Distributions of Weekly Incomes
(survey of 66 incomes)

Weekly income	Cumulative frequency	Cumulative relative frequency
No more than $149	1	.02
No more than $174	4	.07
No more than $199	14	.21
No more than $224	22	.33
No more than $249	33	.50
No more than $274	47	.71
No more than $299	59	.89
No more than $324	64	.97
No more than $349	66	1.00

Exercises

1. The scores received by a certain group of students in a certain examination are:

    1, 1, 5, 5, 6, 9, 10, 9, 7, 8, 4, 9, 10, 0, 2, 5, 6, 8, 7, 6, 5, 5, 2, 8, 3, 3, 4, 3, 4, 6,
    5, 6, 8, 3, 5, 8, 7, 5, 5, 5, 6, 3, 7, 7, 6, 6, 7, 3, 4, 4, 4, 6, 6, 7, 4, 6, 6, 4,

    Let each of the possible scores 0, 1, 2, 3, 4, 5, 6, 7, 8, 9, and 10 be a separate class.

    (a)   Construct a frequency distribution, a relative frequency distribution, and a percentage distribution for this set of data.

    (b)   Construct a cumulative frequency distribution and a cumulative relative frequency distribution for this set of data.

2. The running times, measured in minutes, of a group of runners completing a run around a certain lake are:

    10.10, 12.11, 15.15, 16.05, 11.67, 11.54, 11.86, 15.90, 12.35, 11.65, 12.33,
    10.23, 11.88, 11.75, 11.94, 12.04, 13.66, 13.45, 10.57, 10.75, 12.43, 12.60,
    12.75, 13.00, 13.15, 13.75, 14.00, 14.56, 12.91, 13.45, 14.72, 12.50, 13.25,
    14.60, 14.88, 13.23, 12.09, 13.07, 14.08, 14.92.

    Let the classes be the intervals 10.00–10.99, 11.00–11.99, 12.00–12.99, 13.00–13.99, 14.00–14.99, 15.00–15.99, and 16.00–16.99.

    (a)   Construct a frequency distribution, a relative frequency distribution, and a percentage distribution for this set of data.

    (b)   Construct a cumulative frequency distribution and a cumulative relative frequency distribution for this set of data.

    (c)   Find the class marks of each of the classes.

3. The age in years of the employees of a certain company are given below:

    55, 25, 34, 67, 28, 45, 42, 39, 25, 66, 32, 28, 26, 56, 48, 49, 32, 51, 23, 44,
    47, 61, 25, 24, 31, 33, 35, 47, 42, 30, 24, 28, 44, 37, 51, 57, 24, 39, 49, 44,
    47, 35, 32, 36, 50, 43, 47, 47, 24, 22, 31, 36, 27, 27, 44, 37, 36, 41, 29, 28.

    Let the classes be the intervals 20–24, 25–29, 30–34, 35–39, 40–44, 45–49, 50–54, 55–59, 60–64, 65–69.

    (a)   Construct a frequency distribution, a relative frequency distribution, and a percentage distribution for this set of data.

    (b)   Construct a cumulative frequency distribution and a cumulative relative frequency distribution for this set of data.

    (c)   Find the class marks of each of the classes.

4. Use class boundaries instead of class limits to construct a frequency table for the set of data given in Exercise 2. Choose the class boundaries so that the respective classes of this frequency table will contain the same data as the classes of the frequency table constructed in Exercise 2(a). Let 10 be the lower boundary of the first class.

5. Use class boundaries and thirteen classes of length .5 to construct a frequency table for the set of data given in Exercise 2. Let 10 be the lower boundary of the first class.

9.3   Graphs of Frequency Distributions

The standard graphic forms of presentation of frequency distributions are bar charts, frequency polygons, and histograms. If the frequency distribution is similar in format to the frequency distribution of Example 9.1 (Table 9.1), that is, if each class consists of a single value, then the appropriate graphic presentation is a bar chart. A bar chart of the frequency distribution of Table 9.1 is given in Figure 9.1.

   If the frequency distribution is similar in format to the frequency distribution of Example 9.2 (Table 9.2), that is, if each class consists of an interval of values, then appropriate graphical presentations are frequency polygons and histograms. A frequency polygon of the frequency distribution of Table 9.2 is given in Figure 9.2. The frequency polygon is obtained by plotting the points (x_i, f_i), (where x_i is the midpoint, also called the class mark, of the ith class, and f_i is the corresponding frequency), and then drawing straight-line segments to connect the successive points. It is customary to add

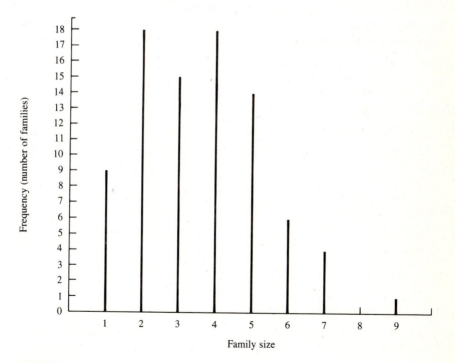

FIGURE 9.1
Bar Chart of Family Size

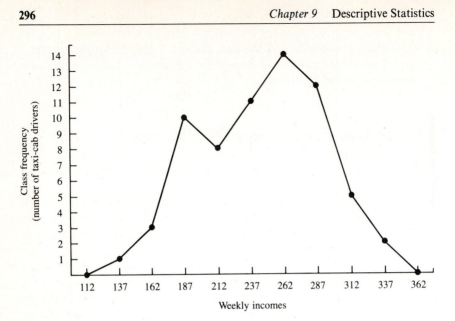

FIGURE 9.2
Frequency Polygon of Weekly Incomes

classes with zero frequencies at each end of the frequency distribution in order to obtain a frequency polygon that begins and ends at the horizontal axis.

A histogram of the frequency distribution of Table 9.2 is given in Figure 9.3. The histogram of a frequency distribution is similar to the histogram of a probability distribution; the midpoint of the base of the ith rectangle is the

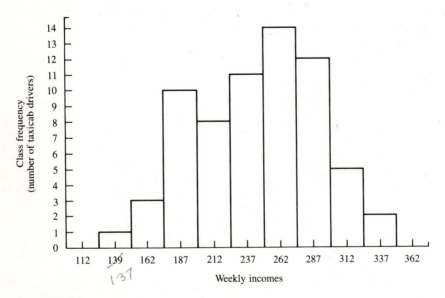

FIGURE 9.3
Histogram of Weekly Incomes

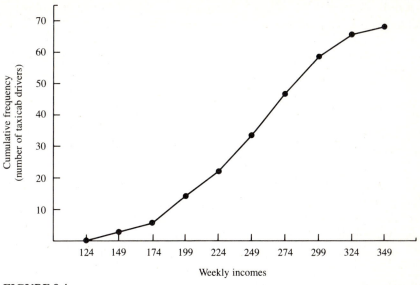

FIGURE 9.4
Ogive of Weekly Income

midpoint (or class mark) of the ith class, and the height of the ith rectangle is the class frequency of the ith class.

Graphs of cumulative frequency distributions are generally presented in terms of **ogives**. An ogive of the cumulative distribution of Table 9.8 is given in Figure 9.4. To construct an ogive, plot the points (x_i, F_i) where x_i is the larger boundary of the ith class (also called the upper class boundary), and F_i is the cumulative frequency for the ith class. Straight-line segments connecting the successive points form the graph of the ogive.

Exercises

1. Construct a bar chart, a frequency polygon, and an ogive for the data of Exercise 1 of Section 9.2.

2. Construct a histogram, a frequency polygon, and an ogive for the data of Exercise 2 of Section 9.2.

3. Construct a histogram, a frequency polygon, and an ogive for the data of Exercise 3 of Section 9.2.

9.4   Descriptive Measures

The mean, median, mode, variance, standard deviation, and various percentiles (as well as fractiles and quartiles) of a set of data are descriptive measures that provide a quick summary of the distribution of the data. The formal definition of these descriptive measures follows. In each case assume that the set of data consists of the numerical values $x_1, x_2, \ldots, x_n$.

Definition 9.1   The **mean** of the set $\{x_1, x_2, \ldots, x_n\}$ of numbers is the quantity

$$\bar{x} = \frac{x_1 + x_2 + \cdots + x_n}{n}$$

It is also called the **arithmetic mean** or the **average** of the set of numbers.

If only one single number may be used to characterize a set of numbers, that number would most likely be the mean (average) of the set. We are all familiar with measurements such as the average score in a test, the average income of high school teachers, or the average life span of white males in the U.S.A. Very often, the mean of a set of data gives us a good idea of the center of the set of data because if the histogram of a set of data is symmetric, then the midpoint of the histogram is the mean of the set of data. In other words, if the histogram is symmetric, then it is symmetric about the point $\bar{x}$ (see Figure 9.5). If the histogram is only more or less symmetric, then the mean is, of course, only an approximate measurement of the midpoint of the histogram.

Another property of the mean may be described through the following examples. If the sum of the incomes of a group of people is to be equally redistributed, then each person in the group should receive an amount that is equal to the mean of the original set of incomes. If every student in a certain class had done equally well on a given test, every student would receive the average of the test scores. In other words, if $x_1 = x_2 = \cdots = x_n$, then $\bar{x} = x_1 = x_2 = \cdots = x_n$.

If the set of data under consideration is the set of observed values of a random sample from a larger set—that is, if the set of data can be considered to be a set of observed values of a random variable—then the mean of the set is

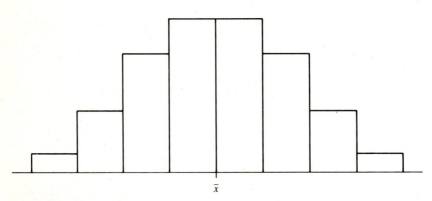

FIGURE 9.5
Location of the Mean when the Histogram is Symmetric

a measurement of great importance. The reason for this is that if the number n of elements in the set of data is large, it is reasonable to assume that the mean of the set is approximately the same as the mean of the random variable (see Chapter 6 on the strong law of large numbers). Thus the mean of the set provides important insights into the nature of the random variable. For example, suppose that the set $\{x_1, x_2, \ldots, x_n\}$ consists of n observations on the number of cars waiting in line at a certain toll booth. If the number of cars waiting in line at this toll booth may be assumed to be a random variable X with a Poisson distribution, then, if n is large, it would be reasonable to assume that the mean $\bar{x}$ of the set of data is approximately the same as the mean λ of the random variable X. Since the mean of a Poisson random variable completely determines the distribution of the random variable, we can appreciate the importance of the mean of the set of data in this example. The use of the mean of a set of data to make inferences concerning the distribution of a random variable or population will be explored in great detail in the chapters to follow.

The formula in Definition 9.1 is used to compute the mean of a set of data when the raw data are available. Sometimes, however, it is necessary to estimate the mean of a set of data from its frequency table, which may be available when the set of raw data is not. In such cases an estimate of the mean may be obtained by finding the product of the frequency of each class with the class mark (i.e., midpoint) of the class, finding the sum of these products, then dividing the sum by the total number of classes in the frequency table. In other words, an approximation of the mean of the set is given by the formula

$$\bar{x} \approx \frac{\sum\limits_{i=1}^{n} f_i m_i}{n}$$

where f_i is the frequency of the ith class, m_i is the class mark of the ith class, and n is the total number of classes in the frequency table.

EXAMPLE 9.5

(a) Use the raw data given in Example 9.2 to find the mean of the set of data.
(b) Use the frequency table given in Table 9.2 to estimate the mean of the set of data of Example 9.2.

SOLUTION

(a) The mean of the set of data is

$$\bar{x} = \frac{225 + 195 + 145 + \cdots + 182}{66} = 243.97$$

(b) To estimate the mean from the frequency table we must first find the class marks for each of the classes. They are: 137, 162, 187, 212, 237, 262, 287,

312, and 337. Therefore, an approximate value of the mean is given by

$$\bar{x} = \frac{1 \cdot 37 + 3 \cdot 162 + 10 \cdot 187 + 8 \cdot 212 + 11 \cdot 237 + 14 \cdot 262 + 12 \cdot 287 + 5 \cdot 312 + 2 \cdot 337}{66}$$

$$\approx 244.58$$

Definition 9.2 Suppose that the numbers in the set $\{x_1, x_2, \ldots, x_n\}$ are arranged so that $x_1 \leq x_2 \leq x_3 \leq \cdots \leq x_n$. The **median** of the set is the number

$$m = \begin{cases} \dfrac{x_{n+1}}{2} & \text{if } n \text{ is odd} \\[2ex] \dfrac{x_{n/2} + x_{(n/2)+1}}{2} & \text{if } n \text{ is even} \end{cases}$$

In other words, the median of a set of n numbers is the number that is in the middle of the arrangement $x_1 \leq x_2 \leq x_3 \leq \cdots \leq x_n$, if there is a single number in the middle. Otherwise, it is the average of the two numbers that are in the middle of the arrangement.

Although the mean is the most frequently used measure of central tendency, it can be a misleading indicator of the center of a set of data if the histogram of the frequency distribution of the data is highly **skewed** (i.e., if much of the area under the histogram is either to the left or to the right of the center of the histogram's baseline). For example, suppose that a billionaire resides in a neighborhood that houses 9 other families of only moderate wealth (e.g., with family incomes of approximately $100,000). The average income of families in this neighborhood is over $10 million — which is a poor description of the distribution of incomes and a terrible measure of the middle income in this neighborhood! The median income, on the other hand, is only approximately $100,000, which is the exact middle among incomes in this neighborhood, thus providing a better description of the distribution of incomes. This example illustrates the importance of the median as a descriptive measure of the distribution of a set of data. Figure 9.6 compares the

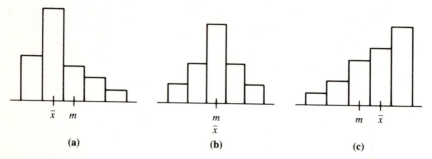

FIGURE 9.6

Comparison of the location of the mean $\bar{x}$ and the median m when the histogram is (a) skewed to the right; (b) symmetric; (c) skewed to the left.

location of the mean and median with a histogram that is skewed to the right, symmetric, and skewed to the left.

Definition 9.3 A number x is called a **mode** of the set if x occurs at least as frequently as any other number in the set. For example, the mode of the set $\{2, -3, 4, -3, 5, -3\}$ is -3, and the modes of the set $\{1, -1, 2, 2, 3, 5, 6, 6\}$ are 2 and 6. If the data are grouped into classes, a modal class is a class which occurs at least as frequently as any other class. A modal class is represented by the class mark (midpoint of the class), and this class mark is also referred to as a mode.

The word "mode" is from the French and means "fashion." Thus, a modal class is a fashionable or popular, class. With reference to Example 9.1, we see that with respect to family sizes, the modes are 2 and 4. Thus, the most fashionable family sizes are 2 and 4. With reference to Example 9.2, we see that with respect to weekly incomes the modal class is defined by the limits 250–274. Thus, more of the taxicab drivers in the survey have weekly incomes between \$250 and \$274 than between any of the other class limits. The incomes between \$250 and \$274 are, so to speak, the fashionable incomes. A comparison of the locations of the mean $\bar{x}$, the median m, and the mode (or modes) for each of three histograms are given in Figure 9.7.

Definition 9.4

(a)   The **variance** of the set $\{x_1, x_2, \ldots, x_n\}$ of numbers is the quantity

$$\sigma^2 = \frac{\sum\limits_{i=1}^{n} (x_i - \bar{x})^2}{n}$$

where $\bar{x}$ is the mean of the set.

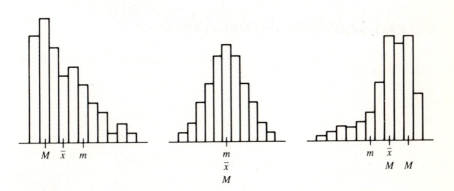

FIGURE 9.7

Location of the mean $\bar{x}$, median m, and mode M

(b)   Let $\{x_1, x_2, \ldots, x_n\}$ be the set of observed values of a random sample of size n. The observed value of the **sample variance** is the quantity

$$s^2 = \frac{\sum\limits_{i=1}^{n} (x_i - \bar{x})^2}{n - 1}$$

where $\bar{x}$ is the mean of the set.

Note: random samples are defined in Chapter 10. For the present, it is sufficient to think of "the set of observed values of a random sample" as a set of data obtained from a larger set of data.

Definition 9.5
(a)   The positive square root of the variance is called the **standard deviation of the set** and is denoted by σ.
(b)   The positive square root of the observed value of the sample variance is called the **observed value of the sample standard deviation** and is denoted by s.

The variance is the average of the terms $(x_1 - \bar{x})^2$, $(x_2 - \bar{x})^2, \ldots$, and $(x_n - \bar{x})^2$. Each of these terms is the square of the distance of one of the elements in the set from the mean of the set. Thus, the variance is the average of the squares of the elements' distances from the mean. In other words, it is a measure of the average deviation from the mean. If the variance is large, the data tend to be spread out from the mean; if it is small the data tend to be clustered about the mean. For instance, in Example 9.6 the variance of the set of data in (a) is considerably less than the variance of the set of data in (b), and we can observe that the data in (a) is clustered about the mean more than the data in (b).

The sample variance and sample standard deviation are used in problems of statistical inference. The special properties of these quantities will be discussed in Chapter 10.

EXAMPLE 9.6

Find the mean, median, mode (or modes), variance, and standard deviation of each of the following sets of data:
(a) $\{2, 3, 3, 1.5, 0\}$
(b) $\{-10, 9, 3, 3, 9, 16\}$
(c)   the data of Example 9.1

SOLUTION

(a)   The mean is

$$\bar{x} = \frac{2 + 3 + 3 + 1.5 + 0}{5} = 1.9$$

The median is $m = 2$. There is one mode, and it is at $x = 3$. The variance is

$$\sigma^2 = \frac{(2 - 1.9)^2 + (3 - 1.9)^2 + (3 - 1.9)^2 + (1.5 - 1.9)^2 + (0 - 1.9)^2}{5}$$

$$= 1.24$$

The standard deviation is $\sigma = 1.24^{1/2} = 1.1136$.

(b) The mean is

$$\bar{x} = \frac{-10 + 9 + 3 + 3 + 9 + 16}{6} = 5$$

The median is $m = 6$. Modes are 3 and 9. The variance is

$$\sigma^2 = \frac{(-10 - 5)^2 + (9 - 5)^2 + (3 - 5)^2 + (3 - 5)^2 + (9 - 5)^2 + (16 - 5)^2}{6}$$

$$= 64.3333$$

The standard deviation is $\sigma = 64.3333^{1/2} = 8.0208$

(c) $\bar{x} = 3.59$, $m = 4$, modes are 2 and 4, $\sigma^2 = 2.97$, $\sigma = 1.72$.

If the variance, standard deviation, sample variance, or sample standard deviation have to be calculated by hand or with the use of a calculator that does not have statistical functions, the use of the following shortcut formulas may simplify the computation.

Shortcut formulas for the variance, standard deviation, sample variance, and sample standard deviation

$$\sigma^2 = \frac{\sum_{i=1}^{n} x_i^2 - n(\bar{x})^2}{n}$$

$$\sigma = \sqrt{\frac{\sum_{i=1}^{n} x_i^2 - n(\bar{x})^2}{n}}$$

$$s^2 = \frac{\sum_{i=1}^{n} x_i^2 - n(\bar{x})^2}{n - 1}$$

$$s = \sqrt{\frac{\sum_{i=1}^{n} x_i^2 - n(\bar{x})^2}{n - 1}}$$

The verification of these formulas is considered in Exercise 5 of Section 9.4.

When the frequency table of a set of data is available, but the raw data are not, an estimate of the variance or sample variance may be computed by using the following formulas:

$$\sigma^2 \approx \frac{\sum\limits_{i=1}^{n} f_i(m_i - \bar{x})^2}{n} \quad \text{and} \quad s^2 \approx \frac{\sum\limits_{i=1}^{n} f_i(m_i - \bar{x})^2}{n}$$

where f_i is the frequency of the ith class, m_i is the class mark of the ith class, and n is the total number of classes in the frequency table. The value of $\bar{x}$ may be estimated with the formula $\bar{x} \approx \left(\sum_{i=1}^{n} f_i m_i\right)/n.$

EXAMPLE 9.7

(a) Use the raw data given in Example 9.2 to find the variance and standard deviation of the set of data.

(b) Use the frequency table in Table 9.2 to estimate the variance and standard deviation of the same set of data.

  SOLUTION

(a) The value of $\bar{x} = 243.97$ was computed in Example 9.5(a). We will use the rounded off value of $\bar{x} = 244$ since the values in the set of data are also rounded off to the nearest unit. Thus

$$\sigma^2 = \frac{\sum\limits_{i=1}^{n} (x_i - \bar{x})^2}{n} = \frac{(225 - 244)^2 + (195 - 244)^2 + \cdots + (182 - 244)^2}{66}$$

$$= 2127$$

and

$$\sigma = 46.12$$

(b) The value of $\bar{x} \approx 244.58$ was obtained in Example 9.5(b). We will use the rounded off value of $\bar{x} \approx 245$. The class marks of the classes in the frequency table are: 137, 162, 187, 212, 237, 262, 287, 312, and 337. Therefore, an approximate value of the variance is given by

$$\sigma^2 \approx \frac{\sum\limits_{i=1}^{n} f_i(m_i - \bar{x})^2}{n}$$

$$= \frac{1 \cdot (137 - 245)^2 + 3 \cdot (162 - 245)^2 + \cdots + 2 \cdot (337 - 245)^2}{66}$$

$$= 2120.64$$

and

$$\sigma \approx 46.05$$

Definition 9.6   A number x is called the **kth percentile** of a set of numbers if k percent of the numbers in the set are less than x. The 25th, 50th, and 75th percentiles have special names, and are called the **first, second**, and **third quartiles**, respectively. **Fractiles** are percentiles with specified percents expressed as decimals. For example, the .85 fractile is the 85th percentile.

EXAMPLE 9.8

Use the cumulative frequency distribution of weekly incomes in Table 9.8 to find (a) the 90th percentile, (b) the first, second, and third quartiles, and (c) the .35 fractile of the set of data given in Example 9.2 (These descriptive measures may also be obtained from the ogive of Figure 9.4. In either case, the numbers obtained will be only estimates of the actual values because of the need to interpolate for these values.)

SOLUTION

(a)   The set of data consists of 66 items. Ninety percent of 66 is 59.4. From Table 9.8 we see that 59 incomes are no greater than $299, and 64 incomes are no greater than $324. Therefore

$$\text{The 90th percentile} = \$299 + \left(\frac{59.4 - 59}{64 - 59}\right) \cdot (\$324 - \$299)$$

$$= \$301$$

(b)   Follow the method of (a).

The first quartile = 25th percentile

$$= \$199 + \left(\frac{16.5 - 14}{22 - 14}\right) \cdot (\$224 - \$199)$$

$$= \$207$$

The second quartile = 50th percentile = $249

The third quartile = 75th percentile

$$= \$274 + \left(\frac{49.5 - 47}{59 - 47}\right) \cdot (\$299 - \$274)$$

$$= \$279$$

(c)

$$.35 \text{ fractile} = 35\text{th percentile} = \$224 + \left(\frac{23.1 - 22}{33 - 22}\right)(\$249 - \$224)$$

$$= \$226.50$$

Exercises

1. The systolic blood pressures of members of a certain health club are given below. Find the mean, median, mode (or modes), variance, and standard deviation of this set of data: 118, 120, 120, 125, 130, 132, 119, 124, 125, 135, 120.

2. The daily receipts of a certain store for a given week are listed below. Find the mean, median, variance, and standard deviation of this set of data: $554.25, $725.12, $699.10, $610.17, $701.20, $900.45, $600.95.

3. (a) Use the raw data given in Exercise 2 of Section 9.2 to find the mean, variance, and standard deviation of the set of data.
   (b) Use the frequency table constructed in Exercise 2 of Section 9.2 to estimate the mean, variance, and standard deviation of the same set of data.

4. (a) Use the raw data given in Exercise 3 of Section 9.2 to find the mean, variance, and standard deviation of the set of data.
   (b) Use the frequency table constructed in Exercise 3 of Section 9.2 to estimate the mean, variance, and standard deviation of the same set of data.

5. Verify the shortcut formulas for the variance and sample variance for the case of $n = 2$.

6. The frequency distribution of scores achieved by 1250 students in a certain placement examination follows. Find
   (a) the first, second, and third quartiles,
   (b) the 80th and 95th percentiles,
   (c) the .65 fractile.

Score	Frequency
0–10	1
11–20	7
21–30	10
31–40	24
41–50	170
51–60	420
61–70	426
71–80	160
81–90	28
91–100	4

ELEMENTS OF STATISTICAL INFERENCE

10.1   Introduction

The object of statistical inference is to generalize from properties of a small body of data to statements concerning properties of a larger system of similar data. For the agronomist, the object may be to generalize from the properties of a number of plants of a certain type to statements concerning properties of all plants of that type. For the research physician, the object may be to generalize from the reactions to a particular medical treatment of a number of patients suffering from a certain disease to statements concerning the reactions of all patients suffering from that disease. For the business manager, the object may be to generalize from the observed level of productivity for a number of employees under a certain program to statements concerning the level of productivity for all employees under that program. This chapter prepares the reader for the study of statistical inference through a discussion of the concepts and technical terms that are basic to the subject. Specific methods and applications of statistical inference are discussed in the remaining chapters of the book.

10.2   Population

10.2.1   Definition

The word "population" conjures up a mental image of a large group of people, and—when dealing with statistical surveys—we tend to think of data (or numbers) as being associated with the people in such a group. In the language of mathematical statistics, however, we define a **population** to be a set of objects. The objects may be people, things, events, or data (numbers) associated with the people, things, or events. For example, the set of all plants of a certain type is a population. The set of reactions of all patients suffering from a certain disease is a population. The set of levels of productivity of all employees under a certain program is a population. A frequent object of statistical inference is to generalize from properties of a small body of data to statements concerning properties of a population.

EXAMPLE 10.1

The following sets of objects are examples of populations.
   (a)   The set of all college students in the United States.
   (b)   The set of incomes of all single women in the United States.
   (c)   The set of all scores in the next test for this course.
   (d)   Suppose that the annual incomes in thousands of dollars for John, Mary, David, and Paul are 25, 25, 15, and 40, respectively. Then, the set $\mathscr{P} = \{25, 25, 15, 40\}$ is the population of incomes for John, Mary, David, and Paul.
   (e)   Suppose that the 10 students in Professor James' class received grades of 95, 95, 87, 87, 87, 85, 85, 75, 70, and 70, respectively. Then, the set $\mathscr{P} = \{95, 95, 87, 87, 87, 85, 85, 75, 70, 70\}$ is the population of grades for the students in Professor James' class.

10.2.2   The Distribution of a Population

We may describe the distribution of incomes for John, Mary, David, and Paul of Example 10.1(d) as follows: one-half of them have incomes of $25,000, one-quarter have incomes of $15,000, and another quarter have incomes of $40,000. In terms of the population $\mathscr{P} = \{25, 25, 15, 40\}$ we would say that 1/2 of the elements of $\mathscr{P}$ have numerical values of 25, 1/4 has a numerical value of 15, and 1/4 has a numerical value of 40. These statements completely describe the distribution of the population $\mathscr{P}$. That is, a complete description of the relative frequency of occurrence of each distinct number or type of object in a population constitutes a complete description of the **distribution of the population**. In other words, the relative frequency distribution of a population provides a complete description of the distribution of a population (a detailed discussion of relative frequency distributions was given in Chapter 9.)

We can also define the distribution of a population $\mathscr{P}$ in terms of a random variable associated with $\mathscr{P}$. Consider the experiment of selecting an element at random from $\mathscr{P}$ in such a manner that each element has equal probability of being selected. Let X be the numerical value of the element selected. The random variable X is called the random variable associated with the population $\mathscr{P}$, and the distribution of X is called the distribution of $\mathscr{P}$. It is intuitively obvious that this method of defining the distribution of a population is consistent with the method described in the previous paragraph. For example, it is easy to see that if $\mathscr{P}$ is a population with only a finite number of elements, the frequency distribution of $\mathscr{P}$ completely defines the probability function of the random variable X that is associated with $\mathscr{P}$, and vice versa. (See Example 10.2.)

EXAMPLE 10.2

Let $\mathscr{P} = \{25, 25, 15, 40\}$, that is, the population of Example 10.1(d). Let X be the random variable associated with $\mathscr{P}$ (i.e., X is the value of the number selected at random from $\mathscr{P}$). Show that the probability function of X completely defines the relative frequencies of the elements in $\mathscr{P}$, and vice versa.

SOLUTION

Since 1/2 of the elements of $\mathscr{P}$ have numerical values of 25, 1/4 have a numerical value of 15, and 1/4 have a numerical value of 40, we say that the relative frequencies of the numbers 25, 15, and 40, are 1/2, 1/4, and 1/4, respectively. Now, if X is the random variable associated with $\mathscr{P}$—that is, if X is the value of the number selected at random from $\mathscr{P}$—then the probability function of X is given by

$$f(x) = P(X = x) = \begin{cases} 1/2 & \text{for} \quad x = 25 \\ 1/4 & \quad x = 15 \\ 1/4 & \quad x = 40 \end{cases}$$

Therefore, it is clear that

$$f(x) = (\text{the relative frequency of the number } x)$$

In other words, the probability function of X completely defines the relative frequencies of the elements in $\mathscr{P}$, and vice versa.

The distribution of a population is often used to describe a population. Thus, if the distribution of a particular population is a normal distribution, we say that the population is a normal population; if the distribution is an exponential distribution, we say that the population is an exponential population; if the distribution is a Poisson distribution, we say that the population is a Poisson population.

EXAMPLE 10.3

(a)   Let $\mathscr{P}$ be the population of SAT scores of students in 1982. Consider the experiment of selecting an SAT score at random from the population $\mathscr{P}$. Let X be the value of the SAT score selected. The random variable X is called the random variable associated with $\mathscr{P}$, and the distribution of X is the distribution of $\mathscr{P}$. If the distribution of X is normal, we say that the distribution of $\mathscr{P}$ is also normal.

With reference to this population $\mathscr{P}$, consider now the experiment of selecting a student at random from the student population of 1982. Suppose we define another random variable, Y, to be the SAT score of the student selected in this experiment. It is obvious that the distribution of Y is identical to the distribution of X (as defined). For that reason, we may say that the distribution of $\mathscr{P}$ is the distribution of Y. In other words, this is still another way of defining the distribution of $\mathscr{P}$.

(b)   Let $\mathscr{P}$ be the set of real numbers in the interval $(0, 1)$. Consider the experiment of selecting an element at random from $\mathscr{P}$. Let X be the value of the number selected. The random variable X is the random variable associated with the population $\mathscr{P}$. If the distribution of X is uniform, the distribution of the population is also uniform.

(c)   Consider the experiment of tossing a coin 10 times and recording the number of heads in the 10 tosses. Suppose the experiment is performed one million times. Let $\mathscr{P}$ be the set consisting of the one million recorded numbers from the one million experiments. Now, consider the experiment of selecting an element at random from the population $\mathscr{P}$. Let X be the value of the number selected. Observe that the distribution of X should be very nearly identical to the distribution of the number of heads in 10 tosses of a coin. In other words, the distribution of X should be approximately binomial with parameters $n = 10$ and $p = 1/2$. Therefore, the distribution of the population is approximately binomial with parameters $n = 10$ and $p = 1/2$.

10.2.3   Parameters of a Population or Distribution

The **parameters of a distribution** are the parameters of the probability function or probability density function associated with the distribution. The **parameters of a population** are the parameters of the distribution of the population. For instance, the parameters $n = 10$ and $p = 1/2$ of the distribution of the population in Example 10.3(c) are also called the parameters of the population $\mathcal{P}$.

EXAMPLE 10.4

   (a)   Let $\mathcal{P}$ be a population and let X be the random variable associated with $\mathcal{P}$. Suppose that X is a continuous random variable and the probability density function of X is

$$f(x) = \begin{cases} \dfrac{1}{\theta}\, e^{-x/\theta} & \text{for} \quad x > 0 \\[2mm] 0 & \text{elsewhere} \end{cases}$$

The parameter θ is a parameter of the distribution of X. It is also a parameter of the population $\mathcal{P}$.

   (b)   Suppose that X is the random variable associated with the population $\mathcal{P}$. Furthermore, suppose that the distribution of X is normal with mean μ and variance σ^2. The values μ and σ^2 are parameters of the distribution of X. Therefore, they are also parameters of the population $\mathcal{P}$, and are called the mean and variance of the population $\mathcal{P}$, respectively.

10.2.4   Summary

The object of statistical inference is to generalize from properties of a small body of data to statements concerning properties of a larger system of similar data. The large system of data is often expressed in terms of a population or a distribution. Parameters such as the mean or variance, are important properties of a population or distribution. Therefore, a frequent object of statistical inference is to generalize from properties of a small body of data to statements concerning the parameters of a population or distribution. For example, the agronomist may wish to generalize from measurements of a number of plants of a certain type to statements concerning the mean of the distribution of heights of all plants of that type. The research physician may wish to generalize from reactions to a particular medical treatment of a number of patients suffering from a certain disease to statements concerning the proportion of the population of all patients suffering from this disease who will recover if given the medical treatment. The business manager may wish to generalize from observed levels of productivity for a number of employees under a certain program to statements concerning the variance of the distribution of productivity levels for all employees under that program.

Exercises

1.   Give 3 examples of a population.
2.   Find the distribution of the population of Example 10.1(e).

3. The random variable X associated with the population $\mathscr{P}$ has pdf

$$f(x) = \begin{cases} cx^2 & \text{for} \quad 0 < x < (3/c)^{1/3} \\ 0 & \text{elsewhere} \end{cases}$$

Which of the following is a parameter of the distribution of $\mathscr{P}$? (a) c; (b) x; (c) 3.

10.3   Sampling from a Population or Distribution

10.3.1   Sampling Without and Sampling with Replacement

We say that the ordered set of elements $(a_1, a_2, \ldots, a_n)$ is a sample of size n obtained under sampling without replacement from the set A if both of the following conditions are satisfied:

1. The elements $a_1, a_2, \ldots, a_n$ are the first, second, ..., and nth elements, respectively, selected from A.

2. Once an element a_i is selected from A, it is permanently removed from A so that it is not available for subsequent selections. Consequently, the elements $a_1, a_2, \ldots, a_n$ are distinct elements of A.

We say that the ordered set of elements $(a_1, a_2, \ldots, a_n)$ is a sample of size n obtained under sampling with replacement from the set A if both of the following conditions are satisfied:

1. The first condition is the same as in previous list.

2. When an element is selected from A, only the identity of the element is recorded in the sample $(a_1, a_2, \ldots, a_n)$. The element itself is replaced into the set A so that it is again available for selection in the next step of the process. Consequently, the elements $a_1, a_2, \ldots, a_n$ need not be distinct elements of A.

EXAMPLE 10.5

Let $A = \{1, 2, 3\}$.

(a) List all the distinct samples of size 2 that can be obtained under sampling without replacement from the set A.

(b) List all the distinct samples of size 2 that can be obtained under sampling with replacement from the set A.

SOLUTION

(a) The distinct samples of size 2 that can be obtained under sampling without replacement form the set A are: $(1, 2), (2, 1), (1, 3), (3, 1), (2, 3), (3, 2)$.

(b) The distinct samples of size 2 that can be obtained under sampling with replacement from the set A are: $(1, 2), (2, 1), (1, 3), (3, 1), (2, 3), (3, 2), (1, 1), (2, 2), (3, 3)$.

The distinction between sampling with and sampling without replacement becomes statistically inconsequential when the population is very large. Example 10.6 illustrates this point.

EXAMPLE 10.6

The principal difference between sampling without replacement and sampling with replacement is that in the second of these two methods there is the possibility of repetitions of elements in the sample, whereas in the first method there is no such possibility. The probability p of repetitions in a sample of size r from a population of size n under sampling with replacement is given by the formula

$$p = P(\text{repetitions}) = 1 - P(\text{no repetition}) = 1 - \frac{n(n-1)\cdots(n-r+1)}{n^r}$$

Find p for $r = 3$ and $n = 10, 100, 1{,}000, 10{,}000$. Observe that as n tends to infinity, p tends to zero, so that the distinction between sampling with and sampling without replacement becomes statistically insignificant.

SOLUTION

Case 1: $r = 3, n = 10$.

$$p = P(\text{repetitions}) = 1 - P(\text{no repetition}) = 1 - \frac{10 \cdot 9 \cdot 8}{10^3} = .28$$

Case 2: $r = 3, n = 100$.

$$p = 1 - \frac{100 \cdot 99 \cdot 98}{100^3} = .0298$$

Case 3: $r = 3, n = 1{,}000$.

$$p = 1 - \frac{1{,}000 \cdot 999 \cdot 998}{1{,}000^3} = .002998$$

Case 4: $r = 3, n = 10{,}000$

$$p = 1 - \frac{10{,}000 \cdot 9{,}999 \cdot 9{,}998}{10{,}000^3} = .0003$$

10.3.2 Random Samples from a Population or Distribution

In this discussion we shall restrict our consideration to populations that are sets of numerical data. A random sample from a population of this type may be defined as follows.

Definition 10.1   Random Sample from a Population   Let $\mathscr{P}$ be a population that is a set of numerical data, let $\mathscr{E}$ be the experiment of selecting an element at random from $\mathscr{P}$, and let X be the random variable representing the numerical value of the element selected. Suppose that the experiment is repeated n times under identical conditions, and X_i is the random variable X on the ith repetition of $\mathscr{E}$. The set $(X_1, X_2, \ldots, X_n)$ of random variables is called a **random sample of size n from the population**.

If x_1 is the observed value of the random variable X_1, x_2 is the observed value of the random variable $X_2, \ldots$, and x_n is the observed value of the random variable X_n, then $(x_1, x_2, \ldots, x_n)$ is called the **set of observed values of the random sample**.

Note: readers who read Chapters 7 or 8 will recognize a random sample to be a set of independent and identically distributed random variables. That is, if $(X_1, X_2, \ldots, X_n)$ is a random sample, then $X_1, X_2, \ldots$, and X_n are independent and identically distributed random variables.

According to this definition, a set of observed values of a random sample of size n is equivalent to a sample of size n obtained under sampling with replacement. Therefore, if the size of the population is large compared to the size of the sample, even a sample of size n obtained under sampling without replacement may be considered the set of observed values of a random sample of size n.

Note: when there is no possibility of confusing the observed set of values of a random sample with the random sample itself, then the former is also called a random sample. This is for the sake of simplicity in the written and spoken language of statistical inference.

EXAMPLE 10.7

In a study of the distribution of weekly grocery allowances of households in a certain suburb, a consumer research group randomly selected 10 households in the area and found their weekly grocery allowances (in dollars) to be 75, 48, 67, 33, 62, 71, 81, 92, 80, and 76, respectively. In this example, what is the population, what is the random sample, and what is the set of observed values of the random sample?

SOLUTION

The Population   If it were possible to list the weekly grocery allowance of every household in the area, the data in the list would constitute the population of the example.

The Random Sample   Let X_i be the random variable representing the weekly grocery allowance of the ith household to be contacted in the survey. The set $(X_1, X_2, \ldots, X_{10})$ of random variables constitutes a random sample of size 10 from the population. (We are assuming that the size of the population is large compared to the size of the sample. Therefore, the respective experiments of selecting the 2nd, 3rd, $\ldots$, and 10th households in the survey may be considered identical to the experiment of selecting the 1st household in the survey.)

The Set of Observed Values of the Random Sample Since the weekly grocery allowances, in dollars, of the 10 randomly selected households were $75, 48, 67, \ldots$, and 76, the set of observed values of the random sample is the set $(75, 48, 67, 33, 62, 71, 81, 92, 80, 76)$.

Definition 10.2 Random Sample from a Distribution Let $\mathscr{E}$ be an experiment, let S be a sample space for $\mathscr{E}$, and let X be a random variable defined on S. Suppose that the experiment $\mathscr{E}$ is repeated n times under identical conditions, and X_i is the random variable X on the ith repetition of $\mathscr{E}$. The set $(X_1, X_2, \ldots, X_n)$ of random variables is called a **random sample of size n from the distribution of X**.

If x_1 is the observed value of the random variable X_1, x_2 is the observed value of the random variable $X_2, \ldots$, and x_n is the observed value of the random variable X_n, then $(x_1, x_2, \ldots, x_n)$ is called the **set of observed values of the random sample**.

When we compare these two definitions of random samples, we can see that a random sample from a population is the same as a random sample from the distribution of the population. However, when we are dealing with a random sample from a distribution there need not be a population under consideration.

EXAMPLE 10.8

Suppose that the owner of a certain die wished to determine the true distribution of values—that is, the true probabilities for the respective events of obtaining a 1, 2, 3, 4, 5, or 6—when his die is thrown. He throws the die 15 times and observes the respective outcomes of 1, 2, 4, 3, 5, 6, 3, 5, 6, 1, 4, 4, 2, 1, and 6. In this example, what is the distribution under consideration, what is the random sample, and what is the set of observed values of the random sample?

SOLUTION

The Distribution Consider the experiment of throwing the die. Let X be the outcome (i.e., value on the top face of the die). The distribution under consideration is the distribution of the random variable X.

The Random Sample Let X_i be the random variable representing the outcome of the ith throw of the die. The set $(X_1, X_2, \ldots, X_{15})$ of random variables constitutes a random sample of size 15 from the distribution under consideration.

The Set of Observed Values of the Random Sample Since the observed respective outcomes of the 15 throws are $1, 2, 4, \ldots$, and 6, the set of observed values of the random sample is the set $(1, 2, 4, 3, 5, 6, 3, 5, 6, 1, 4, 4, 2, 1, 6)$.

10.3.3   The Mean, Variance, and Standard Deviation of a Random Sample

Let $(X_1, X_2, \ldots, X_n)$ be a random sample of size n from a population or distribution. Let

$$\bar{X} = \frac{X_1 + X_2 + \cdots X_n}{n}$$

$$S^2 = \frac{(X_1 - \bar{X})^2 + (X_2 - \bar{X})^2 + \cdots + (X_n - \bar{X})^2}{n-1}$$

and

$$S = \sqrt{S^2}$$

The random variables $\bar{X}$, S^2, and S are called the mean (or sample mean), variance (or sample variance), and standard deviation (or sample standard deviation), respectively, of the random sample $(X_1, X_2, \ldots, X_n)$. When there is no possibility of confusion between the random variables $\bar{X}$, S^2, and S with their observed values $\bar{x}$, s^2, and s, then the latter are also called the mean (or sample mean), variance (or sample variance), and standard deviation (or sample standard deviation) of the random sample.

    Steps for the computation of the observed values of the sample mean, sample variance, and sample standard deviation are given in the example that follows.

EXAMPLE 10.9

If $(x_1, x_2, \ldots, x_n)$ is the set of observed values of the random sample $(X_1, X_2, \ldots, X_n)$, then the observed values of the sample mean $\bar{X}$, the sample variance S^2, and the sample standard deviation S are the values $\bar{x}$, s^2, and s that follow:

$$\bar{x} = \frac{x_1 + x_2 + \cdots + x_n}{n}$$

$$s^2 = \frac{(x_1 - \bar{x})^2 + (x_2 - \bar{x})^2 + \cdots + (x_n - \bar{x})^2}{n-1}$$

$$s = \sqrt{s^2}$$

    Find the observed values of the sample mean, sample variance, and sample standard deviation if the set of observed values of the random sample (X_1, X_2, X_3) is $(2, 4, 3)$.

    SOLUTION

The observed values of the sample mean, sample variance and sample standard deviation are the values $\bar{x}$, s^2, and s that follow:

$$\bar{x} = \frac{x_1 + x_2 + x_3}{3} = \frac{2 + 4 + 3}{3} = 3$$

$$s^2 = \frac{(x_1 - \bar{x})^2 + (x_2 - \bar{x})^2 + (x_3 - \bar{x})^2}{2}$$

$$= \frac{(2 - 3)^2 + (4 - 3)^2 + (3 - 3)^2}{2} = 1$$

$$s = \sqrt{s^2} = 1$$

The following is an important theorem concerning sample means and sample variances. A proof of this theorem is considered in Chapter 8 (Exercise 13, Section 8.3).

Theorem 10.1   *If μ and σ^2 are the mean and variance of the distribution of a population $\mathcal{P}$, and if $\bar{X}$ and S^2 are the mean and variance of a random sample of size n from the population $\mathcal{P}$, then*

$$E(\bar{X}) = \mu \quad and \quad E(S^2) = \sigma^2$$

This theorem has important consequences for statistical inference because it implies that a random sample from a population may tell us something about the mean and variance and, therefore, something about the distribution of a population. Exactly how we will use the sample mean and sample variance to obtain this information is the subject of later chapters in the book.

10.3.4   Summary

A frequent object of statistical inference is to generalize from properties of a small body of data to statements concerning the parameters of a population or distribution. Since random samples contain data concerning such parameters, an object of statistical inference is to generalize from properties of a random sample to statements concerning the parameters of a population or distribution.

Exercises

1.   Let $A = \{2, 5, 9\}$.
     (a)   List all distinct samples of size 2 that can be obtained under sampling without replacement from the set A.
     (b)   List all distinct samples of size 2 that can be obtained under sampling with replacement from the set A.

2. Let $B = \{a, b, c, d\}$.
    (a) List all distinct samples of size 3 that can be obtained under sampling without replacement from the set B.
    (b) List all distinct samples of size 3 that can be obtained under sampling with replacement from the set B.

3. The observed set of values of the random sample (X_1, X_2, X_3, X_4) is $(-1, 3, 6, 2)$. Find the observed values of the sample mean, sample variance, and sample standard deviation.

4. Find the sample mean, sample variance, and sample standard deviation of the random sample $(-1, 2, -1, 4, 0)$.

5. If $\bar{X}$ and S^2 are the mean and variance of a random sample of size 4 from a population whose distribution is normal with mean 1 and variance 9, what is $E(\bar{X})$ and $E(S^2)$?

6. If $\bar{X}$ and S^2 are the mean and variance of a random sample of size 9 from a population whose distribution is exponential with a mean of 2, what is $E(\bar{X})$ and $E(S^2)$?

10.4   Sampling Distributions

In the language of statistical inference, a **statistic** is a random variable that is a function of a random sample, and a **sampling distribution** is a distribution of a statistic. For example, if $(X_1, X_2, \ldots, X_n)$ is a random sample from a population with mean μ and variance σ^2, then X, S^2, $(\bar{X} - \mu)/(\sigma/n^{1/2})$, $(\bar{x} - \mu)/(S/n^{1/2})$, and $(n - 1)S^2/\sigma^2$ are statistics. Their distributions are sampling distributions. As we shall see in Chapters 11–15, sampling distributions play a central role in methods of statistical inference.

10.4.1   Sampling Distributions for Inference Concerning the Mean of a Population

The most commonly used sampling distributions for inference concerning the mean of a population are distributions of the statistics $\bar{X}$, $(\bar{X} - \mu)/(\sigma/n^{1/2})$, and $(\bar{X} - \mu)/(S/n^{1/2})$. This is to be expected, because each of these statistics involves the sample mean $\bar{X}$, and the mathematical expectation of $\bar{X}$ is the mean of the population. The first two of these statistics and their distributions were discussed in Chapter 8 on sums of random variables. (Recall that if $(X_1, X_2, \ldots, X_n)$ is a random sample, then $X_1, X_2, \ldots, X_n$ are independent and identically distributed random variables. Therefore, all the theorems in Chapter 8 on independent and identically distributed random variables are applicable to the random variables in a random sample.) Here we shall discuss only those properties of the statistics that are important for statistical inference.

With respect to each of the statistics stated above, there are two cases of particular importance: (a) if the distribution of the population is normal, and

(b) if the sample size is large. We shall discuss them separately for each statistic. In all cases we shall assume that the mean and variance of the population are μ and σ^2.

1.   The sampling distribution of $\bar{X}$

(a)   If the distribution of the population is normal, then the distribution of $\bar{X}$ is normal with mean μ and variance σ^2/n.

(b)   If the sample size is large, then the distribution of $\bar{X}$ is approximately normal with mean μ and variance σ^2/n. This statement constitutes one version of the central limit theorem. The result is valid for all types of distributions for the population. Of course, if the distribution of the population is normal, statement (a) applies and we may remove the word "approximately" from this statement (statement (b)). This rule applies to all cases below on large sample sizes.

Note: a proof of statement (a) is given in Example A.10 of Appendix A. A proof of a special case of statement (b) is given in Example A.11 of Appendix A.

The statistic $\bar{X}$ is used extensively in the later chapters on statistical inference. A preliminary indication of the role of this statistic in problems of inference is given in Example 10.10.

2.   The sampling distribution of $(\bar{X} - \mu)/(\sigma n^{1/2})$

(a)   If the distribution of the population is normal, the distribution of $(\bar{X} - \mu)/(\sigma/n^{1/2})$ is standard normal.

(b)   If the sample size is large, the distribution of $(\bar{X} - \mu)/(\sigma/n^{1/2})$ is approximately standard normal. This statement is another version of the central limit theorem. As in 1(b), if the distribution of the population is normal, we may remove the word "approximately" in the statement.

Note: a proof of statement (a) is considered in Exercise 5 of Appendix A. A proof of a special case of statement (b) is given in Example A.11 of Appendix A.

In problems of statistical inference the statistics $\bar{X}$ and $(\bar{X} - \mu)/(\sigma/n^{1/2})$ often play essentially the same role. The second of these two statistics is simply a conversion of the first to standard form (i.e., mean 0 and variance 1). Frequently, a solution is stated initially in terms of $\bar{X}$, and then is transformed into a statement in terms of $(\bar{X} - \mu)/(\sigma/n^{1/2})$. Such is the case in Example 10.10.

3.   The sampling distribution of $(\bar{X} - \mu)/(S/n^{1/2})$

(a)   If the distribution of the population is normal, the distribution of $(\bar{X} - \mu)/(S/n^{1/2})$ is the t distribution with $n - 1$ degrees of freedom. (A discussion of the t distribution is given in Section 10.4.3.)

(b)   If the sample size is large, the distribution of $(\bar{X} - \mu)/(S/n^{1/2})$ is approximately standard normal. This is because if the sample size is large, then the standard deviation σ of the population may be approximated by the sample standard deviation s, in which case statement 2(b) applies.

This statistic is especially important in problems of inference when the variance of the population is unknown (see Example 10.12).

EXAMPLE 10.10

Professor Lax is reputed to be a generous grader. Rumor has it that his grades are normally distributed with a mean of 85 and a standard deviation of 5. Not feeling quite confident that she can believe this rumor, Dorothy obtains the grades of a random sample of 9 of Professor Lax' students. She finds a sample mean of only 80! On the basis of this information, should Dorothy infer that the rumor is false?

SOLUTION

This example can be interpreted as a problem of inference concerning the mean of the population of grades given by Professor Lax. The question is: If the sample mean is only 80, can the mean grade be 85? To answer the question, let us find the probability that the sample mean $\bar{X}$ will be only 80 if the distribution of grades is normal (as rumored) with a mean of $\mu = 85$ and a standard deviation of $\sigma = 5$. Under these circumstances the distribution of the statistics $Z = (\bar{X} - \mu)/(\sigma/n^{1/2})$ is standard normal. Therefore

$$P(\bar{X} \le 80) = P[(\bar{X} - \mu)/(\sigma/\sqrt{n}) \le (80 - \mu)/(\sigma/\sqrt{n})]$$
$$= P[Z \le (80 - 85)/(5/\sqrt{9})] = P(Z \le -3) = .0013$$

In conclusion, it is highly unlikely (probability of only .0013) that the sample mean will be only 80 or less if the distribution of grades is normal with a mean of 85 and a standard deviation of 5. On the basis of the data, Dorothy should infer that the rumor is false.

10.4.2   A Sampling Distribution for Inference Concerning the Variance of a Population

The sampling distribution of the statistic $(n - 1)S^2/\sigma^2$ is often used for inference concerning the variance of a population.

4.   The sampling distribution of $(n - 1)S^2/\sigma^2$

If $(X_1, X_2, \ldots, X_n)$ is a random sample of size n from a population whose distribution is normal with variance σ^2, then the distribution of $(n - 1)S^2/\sigma^2$ is the chi-square distribution with $n - 1$ degrees of freedom. (A discussion of the chi-square distribution is given in Section 10.4.4.)

10.4.3 The *t* Distribution

The probability density function for the *t* distribution is a complicated affair, but its exact form need not concern us, because in matters of applied problems, only the table of values for the *t* distribution is used. Nevertheless, for the sake of completeness we have included below a definition of the *t* distribution.

Definition 10.3 A continuous random variable X has a **t distribution with n degrees of freedom** if the probability density function of X is

$$f(x) = \frac{\Gamma\left(\dfrac{n+1}{2}\right)}{\sqrt{\pi n}\,\Gamma\left(\dfrac{n}{2}\right)} \cdot \left(1 + \frac{x^2}{n}\right)^{-[(n+1)/2]} \qquad \text{for} \quad -\infty < t < \infty$$

The symbol Γ denotes the gamma function, which is defined by the equation $\Gamma(\alpha) = \int_0^\infty y^{\alpha-1}\,e^{-y}\,dy$ for $\alpha > 0$.

Note: the derivation of the *t* distribution is due to W. S. Gosset, who published his research under the pen name "Student." For that reason, the *t* distribution is also known as the Student *t* distribution or as Student's *t* distribution.

The probability density function $f(x)$ of the *t* distribution is symmetric about $x = 0$. The mean of the distribution is 0 if *n* (the degrees of freedom) is greater than 1. It is undefined if $n = 1$. As *n* (the degrees of freedom) increases, the distribution becomes approximately standard normal (see Figure 10.1). The general practice is to use the standard normal distribution to approximate the *t* distribution when *n* is greater than 29.

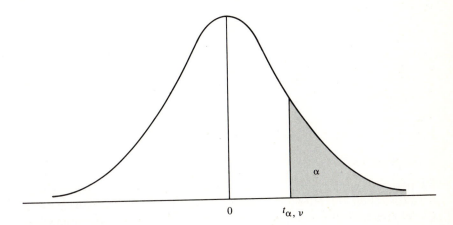

FIGURE 10.1
The Probability Density Function of the *t* Distribution

Because of its importance in statistical inference, the t distribution has been tabulated extensively. Table V in Appendix C contains values of $t_{\alpha,v}$ for $\alpha = .10, .05, .025, .01, .005$ and $v = 1, 2, \ldots, 29$, where $t_{\alpha,v}$ is the value such that the area to its right under the curve of the probability density function of the t distribution with v degrees of freedom is equal to α (see Figure 10.1). That is, if T_v is a random variable with a t distribution with v degrees of freedom, then

$$P(T_v \geq t_{\alpha,v}) = \alpha$$

EXAMPLE 10.11

Let T_v be a random variable with a t distribution with v degrees of freedom.
  (a) Find $P(T_5 \geq 1.476)$, $P(T_{10} < 2.228)$, $P(T_7 \geq -1.895)$, $P(T_{15} \leq -2.602)$.
  (b) Find the value t such that $P(T_{14} \geq t) = .025$; $P(T_9 \leq t) = .05$.
  (c) If a random sample of size 5 is taken from a population whose distribution is normal with mean μ, what is $P[(\bar{X} - \mu)/(S/5^{1/2}) \geq 1.533]$?

SOLUTION

Using the table of values for the t distribution (Appendix Table V) and the fact that the probability density function of the t distribution is symmetric about 0, we obtain the following answers.
  (a)

$$P(T_5 \geq 1.476) = .10$$
$$P(T_{10} < 2.228) = 1 - P(T_{10} \geq 2.228) = 1 - .025 = .975$$
$$P(T_7 \geq -1.895) = 1 - P(T_7 \geq 1.895) = 1 - .05 = .95$$
$$P(T_{15} \leq -2.602) = P(T_{15} \geq 2.602) = .01$$

  (b)

$$\text{If} \quad P(T_{14} \geq t) = .025, \quad \text{then} \quad t = 2.145$$
$$\text{If} \quad P(T_9 \leq t) = .05, \quad \text{then} \quad t = -1.833$$

  (c) From Statement 3(a) of Section 10.4.1 we obtain $(\bar{X} - \mu)/(S/5^{1/2}) = T_4$. Therefore

$$P[(\bar{X} - \mu)/(S/\sqrt{5}) \geq 1.533] = P(T_4 \geq 1.533) = .10$$

EXAMPLE 10.12

Suppose that in Example 10.10 there is no information on the standard deviation of grades. However, suppose that there is a record of each of the grades obtained in the random sample of nine grades and that the grades were 79, 75, 84, 63, 98, 52, 87, 99, 83, respectively. On the basis of this set of data, should Dorothy infer that the rumor of a mean grade of 85 is false?

SOLUTION

Since the variance of the population is unknown, we cannot use the statistic $Z = (\bar{X} - \mu)/(\sigma/n^{1/2})$ of Example 10.10. Let us consider, instead, the statistic $T = (\bar{X} - \mu)/(S/n^{1/2})$. The observed value of $\bar{X}$ is

$$\bar{x} = \frac{79 + 75 + 84 + 63 + 98 + 52 + 87 + 99 + 83}{9} = 80$$

The observed value of S is

$$s = \sqrt{\frac{(79 - 80)^2 + (75 - 80)^2 + \cdots + (83 - 80)^2}{8}} = 15.24$$

Therefore, the observed value of the statistic T is

$$t = \frac{\bar{x} - \mu}{s/\sqrt{n}} = \frac{80 - 85}{15.24/3} = -.984$$

Now, if the population of grades is normally distributed with a mean of 85, then the distribution of the statistic T is the t distribution with 8 degrees of freedom (see Statement 3 of section 10.4.1). The expected value of T is zero. Thus the observed value of $t = -.984$ is less than the expected value. The question is: If the observed value of the statistic T is only $-.984$, can the mean grade be 85? To answer the question, let us find the probability that the statistic T will be only $-.984$ or less if the distribution of grades is normal (as rumored) with a mean of $\mu = 85$. Using the fact that under these circumstances the distribution of T is the t distribution with 8 degrees of freedom, we have

$$P(T \leq -.984) = P(T \geq 0.984) > .10$$

With this set of data, Dorothy may not wish to reject the rumor; there is a substantial probability (a probability greater than .10) that the observed value of the T statistic will be no greater than $-.984$ if the rumor is correct.

10.4.4 The Chi-Square Distribution

Definition 10.4 A continuous random variable X has a **chi-square distribution with n degrees of freedom** if the probability density function of X is

$$f(x) = \begin{cases} \dfrac{1}{2^{n/2}\Gamma(n/2)} x^{(n-2)/2} e^{-(x/2)} & \text{for } x > 0 \\ 0 & \text{elsewhere} \end{cases}$$

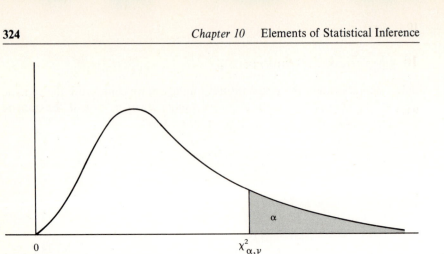

FIGURE 10.2
The Probability Density Function of the Chi-Square Distribution

When n is large, the chi-square distribution with n degrees of freedom is approximately normal. For the purpose of statistical inference one generally uses the normal approximation when $n \geq 30$.

As in the case of the t distribution, the chi-square distribution has been tabulated extensively. Table VI in Appendix C contains values of $\chi^2_{\alpha,\,v}$ for $\alpha = .995, .99, .975, .95, .05, .025, .01, .005$, and $v = 1, 2, 3, \ldots, 29$, where $\chi^2_{\alpha,\,v}$ is the value such that the area to its right under the curve of the density function of the chi-square distribution with v degrees of freedom is equal to α (see Figure 10.2). That is, if X_v is a random variable with a chi-square distribution with v degrees of freedom, then

$$P(X_v \geq \chi^2_{\alpha,v}) = \alpha$$

EXAMPLE 10.13

Let X_v be a random variable with a chi-square distribution with v degrees of freedom.
   (a)   Find $P(X_5 \geq 11.070)$ and $P(X_{10} < 18.307)$.
   (b)   Find the value χ^2 such that $P(X_{20} \geq \chi^2) = .01$; $P(X_{15} < \chi^2) = .99$.

   SOLUTION

   (a)   The answers are:

$$P(X_5 \geq 11.070) = .05$$
$$P(X_{10} < 18.307) = .95$$

   (b)

$$\text{If}\quad P(X_{20} \geq \chi^2) = .01, \quad \text{then}\quad \chi^2 = 37.566$$
$$\text{If}\quad P(X_{15} < \chi^2) = .99, \quad \text{then}\quad \chi^2 = 30.578$$

10.4.5   The *F* Distribution

Definition 10.5   A continuous random variable X has an F distribution with n_1 and n_2 degrees of freedom if the probability density function of X is

$$f(x) = \begin{cases} \dfrac{\Gamma\left(\dfrac{n_1 + n_2}{2}\right)}{\Gamma\left(\dfrac{n_1}{2}\right)\Gamma\left(\dfrac{n_2}{2}\right)} \cdot \left(\dfrac{n_1}{n_2}\right)^{(n_1/2)} x^{(n_1/2)-1}\left(1 + \dfrac{n_1}{n_2}x\right)^{-(1/2)(n_1+n_2)} & \text{for} \quad x > 0 \\ \\ 0 \qquad \text{elsewhere} \end{cases}$$

(See Figure 10.3.) An important use of the F distribution concerns random samples from two distinct populations. It can be shown that if S_1^2 and S_2^2 are the variances of random samples of size n_1 and n_2 from two distinct normal populations with variances σ_1^2 and σ_2^2, respectively, then the statistic

$$F = \frac{S_1^2/\sigma_1^2}{S_2^2/\sigma_2^2}$$

has an F distribution with $n_1 - 1$ and $n_2 - 1$ degrees of freedom. This statistic is used in inference concerning the comparison of two variances (see Chapter 13).

Note: Table VII in Appendix C contains values of F_{α,v_1,v_2}.

10.4.6   The Values z_α, $t_{\alpha,v}$, $\chi^2_{\alpha,v}$, F_{α,v_1,v_2}

The definitions of $t_{\alpha,v}$, $\chi^2_{\alpha,v}$, and F_{α,v_1,v_2} were given in Sections 10.4.3, 10.4.4, and 10.4.5, respectively. Each of these values gives the location such that the area to its right under the curve of the associated probability density function

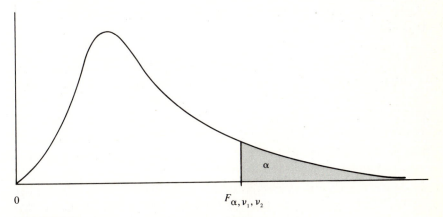

FIGURE 10.3
The Probability Density Function of the *F* Distribution

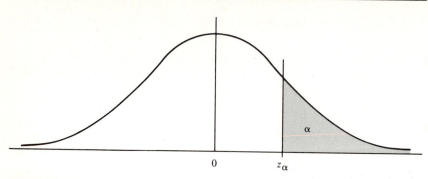

FIGURE 10.4
The Standard Normal Probability Density Function

is α. The notation for the corresponding value for the standard normal distribution is z_α. In other words, z_α is the value such that the area to its right under the curve of the standard normal density function is α. (See Figure 10.4) These values play an important role in problems of statistical inference. (See Chapters 11, 12, 13, 14, and 15.)

EXAMPLE 10.14

(a) Find $z_{.05}, z_{.025}, z_{.01}$, and $z_{.005}$.
(b) Let Z be a standard normal random variable. Find $P(Z > z_{.05})$ and $P(Z < -z_{.01})$.

SOLUTION

(a) The answers are 1.645, 1.96, 2.327, and 2.575.
(b) The answers are

$$P(Z > z_{.05}) = .05 \quad \text{and} \quad P(Z < -z_{.01}) = P(Z > z_{.01}) = .01$$

Exercises

1. If $\bar{X}$ is the sample mean of a random sample of size 9 from a population whose distribution is normal with mean 1 and variance 4, then what is $E(\bar{X})$? What is $Var(\bar{X})$? What is the distribution of $\bar{X}$? Find $P(\bar{X} > 1.5)$.

2. Let $\bar{X}$ be the sample mean of a random sample of size 4 from a population whose distribution is normal with mean 2 and variance 9. Find $P(\bar{X} < 3)$.

3. Let $\bar{X}$ be the sample mean of a random sample of size 100 from a population whose mean is 3 and whose variance is 4. Find $P(\bar{X} < 2.9)$.

4. The owner of a certain gasoline station claims that weekday sales average 1,200 gallons of gasoline (business being less brisk on weekends) and that the distribution of sales is normal with a standard deviation of 100 gallons. A random sample of 16 weekdays yielded an average sale of 1,100 gallons

per day. Is there reason to question the owner's claim? (*Hint:* follow the method of Example 10.8.)

5. Let T_n be a random variable with a t distribution with n degrees of freedom.
   (a)   Find $P(T_6 \geq 1.945)$, $P(T_{15} < 2.131)$, $P(T_3 < -9.925)$,   and $P(T_{21} > -1.323)$.
   (b)   Find the value t such that $P(T_{13} \geq t) = .025$, $P(T_7 \leq t) = .05$.

6. The administration of a certain college claims that on the average it should take a student 75 minutes to complete the registration process. A random sample of 16 students showed that the time required for them to complete the registration process was 75, 78, 74, 82, 90, 82, 69, 70, 69, 75, 85, 79, 66, 68, 72, 85 minutes, respectively. Is there reason to question the administration's claim of 75 minutes? (*Hint:* follow the method of Example 10.10.)

7. Let X_n be a random variable with a chi-square distribution with n degrees of freedom.
   (a)   Find $P(X_7 \geq .989)$, $P(X_{15} < 30.578)$, $P(X_{18} > 9.390)$.
   (b)   Find the value χ^2 such that $P(X_{21} \geq \chi^2) = .01$, $P(X_3 \leq \chi^2) = .95$, $P(X_{13} \geq \chi^2) = .05$.

8. Let $\mathscr{P}_1$ and $\mathscr{P}_2$ be two different normal populations with means equal to μ_1 and μ_2, respectively, and variances equal to σ_1^2, and σ_2^2, respectively. Let $\bar{X}_1$ and $\bar{X}_2$ be the sample means of random samples of size n_1 and n_2, respectively, from populations $\mathscr{P}_1$ and $\mathscr{P}_2$, respectively (so that $\bar{X}_1$ and $\bar{X}_2$ are independent random variables). Show that the distribution of $[(\bar{X}_1 - \bar{X}_2) - (\mu_1 - \mu_2)]/[(\sigma_1^2/n_1) + (\sigma_2^2/n_2)]^{1/2}$ is standard normal. (*Hint:* see Section 8.4, Chapter 8.)

9. Show that if the populations of Exercise 8 are not normally distributed, then the distribution of $[(X_1 - X_2) - (\mu_1 - \mu_2)]/[(\sigma_1^2/n_1) + (\sigma_2^2/n_2)]^{1/2}$ is approximately standard normal if n_1 and n_2 are both large values. (*Hint:* see the central limit theorem in Chapter 8.)

References

Derivations of the sampling distributions discussed in this chapter may be found in advanced texts on mathematical statistics, including the following:

1. Cramér, H. 1946. *Mathematical methods of statistics*, Princeton: Princeton University Press.
2. De Groot, M. H. 1975. *Probability and statistics*, Menlo Park, CA: Addison-Wesley Publishing Co.
3. Hogg, R. V., and A. T. Craig. 1970. *Introduction to mathematical statistics*, 3d ed. London: The Macmillan Co.
4. Lindgren, B. W. 1968. *Statistical theory*, 2d ed. London: The Macmillan Co.

Chapter 11

ESTIMATION

Estimating the Mean
   Point estimates
   Confidence interval estimates
   Error of estimation
   Relating sample size with size of error
Estimating the Difference Between Two Means
Estimating a Proportion
Estimating the Variance
Criteria of Goodness of an Estimator
The General Concept of Confidence Interval Estimates

11.1   Estimating the Mean

There are many occasions when one may wish to estimate the mean of a population or distribution: an agronomist may wish to estimate the mean yield per acre of a certain new type of corn; an industrial engineer may wish to estimate the mean time that will be required to assemble a certain new piece of equipment; a business manager may wish to estimate the mean increase in productivity if a certain new program is adopted. The following is a discussion of the standard methods for obtaining such estimates.

11.1.1   Point Estimates

We know from Theorem 10.1 of Chapter 10 that if a random sample is obtained from a population or distribution, the expected value of the sample mean is the mean of the population or distribution. In other words, if $(X_1, X_2, \ldots, X_n)$ is a random sample of size n from a population or distribution whose mean is μ, and if $\bar{X} = (X_1 + X_2 + \cdots + X_n)/n$ is the random variable that is the sample mean, then $E(\bar{X}) = \mu$. This means that if repeated random samples of size n are taken from the population, then the values of the sample means should, on the average, be equal to μ. Furthermore, if the sample size is large, the strong law of large numbers tells us that there is a great probability that the sample mean $\bar{X}$ will be approximately equal to μ. (See supplementary exercise at the end of the chapter.) Thus, if we are looking for an estimate of the mean of a population it is reasonable to use the observed value $\bar{x}$ of the sample mean $\bar{X}$ as an estimate. For example, suppose that the respective yield measured in bushels of 100 acres each of a certain new type of corn is $x_1, x_2, \ldots,$ and x_{100}. Furthermore, suppose that the mean of this random sample is $\bar{x} = (x_1 + x_2 + \cdots + x_{100})/100 = 82.3$. Then, it would be reasonable to say that 82.3 is an estimate of the mean yield per acre of this type of corn. In the formal language of statistical inference, if the observed value $\bar{x}$ of the sample mean $\bar{X}$ of a random sample of size n from a population is used to estimate the mean μ of the population, we say that $\bar{x}$ is a **point estimate** of the mean, and $\bar{X}$ is an **estimator** of the mean.

Summary

If an observed value $\bar{x}$ of the sample mean $\bar{X}$ is used to estimate the mean μ of a population, then $\bar{x}$ is called a point estimate of μ, and the random variable $\bar{X}$ is called an estimator.

We have discussed only one method of estimating the mean of a population. There are, of course, other methods. For example, we could have used the sample median instead of the sample mean. In that case the random variable that is the sample median would be called the estimator, and the observed value of the sample median would be the point estimate of μ. The sample mean is, however, the most commonly used point estimate of the mean of a population. Why this is so is discussed in Section 11.5, where a formal definition of point estimates and estimators, as well as criteria for judging the goodness of the estimators, are also given.

EXAMPLE 11.1

A survey of 100 randomly selected teenage smokers provided the following set of data on the number of cigarettes smoked per day per teenager.

Number of teenagers	30	21	14	13	8	3	3	2	4	2
Number of cigarettes	1	2	3	4	5	6	7	8	9	10

On the average, how many cigarettes per day did the teenagers smoke? Give point estimates using (a) the sample mean, and (b) the sample median.

SOLUTION

(a) The point estimate using the sample mean is

$$\bar{x} = \frac{1.30 + 2.21 + 3.14 + 4.13 + 5.8 + 6.3 + 7.3 + 8.2 + 9.4 + 10.2}{100}$$

$$= 3.17$$

(b) The point estimate using the sample median is

$$m = 2$$

(The median was obtained by looking at the table of values and then observing that 50 teenagers smoked no more than 2 cigarettes per day, and 50 teenagers smoked no fewer than 2 cigarettes per day.)

EXAMPLE 11.2

Some insight into the relative merits of the sample mean and the sample median as point estimates of the mean μ of a population may be obtained by examining the solution to the following problem.

Consider the population $\mathscr{P}$ that is the set $\{1, 3, 4, 10\}$ of numbers.

(a) Find the mean of the population.

(b) List all possible random samples of size 3 from the population. Find the sample mean and sample median of each random sample. Find the magnitude of the difference between each sample mean and the mean of the population. Find the magnitude of the difference between each sample median and the mean of the population.

(c) Find the average of the magnitudes of the differences between sample means and the mean of the population. Find the average of the magnitudes of the differences between sample medians and the mean of the population. Compare these two averages.

(d)   Is the sample mean, on the average, a "better" estimator of the mean of the population than the sample median (for this example)?

(e)   Is the sample mean always a better estimator of the mean of the population than the sample median?

SOLUTION

(a)   The mean of the population is

$$\mu = \frac{1 + 3 + 4 + 10}{4} = 4.5$$

(b)   See the table of values that follows. The first column lists the elements in the random samples. The second column gives the number of random samples that have the elements listed in the first column. For example, there are 6 distinct random samples that consist of the elements $1, 3, 4$ (i.e., the different permutations of the elements). The sample median is denoted by m and is given in the third column; the sample mean is denoted by $\bar{x}$ and is given in the fourth column.

TABLE 11.1
Comparing the Mean and Median of Random Samples

Elements in sample	Number of samples	m	$\bar{x}$	$|m - \mu|$	$|\bar{x} - \mu|$
$\{1,3,4\}$	6	3	2.67	1.5	1.83
$\{1,3,10\}$	6	3	4.67	1.5	.17
$\{1,4,10\}$	6	4	5.00	.5	.50
$\{3,4,10\}$	6	4	5.67	.5	1.17
$\{1,1,3\}$	3	1	1.67	3.5	2.83
$\{1,1,4\}$	3	1	2.00	3.5	2.50
$\{1,1,10\}$	3	1	4.00	3.5	.50
$\{3,3,1\}$	3	3	2.33	1.5	2.17
$\{3,3,4\}$	3	3	3.33	1.5	1.17
$\{3,3,10\}$	3	3	5.33	1.5	.83
$\{4,4,1\}$	3	4	3.00	.5	1.50
$\{4,4,3\}$	3	4	3.67	.5	.83
$\{4,4,10\}$	3	4	6.00	.5	1.50
$\{10,10,1\}$	3	10	7.00	5.5	2.50
$\{10,10,3\}$	3	10	7.67	5.5	3.17
$\{10,10,4\}$	3	10	8.00	5.5	3.50
$\{1,1,1\}$	1	1	1.00	3.5	3.50
$\{3,3,3\}$	1	3	3.00	1.5	1.50
$\{4,4,4\}$	1	4	4.00	.5	.50
$\{10,10,10\}$	1	10	10.00	5.5	5.50
	64				

m = sample median $\bar{x}$ = sample mean μ = mean of the population

(c)

(the average of the magnitudes of the differences between
sample medians and the mean of the population)

$$= \frac{(6)(1.5) + (6)(1.5) + (6)(.5) + (6)(.5) + (3)(3.5) + \cdots + (1)(5.5)}{64}$$

$$= 2.09375$$

(the average of the magnitudes of the differences between
sample means and the mean of the population)

$$= \frac{(6)(1.83) + (6)(.17) + (6)(.5) + (6)(1.17) + (3)(2.83) + \cdots + (1)(5.5)}{64}$$

$$= 1.59406$$

(d) Since the sample mean is, on the average, 1.59406 away from the
mean of the population and the sample median is, on the average, 2.09375
away from the mean of the population, we would conclude that the sample
mean is, on the average, a better estimator of the mean of the population than
the sample median. (The numbers are obtained in (c).)

(e) The sample mean is not always a better estimator of the mean of the
population than the sample median. We can see from the table above that the
value of $|m - \mu|$ is less than the value of $|\bar{x} - \mu|$ for 24 of the 64 random
samples. The values of $|m - \mu|$ is no greater than the value of $|\bar{x} - \mu|$ for 34 of
the 64 random samples.

11.1.2   Confidence Interval Estimates

Although the agronomist in the example of Section 11.1.1 may feel that 82.3 is
a reasonable point estimate of the mean yield in bushels of the new type of corn
under consideration, he may nevertheless have reservations about making a
statement to the effect that the mean yield will be exactly 82.3 bushels. He may
prefer to allow for some margin of error. That is, he may prefer to say that he
estimates that the mean yield will be some number in the range of $82.3 - e$ to
$82.3 + e$, e being the margin of error. The following is a discussion of how one
may determine estimates of this sort.

Suppose we wish to estimate the mean μ of a normal population. If $\bar{X}$ is
the sample mean of a random sample of size n from the population, then—
according to Statement 1(a), Section 10.4.1, Chapter 10—the distribution of $\bar{X}$
is normal with mean μ and variance σ^2/n (σ^2 being the variance of the
population under consideration), and the distribution of $(\bar{X} - \mu)/(\sigma/n^{1/2})$ is

standard normal. Therefore, for any value α such that $0 < \alpha < 1$, we have

$$P\left(-z_{\alpha/2} < \frac{\bar{X} - \mu}{\sigma/\sqrt{n}} < z_{\alpha/2}\right) = 1 - \alpha$$

Note: z_α is defined by the equation $\int_{z_\alpha}^\infty [1/(2\pi)^{1/2}] \cdot e^{-x^2/2}\, dx = \alpha$. In other words, if we examine the area under the curve of the standard normal probability density function, then the area to the right of z_α is α. See Figure 10.4. in Chapter 10.

Then it follows that

$$P\left(-z_{\alpha/2} \cdot \frac{\sigma}{\sqrt{n}} < \bar{X} - \mu < z_{\alpha/2} \cdot \frac{\sigma}{\sqrt{n}}\right) = 1 - \alpha$$

and

$$P\left(-\bar{X} - z_{\alpha/2} \cdot \frac{\sigma}{\sqrt{n}} < -\mu < -\bar{X} + z_{\alpha/2} \cdot \frac{\sigma}{\sqrt{n}}\right) = 1 - \alpha$$

Now, if we multiply each term in the inequality by -1 (reversing the sense of the inequalities), we obtain

$$P\left(\bar{X} - z_{\alpha/2} \cdot \frac{\sigma}{\sqrt{n}} < \mu < \bar{X} + z_{\alpha/2} \cdot \frac{\sigma}{\sqrt{n}}\right) = 1 - \alpha$$

The above statement tells us that if the random variable $\bar{X}$ is used to provide a point estimate $\bar{x}$ of μ, then there is a $1 - \alpha$ probability that μ will lie in the random interval from $\bar{X} - z_{\alpha/2}(\sigma/n^{1/2})$ to $\bar{X} + z_{\alpha/2}(\sigma/n^{1/2})$. This means that if α is small so that $1 - \alpha$ is large, then there is a great probability that the inequality $\bar{X} - z_{\alpha/2}(\sigma/n^{1/2}) < \mu < \bar{X} + z_{\alpha/2}(\sigma/n^{1/2})$ will be satisfied. Now, it seems reasonable to assume that if there is a great probability that the inequality

$$\bar{X} - z_{\alpha/2} \cdot \frac{\sigma}{\sqrt{n}} < \mu < \bar{X} + z_{\alpha/2} \cdot \frac{\sigma}{\sqrt{n}}$$

will be satisfied, we can have great confidence that if $\bar{x}$ is an observed value of the sample mean $\bar{X}$, the inequality

$$\bar{x} - z_{\alpha/2} \cdot \frac{\sigma}{\sqrt{n}} < \mu < \bar{x} + z_{\alpha/2} \cdot \frac{\sigma}{\sqrt{n}}$$

will also be satisfied. The value $1 - \alpha$ is a measure of the degree of confidence we have in the statement

$$\bar{x} - z_{\alpha/2} \cdot \frac{\sigma}{\sqrt{n}} < \mu < \bar{x} + z_{\alpha/2} \cdot \frac{\sigma}{\sqrt{n}}$$

For this reason, we call the value $1 - \alpha$ a **confidence coefficient**, and we call the inequality $\bar{x} - z_{\alpha/2}(\sigma/n^{1/2}) < \mu < \bar{x} + z_{\alpha/2}(\sigma/n^{1/2})$ a **$1 - \alpha$ confidence interval** estimate or a **$(1 - \alpha) \cdot 100\%$ confidence interval** estimate for μ.

With respect to this discussion, it is important to realize that although we can say that $P[\bar{X} - z_{\alpha/2}(\sigma/n^{1/2}) < \mu < \bar{x} + z_{\alpha/2}(\sigma/n^{1/2})] = 1 - \alpha$ we cannot say that $P[\bar{x} - z_{\alpha/2}(\sigma/n^{1/2}) < \mu < \bar{x} + z_{\alpha/2}(\sigma/n^{1/2})] = 1 - \alpha$. This is because in the first case we are dealing with an inequality involving a random variable $\bar{X}$, whereas in the second case we are dealing with an inequality involving only the fixed numbers μ, $\bar{x}$, and $z_{\alpha/2}(\sigma/n^{1/2})$. The $1 - \alpha$ confidence interval $\bar{x} - z_{\alpha/2}(\sigma/n^{1/2}) < \mu < \bar{x} + z_{\alpha/2}(\sigma/n^{1/2})$ is either true or false. It is meaningless to say that it is true or false with a probability that is strictly greater than zero and less than 1. We must not confuse a $1 - \alpha$ confidence with a $1 - \alpha$ probability. We can, however, make the following statement. If repeated random samples of size n are taken from the population, and if with each random sample the sample mean is computed and a $1 - \alpha$ confidence interval is constructed, then we should expect approximately $(1 - \alpha) \cdot 100\%$ of these confidence intervals to be correct statements.

Formula 1   Confidence Interval for the Mean of a Normal Population with Known Variance   If $\bar{x}$ is the observed value of the sample mean of a random sample of size n from a normal population with known variance equal to σ^2, then the inequality

$$\bar{x} - z_{\alpha/2} \cdot \frac{\sigma}{\sqrt{n}} < \mu < \bar{x} + z_{\alpha/2} \cdot \frac{\sigma}{\sqrt{n}}$$

is a $1 - \alpha$ confidence interval estimate for the mean μ of the population.

EXAMPLE 11.3

A certain firm manufactures a type of compressor whose length of life is approximately normally distributed with a standard deviation of 6 months. If a random sample of 16 such compressors has an average life of 65 months, find a .90 confidence interval estimate of the mean life of all compressors of this type manufactured by the firm.

SOLUTION

The data for this example are: $\bar{x} = 65$, $\sigma = 6$, $n = 16$, $1 - \alpha = .90$, $\alpha = .10$, $z_{\alpha/2} = 1.645$. The appropriate formula for a $1 - \alpha$ confidence interval estimate of μ (the mean life) is

$$\bar{x} - z_{\alpha/2} \cdot \frac{\sigma}{\sqrt{n}} < \mu < \bar{x} + z_{\alpha/2} \cdot \frac{\sigma}{\sqrt{n}}$$

By substituting the data into the formula we see that a .90 confidence interval estimate of μ is

$$65 - (1.645) \cdot \frac{6}{4} < \mu < 65 + (1.645) \cdot \frac{6}{4}$$

or

$$62.5 < \mu < 67.5$$

When the sample size is large (for most populations, this means $n \geq 30$), then—regardless of the specific form of the distribution of the population—the distribution of the sample mean $\bar{X}$ may be assumed to be approximately normal with mean μ and variance σ^2/n (μ and σ^2 being the mean and variance of the population under consideration). (See Section 8.5 of Chapter 8 on the central limit theorem, and Statement 1(b), Section 10.4.1 of Chapter 10 on the sampling distribution of $\bar{X}$.) Therefore, we may assume that the format for the confidence interval stated above for the mean of a normal population with known variance is also the format for the confidence interval for the mean of any population when the sample size is large and the variance is known. If the variance of the population is also unknown, the accepted procedure is to replace σ (the standard deviation of the population) with s (the observed value of the sample standard deviation) in the format for the $1 - \alpha$ confidence interval estimate for μ.

Formula 2   Confidence Interval for the Mean of a Population When the Sample Size Is Large   When the sample size is large and the variance of the population is known to be σ^2, then the inequality

$$\bar{x} - z_{\alpha/2} \cdot \frac{\sigma}{\sqrt{n}} < \mu < \bar{x} + z_{\alpha/2} \cdot \frac{\sigma}{\sqrt{n}}$$

is a $1 - \alpha$ confidence interval estimate for the mean μ of the population.

When the sample size is large and the variance of the population is unknown, then the inequality

$$\bar{x} - x_{\alpha/2} \cdot \frac{s}{\sqrt{n}} < \mu < \bar{x} + z_{\alpha/2} \cdot \frac{s}{\sqrt{n}}$$

is a $1 - \alpha$ confidence interval estimate for the mean μ of the population.

EXAMPLE 11.4

The sample mean and sample standard deviation for the grade point averages of a random sample of 100 college freshmen at a certain university are 2.5 and .2, respectively.

(a)   What is a reasonable point estimate for the mean grade point average for the entire freshmen class?

(b) Show that $2.4485 < \mu < 2.5515$ is a .99 confidence interval estimate for the mean μ of grade point averages for the entire freshmen class.

(c) Can we say that $P(2.4485 < \mu < 2.5515) = .99$?

(d) Can we say that we used a method for deriving a .99 confidence interval for μ that should be correct, in the long run, approximately 99% of the time?

SOLUTION

(a) A reasonable point estimate for the mean grade point average for the entire freshman class is the sample mean

$$\bar{x} = 2.5$$

(b) Since the sample size of $n = 100$ can be considered large, and since the variance of the population is not given, a $1 - \alpha$ confidence interval estimate of the mean μ of the population of grade point averages for the entire freshmen class is

$$\bar{x} - z_{\alpha/2} \cdot \frac{s}{\sqrt{n}} < \mu < \bar{x} + z_{\alpha/2} \cdot \frac{s}{\sqrt{n}}$$

In this case, $\bar{x} = 2.5$, $s = 0.2$, $n = 100$, $\alpha = .01$, $z_{\alpha/2} = z_{.005} = 2.575$. Therefore, by substituting these numbers into the format given above, we obtain

$$2.4485 < \mu < 2.5515$$

as a .99 confidence interval estimate for μ.

(c) No, we cannot say that $P(2.4485 < \mu < 2.5515) = .99$ because μ is a fixed number, not a random variable. It is no less senseless to say that $P(2.4485 < \mu < 2.5515) = .99$ than it is to say that $P(3 < 4 < 0) = .99$.

(d) Yes, we can say that we used a method for deriving a .99 confidence interval for μ that should be correct in the long run approximately 99% of the time. This is because the method consists of replacing the random variable $\bar{X}$ with the observed value $\bar{x}$ in the inequality

$$\bar{X} - z_{.005} \cdot \frac{\sigma}{\sqrt{n}} < \mu < \bar{X} + z_{.005} \cdot \frac{\sigma}{\sqrt{n}}$$

Since $P[\bar{X} - z_{.005} \cdot (\sigma/n^{1/2}) < \mu < \bar{X} + z_{.005} \cdot (\sigma/n^{1/2})] = .99$, it follows that if the method is used over and over again in the same type of situations, it should give the correct answer approximately 99% of the time. For this reason, a .99 confidence interval is sometimes called a 99% confidence interval.

When the sample size is small, the variance of the population unknown, and the distribution of the population normal, then a $1 - \alpha$ confidence interval estimate for the mean μ of the population may be obtained as follows. Under these circumstances, if $\bar{X}$ is the sample mean, S is the sample standard deviation, and n is the sample size, then it is known that the distribution of $(\bar{X} - \mu)/(S/n^{1/2})$ is the t distribution with $n - 1$ degrees of freedom (see Statement 3(a), Section 10.4.1, Chapter 10). Therefore, for any value α such that $0 < \alpha < 1$, we have

$$P\left(-t_{\alpha/2,n-1} < \frac{\bar{X} - \mu}{S/\sqrt{n}} < t_{\alpha/2,n-1}\right) = 1 - \alpha$$

(See Section 10.4.3, Chapter 10, for definitions of $t_{\alpha,v}$ and $t_{\alpha/2,v}$.) Now, let us multiply each term in the inequality by $S/n^{1/2}$, then subtract $\bar{X}$ from each term, and finally multiply each term by -1. We then obtain

$$P\left(\bar{X} - t_{\alpha/2,n-1} \cdot \frac{S}{\sqrt{n}} < \mu < \bar{X} + t_{\alpha/2,n-1} \cdot \frac{S}{\sqrt{n}}\right) = 1 - \alpha$$

In other words, there is a $1 - \alpha$ probability that μ will lie in the random interval from $\bar{X} - t_{\alpha/2,n-1}(S/n^{1/2})$ to $\bar{X} + t_{\alpha/2,n-1}(S/n^{1/2})$. If we now apply the same principles that were previously put to use to define a confidence interval estimate for the mean of a population whose variance is known to be σ^2, then we should now conclude that a $1 - \alpha$ confidence interval estimate for the mean μ of a normal population whose variance is unknown is

$$\bar{x} - t_{\alpha/2,n-1} \cdot \frac{S}{\sqrt{n}} < \mu < \bar{x} + t_{\alpha/2,n-1} \cdot \frac{S}{\sqrt{n}}$$

Formula 3   Small Sample Confidence Interval for the Mean of a Normal Population When the Variance Is Unknown   When the sample size is small, the distribution of the population is normal, and the variance is unknown, then the inequality

$$\bar{x} - t_{\alpha/2,n-1} \cdot \frac{S}{\sqrt{n}} < \mu < \bar{x} + t_{\alpha/2,n-1} \cdot \frac{S}{\sqrt{n}}$$

is a $1 - \alpha$ confidence interval estimate for the mean of the population.

EXAMPLE 11.5

Nine randomly selected mice of a certain type completed a maze in 3.5, 3.7, 4.5, 3.9, 3.9, 4.1, 4.3, 4.6, and 3.8 minutes, respectively.

    (a)   Find a point estimate of the mean time required for the mice to complete the maze;

    (b)   Find a .95 confidence interval estimate of the mean;

    (c)   Find a .99 confidence interval estimate of the mean;

(d)   As the confidence coefficient increases, does the width of the confidence interval increase or decrease?

SOLUTION

(a)   A point estimate of the mean is the observed value of the sample mean. That is, it is

$$\bar{x} = \frac{3.5 + 3.7 + 4.5 + 3.9 + 3.9 + 4.1 + 4.3 + 4.6 + 3.8}{9} = 4.03$$

(b)   The formula for a $1 - \alpha$ confidence interval estimate for the mean μ in this example is

$$\bar{x} - t_{\alpha/2, n-1} \cdot \frac{s}{\sqrt{n}} < \mu < \bar{x} + t_{\alpha/2, n-1} \cdot \frac{s}{\sqrt{n}}$$

In this case, we have

$$\bar{x} = 4.03 \text{ (computed in (a))}$$

$$s = \sqrt{\frac{(3.5 - 4.03)^2 + (3.7 - 4.03)^2 + \cdots + (3.8 - 4.03)^2}{8}} = .37$$

$$n = 9, \qquad 1 - \alpha = .95, \qquad t_{\alpha/2, n-1} = t_{.025, 8} = 2.306$$

Therefore, a .95 confidence interval estimate for μ is

$$3.75 < \mu < 4.30$$

(c)   For a .99 confidence interval, $t_{\alpha/2, n-1} = t_{.005, 8} = 3.355$. Therefore, a .99 confidence interval for μ is

$$3.62 < \mu < 4.45$$

(d)   For fixed values of $\bar{x}$, s, and n, we can see that as the confidence coefficient increases, the width of the confidence interval as defined by the formula $\bar{x} - t_{\alpha/2, n-1} (s/n^{1/2}) < \mu < \bar{x} + t_{\alpha/2, n-1} (s/n^{1/2})$ increases, because as the value of $1 - \alpha$ increases, the corresponding value of $t_{\alpha/2, n-1} (s/n^{1/2})$ also increases. For example, compare the .99 confidence interval above with the .95 confidence interval above.

In concluding this discussion on confidence interval estimates for the mean of a population, we would like to bring to the attention of the reader a class of problems that properly belongs in this section (Section 11.1) on estimating the mean but is often confused with problems of the next section (Section 11.2) on estimating the difference between two means. Consider the following example.

EXAMPLE 11.6

An experiment was conducted to estimate the average weight loss of grossly overweight men on a special 2-week diet. The observed data for 9 such men are shown in the following table.

Man	1	2	3	4	5	6	7	8	9
Weight before diet	210	198	212	230	199	206	201	225	216
Weight after diet	200	191	200	215	185	200	195	215	211

Find a .95 confidence interval for the mean weight loss for grossly overweight men on this diet. Assume that the distribution of weight losses is approximately normal.

SOLUTION

By computing the difference in weight (i.e., weight before diet minus weight after diet) for each man, we compile the following table of values

Man	1	2	3	4	5	6	7	8	9
Weight loss	10	7	12	15	14	6	6	10	5

If we let x_i denote the weight loss of the ith man, then $(x_1, x_2, \ldots, x_{10}) = (10, 7, 12, 15, 14, 6, 6, 10, 5)$ is the observed value of a random sample of size 10 from a normal population. The mean of the random sample is

$$\bar{x} = \frac{10 + 7 + 12 + 15 + 14 + 6 + 6 + 10 + 5}{9} = 9.44$$

and the standard deviation is

$$s = \sqrt{\frac{(10 - 9.44)^2 + (7 - 9.44)^2 + \cdots + (5 - 9.44)^2}{8}} = 3.68$$

With this information, the appropriate formula for a $1 - \alpha$ confidence interval for the mean μ of weight losses is Formula 3. That is

$$\bar{x} - t_{\alpha/2, n-1} \cdot \frac{s}{\sqrt{n}} < \mu < \bar{x} + t_{\alpha/2, n-1} \cdot \frac{s}{\sqrt{n}}$$

Therefore, a .95 confidence interval for the mean weight loss for grossly overweight men on this diet is

$$9.44 - (2.306) \cdot \frac{3.68}{\sqrt{9}} < \mu < 9.44 + (2.306) \cdot \frac{3.68}{\sqrt{9}}$$

or

$$6.61 < \mu < 12.27$$

A problem of the sort considered in Example 11.6 is sometimes referred to as a problem of estimating the mean of differences between paired measurements. In our example the measurement of the weight before the diet is paired with the measurement of the weight after the diet for each of the men in the sample. In general, the data for a problem of this type can be presented in a table of values such as the one that follows:

TABLE 11.2
Data for the Mean of Differences between Paired Measurements

Pair	First measurement	Second measurement	Difference (1st measurement) − (2nd measurement)
1	m_1	m_1^*	x_1
2	m_2	m_2^*	x_2
3	m_3	m_3^*	x_3
$\vdots$	$\vdots$	$\vdots$	$\vdots$
n	m_n	m_n^*	x_n

For each i, the value of x_i is computed from the equation $x_i = m_i - m_i^*$. The set $(x_1, x_2, \ldots, x_n)$ of values constitutes the set of observed values of a random sample $(X_1, X_2, \ldots, X_n)$ from the population whose mean μ we wish to estimate. However, now the problem has been reduced to the familiar one of estimating the mean of a population. We have already shown that—depending on the answers to questions such as 'Is the sample size large?', 'Is the population normal?', and 'Is the variance of the population a known quantity?'—the appropriate formula for a $1 - \alpha$ confidence interval estimate of μ is one of those already given (i.e., Formulas 1, 2, and 3).

11.1.3   Error of Estimation

In the third paragraph of Section 11.1.2 we gave an interpretation of the statement

$$P\left(\bar{X} - z_{\alpha/2} \cdot \frac{\sigma}{\sqrt{n}} < \mu < \bar{X} + z_{\alpha/2} \cdot \frac{\sigma}{\sqrt{n}} \right) = 1 - \alpha$$

and from it we defined a $1 - \alpha$ confidence interval estimate of the mean μ of a population. Another interpretation of the statement

$$P\left(\bar{X} - z_{\alpha/2} \cdot \frac{\sigma}{\sqrt{n}} < \mu < \bar{X} + z_{\alpha/2} \cdot \frac{\sigma}{\sqrt{n}}\right) = 1 - \alpha$$

is that if we estimate the mean μ of a population with the sample mean $\bar{X}$, then there is a $1 - \alpha$ probability of obtaining an error that is less than $z_{\alpha/2}\,(\sigma/n^{1/2})$. Thus, if the value of $1 - \alpha$ is large, it is reasonable to place great confidence in the statement that if $\bar{x}$ is an observed value of the sample mean $\bar{X}$, then the error in estimating μ with $\bar{x}$ will be less than $z_{\alpha/2}\,(\sigma/n^{1/2})$. The value $1 - \alpha$ is a measure of the degree of confidence we have in the statement.

When we wish to emphasize the interpretation given above, the $1 - \alpha$ confidence interval

$$\bar{x} - z_{\alpha/2} \cdot \frac{\sigma}{\sqrt{n}} < \mu < \bar{x} + z_{\alpha/2} \cdot \frac{\sigma}{\sqrt{n}}$$

is expressed as

$$\mu = \bar{x} \pm z_{\alpha/2} \cdot \frac{\sigma}{\sqrt{n}}$$

The focus is then centered on the value of $\bar{x}$ as a point estimate of μ. The value $z_{\alpha/2}\,(\sigma/n^{1/2})$ provides a measure of the maximum size of the error that is to be expected with a confidence of $1 - \alpha$.

This interpretation and this form of expression is of course applicable to all other confidence intervals given in this chapter. For example, the $1 - \alpha$ confidence interval

$$\bar{x} - t_{\alpha/2} \cdot \frac{s}{\sqrt{n}} < \mu < \bar{x} + t_{\alpha/2} \cdot \frac{s}{\sqrt{n}}$$

under this interpretation, would be written as

$$\mu = \bar{x} \pm t_{\alpha/2} \cdot \frac{s}{\sqrt{n}}$$

EXAMPLE 11.7

To estimate the mean time required to assemble a certain television component, an industrial engineer timed 64 technicians in the performance of this task and found a mean of 15.5 minutes and a standard deviation of 3.2 minutes. What can this engineer assert with a confidence of .95 about the possible size of error in estimating the mean time to be 15.5 minutes?

SOLUTION

In this example the appropriate formula for a $1 - \alpha$ confidence interval estimate of the mean μ is

$$\bar{x} - z_{\alpha/2} \cdot \frac{s}{\sqrt{n}} < \mu < \bar{x} + z_{\alpha/2} \cdot \frac{s}{\sqrt{n}}$$

Consequently, the value $z_{\alpha/2} (s/n^{1/2})$ is the maximum size of the error that is to be expected with a confidence of $1 - \alpha$ if the sample mean $\bar{x}$ is used to estimate μ. With the values given above for α, s, and n, the engineer can assert with a confidence of .95 that the maximum error will be

$$z_{\alpha/2} \cdot \frac{s}{\sqrt{n}} = (1.96) \cdot \frac{3.2}{\sqrt{64}} = .78 \text{ minutes}$$

11.1.4   Relating Sample Size with Size of Error

Suppose we wish to determine the sample size n that will give us a $1 - \alpha$ confidence that the error in estimating μ with $\bar{x}$ will be less than a specified amount e. If the population is normal (or if the sample size is expected to be large), and the variance is known to be σ^2, then we know that a $1 - \alpha$ confidence interval for μ is

$$\mu = \bar{x} \pm z_{\alpha/2} \cdot \frac{\sigma}{\sqrt{n}}$$

In other words, under these circumstances, we can have a $1 - \alpha$ confidence that the error in estimating μ with $\bar{x}$ will be less than $z_{\alpha/2} (\sigma/n^{1/2})$. Therefore, if we want the error to be less than e, we want n such that

$$z_{\alpha/2} \cdot \frac{\sigma}{\sqrt{n}} = e$$

This means that

$$z_{\alpha/2} \cdot \frac{\sigma}{e} = \sqrt{n}$$

or

$$n = \left(z_{\alpha/2} \cdot \frac{\sigma}{e} \right)^2$$

If the variance of the population is unknown, the accepted method of finding n in the formula $n = [z_{\alpha/2}(\sigma/e)]^2$ is as follows: take a preliminary random sample of size $n \geq 30$; compute the sample standard deviation s;

substitute s for σ in the formula $n = [z_{\alpha/2}(\sigma/e)]^2$. The number n obtained in this manner will provide an approximate sample size for the desired margin of error. In fact, because of the central limit theorem, this method of finding n is always applicable when the sample size is expected to be large.

Summary

When the variance of a normal population is known to be σ^2, we can have a $1 - \alpha$ confidence that the error in estimating μ with $\bar{x}$ will be less than a specified amount e if the sample size is

$$n = \left(z_{\alpha/2} \cdot \frac{\sigma}{e} \right)^2$$

If the sample size is expected to be large, the formula is applicable even if the distribution of the population is unknown.

If the variance of the population is unknown, and the sample size is expected to be large, then the value σ may be replaced by a preliminary value of s obtained from a random sample of size $n \geq 30$.

EXAMPLE 11.8

At a certain fish hatchery, the mean weight of trouts in a certain pond is to be estimated from a random sample of n trouts. A preliminary random sample of 30 trouts from the pond showed a sample standard deviation of 3 ounces. How large should n be if in estimating the mean weight with the sample mean there is to be a .99 confidence that the error will be less than 1 ounce?

SOLUTION

In this case, $\alpha = .01$, $e = 1$, and σ is to be replaced by $s = 3$. Therefore

$$n = \left(z_{\alpha/2} \cdot \frac{\sigma}{e} \right)^2 = [(2.575)(3/1)]^2 = 60$$

Exercises

1. Consider the population $\mathscr{P}$ that is the set $\{1, 3, 4, 5\}$ of numbers. Answer all the questions of Example 11.2 with this population.

2. Does the sample mean always give a better estimate of the population mean than the sample median? Explain. (See Example 11.2)

3. The set of values from 4 throws of a die constitutes a random sample of size 4 from the population $\mathscr{P}$ that is the set $\{1, 2, 3, 4, 5, 6\}$ of numbers.
    (a) Use this principle to obtain 5 random samples of size 4 from $\mathscr{P}$.
    (b) Answer the questions of Example 11.2 for this set of random samples. Report your findings to the class.

(c)   Answer the questions of Example 11.2 for the set of random samples obtained by all students in the class.

4.   The sound intensity of a certain type of food processors is normally distributed with a standard deviation of 2.9 decibels. If the measurements of the sound intensities of a random sample of 9 such food processors showed a sample mean of 50.3 decibels, find a .95 confidence interval estimate of the mean sound intensity of all food processors of this type.

5.   The number of errors in 81 randomly selected pages of a certain size from a certain printer yielded a sample mean of .3 errors per page and a sample standard deviation of .01. Find a .99 confidence interval estimate of the mean number of error per page of this size for this printer.

6.   What is the answer to Exercise 4 if the standard deviation of the population is unknown but the sample standard deviation of the random sample of 9 food processors is $s = 2.9$ decibels? Compare the length of the confidence interval obtained here with the length obtained in Exercise 4. For what value of s would the length of the confidence interval computed in this exercise be the same as the length of the confidence interval of Exercise 4? Does this seem to indicate that in general the length of the confidence interval will be shorter if we know the value of the standard deviation of the population?

7.   In reference to Example 11.1, find a .95 confidence interval estimate of the mean number of cigarettes smoked per day by teenage smokers if we can assume that the distribution under consideration is normal.

8.   The marketing research department of a certain breakfast cereal company wished to estimate the effect of increased television advertising on the sales of one of their more popular cereals whose sales, though high, had shown no increase in volume in the last five years. They decided to increase substantially the television advertising of the cereal in 8 cities for the two-month period of September and October. The average weekly sales (in units) of the cereal for the months of September and October of the last five years, and for the months of September and October during the sales campaign, are given in the table.

City	1	2	3	4	5	6	7	8
Average weekly sales before campaign	60	75	42	80	35	51	22	19
Average weekly sales during campaign	65	82	45	90	37	57	25	22

(a)   Compute the increase in sales for each city. Use this information to find a .95 confidence interval for the mean increase for all cities.
(b)   Compute the percentage of increase in sales for each city. Use this information to find a .90 confidence interval for the mean increase

measured in percents for all cities. Assume that the distribution of increases can be assumed to be approximately normal.

9. A research team whose project tasks included weighing sea shells, acquired a second scale to verify the accuracy of measurement of their first scale. To check the mean difference in weights obtained by the two available scales, a random sample of 10 sea shells were weighed on each of them. The results were:

Shell	Weight in grams	
	Old scale	New scale
1	5.13	5.12
2	4.89	4.87
3	6.14	6.13
4	7.85	7.86
5	4.99	4.99
6	5.73	5.72
7	6.80	6.77
8	7.14	7.12
9	6.25	6.24
10	5.96	5.96

Assuming that the distribution of the differences in weights obtained using the two scales can be assumed to be approximately normal, can we assert with a confidence of approximately .99 that the mean difference is less than .02 grams?

10. In Example 11.1, what can we assert, with a confidence of .95, about the possible size of error in estimating the mean with the sample mean?

11. What can we assert, with a confidence of .99, about the possible size of error in estimating the mean number of errors per page with the value of .3 in Exercise 5?

12. If the mean number of sales contacts made by agents of a certain large insurance company in December is to be estimated from the sample mean of a random sample of n agents, how large should n be to ensure, with a confidence of .95, that the error of the estimation will be less than 2? Assume that a preliminary random sample of 30 agents yielded a sample standard deviation of 10 sales contacts per agent.

13. How large a sample size is required in Exercise 4 if we want a confidence of .99 that an estimation of the mean intensity with the sample mean will have an error of less than .5 decibels?

11.2   Estimating the Difference Between Two Means

Point estimates and confidence interval estimates for the differences between two means may be obtained as follows. Let μ_1 and μ_2 be the respective means, and let σ_1^2 and σ_2^2 be the respective variances of two distinct populations $\mathscr{P}_1$

and $\mathcal{P}_2$. Furthermore, let $\bar{X}_1$ and $\bar{X}_2$ be the sample means of independent random samples of sizes n_1 and n_2, respectively, from the populations $\mathcal{P}_1$ and $\mathcal{P}_2$, respectively.

Note: random samples $(X_1, X_2, \ldots, X_{n_1})$ and $(X_1^*, X_2^*, \ldots, X_{n_2}^*)$ are said to be independent random samples if and only if the random variables $X_1, X_2, \ldots, X_{n_1}, X_1^*, \ldots, X_{n_2}^*$, taken together in one group, are independent random variables.

From Exercise 14, Section 8.3, Chapter 8, we see that $E(\bar{X}_1 - \bar{X}_2) = \mu_1 - \mu_2$. Therefore, a reasonable estimate of $\mu_1 - \mu_2$ is $\bar{x}_1 - \bar{x}_2$, where $\bar{x}_1$ and $\bar{x}_2$ are the observed values of the sample means $\bar{X}_1$ and $\bar{X}_2$, respectively. The value $\bar{x}_1 - \bar{x}_2$ is called a point estimate of $\mu_1 - \mu_2$, and the random variable $\bar{X}_1 - \bar{X}_2$ is called an estimator of $\mu_1 - \mu_2$.

Summary   Point Estimate of the Difference Between Two Means   The value $\bar{x}_1 - \bar{x}_2$ is a point estimate of $\mu_1 - \mu_2$, and the random variable $\bar{X}_1 - \bar{X}_2$ is an estimator of $\mu_1 - \mu_2$.

To obtain confidence interval estimates of $\mu_1 - \mu_2$, we may proceed as follows. If the distributions of both populations are normal, and if the respective variances are known to be σ_1^2 and σ_2^2, then the distribution of $[(\bar{X}_1 - \bar{X}_2) - (\mu_1 - \mu_2)]/\sqrt{(\sigma_1^2/n_1) + (\sigma_2^2/n_2)}$ is standard normal (see Exercise 8 of Section 10.4, Chapter 10). Therefore, for any value α such that $0 < \alpha < 1$, we have

$$P\left[-z_{\alpha/2} < \frac{(\bar{X}_1 - \bar{X}_2) - (\mu_1 - \mu_2)}{\sqrt{(\sigma_1^2/n_1) + (\sigma_2^2/n_2)}} < z_{\alpha/2}\right] = 1 - \alpha$$

It follows that

$$P\left[(\bar{X}_1 - \bar{X}_2) - z_{\alpha/2}\sqrt{\frac{\sigma_1^2}{n_1} + \frac{\sigma_2^2}{n_2}} < \mu_1 - \mu_2 < (\bar{X}_1 - \bar{X}_2) + z_{\alpha/2}\sqrt{\frac{\sigma_1^2}{n_1} + \frac{\sigma_2^2}{n_2}}\right]$$
$$= 1 - \alpha$$

Now, if we apply the principles that were used previously to define confidence intervals, we may conclude that the inequality

$$(\bar{x}_1 - \bar{x}_2) - z_{\alpha/2}\sqrt{\frac{\sigma_1^2}{n_1} + \frac{\sigma_2^2}{n_2}} < \mu_1 - \mu_2 < (\bar{x}_1 - \bar{x}_2) + z_{\alpha/2}\sqrt{\frac{\sigma_1^2}{n_1} + \frac{\sigma_2^2}{n_2}}$$

is a $1 - \alpha$ confidence interval estimate for $\mu_1 - \mu_2$.

Formula 4   Confidence Interval for the Difference Between the Means of Two Normal Populations with Known Variances   When the variances are known, and the distributions of the populations are normal, then the inequality

$$(\bar{x}_1 - \bar{x}_2) - z_{\alpha/2}\sqrt{\frac{\sigma_1^2}{n_1} + \frac{\sigma_2^2}{n_2}} < \mu_1 - \mu_2 < (\bar{x}_1 - \bar{x}_2) + z_{\alpha/2}\sqrt{\frac{\sigma_1^2}{n_1} + \frac{\sigma_2^2}{n_2}}$$

is a $1 - \alpha$ confidence interval estimate for $\mu_1 - \mu_2$.

When the sample sizes are large (for most populations, this means $n_1 \geq 30$ and $n_2 \geq 30$), then—regardless of the specific forms of the distributions of the two populations—the distribution of the estimator $\bar{X}_1 - \bar{X}_2$ may be assumed to be approximately normal with mean $\mu_1 - \mu_2$ and variance $(\sigma_1^2/n_1 + \sigma_2^2/n_2)$. (See Exercise 9, Section 10.4, Chapter 10.) Therefore, we may assume that the formula for the confidence interval of $\mu_1 - \mu_2$, when the sample sizes are large and when the variances are known, is the same as that stated above for normal populations. If the variances of the populations are also unknown, the accepted procedure is to replace σ_1 and σ_2 (the standard deviations of the respective populations) with s_1 and s_2 (the observed values of the respective sample standard deviations), respectively, in the formula for the $1 - \alpha$ confidence interval estimate for $\mu_1 - \mu_2$.

Formula 5   Confidence Interval for the Difference Between Two Means When the Sample Sizes Are Large
(a)   When the sample sizes are large and the variances of the two populations are known to be σ_1^2 and σ_2^2, respectively, then the inequality

$$(\bar{x}_1 - \bar{x}_2) - z_{\alpha/2}\sqrt{\frac{\sigma_1^2}{n_1} + \frac{\sigma_2^2}{n_2}} < \mu_1 - \mu_2 < (\bar{x}_1 - \bar{x}_2) + z_{\alpha/2}\sqrt{\frac{\sigma_1^2}{n_1} + \frac{\sigma_2^2}{n_2}}$$

is a $1 - \alpha$ confidence interval estimate for $\mu_1 - \mu_2$.
(b)   When the sample sizes are large and the variances of the populations are unknown, then the inequality

$$(\bar{x}_1 - \bar{x}_2) - z_{\alpha/2}\sqrt{\frac{s_1^2}{n_1} + \frac{s_2^2}{n_2}} < \mu_1 - \mu_2 < (\bar{x}_1 - \bar{x}_2) + z_{\alpha/2}\sqrt{\frac{s_1^2}{n_1} + \frac{s_2^2}{n_2}}$$

is a $1 - \alpha$ confidence interval estimate for $\mu_1 - \mu_2$.

EXAMPLE 11.9

In a certain experiment, the mean response time to a certain signal of 50 subjects under treatment A was $\bar{x}_1 = 40$ seconds; the standard deviation was $s_1 = 5$ seconds. The mean response time of 45 subjects under treatment B was $\bar{x}_2 = 35$ seconds; the standard deviation was $s_2 = 3$ seconds.
(a)   Find a point estimate of the difference between the mean response time of all subjects under treatment A and the mean response time of all subjects under treatment B.

(b)   Find a .98 confidence interval estimate for the difference in the two mean response times.

SOLUTION

(a)   A point estimate of the difference between μ_1 (the mean response time under treatment A) and μ_2 (the mean response time under treatment B) is

$$\bar{x}_1 - \bar{x}_2 = 40 - 35 = 5 \text{ seconds}$$

(b)   Since the sample sizes are large and the variances of the populations are unknown, a $1 - \alpha$ confidence interval estimate of $\mu_1 - \mu_2$ is

$$(\bar{x}_1 - \bar{x}_2) - z_{\alpha/2}\sqrt{\frac{s_1^2}{n_1} + \frac{s_2^2}{n_2}} < \mu_1 - \mu_2 < (\bar{x}_1 - \bar{x}_2) + z_{\alpha/2}\sqrt{\frac{s_1^2}{n_1} + \frac{s_2^2}{n_2}}$$

In this example, we have

$$\bar{x}_1 - \bar{x}_2 = 5, \qquad z_{\alpha/2} = z_{.01} = 2.327, \qquad s_1 = 5, \qquad s_2 = 3,$$
$$n_1 = 50, \qquad n_2 = 45$$

Therefore, a .98 confidence interval estimate is

$$3.1 < \mu_1 - \mu_2 < 6.9$$

Confidence interval estimates for the difference between two means when the populations are normal, the sample sizes are small, and the variances are unknown are given in the summary statement below. Derivations of these formulas are similar to those given and are considered in the exercises.

Formula 6   Small Sample Confidence Intervals for the Difference of Two Means of Normal Populations with Unknown Variances

(a)   When the populations are normal, the sample sizes small, and the variances unknown but equal to each other (i.e., $\sigma_1^2 = \sigma_2^2$), then an approximate $1 - \alpha$ confidence interval for $\mu_1 - \mu_2$ is

$$(\bar{x}_1 - \bar{x}_2) - t_{\alpha/2, v} \cdot s^* < \mu_1 - \mu_2 < (\bar{x}_1 - \bar{x}_2) + t_{\alpha/2, v} \cdot s^*$$

where $v = n_1 + n_2 - 2$, and

$$(s^*)^2 = \left[\frac{(n_1 - 1)s_1^2 + (n_2 - 1)s_2^2}{n_1 + n_2 - 2}\right]\left[\frac{1}{n_1} + \frac{1}{n_2}\right]$$

(b)   When the populations are normal, the sample sizes small, and the variances unknown and not equal to each other (i.e., $\sigma_1^2 \neq \sigma_2^2$), then an approximate $1 - \alpha$ confidence interval estimate for $\mu_1 - \mu_2$ is

$$(\bar{x}_1 - \bar{x}_2) - t_{\alpha/2, v}\sqrt{\frac{s_1^2}{n_1} + \frac{s_2^2}{n_2}} < \mu_1 - \mu_2 < (\bar{x}_1 - \bar{x}_2) + t_{\alpha/2, v}\sqrt{\frac{s_1^2}{n_1} + \frac{s_2^2}{n_2}}$$

where

$$v = \frac{(s_1^2/n_1 + s_2^2/n_2)^2}{[(s_1^2/n_1)^2/(n_1 - 1)] + [(s_2^2/n_2)^2/(n_2 - 1)]}$$

Note: let v be the nearest whole number.

Finally, we should mention that if the sample sizes n_1 and n_2 are equal, the problem of estimating $\mu_1 - \mu_2$ may be reduced to the problem of estimating the mean of the differences between paired measurements (Section 11.1.2 and Example 11.6). The procedure is as follows. Suppose that $(x_{11}, x_{12}, \ldots, x_{1n})$ and $(x_{21}, x_{22}, \ldots, x_{2n})$ are the respective random samples from the populations with means μ_1 and μ_2. Let $\mu = \mu_1 - \mu_2$ and let $x_i = x_{1i} - x_{2i}$ for $i = 1, 2, \ldots, n$. Then $(x_1, x_2, \ldots, x_n)$ is a random sample of size n from a population whose mean is μ. We can use any of the formulas given in Section 11.1 to estimate μ with the result that an estimate of μ is also an estimate of $\mu_1 - \mu_2$. This procedure may be justified by observing that if X_1 and X_2 are random variables with means μ_1 and μ_2, respectively, then $\mu_1 - \mu_2 = E(X_1) - E(X_2) = E(X_1 - X_2)$. Consequently, the difference between two population means is the mean of the differences between paired elements of the two populations.

EXAMPLE 11.10

The average daily calorie intakes of 9 randomly selected 20-year-old men and 9 randomly selected 20-year-old women are:

Men:	4515	5200	3515	2798	3808	1979	3506	4101	3414
Women:	2515	3200	1915	1500	1850	4115	2316	1995	2988

Find a .95 confidence interval estimate of the difference between the mean calorie intake of 20-year-old men and the mean calorie intake of 20-year-old women.

SOLUTION

Following the procedure described, we obtain this table of values.

x_{1_i}:	4515	5200	3515	2798	3808	1979	3506	4101	3414
x_{2_i}:	2515	3200	1915	1500	1850	4115	2316	1995	2988
x_i :	2000	2000	1600	1298	1958	-2136	1190	2106	426

(x_{1i} is the calorie intake of the ith man, x_{2i} is the calorie intake of the ith woman, and $x_i = x_{1i} - x_{2i}$.)

Let μ_1 and μ_2 be the mean calorie intake of 20-year-old men and 20-year-old women, respectively. If we can now assume that $(x_1, x_2, \ldots, x_9)$ is a random sample from a normal population, then the procedure for finding a .95 confidence interval for $\mu_1 - \mu_2$ would be to follow Formula 3 of Section 11.1.2. In other words, a .95 confidence interval for $\mu_1 - \mu_2$ would be

$$\bar{x} + t_{.025,8} \frac{s}{\sqrt{9}} < \mu_1 - \mu_2 < \bar{x} + t_{.025,8} \frac{s}{\sqrt{9}}$$

where $\bar{x}$ is the mean and s is the standard deviation of the random sample $(x_1, x_2, \ldots, x_9)$.

Computing for the mean and standard deviation of the random sample, we find

$$\bar{x} = \frac{2000 + 2000 + 1600 + \cdots + 426}{9} = 1160.22$$

and

$$s = \sqrt{\frac{(2000 - 1160.22)^2 + (2000 - 1160.22)^2 + \cdots + (426 - 1160.22)^2}{8}}$$

$$= 1349.0$$

Therefore, a .95 confidence interval for $\mu_1 - \mu_2$ is

$$1160.22 - 2.306 \cdot (1349.0/3) < \mu_1 - \mu_2 < 1160.22 + 2.306 \cdot (1349.0/3)$$

or

$$123.29 < \mu_1 - \mu_2 < 2197.15$$

Exercises

1. A standardized basic arithmetic test given to 60 randomly selected college students, who were told of a $100 cash award for excellence, yielded a mean test score of 84 and a standard deviation of 5. The same test given to 63 randomly selected college students without any mention of a cash award for excellence yielded a mean test score of 78 and a standard deviation of 10. Find a .96 confidence interval for the difference $\mu_1 - \mu_2$, where μ_1 is the mean score for all college students who are promised a cash award for excellence, and μ_2 is the mean score for all college students who are not.

2. A random sample of 49 brand A cigarettes has a mean nicotine content of 18.4 milligrams and a standard deviation of 1.9 milligrams. A random sample of 53 brand B cigarettes has a mean nicotine content of 17.1

milligrams and a standard deviation of 1.2 milligrams. Construct a .99 confidence interval estimate for $\mu_1 - \mu_2$, where μ_1 is the mean nicotine content of brand A cigarettes, and μ_2 is the mean nicotine content of brand B cigarettes.

3. If a random sample of size $n_1 = 9$ taken from a normal population with a standard deviation $\sigma_1 = 4$ has a sample mean of $\bar{x}_1 = 35$, and if a second random sample of size $n_2 = 16$ taken from a different normal population with a standard deviation $\sigma_2 = 2$ has a mean of $\bar{x}_2 = 27$, then what is a .93 confidence interval for $\mu_1 - \mu_2$, where μ_1 and μ_2 are the respective means of the first and second populations described above.

4. The weight gains measured in ounces of 9 rabbits on diet A and 12 rabbits on diet B are given in the following table. Construct a .90 confidence interval for the difference between the mean weight gains for rabbits on the two diets. Assume that the distribution of weight gains for each of the two diets is approximately normal and that the two variances are equal.

Diet A	12	11	15	13	12	9	13	16	11			
Diet B	7	8	6	9	7	7	9	9	6	10	8	7

5. What is the answer to Exercise 4 if we do not assume equal variances for the two populations?

6. Following the principles discussed in Section 11.1.3, what can we assert, with a confidence of $1 - \alpha$, about the maximum size of the error in estimating $\mu_1 - \mu_2$ with $\bar{x}_1 - \bar{x}_2$ if
   (a) the populations are normal and the variances are known;
   (b) the sample sizes are large and the variances are known;
   (c) the sample sizes are large and the variances are unknown;
   (d) the populations are normal, the sample sizes small, and the variances are unknown but equal to each other;
   (e) the populations are normal, the sample sizes small, and the variances are unknown and not equal to each other?

7. If 45 boxes of sugar packaged by a certain machine had a mean weight of 1.04 pounds and a standard deviation of .03 pounds, and 49 boxes of sugar packaged by a second machine had a mean weight of .99 pounds and a standard deviation of .04 pounds, then what can we assert with a confidence of .95 concerning the maximum size of the error in estimating that the average weight of sugar packed by the first machine is .05 pounds heavier than the average weight of sugar packed by the second machine?

8. Let $\mathscr{P}_1$ and $\mathscr{P}_2$ be normal populations with means μ_1 and μ_2, respectively, and variances σ_1^2 and σ_2^2, respectively. Let $\bar{x}_1$ and $\bar{s}_1$, and $\bar{x}_2$ and $\bar{s}_2$ be the sample means and sample standard deviations of respective independent random samples of sizes n_1 and n_2 from populations $\mathscr{P}_1$ and $\mathscr{P}_2$.
   (a) Find the length of a $1 - \alpha$ confidence interval for $\mu_1 - \mu_2$ if we assume that $\sigma_1^2 = \sigma_2^2$ so that Formula 6(a) is applicable.

(b)   Find the length if we assume that $\sigma_1^2 \neq \sigma_2^2$ so that Formula 6(b) is applicable.

(c)   Suppose that $n_1 = 9, n_2 = 15, s_1 = 3$, and $s_2 = 3$. Find the length of a .95 confidence interval for $\mu_1 - \mu_2$ if we assume $\sigma_1^2 = \sigma_2^2$. Find the length if we assume $\sigma_1^2 \neq \sigma_2^2$.

(d)   Suppose that $n_1 = 9, n_2 = 15, s_1 = 3$, and $s_2 = 4$. Find the length of a .95 confidence interval for $\mu_1 - \mu_2$ if we assume $\sigma_1^2 = \sigma_2^2$. Find the length if we assume $\sigma_1^2 \neq \sigma_2^2$.

(e)   Suppose that $n_1 = 9, n_2 = 15, s_1 = 3$, and $s_2 = 10$. Find the length of a .95 confidence interval for $\mu_1 - \mu_2$ if we assume $\sigma_1^2 = \sigma_2^2$. Find the length if we assume $\sigma_1^2 \neq \sigma_2^2$.

(f)   Judging from the results of (c), (d), and (e), would it be better to assume that $\sigma_1^2 = \sigma_2^2$ than to assume that $\sigma_1^2 \neq \sigma_2^2$ if $|s_1 - s_2| = 0$; if $|s_1 - s_2|$ is very small; if $|s_1 - s_2|$ is large?

11.3   Estimating a Proportion

There are many occasions when one may wish to estimate a proportion. For example, one may wish to estimate the proportion of voters who will vote for a certain candidate; or, one may wish to estimate the proportion of patients under a certain treatment who will experience unpleasant side effects; or, one may wish to estimate the proportion of TV viewers who will respond favorably to a certain commercial.

Suppose that p is the proportion of elements in a population that possesses property A. To estimate p, we may wish to obtain a random sample of size n from the population. Suppose we let $\hat{P}$ be the random variable that is the proportion of elements in the random sample that possesses property A, and we let $\hat{p}$ denote an observed value of $\hat{P}$. It is intuitively obvious that $\hat{p}$ is a reasonable estimate of p. We call $\hat{p}$ a point estimate of p and we call $\hat{P}$ an estimator of p.

Summary

If p is the proportion of elements in a population that possesses property A, $\hat{P}$ is the random variable that is the proportion of elements in the random sample that possesses property A, and $\hat{p}$ is an observed value of $\hat{P}$, then $\hat{p}$ is a point estimate of p, and $\hat{P}$ is an estimator of p.

If we can be certain that the random sample is obtained under sampling with replacement, or—failing that—if we can assume that the population is very large, then a large sample $1 - \alpha$ confidence interval estimate of p may be obtained as follows. Consider n Bernoulli trials where the event of a success on the ith trial is equivalent to the event that the ith element in the random sample possesses property A. It is easy to see that under these circumstances the distribution of the estimator P is identical (or approximately equal) to the distribution of S_n/n, where S_n is the binomial random variable of the number of successes in the n Bernoulli trials. Now, since the distribution of $(S_n - np)/[np(1 - p)]^{1/2}$ is approximately standard normal for n large (see Chapter 8), it follows that the distribution of $(n\hat{P} - np)/[np(1 - p)]^{1/2}$ will also be approximately standard normal for n large. Therefore, for any value α such that $0 < \alpha < 1$, we have

$$P\left(-z_{\alpha/2} < \frac{n\hat{P} - np}{\sqrt{np(1 - p)}} < z_{\alpha/2}\right) \approx 1 - \alpha$$

This means that

$$P\left(\hat{P} - z_{\alpha/2}\sqrt{\frac{p(1 - p)}{n}} < p < \hat{P} + z_{\alpha/2}\sqrt{\frac{p(1 - p)}{n}}\right) \approx 1 - \alpha$$

Now, if we replace p with the point estimate $\hat{p}$ on the extreme left and extreme right sides of the inequality, then

$$P\left(\hat{P} - z_{\alpha/2}\sqrt{\frac{\hat{p}(1 - \hat{p})}{n}} < p < \hat{P} + z_{\alpha/2}\sqrt{\frac{\hat{p}(1 - \hat{p})}{n}}\right) \approx 1 - \alpha$$

We then apply the principles that were used previously to define confidence intervals and conclude that the inequality

$$\hat{p} - z_{\alpha/2}\sqrt{\frac{\hat{p}(1 - \hat{p})}{n}} < p < \hat{p} + z_{\alpha/2}\sqrt{\frac{\hat{p}(1 - \hat{p})}{n}}$$

is an approximate $1 - \alpha$ confidence interval estimate of p.

Formula 7   Large Sample Confidence Interval for Proportion   Let p be the proportion of elements in a population that possesses property A. Let $\hat{p}$ be the proportion of elements in a large random sample of size n that possesses property A. An approximate $1 - \alpha$ confidence interval estimate of p is

$$\hat{p} - z_{\alpha/2}\sqrt{\frac{\hat{p}(1 - \hat{p})}{n}} < p < \hat{p} + z_{\alpha/2}\sqrt{\frac{\hat{p}(1 - \hat{p})}{n}}$$

In actual practice, confidence intervals for proportions are usually obtained from tables of values such as the ones given in *Biometrika Tables for Statisticians*, Vol. I, John Wiley & Sons, Inc., 1968.

EXAMPLE 11.11

A survey of 100 randomly selected residents of a certain city showed that $\frac{1}{5}$ of them were born in that city. Construct a .95 confidence interval estimate of the proportion p of residents in that city who were born there.

SOLUTION

With $\hat{p} = 1/5$, and $z_{\alpha/2} = z_{.025} = 1.96$ substituted into the formula

$$\hat{p} - z_{\alpha/2}\sqrt{\frac{\hat{p}(1-\hat{p})}{n}} < p < \hat{p} + z_{\alpha/2}\sqrt{\frac{\hat{p}(1-\hat{p})}{n}}$$

we obtain

$$\frac{1}{5} - 1.96\sqrt{\frac{\frac{1}{5}\cdot\frac{4}{5}}{100}} < p < \frac{1}{5} + 1.96\sqrt{\frac{\frac{1}{5}\cdot\frac{4}{5}}{100}}$$

or

$$.12 < p < .28$$

as a .95 confidence interval estimate of p.

Exercises

1. In a random sample of $n = 500$ coffee drinkers, 50 of them will not drink instant coffee. Find a .95 confidence interval for the actual proportion of coffee drinkers who will not drink instant coffee.

2. If 15 of 45 randomly selected office workers in a certain very large firm disapprove of the new policy issued by the firm's president, what is a .99 confidence interval estimate of the actual proportion of office workers who disapprove of the new policy?

3. Show that if $\hat{p}$ is the proportion of elements in a random sample of size n that possesses a certain property and p is the actual proportion of elements in the population that possesses that property, then we can have a confidence of $1 - \alpha$ that the error in estimating p with $\hat{p}$ will be less than $z_{\alpha/2}[\hat{p}(1-\hat{p})/n]^{1/2}$. (Follow the principles discussed in Section 11.1.3.)

4. In testing a certain new drug, it was found that 45 of 50 randomly selected patients with a certain disease were cured after a one-week treatment with the drug. What can we assert, with a confidence of .95, about the possible size of error in estimating that 90% of patients with this disease will be cured after a one-week treatment with the drug? (See Exercise 3.)

5. Show that if we can obtain a preliminary value of $\hat{p}$ from a random sample that is relatively large (for example, at least 30), then we can use the formula

$$n = [z_{\alpha/2}^2 \cdot \hat{p}(1-\hat{p})]/e^2$$

to obtain a sample size n that will provide a confidence of approximately

$1 - \alpha$ to the statement that p (obtained from the random sample of size n) estimates p with an error that is less than e. (Follow the procedure used in Section 11.1.4.)

6. Suppose that the manager of the office workers of Exercise 2 wishes to obtain a fairly accurate estimate of the proportion of office workers who disapprove of the new policy. Find the required sample size for a .99 confidence that the error of the estimate $\hat{p}$ will be less than .1.

7. Find the required number of patients that would have to be treated with the new drug in Exercise 4 if we want a .95 confidence that the error of the estimate will be less than 1%.

11.4 Estimating the Variance

If S^2 is the sample variance of a random sample of size n from a population with variance equal to σ^2, then $E(S^2) = \sigma^2$. (See Chapter 10, Section 10.3.3.) Therefore, it is reasonable to estimate σ^2 with an observed value s^2 of the random variable S^2. Furthermore, if the population has a normal distribution, the random variable $(n - 1)S^2/\sigma^2$ also has a chi-square distribution with $n - 1$ degrees of freedom. Therefore, a $1 - \alpha$ confidence interval estimate of σ^2 may be obtained as follows. For any α such that $0 < \alpha < 1$, we have

$$P(\chi^2_{1-\alpha/2,\, n-1} < \frac{(n-1)S^2}{\sigma^2} < \chi^2_{\alpha/2,\, n-1}) = 1 - \alpha$$

Therefore

$$P\left(\frac{(n-1)S^2}{\chi^2_{\alpha/2,\, n-1}} < \sigma^2 < \frac{(n-1)S^2}{\chi^2_{1-\alpha/2,\, n-1}}\right) = 1 - \alpha$$

It follows that a $1 - \alpha$ confidence interval estimate of σ^2 is

$$\frac{(n-1)s^2}{\chi^2_{\alpha/2,\, n-1}} < \sigma^2 < \frac{(n-1)s^2}{\chi^2_{1-\alpha/2,\, n-1}}$$

Formula 8   Point and Confidence Interval Estimates for Variance If σ^2 is the variance of a population, S^2 is the sample variance of a random sample of size n from the population, and s^2 is an observed value of S^2, then s^2 is a **point estimate** of σ^2 and S^2 is an estimator of σ^2.

If the distribution of the population is normal, then a $1 - \alpha$ confidence interval estimate of σ^2 is

$$\frac{(n-1)s^2}{\chi^2_{\alpha/2,\, n-1}} < \sigma^2 < \frac{(n-1)s^2}{\chi^2_{1-\alpha/2,\, n-1}}$$

EXAMPLE 11.12

Test runs for the operating time before overheating of 9 randomly selected motors of a certain type showed a sample mean of $\bar{x} = 39.7$ hours and a sample standard deviation of $s = 2.5$ hours. Find a .95 confidence interval estimate for the standard deviation σ in the length of time a motor of this type will operate before overheating.

SOLUTION

Substituting $s^2 = (2.5)^2$, $n = 9$, $\chi^2_{\alpha/2, n-1} = \chi^2_{.025, 8} = 17.535$, and $\chi^2_{1-\alpha/2, n-1} = \chi^2_{.975, 8} = 2.180$ into the formula

$$\frac{(n-1)s^2}{\chi^2_{\alpha/2, n-1}} < \sigma^2 < \frac{(n-1)s^2}{\chi^2_{1-\alpha/2, n-1}}$$

we obtain

$$\frac{8 \cdot (2.5)^2}{17.535} < \sigma^2 < \frac{8 \cdot (2.5)^2}{2.180}$$

or

$$2.85 < \sigma^2 < 22.94$$

as a .95 confidence interval estimate of σ^2. Taking square roots we obtain

$$1.69 < \sigma < 4.79$$

as a .95 confidence interval estimate of σ.

It should also be noted that the above discussion assumes that either the random samples from a population are obtained under the rules of sampling with replacement, or that the population is very large (so that sampling without replacement differs very little from sampling with replacement). If the population is finite (and not very large) and random samples are obtained under sampling without replacement, the distribution of $(n-1)S^2/\sigma^2$ cannot be assumed to be a chi-square distribution, nor can we assume that $E(S^2) = \sigma^2$.

Exercises

1. The systolic blood pressure of 9 randomly selected Army volunteers was measured after each of them had run 100 yards. Individual measurements were found to be 120, 130, 125, 120, 140, 125, 135, 130, and 145, respectively. Find a .95 confidence interval for the variance of systolic blood pressures among all volunteers. Assume that the distribution of systolic blood pressures is normal.

2. Find a .99 confidence interval for the variance in the sound intensity of the food processors described in Exercise 4 of Section 11.1. (Assume that the sample standard deviation is 2.9 decibels.)

3. Construct a .90 confidence interval for the standard deviation of mean grade point averages of college freshmen in Example 11.4.

4. Find a point estimate of the standard deviation of the distribution of times required of mice, of the type in Example 11.5, to complete the maze. Can we assert with a confidence of at least .95 that the population standard deviation will be within .5 minutes of the sample standard deviation? Explain.

11.5   Criteria of Goodness of an Estimator

In the preceding sections we called $\bar{X}$ an estimator of μ, we called S^2 an estimator of σ^2, and we called S_n/n an estimator of p. In each case, the estimator is a random variable that is also a function of the random variables in a random sample $(X_1, X_2, \ldots, X_n)$ of size n from a population. For example, $\bar{X} = (\bar{X}_1 + X_2 + \cdots + X_n)/n$;   and $S^2 = [(X_1 - \bar{X})^2 + \cdots + (X_n - \bar{X})^2]/(n-1)$. In other words, each of these estimators is a statistic. (See the definition of statistics in Section 10.4 of Chapter 10.) In the language of statistical inference, an estimator is a statistic that is used to obtain estimates of a population parameter. A value of an estimator is called a point estimate of the parameter. With this definition, we see that any one parameter may give rise to any number of estimators of that parameter. How do we compare two different estimators of the same parameter? How do we choose the 'best' estimator among all estimators of the same parameter? Numerous criteria have been developed to judge the goodness of estimators. The more important of these are stated in terms of the properties of being unbiased, consistent, relatively efficient, and having minimum variance. A discussion of these properties follows.

Since an estimator is a random variable whose values are unpredictable (even though they are statistically governed by the distribution of the random variable), we cannot expect it to provide the correct answer at all times. For example, if the mean of a population is $\mu = 5$, then one random sample of size $n = 30$ might have a sample mean of $\bar{x} = 4.9$; another random sample of size $n = 30$ might have a sample mean of $\bar{x} = 5.2$. However, if it can be shown that the expected value of the estimator is equal to the actual value of the parameter to be estimated, at least we can expect the estimator to give us, on the average, the correct answer. Such estimators are called unbiased estimators.

Definition 11.1   An estimator θ is said to be an **unbiased estimator** of the parameter θ if $E(\theta) = \theta$. Otherwise, it is said to be **biased**.

According to this definition, $\bar{X}$ is an unbiased estimator of μ (see Section 11.1), $\bar{X}_1 - \bar{X}_2$ is an unbiased estimator of $\mu_1 - \mu_2$ (see Section 11.2), $\hat{P}$ is an unbiased estimator of p (under sampling with replacement—see Section 11.3), and S^2 is an unbiased estimator of σ^2 (under sampling with replacement—see Section 11.4). All of these estimators have the desirable property of providing, on the average, a correct estimate of the parameter.

The variance and the standard deviation of the distribution of a random variable provide a measure of the average difference between values of the random variable and the mean of the random variable. (See Chapter 6 on variance and standard deviation.) This means that if an estimator is an unbiased estimator, the variance and the standard deviation of the estimator provide a measure of the average difference between the values of the estimator and the value of the parameter to be estimated. In other words, the variance and the standard deviation of an unbiased estimator provide a measure of the average size of error in the estimations. For this reason, we call the standard deviation of the distribution of the estimator the standard error of the estimator.

Definition 11.2   The standard deviation of the distribution of an unbiased estimator is called the **standard error** of the estimator.

Now, suppose that θ_1 and θ_2 are both unbiased estimators of the same population parameter θ, and that their respective variances are $\sigma_{\theta_1}^2$ and $\sigma_{\theta_2}^2$. If we are trying to choose between the two estimators, and if $\sigma_{\theta_1}^2 < \sigma_{\theta_2}^2$, then we would surely choose θ_1 over θ_2 because, according to the discussion above, the average size of errors in using θ_1 will be smaller than the average size of errors in using θ_2. We call θ_1 a more efficient estimator than θ_2. More generally, if we consider all possible unbiased estimators of some parameter θ, then the one with the smallest variance is called the most efficient or the minimum variance unbiased estimator.

Definition 11.3   If θ_1 and θ_2 are both unbiased estimators of the same population parameter θ, θ_1 is said to be a **more efficient** estimator than θ_2 if the variance of θ_1 is less than the variance of θ_2.

If we consider all possible unbiased estimators of some parameter θ, then the one with the smallest variance is called the **most efficient** or the **minimum variance** unbiased estimator of θ.

We would all agree that minimum variance is a desirable property. In fact, minimum variance unbiased estimators are also called **best estimators**. The problem is, how can we know if an estimator has this property? The Cramer-Rao Theorem can sometimes be used to answer this question.

Theorem 11.1   The Cramer-Rao Theorem   *If θ is an unbiased estimator of θ and*

$$\mathrm{Var}(\theta) = \cfrac{1}{n \cdot E\left[\left(\dfrac{\partial \ln f(X)}{\partial \theta}\right)^2\right]}$$

where f is the pdf of the population, and n is the sample size, then θ is a minimum variance unbiased estimator of θ.

EXAMPLE 11.13

Show that the sample mean is a minimum variance unbiased estimator of the mean of a normal population.

SOLUTION

The pdf of a normal population is

$$f(x) = \frac{1}{\sqrt{2\pi}\,\sigma}\, e^{-(x-\mu)^2/(2\sigma^2)} \qquad \text{for} \quad -\infty < x < \infty$$

Therefore

$$\ln f(x) = -\ln \sigma\sqrt{2\pi} \;-\; \frac{1}{2}\left(\frac{x-\mu}{\sigma}\right)^2$$

so that

$$\frac{\partial \ln f(x)}{\partial \mu} = \frac{1}{\sigma}\left(\frac{x-\mu}{\sigma}\right)$$

and hence

$$E\left[\left(\frac{\partial \ln f(X)}{\partial \mu}\right)^2\right] = \frac{1}{\sigma^2}\, E\left[\left(\frac{X-\mu}{\sigma}\right)^2\right] = \frac{1}{\sigma^2}\cdot 1 = \frac{1}{\sigma^2}$$

Thus

$$\frac{1}{n\cdot E\left[\left(\dfrac{\partial \ln f(x)}{\partial \mu}\right)^2\right]} = \frac{1}{n\cdot \dfrac{1}{\sigma^2}} = \frac{\sigma^2}{n}$$

Now, $\bar{X}$ is an unbiased estimator of μ, and $\text{Var}(\bar{X}) = \sigma^2/n$. Therefore, it follows that $\bar{X}$ is a minimum variance unbiased estimator of μ.

A good estimator of a parameter should give us values that are close to the true value of the parameter. Furthermore, as we increase the sample size, we expect a good estimator of a parameter to give us values that are even closer to the true value of the parameter. For example, we think of the sample mean $\bar{X}$ as an estimator of μ that possesses this property, because, as the sample size increases, we expect values $\bar{x}$ of the estimator $\bar{X}$ to move closer to the true value of μ. Loosely speaking, an estimator that possesses this property is called a consistent estimator. The exact definition follows.

Definition 11.4   An estimator θ is said to be a **consistent estimator** of the parameter θ if for each positive constant c

$$\lim_{n \to \infty} P(|\theta - \theta| \geq c) = 0$$

(n denotes the size of the random sample)

Again, how can we know if an estimator has this property? The following theorem is sometimes useful.

Theorem 11.2   *An unbiased estimator θ of the parameter θ is also a consistent estimator of the parameter if*

$$\lim_{n \to \infty} \text{Var}(\theta) = 0$$

EXAMPLE 11.14

Show that
    (a) $\bar{X}$ is a consistent estimator of μ;
    (b) $\bar{X}_1 - \bar{X}_2$ is a consistent estimator of $\mu_1 - \mu_2$; and
    (c) $\hat{P}$ is a consistent estimator of p if the random sample is obtained under sampling with replacement.

    SOLUTION

    (a)   From Chapter 8, we have $E(\bar{X}) = \mu$ and $\text{Var}(\bar{X}) = \sigma^2/n$. Therefore, $\bar{X}$ is an unbiased estimator of μ and $\lim_{n \to \infty} \text{Var}(\bar{X}) = \lim_{n \to \infty} (\sigma^2/n) = 0$. Hence, it follows that $\bar{X}$ is a consistent estimator of μ.
    (b) and **(c)**   See Exercise 3.

Exercises

1.   Which of the following statements are true?
    (a)   The sample median is an unbiased estimator of the population mean for all populations.
    (b)   The sample median is a biased estimator of the population mean for all populations.
    (*Hint:* see Examples 11.1 and 11.2.)

2.   Find the standard error of
    (a)   the unbiased estimator $\bar{X}$,
    (b)   the unbiased estimator $\bar{X}_1 - \bar{X}_2$.

3.   Show that
    (a) $\bar{X}_1 - \bar{X}_2$ is a consistent estimator of $\mu_1 - \mu_2$ and

(b) $\hat{P}$ is a consistent estimator of p if the random sample is obtained under sampling with replacement.

4.   Show that if $(X_1, X_2, \ldots, X_n)$ is a random sample from a normal population with mean $\mu = 0$ and variance $\sigma^2 > 0$, then $(X_1^2 + X_2^2 + \cdots + X_n^2)/n$ is an unbiased estimator of σ^2.

5.   Let X_1 be the first random variable in a random sample $(X_1, X_2, \ldots, X_n)$ from a population. Show that X_1 is an unbiased estimator of the mean of the population, but that it is *not* a minimum variance estimator. Thus, it is not a best estimator.

6.   Give an example (different from those already mentioned) of an unbiased estimator that is not a minimum variance estimator.

11.6   The General Concept of Confidence Interval Estimates

In deriving a $1 - \alpha$ confidence interval estimate for the mean μ of a normal population, we saw that

$$P\left(\bar{X} - z_{\alpha/2} \cdot \frac{\sigma}{\sqrt{n}} < \mu < \bar{X} + z_{\alpha/2} \cdot \frac{\sigma}{\sqrt{n}}\right) = 1 - \alpha$$

We then called

$$\bar{x} - z_{\alpha/2} \cdot \frac{\sigma}{\sqrt{n}} < \mu < \bar{x} + z_{\alpha/2} \cdot \frac{\sigma}{\sqrt{n}}$$

where $\bar{x} - z_{\alpha/2}(\sigma/n^{1/2})$ and $\bar{x} + z_{\alpha/2}(\sigma/n^{1/2})$ are respective observed values of the random variables $\bar{X} - z_{\alpha/2}(\sigma/n^{1/2})$ and $\bar{X} + z_{\alpha/2}(\sigma/n^{1/2})$, a $1 - \alpha$ confidence interval estimate of μ. In deriving a $1 - \alpha$ confidence interval estimate for the variance σ^2 of a normal population, we saw that

$$P\left(\frac{(n-1)S^2}{\chi_{\alpha/2,\,n-1}^2} < \sigma^2 < \frac{(n-1)S^2}{\chi_{1-\alpha/2,\,n-1}^2}\right) = 1 - \alpha$$

We then called

$$\frac{(n-1)s^2}{\chi_{\alpha/2,\,n-1}^2} < \sigma^2 < \frac{(n-1)s^2}{\chi_{1-\alpha/2,\,n-1}^2}$$

where $(n-1)s^2/\chi_{\alpha/2,\,n-1}^2$ and $(n-1)s^2/\chi_{1-\alpha/2,\,n-1}^2$ are respective observed values of the random variables $(n-1)S^2/\chi_{\alpha/2,\,n-1}^2$ and $(n-1)S^2/\chi_{1-\alpha/2,\,n-1}^2$, a $1 - \alpha$ confidence interval estimate of σ^2.

In each of these cases, we began with a parameter θ of a population. Then we saw that there existed random variables θ_1 and θ_2 such that both of the following conditions are satisfied.

1.   The random variables θ_1 and θ_2 are functions of the random variables in a

random sample from the population of the parameter θ (i.e., they are statistics).

2. $P(\theta_1 < \theta < \theta_2) = 1 - \alpha$.

Next we called

$$\theta_1 < \theta < \theta_2$$

where θ_1 and θ_2 are respective observed values of θ_1 and θ_2, a $1 - \alpha$ confidence interval estimate of θ.

This procedure for finding a $1 - \alpha$ confidence interval estimate for a parameter contains within it the formal definition of a confidence interval estimate. A statement of the formal definition follows.

Definition 11.5   Let θ be a parameter of a population. Let $\hat{\theta}_1$ and $\hat{\theta}_2$ be random variables that satisfy both of the following conditions.

1.   They are functions of the random variables $X_1, X_2, \ldots, X_n$ that are in a random sample $(X_1, X_2, \ldots, X_n)$ from the population of the parameter θ.
2. $P(\hat{\theta}_1 < \theta < \hat{\theta}_2) = 1 - \alpha$.

If $\hat{\theta}_1$ and $\hat{\theta}_2$ are respective observed values of $\hat{\theta}_1$ and $\hat{\theta}_2$, then

$$\hat{\theta}_1 < \theta < \hat{\theta}_2$$

is a $1 - \alpha$ confidence interval estimate of θ. The fraction $1 - \alpha$ is called the **confidence coefficient**, and the end points, $\hat{\theta}_1$ and $\hat{\theta}_2$, are called the lower and upper **confidence limits** or **fiducial limits**, respectively.

Note: a $1 - \alpha$ confidence interval is also called a $(1 - \alpha) \cdot 100\%$ confidence interval.

EXAMPLE 11.15

Let μ be the mean of a normal population with variance $\sigma^2 = 1$. Let $(X_1, X_2, \ldots, X_n)$ be a random sample of size n from the population.
   (a)   Show that

$$x_1 - z_{\alpha/2} < \mu < x_1 + z_{\alpha/2}$$

where x_1 is an observed value of X_1, is a $1 - \alpha$ confidence interval estimate of μ.
   (b)   Suppose that $n = 4$, $x_1 = 1$, $x_2 = 2$, $x_3 = 1$, and $x_4 = 3$. Use these data and the formula of (a) to obtain a .95 confidence interval estimate of μ. Also, use the formula of Section 11.1.2 to obtain a confidence interval estimate of μ.
   (c)   Compare the lengths of the two confidence intervals. Which confidence interval would you say is the better one?
   (d)   For any fixed value of n, what is the length of a $1 - \alpha$ confidence interval for μ, using the formula of (a)? What is the length using the formula of Section 11.1.2.?
   (e)   Which formula provides the better confidence interval estimate?

SOLUTION

(a)   The distribution of X_1 is normal with mean μ and variance $\sigma^2 = 1$. Therefore, the distribution of $X_1 - \mu$ is standard normal. This means that for any α such that $0 < \alpha < 1$, we have

$$P(-z_{\alpha/2} < X_1 - \mu < z_{\alpha/2}) = 1 - \alpha$$

Consequently

$$P(X_1 - z_{\alpha/2} < \mu < X_1 + z_{\alpha/2}) = 1 - \alpha$$

Now, observe that both $X_1 - z_{\alpha/2}$ and $X_1 + z_{\alpha/2}$ are random variables that satisfy conditions 1 and 2 in the definition of a confidence interval. Therefore, if x_1 is an observed value of X_1, then

$$x_1 - z_{\alpha/2} < \mu < x_1 + z_{\alpha/2}$$

is a $1 - \alpha$ confidence interval estimate of μ.

(b)   Using the data $n = 4$, $x_1 = 1$, $x_2 = 2$, $x_3 = 1$, and $x_4 = 3$, a .95 confidence interval estimate of μ using the formula in (a) is

$$1 - z_{.025} < \mu < 1 + z_{.025}$$

or

$$-.96 < \mu < 2.96$$

whereas, if we use the formula of Section 11.1.2, a .95 confidence interval estimate of μ is

$$\bar{x} - z_{\alpha/2} \cdot \frac{\sigma}{\sqrt{n}} < \mu < \bar{x} + z_{\alpha/2} \cdot \frac{\sigma}{\sqrt{n}}$$

or

$$.77 < \mu < 2.73$$

(c)   The length of the .95 confidence interval using the formula of (a) is 3.92; the length of the other is 1.96. Because we can have just as much confidence in the correctness of the shorter interval as we do in the longer one (.95 in both cases), the shorter interval, of course, is the better one—it implies greater precision in the estimation.

(d)   The length of a $1 - \alpha$ confidence interval for μ using the formula of (a) is

$$(x_1 + z_{\alpha/2}) - (x_1 - z_{\alpha/2}) = 2z_{\alpha/2}$$

(i.e., the difference between the two confidence limits). The length of a $1 - \alpha$

confidence interval for μ using the formula of Section 11.12 is

$$\left(\bar{x} + z_{\alpha/2} \cdot \frac{1}{\sqrt{n}}\right) - \left(\bar{x} - z_{\alpha/2} \cdot \frac{1}{\sqrt{n}}\right) = 2z_{\alpha/2} \cdot \frac{1}{\sqrt{n}}$$

(e)   For any sample size greater than 1, the formula of Section 11.1.2 provides a shorter and hence better confidence interval estimate.

From the definition of confidence intervals and from Example 11.15 we can see that for any one parameter there are any number of formulas for $1 - \alpha$ confidence interval estimates for that parameter. The best formula is of course the one that provides the shortest interval for a given confidence coefficient, because that is the formula that provides the greatest precision at a given level of confidence. Since a confidence interval is essentially an interval about a point estimate, the problem of obtaining a best confidence interval is related to the problem of obtaining a best point estimator. For example, it is easy to see that a minimum variance unbiased estimator can be used to obtain a better confidence interval than most estimators that do not have these properties.

Exercises

1. Let X be a random sample of size $n = 1$ from a population whose distribution is exponential, with mean equal to θ. Demonstrate that $0 < \theta < -x/\ln(1 - \alpha)$ is a $1 - \alpha$ confidence interval estimate of θ.

2. Suppose that if X is a random sample of size 1 from a certain population, then the pdf of X is

$$f(x) = \begin{cases} \dfrac{1}{\beta} & \text{for} \quad 0 < x < \beta \\ 0 & \text{elsewhere} \end{cases}$$

   Find K so that $0 < \beta < Kx$ is a $1 - \alpha$ confidence interval estimate of β.

3. Suppose that if X is a random sample of size 1 from a certain population, then the pdf of X is

$$f(x) = \begin{cases} 1 & \theta - .5 < x < \theta + .5 \\ 0 & \text{elsewhere} \end{cases}$$

   Show that

$$x - \frac{1 - \alpha}{2} < \theta < x + \frac{1 - \alpha}{2}$$

   is a $1 - \alpha$ confidence interval estimate of θ.

Supplementary Exercise

The strong law of large numbers states that if $X_1, X_2, \ldots, X_n$ are independent, identically distributed random variables with means equal to μ, then $P[\lim_{n \to \infty} (\sum_{i=1}^{n} X_i/n) = \mu] = 1$. Use this theorem to show that if the mean of a population is μ, there is a great probability that the sample mean $\bar{X}$ will be approximately equal to μ if the sample size is large. (*Hint:* if $(X_1, X_2, \ldots, X_n)$ is a random sample of size n from a population, the random variables $X_1, X_2, \ldots, X_n$ are independent and identically distributed.)

Chapter 12

UNDERSTANDING TESTING OF STATISTICAL HYPOTHESES

12.1   Two Informal Examples of Tests of Hypotheses

Let us introduce the topic of testing of statistical hypotheses with two informal examples. (The reader who prefers a formal initial approach may omit this section and proceed directly to Section 12.2. of this chapter. On the other hand, readers primarily interested in applications may wish to read only this first section, omitting Sections 12.2, 12.3, and 12.4, and then proceed directly to Chapter 13.)

12.1.1   Example I

The following situation, which leads to a testing of statistical hypotheses, is somewhat artificial. Nevertheless, because of its simplicity, it is valuable as an illustration of the basic concepts.

Let us assume that in flipping an ordinary coin the probabilities of heads and tails are both 1/2. Suppose that Mr. Gold, a coin collector, has a rare dime whose probabilities of heads and tails are 3/4 and 1/4, respectively. Imagine, now, the following situation. Mr. Gold misplaces his rare dime. Later he finds a dime that is identical in appearance and wonders if it is his rare dime. To help him make a decision he sets up a test of hypotheses.

12.1.1.1   Setting up the test

Let θ_0, θ_1, and θ be the probabilities of heads using the rare dime, an ordinary dime, and the dime that Mr. Gold found, respectively. Then, θ_0 is 3/4, θ_1 is 1/2, and θ is unknown. There are two possible hypotheses concerning θ. Either $\theta = \theta_0 = 3/4$ or $\theta = \theta_1 = 1/2$. We call the first hypothesis the null hypothesis (null because it indicates no changing of coins in Mr. Gold's possession), and we label it H_0. We call the second hypothesis the alternative hypothesis and label it H_1. Thus we have

$$H_0: \theta = \theta_0 = 3/4 \quad \text{versus} \quad H_1: \theta = \theta_1 = 1/2$$

Let us assume that Mr. Gold decides to flip the coin 10 times. Furthermore, suppose he decides that if the 10 flips yield 7 or more heads, he will assume that the coin he found is his rare dime, whereas if the 10 flips yield fewer than 7 heads, he will assume that the coin he found is an ordinary coin. We say, then, that Mr. Gold has designed a test of the null hypothesis H_0 versus the alternative hypothesis H_1, and that the test (or decision rule) is the following.

Test (Decision Rule)

Reject H_0 and accept H_1 if 10 flips yield fewer than 7 heads.

Accept H_0 and reject H_1 if 10 flips yield at least 7 heads.

12.1.1.2   How good is the test?

To evaluate the test (or decision rule) we define two types of errors.

Type I error: A type I error is the error of rejecting the null hypothesis when the null hypothesis is true.

Type II error: A type II error is the error of accepting the null hypothesis when the null hypothesis is false.

Next we compute the probability that the test will lead to a type I error and the probability that the test will lead to a type II error. We call these probabilities α and β, respectively.

$$\alpha = P(\text{test will lead to a type I error})$$
$$\beta = P(\text{test will lead to a type II error})$$

Let us compute α and β for this example.

Let S_n be the number of heads in n tosses of the coin that Mr. Gold found. With $n = 10$, we have

$$\alpha = P(\text{test will lead to a type I error})$$
$$= P(\text{test will reject } H_0 \text{ when } H_0 \text{ is true})$$
$$= P(S_n < 7 \qquad \text{when} \quad \theta = \theta_0 = 3/4)$$
$$= \sum_{x=0}^{6} b(x; 10, 3/4) = .2241$$

$$\beta = P(\text{test will lead to a type II error})$$
$$= P(\text{test will accept } H_0 \text{ when } H_0 \text{ is false})$$
$$= P(S_n \geq 7 \qquad \text{when} \quad \theta = \theta_1 = 1/2)$$
$$= \sum_{x=7}^{10} b(x; 10, 1/2) = .1719$$

12.1.1.3   Interpretation of the test results

Now, suppose that Mr. Gold proceeds to test the hypotheses H_0 versus H_1 as outlined above. He flips the coin 10 times; let us suppose he obtains 6 heads. Following the rules of the test, he rejects H_0.

Question: What is the probability that he made an error?   The casual reader would say .2241, since that is the probability that the test will lead to a type I error (an error of rejecting H_0 when H_0 is true). However, careful readers will note that the coin that Mr. Gold found either is or is not the rare dime, so that once Mr. Gold makes a decision concerning H_0, he is either completely correct or completely incorrect. It is meaningless, then, to talk about a probability of his being wrong. The α and β computed above are related to the nature of the test (or decision rule) that Mr. Gold had set up to make a decision about the coin he found; they cannot be interpreted as probabilities of error in relation to a decision made after the 10 flips of the coin.

To understand the meaning of α and β, it may help to analyze the following situation. Let us suppose that Mr. Gold is infinitely careless and misplaces the rare dime, not once, but many, many times. Each time he misplaces the rare dime, he finds a dime and conducts a test of hypotheses according to the rules set up above. Suppose that after each test, and after each decision, the true rare dime is returned to him, that he proceeds to misplace it again, and that the whole round of activities is repeated. In the context of this sequence of tests (and of decisions), and assuming the law of large numbers, α and β indicate the expected frequencies with which Mr. Gold will make a type I or a type II error, respectively. For example, suppose that the test of hypotheses is carried out 10,000 times; then, it is very likely that Mr. Gold will make type I errors approximately $\alpha \times 10{,}000 = .2241 \times 10{,}000 = 2{,}241$ times; that he will make type II errors approximately $\beta \times 10{,}000 = .1719 \times 10{,}000 = 1{,}719$ times; and that he will make the correct decision approximately $10{,}000 - 2{,}241 - 1{,}719 = 6{,}040$ times.

12.1.1.4   Reducing error probabilities

The test described above is associated with probabilities of .2241 and .1719 of making errors of type I and type II, respectively. If these probabilities are considered too large, how might we design a test with smaller probabilities of error? One obvious answer is that we might design a test that is based on a larger number of flips of the coin. For example, consider the following test for the same set of hypotheses.

Test (Decision Rule)

Reject H_0 and accept H_1 if 100 flips yield fewer than 65 heads.

Accept H_0 and reject H_1 if 100 flips yield at least 65 heads.

Let us find α and β for this test. Since $n = 100$ is a relatively large number, we may use the normal approximation of the binomial distribution (see Section 8.6 of Chapter 8). Therefore, letting S_n be the number of heads in n flips of the coin (with $n = 100$), letting $Z = (S_n - n\theta)/[n\theta(1 - \theta)]^{1/2}$ (so that the distribution of Z is approximately standard normal), and using a correction for continuity, we have

$$
\begin{aligned}
\alpha &= P(\text{reject } H_0 \text{ when } H_0 \text{ is true}) \\
&= P(S_n < 65) \quad \text{when} \quad \theta = \theta_0 = 3/4 \\
&= P(S_n < 64.5) \quad \text{when} \quad \theta = \theta_0 = 3/4 \\
&= P\left(\frac{S_n - n\theta}{\sqrt{n\theta(1 - \theta)}} < \frac{64.5 - 75}{\sqrt{18.75}}\right) \\
&= P(Z < -2.425) \\
&= .008
\end{aligned}
$$

$$\beta = P(\text{accept } H_0 \text{ when } H_0 \text{ is false})$$

$$= P(S_n \geqslant 65) \qquad \text{when} \quad \theta = \theta_1 = 1/2)$$

$$= P(S_n > 64.5) \qquad \text{when} \quad \theta = \theta_1 = 1/2)$$

$$= P\left(\frac{S_n - n\theta}{\sqrt{n\theta(1 - \theta)}} > \frac{64.5 - 50}{5}\right)$$

$$= P(Z > 2.9) = .002$$

Thus, we have successfully designed a second test (for the same set of hypotheses) with much smaller error probabilities. When the decision rule is appropriately chosen, this phenomenon of a reduction of probabilities of errors through an increase in the size of the random sample is applicable to nearly all test situations.

12.1.2   Example II

Suppose that the life, measured in hours, of TV tubes manufactured by a certain standard process may be assumed to be normally distributed with a mean of 1,000 hours and a standard deviation of 200 hours. Now, suppose that the engineering department of a certain firm has developed a new process for manufacturing TV tubes. Will the TV tubes manufactured by this new process have a longer mean life? A discussion follows on ways in which we might use the methods of testing of statistical hypotheses to help us answer this question with a certain sense of confidence.

12.1.2.1   Setting up the test

First of all, let us make several assumptions that will make this informal example a simple one. Let us assume that, because of the nature of the process, TV tubes manufactured by the new process will have a mean life that is at least as great as the mean life of TV tubes manufactured by the old process, but that the standard deviation remains the same and the distribution remains normal. That is, let us assume that if X is the life (measured in hours) of a TV tube manufactured by the new process, then X is normally distributed with a mean μ that is greater or equal to 1,000 and with a standard deviation that is $\sigma = 200$. Now, if the TV tubes manufactured by the new process do not have a longer mean life, then the mean of X is $\mu = 1,000$, whereas, if the TV tubes manufactured by the new process do have a longer mean life, then $\mu > 1,000$. These two statements specify two possible hypotheses concerning the distribution of the life of TV tubes manufactured by the new process. Let us call the hypothesis of the first statement, that $\mu = 1,000$, the null hypothesis (null, because no difference from the old process is expected) and label it H_0. Let us call the hypothesis of the second statement, that $\mu > 1,000$, the alternative hypothesis and label it H_1. Then we have

$$H_0\colon \mu = 1,000 \quad \text{versus} \quad H_1\colon \mu > 1,000$$

To test these hypotheses, suppose we decide to examine a random sample of 25 TV tubes manufactured by the new process. Furthermore, suppose we decide that the only condition under which we are willing to reject the null hypothesis is if the 25 TV tubes in the random sample have an average life greater than 1,080 hours. If the average life of the 25 TV tubes does not exceed 1,080 hours, we will assume that the null hypothesis is true. These statements define a test (or decision rule) that may be summarized as follows.

Test (Decision Rule)

Let $x_1, x_2, \ldots,$ and x_n be the respective life (measured in hours) of 25 randomly selected TV tubes manufactured by the new process. Let $\bar{x} = (x_1 + x_2 + \cdots + x_{25})/25$.

Reject H_0 and accept H_1 if $\bar{x} > 1,080$.

Reject H_1 and accept H_0 if $\bar{x} \leq 1,080$.

Again, we may wish to evaluate this test by computing the probabilities of type I and type II errors as defined in Example I. In this case

$$\alpha = P(\text{test will lead to a type I error})$$

$$= P(\text{test will reject } H_0 \text{ when } H_0 \text{ is true})$$

$$= P(\bar{X} > 1,080 \qquad \text{when} \quad \mu = 1,000)$$

$$= P\left(\frac{\bar{X} - \mu}{\sigma/\sqrt{n}} > \frac{1,080 - 1,000}{200/5}\right)$$

$$= P\left(\frac{\bar{X} - \mu}{\sigma/\sqrt{n}} > 2\right) = .02$$

The last step uses Statement 2(a) of Section 10.4.1 of Chapter 10, which states that $(\bar{X} - \mu)/(\sigma/n^{12})$ is standard normal if $\bar{X}$ is the sample mean from a normal distribution with mean μ and standard deviation σ. Figure 12.1 illustrates the basic elements of the test.

Now, let us try to find β. We have

$$\beta = P(\text{test will lead to a type II error})$$

$$= P(\text{test will accept } H_0 \text{ when } H_0 \text{ is false})$$

$$= P(\bar{X} \leq 1,080 \qquad \text{when} \quad \mu > 1,000)$$

$$= P\left(\frac{\bar{X} - \mu}{\sigma/\sqrt{n}} \leq \frac{1,080 - \mu}{200/5} \qquad \text{when} \quad \mu > 1,000\right) = ?$$

At this point we cannot complete the calculation for β because μ is not a specific value. Therefore, in this example, and in all examples where the alternative hypothesis is not a single value for a parameter, it is not possible to

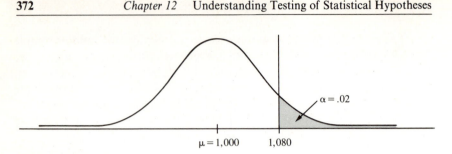

FIGURE 12.1

Elements of the Test for Example II

1.   The hypotheses are: H_0: $\mu = 1{,}000$ versus H_1: $\mu > 1{,}000$.
2.   The curve is for the pdf of $\bar{X}$, assuming that the null hypothesis $\mu = 1{,}000$ is true.
3.   The decision rule is: reject H_0 if $\bar{x} > 1{,}080$. (Values of $\bar{X}$ are more likely to be greater than 1,080 if H_1 is true than if H_0 is true.)
4.   The area under the curve from 1,080 to $+\infty$ is $\alpha = P(\text{type I error}) = .02$.

find β. It is possible, however, to define what is called a power function for such situations. This will be discussed in Section 12.4. Situations of this type are also dealt with in terms of *significance tests* in Section 12.2.4.

12.1.2.2   Interpretation of the test results

Let us return to our original question: will the TV tubes manufactured by the new process have a longer mean life? Suppose we now obtain a random sample of 25 TV tubes manufactured by the new process and find that the average life, measured in hours, of the 25 TV tubes is

$$\bar{x} = 1{,}100$$

According to the decision rule of the test we should reject H_0 because $\bar{x}$ is greater than 1,080. In other words, if we were to follow the decision rule of the test, we should assume that TV tubes manufactured by the new process have a longer mean life.

   If we do reject H_0, how sure can we be that we have made the right decision? In this case, the only possible error is a type I error (the error of rejecting H_0 when H_0 is true). We have based our decision on a test or decision rule whose probability of type I errors is .02. Therefore, we can feel rather confident about our decision—a feeling somewhat akin to having consulted a very reliable expert on the matter.

12.1.3   Unanswered Questions

The two informal examples, while giving an idea of what we mean by a test of hypotheses, have left many questions unanswered. For example:

1.   How do we design a test that will have a given probability of type I errors, or a given probability of type II errors?

2.   Is there more than one test for a given probability of type I errors—or for a given probability of type II errors?

3.   If so, which test is the best test?

4.   What are significance tests?

5.   What is the power function of a test?

These questions are answered in the remaining sections of this chapter. Question 1 is answered in Section 12.3. Questions 2, 3, and 5 are answered in Section 12.4. Question 4 is answered in Section 12.2.4.

Readers who are less interested in the answers to these questions than they are in learning the standard tests of statistical hypotheses may omit the remaining sections of this chapter and proceed directly to Chapter 13.

Exercises

1.   With reference to Example I, find α and β if the decision rule is changed to: reject H_0 and accept H_1 if 15 flips yield fewer than 11 heads. Accept H_0 and reject H_1 if 15 flips yield at least 11 heads.

2.   With reference to Example II, find α if the decision rule is changed to: reject H_0 and accept H_1 if $\bar{x} > 1,060$.

12.2   Basic Concepts in Testing of Statistical Hypotheses

This section discusses the basic concepts in testing of statistical hypotheses.

12.2.1   Statistical Hypotheses: Null, Alternative, Simple, Composite

A **statistical hypothesis** (except in nonparametric tests) is a hypothesis about a parameter of a population or distribution. In Example I, the distribution under consideration was the distribution of heads in one toss of the coin that Mr. Gold found. The parameter under consideration was θ—the probability of heads, using the coin that Mr. Gold found. Both statistical hypotheses under consideration were concerned with the parameter θ. One hypothesized that θ is equal to 3/4; the other hypothesized that θ is equal to 1/2. In Example II, the population under consideration was the population of the lengths of life of TV tubes manufactured by the new process. The parameter under consideration was μ, the mean of the population. Both statistical hypotheses under consideration were concerned with the parameter μ. One hypothesized that μ is equal to 1,000; the other hypothesized that μ is greater than 1,000.

In both Examples I and II we were faced with the problem of having to choose between two statistical hypotheses. When such is the case we call one hypothesis the **null hypothesis** and label it H_0. We call the other hypothesis the

alternative hypothesis and label it H_1. There is no rigid rule concerning the labeling of null and alternative. However, if the test is devised to detect a change in some standard or prevailing value for the parameter under question, then the hypothesis of no change is generally labeled the *null hypothesis*. For instance, in Example I the hypothesis $\theta = 3/4$ was labeled the null hypothesis because it indicated no change in Mr. Gold's possession of the rare dime. In Example II, the hypothesis $\mu = 1,000$ was labeled the null hypothesis because it indicated no difference between the mean life of TV tubes manufactured by the old process and the mean life of TV tubes manufactured by the new process. This general rule for the labeling of hypotheses becomes particularly meaningful when we consider significance tests in Section 12.2.4.

A hypothesis is called **simple** if it specifies a single value for the parameter in question. For example, the hypotheses $\theta = 3/4, \theta = 1/2$, and $\mu = 1,000$ of Examples I and II are all simple hypotheses. A hypothesis is called **composite** if it indicates more than one possible value for the parameter in question. For example, the hypothesis $\mu > 1,000$ of Example II is a composite hypothesis.

It is possible to have tests of statistical hypotheses involving more than two hypotheses. However, such tests are beyond the scope of this book. We shall limit our discussion to tests of two hypotheses.

12.2.2   Test of Statistical Hypotheses (Classical)

In the classical theory of mathematical statistics a test of a null hypothesis H_0 versus an alternative hypothesis H_1 is a decision rule that defines the conditions under which the null hypothesis should be rejected. The understanding is that if the null hypothesis is not rejected, then it is accepted. Also, it is understood that to reject H_0 is to accept H_1. A convenient (and useful) variation of these rules is offered in what is called a significance test (see Section 12.2.4).

12.2.3   Type I and Type II Errors, α and β

In the testing of statistical hypotheses a type I error is the error of rejecting the null hypothesis H_0 when H_0 is true. A type II error is the error of accepting H_0 when H_0 is false. The letter α denotes the probability that the test will lead to a type I error, and β denotes the probability that the test will lead to a type II error.

EXAMPLE 12.1

A box contains 10 balls, some of which are red and some blue. Let p denote the proportion of balls that are red. A random sample (sampling with replacement) of 4 balls is obtained from the box to test the hypotheses:

$$H_0: \rho = 1/2 \quad \text{versus} \quad H_1: \rho = 1/5$$

Find α and β if the decision rule is: reject H_0 if the sample consists of fewer than 2 red balls.

SOLUTION

Let S_n denote the number of red balls in the random sample of $n = 4$ balls. Then

$$\alpha = P(\text{reject } H_0 \text{ when } H_0 \text{ is true})$$
$$= P(S_n < 2 \mid p = 1/2)$$
$$= b(0; 4, 1/2) + b(1; 4, 1/2)$$
$$= .0625 + .2500$$
$$= .3125$$
$$\beta = P(\text{accept } H_0 \text{ when } H_0 \text{ is false})$$
$$= P(S_n \geq 2 \mid p = 1/5)$$
$$= b(2; 4, 1/5) + b(3; 4, 1/5) + b(4; 4, 1/5)$$
$$= .1536 + .0256 + .0016$$
$$= .1808$$

EXAMPLE 12.2

If the sample of size r in Example 12.1 was obtained under sampling without replacement, what are α and β?

SOLUTION

$$\alpha = P(\text{fewer than 2 red} \mid p = 1/2)$$

$$= \frac{\binom{5}{0} \cdot \binom{5}{4}}{\binom{10}{4}} + \frac{\binom{5}{1} \cdot \binom{5}{3}}{\binom{10}{4}}$$

$$= .2619$$

$$\beta = P(\text{2 or more red} \mid p = 1/5)$$

$$= 1 - P(\text{fewer than 2 red} \mid p = 1/5)$$

$$= 1 - \frac{\binom{2}{0} \cdot \binom{8}{4}}{\binom{10}{4}} - \frac{\binom{2}{1} \cdot \binom{8}{3}}{\binom{10}{4}}$$

$$= .1334$$

12.2.4   Significance Tests

Frequently a test of a simple null hypothesis versus a composite alternative hypothesis is designed and performed mainly because there is a need to determine if the null hypothesis should be rejected, rather than a need to ascertain the truth of the null hypothesis. For example, suppose that a chemist of a certain pharmaceutical company is anxious to test the effectiveness of a new drug in the treatment of a certain disease. If the standard medication before the development of this new drug provides a recovery rate of $\theta = \theta_0$, then there may be an interest in testing a null hypothesis $H_0: \theta = \theta_0$, against an alternative hypothesis $H_1: \theta > \theta_0$ on a random sample of patients receiving treatment with the new drug. If the test result does not indicate a rejection of H_0, the chemist may decide to withhold judgment concerning H_0 rather than accept H_0. Such tests, where the alternative to rejecting H_0 is a decision to withhold judgment concerning H_0, are called **significance tests**. These tests have tremendous appeal and enjoy great popularity among research workers in many fields. Because there is no possibility of a type II error (error of accepting H_0 when H_0 is false) in these tests, there appears to be, no need to be concerned about the size of β. Of course, this is misleading. The reasons why we should be concerned with the size of β, even for significance tests are discussed in Section 12.4 on evaluating tests of statistical hypotheses and in Section 13.6 of Chapter 13 on the power and required sample size of a test. There is a vast literature on the controversial nature of significance tests. A reference—and a source for references—is the book *The Significance Test Controversy*, edited by Morrison and Henkel, 1970, Aldine Publishing Company.

12.2.5   Level of Significance of a Test

If the probability of a type I error is α for a certain test, we say that the level of significance of the test is α. The term was originally used in designating the probability of rejecting the null hypothesis when the null hypothesis is true for significance tests. However, the term is now commonly used to specify the size of α for all tests.

Two levels of significance, namely $\alpha = .01$ and $\alpha = .05$, are most frequently used by statisticians and research workers. Generally, the level of .01 (and sometimes .001) is chosen when there is great reluctance to reject the null hypothesis unless there is overwhelming statistical evidence to support the alternative hypothesis. For example, if the consequence of a type I error is something in the nature of increased risk of death for a patient, or substantial financial losses for a company, then it would be prudent to use a level of significance that is no greater than .01. On the other hand, if the consequence of the alternative hypothesis is extremely attractive and if the results of a type I error are not catastrophic, it may be advisable to increase the risk of making a type I error and use a level of significance that is .05 or higher. With respect to these levels of significance, it is common practice among users of significance tests to report that a test result is significant if the null hypothesis is rejected at the .05 level of significance, and that it is highly significant if the null hypothesis is rejected at the .01 level of significance. We must understand, however, that the word "significant," when used in this sense, is not necessarily equivalent to the usual sense of the word "important." For instance, a difference of 2 IQ points between two groups of children may prove to be statistically significant, yet it is doubtful that any educator would consider the difference to be of any importance.

12.2.6   Statistics and Test Statistics

In the testing of statistical hypotheses we refer to a random variable as a **statistic** if it is a function of the random variables in a random sample. For example, $\bar{X}$, S^2, and $(\bar{X} - \mu)/(\sigma/n^{1/2})$ are statistics (see Section 10.4, Chapter 10). A statistic is called a **test statistic** if its observed values are determining factors in the decision rule of a test. For instance, in Example II, the decision rule specifies rejection of H_0 if the observed value $\bar{x}$ of the random variable $\bar{X}$ is greater or equal to 1,080. The random variable $\bar{X}$ is the test statistic of that decision rule. In Example I, the decision rule specifies rejection of H_0 if 10 flips yield fewer than 7 heads. Thus, if we let S_n be the random variable representing the number of heads in n flips of the coin, and if we set $n = 10$, then S_n is the test statistic of the test of Example I.

12.2.7   Sample Space for Tests of Hypotheses

If a test of a null hypothesis versus an alternative hypothesis is based on a random sample $(X_1, X_2, \ldots, X_n)$ of size n from a population or distribution, then the **sample space** for the test is defined to be the collection Ω of all possible values $(x_1, x_2, \ldots, x_n)$ of the random sample $(X_1, X_2, \ldots, X_n)$.

EXAMPLE 12.3

(a)   What is the sample space for Example I?
(b)   What is the sample space for Example II?

SOLUTION

(a)   In Example I, the distribution under consideration was the distribution of heads in one toss of the coin that Mr. Gold found. If we define

$$X_i = \begin{cases} 1 & \text{if the } i\text{th toss of the coin is heads} \\ 0 & \text{if the } i\text{th toss of the coin is tails} \end{cases}$$

then the test was based on the random sample $(X_1, X_2, \ldots, X_{10})$ of size 10 from the distribution. For example, if the observed set of values of the random sample $(X_1, X_2, \ldots, X_n)$ is $(1, 0, 1, 0, 0, 0, 0, 0, 0, 0)$, then there are exactly 2 heads in 10 tosses, and we should, according to the decision rule, reject the null hypothesis. Therefore, the sample space for the test is the set

$$\Omega = \{(x_1, x_2, \ldots, x_{10}) \mid x_i = 0 \quad \text{or} \quad 1 \quad \text{for} \quad i = 1, 2, \ldots, 10\}$$

(b) The test of Example II was based on the average life of the 25 TV tubes in a random sample. If $X_1, X_2, \ldots,$ and X_{25} are the respective lengths of life of the 25 TV tubes, then the sample space for the test is

$$\Omega = \{(x_1, x_2, \ldots, x_{25}) \mid 0 < x_i < \infty \quad \text{for} \quad i = 1, 2, \ldots, 25\}$$

12.2.8   Critical Region, Region of Rejection

As stated in Section 12.2.2, a test is a decision rule that specifies conditions under which the null hypothesis should be rejected. If the test is based on a random sample of size n from the population, the decision rule must specify the values of the random sample for which the null hypothesis should be rejected. With the exception of significance tests, the understanding is that the null hypothesis should be accepted for all other values of the random sample. In other words, a test, according to the classical definition, must partition the sample space Ω into a **region of rejection** of H_0 and a region of acceptance of H_0. The region of rejection is also called the **critical region**. An element $(x_1, x_2, \ldots, x_n)$ of the sample space Ω is said to be in the critical region if and only if the observed values $x_1, x_2, \ldots, x_n$ of the random sample $(X_1, X_2, \ldots, X_n)$ imply rejection of H_0.

In most tests of statistical hypotheses the decision rule is stated in terms of a test statistic. Consequently, the critical region is often most conveniently stated in terms of the test statistic. When this is the case, the term critical region may also be used to denote the set of those values of the test statistic for which the decision rule specifies a rejection of H_0. If the decision rule is a statement to the effect that the null hypothesis should be rejected for all values (or absolute values) of the test statistic that are greater than some number c, or less than some number c, then the number c is called the critical value of the test statistic. In Example II the decision rule specifies rejection of H_0 if $\bar{x} > 1,080$; therefore, $1,080$ is the critical value of the test statistic $\bar{X}$.

A critical region is said to be of size α if the test defined by the region will lead to a type I error with probability α.

EXAMPLE 12.4

(a) With reference to the test of Example I, find the critical region and the critical value of the test statistic.

(b) With reference to the test of Example II, find the critical region and the critical value of the test statistic.

SOLUTION

(a) The decision rule in Example I was: reject H_0 if 10 flips yield fewer than 7 heads. The sample space for Example I was given in Example 12.3(a) and is the set

$$\Omega = \{(x_1, x_2, \ldots, x_{10}) \mid x_i = 0 \quad \text{or} \quad 1 \quad \text{for} \quad i = 1, 2, \ldots, 10\}$$

where $x_i = 1$ if the ith flip is heads and $x_i = 0$ if tails. Therefore the critical region is the set

$$C = \{(x_1, x_2, \ldots, x_{10}) \mid x_i = 0 \quad \text{or} \quad 1 \quad \text{and} \quad \sum_{i=1}^{10} x_i < 7\}$$

The test statistic is the random variable

$$S_n = X_1 + X_2 + \cdots + X_n$$

where $n = 10$ and $X_i = 1$ if the ith flip is heads and $X_i = 0$ if the ith flip is tails. In other words, S_n is the number of heads in n flips of the coin (with $n = 10$). Stated in terms of the observed value s_n of the test statistic S_n, the decision rule is: reject H_0 if $s_n < 7$. Therefore, the critical value of the test statistic is the value

$$s_n = 7$$

(b)   The test statistic in Example II was the random variable $\bar{X}$, the average life of the 25 TV tubes in a random sample from the population. Stated in terms of the observed value $\bar{x}$ of the test statistic $\bar{X}$, the decision rule was: Reject H_0 if $\bar{x} > 1{,}080$. Therefore, the critical value of the test statistic is the value

$$\bar{x} = 1{,}080$$

The critical region is the set

$$C = \left\{(x_1, x_2, \ldots, x_{25}) \,\middle|\, \bar{x} = \frac{x_1 + x_2 + \cdots + x_{25}}{25} > 1{,}080\right\}$$

where x_i is the length of life of the ith TV tube in the sample.

Exercises

1.   Fill in the missing words in the following statements.

(a)   A statistical hypothesis (except in nonparametric tests) is a hypothesis concerning a _____ of a population or distribution.

(b)   If a test is devised to detect a change in some standard or prevailing value for the parameter in question, then the hypothesis of no change is generally labeled the _____ hypothesis.

(c)   A hypothesis is called _____ if it specifies a single value for the parameter in question.

(d)   A hypothesis is called _____ if it indicates more than one possible value for the parameter in question.

(e) A test of a null hypothesis versus an alternative hypothesis is a _____ _____ that defines the conditions under which the null hypothesis should be rejected.

(f) The error of rejecting the null hypothesis when the null hypothesis is true is called a _____ error; the Greek letter _____ is used to denote the probability of this type of error. The error of accepting the null hypothesis when the null hypothesis is false is called a _____ error; the Greek letter _____ is used to denote the probability of this type of error.

(g) Tests where the alternative to rejecting the null hypothesis is a decision to withhold judgment concerning the null hypothesis are called _____ tests.

(h) The level of significance of a test is equivalent to _____.

(i) In the testing of statistical hypotheses we refer to a random variable as a _____ if it is a function of the random variables in a random sample and if its observed values are determining factors in the decision rule of the test.

(j) The collection of all possible values of the random sample for a test is called the _____ space.

(k) The region of the sample space that indicates rejection of the null hypothesis is called the _____ region.

2. A box contains 4 balls of which r are red and $4 - r$ are black. A random sample (under sampling with replacement) of 2 balls is to be obtained from the box to test the hypotheses

$$H_0: r = 2 \quad \text{versus} \quad H_1: r = 1$$

The decision rule is: reject H_0 and accept H_1 if the sample of size 2 contains fewer than 2 red balls. Accept H_0 and reject H_1 if the sample of size 2 contains 2 red balls.

(a) Find α and β.
(b) Define the sample space for the test.
(c) What is the critical region?
(d) Is the null hypothesis simple or composite?
(e) Is the alternative hypothesis simple or composite?

12.3 How to Design a Test of Statistical Hypotheses

Suppose we wish to design a test concerning the value of a parameter θ of a population or distribution. Let us suppose that the test can be formulated in terms of a null hypothesis H_0 versus an alternative hypothesis H_1. Furthermore, suppose we wish the test to be based on a random sample of size

n from the population or distribution, and we want the probability of a type I error to be a specific value of α. How can we design a test to meet these requirements? To answer this question, let us begin with a consideration of the hypotheses of Example II.

In Example II we examined a test based on a random sample of size 25 that had a .02 probability of committing type I errors. Suppose we want to design another test of the same hypotheses based on a random sample of size 25 that has only a .01 probability of committing type I errors. The following is a discussion of how this may be done.

The hypotheses of Example II were:

$$H_0: \mu = 1,000 \quad \text{versus} \quad H_1: \mu > 1,000$$

where μ is the mean of a normal population (and where $\sigma = 200$ is the standard deviation of the population). Suppose we let $(X_1, X_2, \ldots, X_{25})$ be a random sample of size 25 from the population (in this case, X_i is the length of life of the ith TV tube in the sample) and let $\bar{X}$ be the sample mean (i.e., $\bar{X} = (X_1 + X_2 + \cdots + X_{25})/25$ is the average life of the 25 TV tubes in the sample). Since the observed value of $\bar{X}$ is likely to be large if the mean μ of the population is large, and is likely to be small if the mean μ of the population is small, it makes sense to reject H_0 (and possibly accept H_1) if the observed value of $\bar{X}$ is substantially larger than 1,000. Thus, it makes sense to use a decision rule of the following form:

Test (Decision Rule)

Reject H_0 if $\bar{x} > c$

Can we determine a value of c so that α (the probability of type I error) is .01? Consider the following:

$$\alpha = P(\text{the test will lead to a type I error})$$
$$= P(\text{the test will reject } H_0 \text{ when } H_0 \text{ is true})$$
$$= P(\bar{X} > c) \quad \text{when} \quad \mu = 1,000)$$
$$= P\left(\frac{\bar{X} - \mu}{\sigma/\sqrt{n}} > \frac{c - 1,000}{200/5}\right)$$
$$= P\left(\frac{\bar{X} - \mu}{\sigma/\sqrt{n}} > \frac{c - 1,000}{40}\right)$$

Note: we have previously established in Example II that the distribution of $(\bar{X} - \mu)/(\sigma/n^{1/2})$ is standard normal.

Now, we can use the table of values for the standard normal distribution to conclude that if the value of α is .01, then the value of c must satisfy the equation

$$\frac{c - 1,000}{40} = 2.327$$

Therefore

$$c = 1,093$$

Thus we have found a test (decision rule) based on a random sample of size 25 that has a .01 probability of committing a type I error. It is

Test (Decision Rule)

Reject H_0 if $\bar{x} > 1,093$.

The method used in this example may be summarized into the following 3 steps that are applicable to all tests of a simple null hypothesis versus an alternative hypothesis (the alternative hypothesis may be either simple or composite).

The method

Step 1: Find a statistic (i.e., function of the random variables in the random sample) whose distribution is completely determined if H_0 is true. (This enables us to compute α if we use this statistic as the test statistic.)

In the example discussed above, the statistic was the sample mean $\bar{X}$.

Step 2: Use the statistic of Step 1 as the test statistic. In other words, use it to determine a decision rule whose probability of type I errors is α. (Students who have read Sections 12.2.7 and 12.2.8 will recognize this to be equivalent to a determination of a critical region of size α in terms of the test statistic.)

In the example discussed above we determined the decision rule when we found c so that $P(\bar{X} > c$ when H_0 is true$) = .01$.

Step 3: Make a formal statement of the decision rule, determined in Step 2, in terms of observed values of the test statistic.

In the example discussed above, the formal statement became:

$$\text{Reject } H_0 \quad \text{if} \quad \bar{x} > 1,093$$

EXAMPLE 12.5

Suppose it is known that the mean μ of a certain normal population is either 1 or 2, and that the standard deviation is .5. Use the 3 steps outlined above to design a test of the hypotheses

$$H_0 : \mu = 1 \quad \text{versus} \quad H_1 : \mu = 2$$

so that it uses a random sample of size $n = 9$, and has a probability of type I errors equal to .05.

SOLUTION

Step 1:   (Find a statistic whose distribution is completely determined if H_0 is true.)

Consider the sample mean $\bar{X}$. Can we use this random variable as a test statistic? The answer is yes, because if H_0 is true (i.e., if $\mu = 1$), then $\bar{X}$ has a normal distribution with mean $\mu = 1$ and standard deviation $\sigma/n^{1/2} = .5/3$. In other words, we may use this random variable as the test statistic because the distribution of this random variable is completely determined if H_0 is assumed to be true.

Step 2:   (Use the statistic of Step 1 to determine a decision rule whose probability of type I errors is α.)

Since the value of μ is greater under H_1 than under H_0, we would naturally expect to observe greater values for $\bar{X}$ under H_1 than under H_0. Hence, it would seem reasonable to reject H_0 if values of the test statistic $\bar{X}$ are relatively large. In other words, it would seem reasonable to use a decision rule of the form: reject H_0 when $\bar{x} > k$, where k is some constant. To find the constant k so that the resulting decision rule will have a .05 probability of committing type I errors, observe that

$$\alpha = P(\text{the test will lead to a type I error})$$

$$= P(\text{the test will reject } H_0 \text{ when } H_0 \text{ is true})$$

$$= P(\bar{X} > k \qquad \text{when} \quad \mu = 1)$$

$$= P\left(\frac{\bar{X} - \mu}{\sigma/\sqrt{n}} > \frac{k - 1}{.5/3}\right)$$

We have already established in previous discussions that the distribution of $(\bar{X} - \mu)/(\sigma/n^{1/2})$ is standard normal. Therefore, we may use the table of values for the standard normal distribution to find that if the value of α is .05, then the value of $(k - 1)/(.5/3)$ must be 1.645. Therefore, the value of k must be

$$k = 1.2742$$

Step 3:   (Make a formal statement of the decision rule determined in step 2, in terms of observed values of the test statistic.) The test (or decision rule) determined in Step 2 is

Test (Decision Rule)

Reject H_0 if $\bar{x} > 1.2742$.

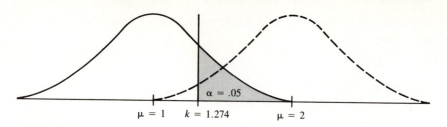

FIGURE 12.2

Elements of the Test for Example 12.5

1. The hypotheses are: H_0: $\mu = 1$ versus H_1: $\mu = 2$.
2. The curve with the solid line is for the pdf of $\bar{X}$ if the null hypothesis $\mu = 1$ is true.
3. The curve with the broken line is for the pdf of $\bar{X}$ if the alternative hypothesis $\mu = 2$ is true.
4. The decision rule is: reject H_0 if $\bar{x} > 1.2742$ (Values of $\bar{X}$ are more likely to be greater than 1.2742 if H_1 is true than if H_0 is true.)
5. The area under the curve with solid line from 1.2742 to $+\infty$ is $\alpha = P(\text{type I error}) = .05$.

Figure 12.2 illustrates the basic elements of this test.

EXAMPLE 12.6

Use the 3 steps outlined to design a test of the hypotheses

$$H_0: \mu = \mu_0 \quad \text{versus} \quad H_1: \mu < \mu_0$$

where μ is the unknown mean of a normal population whose standard deviation is the known quantity σ.

   (a) Design the test so that it uses a random sample of size n and has a probability of type I errors equal to .05.

   (b) Design the test so that the probability of type I errors is equal to α.

   SOLUTION

   (a) *Step 1:* (Find a statistic whose distribution is completely determined if H_0 is true.)

   Consider the random variable

$$\frac{\bar{X} - \mu_0}{\sigma/\sqrt{n}}$$

where $\bar{X}$ is the sample mean. Can we use this random variable as a test statistic? The answer is yes, because if H_0 is true (i.e., if $\mu = \mu_0$), then the distribution of this random variable is the standard normal distribution (see Statement 2(a) of Section 10.4.1 of Chapter 10). In other words, we may use this random variable as the test statistic because the distribution of this random variable is completely determined if H_0 is assumed to be true.

Step 2: (Use the statistic of Step 1 to determine a decision rule whose probability of type I errors is α.)

Since the observed value of $\bar{X}$ is likely to be near the true value of μ, it makes sense to reject H_0 (and possibly accept H_1) when the observed value of $\bar{X}$ is substantially less than μ_0. Thus, it makes sense to reject H_0 when the observed value of $(\bar{X} - \mu_0)/(\sigma/n^{1/2})$ is substantially less than 0. In other words, it is reasonable to use a decision rule of the form: reject H_0 when $(\bar{x} - \mu_0)/(\sigma/n^{1/2}) < c$, where c is some constant less than 0. To find the constant c so that the resulting decision rule will have a .05 probability of committing type I errors, observe that

$$\alpha = P(\text{the test will lead to a type I error})$$

$$= P(\text{the test will reject } H_0 \text{ when } H_0 \text{ is true})$$

$$= P\left(\frac{\bar{X} - \mu_0}{\sigma/\sqrt{n}} < c \quad \text{when} \quad \mu = \mu_0\right)$$

$$= P\left(\frac{\bar{X} - \mu}{\sigma/\sqrt{n}} < c\right)$$

Now, we can use the table of values for the standard normal distribution to conclude that if the value of α is .05, then the value of c must be

$$c = -1.645$$

Step 3: (Make a formal statement of the decision rule determined in Step 2, in terms of observed values of the test statistic.)

The test (or decision rule) determined in Step 2 is:

Test (Decision Rule)

Reject H_0 if $\dfrac{\bar{x} - \mu_0}{\sigma/\sqrt{n}} < -1.645$.

(b) To determine the decision rule for this part of the problem we need only replace the number 1.645 in the decision rule with the general expression of $-z_\alpha$, because

$$P\left(\frac{\bar{X} - \mu}{\sigma/\sqrt{n}} < -z_\alpha\right) = \alpha$$

Note: the value of z_α is determined by the equation

$$\int_{z_\alpha}^{\infty} \frac{1}{\sqrt{2\pi}} e^{-x^2/2} \, dx = \alpha$$

In other words, if we examine the area under the curve of the standard normal

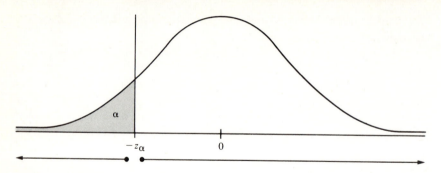

FIGURE 12.3

Elements of the Test for Example 12.6(b)

1. The hypotheses are: H_0: $\mu = \mu_0$ versus H_1: $\mu < \mu_0$
2. The curve is for the pdf of $(\bar{X} - \mu_0)/(\sigma/n^{1/2})$
3. The decision rule is: reject H_0 if $(\bar{x} - \mu_0)/(\sigma/n^{1/2}) < -z_\alpha$. (Values of $(\bar{X} - \mu)/(\sigma/n^{1/2})$ are more likely to be less than z_α if H_1 is true than if H_0 is true.)
4. The area under the curve from $-\infty$ to $-z_\alpha$ is $\alpha = P(\text{type I error})$.

density function, then the area to the right of z_α is α. See Figure 10.4 in Chapter 10.

Therefore, a test with a probability of type I errors equal to α is given by the following decision rule:

Test (Decision Rule)

Reject H_0 if $\dfrac{\bar{x} - \mu_0}{\sigma/\sqrt{n}} < -z_\alpha.$

Figure 12.3 illustrates the basic elements of this test.

EXAMPLE 12.7

A random sample of 25 cups from a certain coffee dispensing machine yields a mean of $\bar{x} = 6.9$ ounces per cup. Use the test designed in Example 12.6 to test (at the .05 level of significance) the null hypothesis that, on the average, the machine dispenses $\mu = 7.0$ ounces against the alternative hypothesis that, on the average, the machine dispenses $\mu < 7.0$ ounces. Assume that the distribution of ounces per cup is normal, and that the variance is the known quantity $\sigma^2 = .01$ ounces.

SOLUTION

The test designed in Example 12.6 is given by the decision rule:

$$\text{Reject } H_0 \quad \text{if} \quad \frac{\bar{x} - \mu_0}{\sigma/\sqrt{n}} < 1.645$$

In this example $\sigma = .1$, $n = 25$, and $\mu_0 = 7.0$. Therefore, the value of the test statistic is

$$\frac{\bar{x} - \mu_0}{\sigma/\sqrt{n}} = \frac{6.9 - 7.0}{.1/5} = -5$$

Since -5 is less than -1.645, we should reject the null hypothesis. If we are treating the test as a significance test, then we might report that statistical data show that the average amount of coffee dispensed by the machine is significantly less than 7.0 ounces (because we were able to reject the null hypothesis at the .05 level of significance).

EXAMPLE 12.8

Use the preceding 3 steps outlined to design a test of the hypotheses

$$H_0: \mu = \mu_0 \quad \text{versus} \quad H_1: \mu \neq \mu_0$$

where μ is the unknown mean of a normal population whose standard deviation is the known quantity σ. Design the test so that it uses a random sample of size n and has a probability of type I errors equal to α.

SOLUTION

Step 1: (Find a statistic whose distribution is completely determined if H_0 is true.)

Since the population and the null hypothesis of this example are identical to the population and the null hypothesis of Example 12.6, we may use the same test statistic in both examples. Therefore, we may use the random variable

$$\frac{\bar{X} - \mu_0}{\sigma/\sqrt{n}}$$

as the test statistic.

Step 2: (Use the statistic of Step 1 to determine a decision rule whose probability of type I errors is α.)

Since the observed value of $\bar{X}$ is likely to be near the true value of μ, it makes sense to reject H_0 (and possibly accept H_1) when the observed value of $\bar{X}$ is substantially different from μ_0. Thus, it makes sense to reject H_0 (and possibly accept H_1) when the observed value of the test statistic $(\bar{X} - \mu_0)/(\sigma/n^{1/2})$ is substantially different from 0. In other words, it is reasonable to use a decision rule of the form: reject H_0 when $|(\bar{x} - \mu_0)/(\sigma/n^{1/2})| > c$, where c is some constant greater than 0. To find the constant c observe that

$$\alpha = P(\text{the test will lead to a type I error})$$

$$= P(\text{the test will reject } H_0 \text{ when } H_0 \text{ is true})$$

$$= P\left(\left|\frac{\bar{X} - \mu_0}{\sigma - \sqrt{n}}\right| > c \quad \text{when} \quad \mu = \mu_0\right) = P\left(\left|\frac{\bar{X} - \mu}{\sigma\sqrt{n}}\right| > c\right)$$

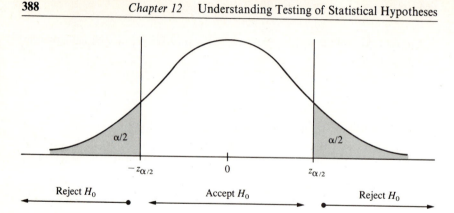

FIGURE 12.4

Elements of the Test for Example 12.8

1. The hypotheses are: H_0: $\mu = \mu_0$ versus H_1: $\mu \neq \mu_0$.
2. The curve is for the pdf of $(\bar{X} - \mu_0)/(\sigma/n^{1/2})$. That is, the pdf of $(\bar{X} - \mu)/(\sigma/n^{1/2})$ when $\mu = \mu_0$.
3. The decision rule is: reject H_0 if $|(\bar{x} - \mu_0)/(\sigma/n^{1/2})| > z_{\alpha/2}$
   (If H_0 is true, values of $(\bar{X} - \mu)/(\sigma/n^{1/2})$ are more likely to be near 0.)
4. The area under the curve from $-\infty$ to $-z_{\alpha/2}$ and from $z_{\alpha/2}$ to $+\infty$ is $\alpha = P$(type I error).

The random variable $(\bar{X} - \mu)/(\sigma/n^{1/2})$ is standard normal; therefore, if we use the standard notation introduced in Section 10.4.6 of Chapter 10, the value of c that will satisfy the above equation is

$$c = z_{\alpha/2}$$

Step 3: (Make a formal statement of the decision rule determined in Step 2, in terms of observed values of the test statistic.)

The test (or decision rule) determined in Step 2 is:

Test (Decision Rule)

Reject H_0 if $\left|\dfrac{\bar{x} - \mu_0}{\sigma/\sqrt{n}}\right| > z_{\alpha/2}$.

Figure 12.4 illustrates the basic elements of this test.

EXAMPLE 12.9

Suppose that the nationwide average reading score of second-graders tested at the midpoint of the academic year is 2.5. If 100 second-graders selected at random from a certain large metropolitan school district were tested at the midpoint of the academic year and were found to have a mean score of 2.7, can we conclude, at the .01 level of significance, that the mean score of second-graders from the metropolitan school district is different from the national mean? Use the test designed in Example 12.8 and assume that the distribution of scores is normal with a known variance of $\sigma^2 = 1$. What can we conclude at the .05 level of significance?

SOLUTION

The appropriate hypotheses for this problem are

$$H_0: \mu = 2.5 \quad \text{versus} \quad H_1: \mu \neq 2.5$$

where μ is the mean score for all second-graders in the metropolitan school district. The decision rule determined in Example 12.8 for these hypotheses is

Test (Decision Rule)

Reject H_0 if $\left| \dfrac{\bar{x} - \mu_0}{\sigma / \sqrt{n}} \right| > z_{\alpha/2}$.

In this example, $\bar{x} = 2.7, \mu_0 = 2.5, \sigma = 1$, and $n = 100$. Therefore, the value of the test statistic is

$$\frac{\bar{x} - \mu_0}{\sigma / \sqrt{n}} = \frac{2.7 - 2.5}{1/10} = 2.0$$

At $\alpha = .01$, the value of $z_{\alpha/2}$ is 2.573, whereas, at $\alpha = .05$, the value of $z_{\alpha/2}$ is 1.96. Therefore, we cannot conclude, at the .01 level of significance, that the mean score of second-graders from the metropolitan school district is different from the national mean. However, at the .05 level of significance we can conclude that the mean score is different from the national mean.

The test of Example 12.8 (see Figure 12.4) is called a **two-tailed test**. In general, a test is called two-tailed if the decision rule specifies rejection of the null hypothesis for all values of the test statistic to the right of some number b and to the left of some number $a < b$. Two-tailed tests are usually tests of hypotheses of the form

$$H_0: \theta = \theta_0 \quad \text{versus} \quad H_1: \theta \neq \theta_0$$

The tests of Example II and Example 12.5 (see Figures 12.1 and 12.2) are called **one-tailed tests**. One-tailed tests have decision rules that specify rejection of the null hypothesis only if values of the test statistic are to the right of some number, or else only if values of the test statistic are to the left of some number. These tests are usually tests of hypotheses of the form

$$H_0: \theta = \theta_0 \quad \text{versus} \quad H_1: \theta > \theta_0$$

or of the form

$$H_0: \theta = \theta_0 \quad \text{versus} \quad H_1: \theta < \theta_0$$

Exercises

1. Let $\mathscr{P}$ be a population whose distribution is normal with known variance $\sigma^2 = 4$ and unknown mean μ. Use the method of Section 12.3 to design a test with $\alpha = .01$, based on a random sample of size $n = 16$, for testing the hypotheses

$$H_0: \mu = 5 \quad \text{versus} \quad H_1: \mu = 8$$

   (*Hint:* follow the steps of Example 12.5.)

2. Before the increase from \$10 to \$25 in the fine for parking violations, the average number of parking violations per day in a certain downtown area was 20.5. The standard deviation was 5. After the increase a random sample of 25 days showed an average of 19 parking violations per day. Can we conclude, at the .05 level of significance, that the average number of parking violations per day has decreased? Use the test that was designed in Example 12.6. Assume that the distribution is normal with a known standard deviation of 5. Let μ denote the new average number of parking violations per day (i.e., since the increase in the fine). What is the answer if we increase the level of significance to .10?

3. Consider the hypotheses

$$H_0: \mu = \mu_0 \quad \text{versus} \quad H_1: \mu > \mu_0$$

   where μ is the mean of a normal population with a variance that is known to be σ^2, and consider the test whose decision rule is given by: reject H_0 if $[(\bar{x} - \mu_0)/(\sigma/n^{1/2})] > z_\alpha$. Show that α is the probability of a type I error with this test.

4. A random sample of 100 families in a very large suburban area gave the result of a sample mean of $\bar{x} = 2.25$ TVs per family. If we can assume that the distribution of television sets per family is a normal distribution with a standard deviation equal to 1, can we conclude, at the .03 level of significance, that in this area the average number of TVs per family is greater than 2? Use the test given in Exercise 3.

5. The production engineer of a certain firm estimated that it would require technicians an average of 30 minutes to assemble a certain newly designed computer component. Twenty-five technicians were randomly selected from the company's work force, and after an initiation period, it was found that among this group of technicians, an average of 32 minutes was required to assemble the component. If we assume that the distribution of required time is normal with a standard deviation equal to 5 minutes, can we conclude, at the .05 level of significance, that the production engineer has made an incorrect estimate of the required time? Use the test designed in Example 12.8.

6. Consider a test for the mean μ of a population whose distribution is normal and whose variance is unknown. Verify that if the null hypothesis is $H_0: \mu = \mu_0$, then the statistic $(\bar{X} - \mu_0)/(S/n^{1/2})$ may be used as the test

statistic (i.e., verify that its distribution is completely determined if H_0 is true). Show that if the alternative hypothesis is $H_1: \mu \neq \mu_0$, then the decision rule "Reject H_0 if $|(\bar{x} - \mu_0)/(s/n^{1/2})| > t_{\alpha/2, n-1}$" defines a test of size α. Show that if the alternative hypothesis is $H_1: \mu > \mu_0$, then the decision rule "Reject H_0 if $[(\bar{x} - \mu_0)/(s/n^{1/2})] > t_{\alpha, n-1}$" defines a test of size α. Show that if the alternative hypothesis is $H_1: \mu < \mu_0$, then the decision rule "Reject H_0 if $[(\bar{x} - \mu_0)/(s/n^{1/2})] < -t_{\alpha, n-1}$" defines a test of size α. (*Hint:* see Statement 3, Section 10.4.1 of Chapter 10.)

7. Consider a test for the mean μ of a population whose distribution is unknown (hence not necessarily normal). Suppose that the variance of the population is σ^2, and the null hypothesis is $H_0: \mu = \mu_0$. Explain why we may use the statistic $(\bar{X} - \mu_0)/(\sigma/n^{1/2})$ as the test statistic if the sample size is large. (*Hint:* see Statement 2, Section 10.4.1 of Chapter 10.) Show that if the alternative hypothesis is $H_1: \mu \neq \mu_0$, then the decision rule "Reject H_0 if $|(\bar{x} - \mu_0)/(\sigma/n^{1/2})| > z_{\alpha/2}$" defines a test of size α. Show that if the alternative hypothesis is $H_1: \mu > \mu_0$, then the decision rule, reject H_0 if $[(\bar{x} - \mu_0)/(\sigma/n^{1/2})] > z_{\alpha}$, defines a test of size α. Show that if the alternative hypothesis is $H_1: \mu < \mu_0$, then the decision rule, "Reject H_0 if $[(\bar{x} - \mu_0)/(\sigma/n^{1/2})] < -z_{\alpha}$," defines a test of size α. (*Hint:* see Statement 2, Section 10.4.1. of Chapter 10.)

8. Consider the problem of Exercise 7. Suppose that the variance of the population is also unknown. Explain why we may use the statistic $(\bar{X} - \mu_2)/(S/n^{1/2})$ as the test statistic if the sample size is large. (*Hint:* see Statement 3, Section 10.4.1 of Chapter 10.) Formulate the appropriate decision rules for the three sets of hypotheses given in Exercise 7.

12.4   Evaluating Tests of Statistical Hypotheses

How good is a particular test of statistical hypotheses? Is it better than a different test of the same set of statistical hypotheses? Is it the best among all tests of the same set of statistical hypotheses? What are the standard criteria for evaluating tests of statistical hypotheses? The answers to these questions depend on the type of test under consideration. Tests of simple null hypotheses versus simple alternative hypotheses are evaluated in terms of their *power*, whereas, tests of simple hypotheses versus composite alternative hypotheses are evaluated in terms of their *power functions* or *operating characteristic curves*. The following is a discussion of these characteristics of a test and includes a discussion of the use of these characteristics to define *best tests* and *uniformly most powerful tests*.

The power of a test is a measure of its ability to reject the null hypothesis when the null hypothesis is false. The formal definition follows.

Definition 12.1   Consider a test of the simple hypotheses:

$$H_0: \theta = \theta_0 \quad \text{versus} \quad H_1: \theta = \theta_1$$

where θ is a parameter of a population of distribution. The **power** of the test is defined to be the probability that the test will lead to a rejection of the null hypothesis, $\theta = \theta_0$, when the alternative hypothesis, $\theta = \theta_1$, is true. That is, the power of the test is the quantity $1 - \beta$, where β is the probability that the test will accept the null hypothesis when the alternative hypothesis is true.

If the test is of a simple hypothesis $H_0: \theta = \theta_0$ versus some composite hypothesis concerning θ, then the power of the test at $\theta = \theta_1$ is defined to be the probability that the test will lead to a rejection of the null hypothesis, $\theta = \theta_0$, when the actual value of the parameter θ is θ_1.

EXAMPLE 12.10

Let $\mathscr{P}$ be a population whose distribution is normal with known variance $\sigma^2 = .25$ and unknown mean μ. Find α, β, and the power of each of the tests given below of the hypotheses

$$H_0: \mu = 1 \quad \text{versus} \quad H_1: \mu = 2$$

In each test, let $\bar{x}$ denote the observed value of the mean of a random sample of size 9 from the population.

Test 1: Reject H_0 when $\bar{x} > 1.3878$.

Test 2: Reject H_0 when $\bar{x} < .6122$.

Test 3: Reject H_0 when $\bar{x} > 2$ or when $\bar{x} < .6122$.

SOLUTION

For Test 1:

$$\alpha = P(\text{reject } H_0 \text{ when } H_0 \text{ is true})$$

$$= P(\bar{X} > 1.3878 \qquad \text{when} \quad \mu = 1)$$

$$= P\left(\frac{\bar{X} - \mu}{\sigma/\sqrt{n}} > \frac{1.3878 - 1}{.5/3}\right) = P\left(\frac{\bar{X} - \mu}{\sigma/\sqrt{n}} > 2.327\right)$$

$$= .01$$

Note: we made use of the fact that the distribution of $(\bar{X} - \mu)/(\sigma/n^{1/2})$ is standard normal.

$$\beta = P(\text{accept } H_0 \text{ when } H_0 \text{ is false})$$

$$= P(\text{accept } H_0 \text{ when } H_1 \text{ is true})$$

$$= P(\bar{X} \leq 1.3878 \qquad \text{when} \quad \mu = 2)$$

$$= P\left(\frac{\bar{X} - \mu}{\sigma/\sqrt{n}} \leq \frac{1.3878 - 2}{.5/3}\right) = P\left(\frac{\bar{X} - \mu}{\sigma/\sqrt{n}} \leq -3.6732\right)$$

$$= .001$$

(the power of Test 1) $= P(\text{reject } H_0 \text{ when } H_1 \text{ is true})$

$$= 1 - P(\text{accept } H_0 \text{ when } H_1 \text{ is true})$$

$$= 1 - \beta$$

$$= 1 - .001$$

$$= .999$$

For Test 2:

$$\alpha = P(\bar{X} < 0.6122 \qquad \text{when} \quad \mu = 1)$$

$$= P\left(\frac{\bar{X} - \mu}{\sigma/\sqrt{n}} < \frac{.6122 - 1}{.5/3}\right)$$

$$= P\left(\frac{\bar{X} - \mu}{\sigma/\sqrt{n}} < -2.3268\right)$$

$$= .01$$

$$\beta = P(\bar{X} \geq 0.6122 \qquad \text{when} \quad \mu = 2)$$

$$= P\left(\frac{\bar{X} - \mu}{\sigma/\sqrt{n}} \geq \frac{.6122 - 2}{.5/3}\right)$$

$$= P\left(\frac{\bar{X} - \mu}{\sigma/\sqrt{n}} \geq -8.3268\right)$$

$$= 1$$

(the power of Test 2) $= 1 - \beta = 1 - 1 = 0$

For Test 3:

$$\alpha = P(\bar{X} > 2 \quad \text{or} \quad \bar{X} < .6122 \qquad \text{when} \quad \mu = 1)$$

$$= P(\bar{X} > 2 \qquad \text{when} \quad \mu = 1) + P(\bar{X} < .6122 \qquad \text{when} \quad \mu = 1)$$

$$= .01$$

$$\beta = P(.6122 \leq \bar{X} \leq 2 \qquad \text{when} \quad \mu = 2) = .5$$

Note: the reader will be asked to verify these calculations in Exercise 1 at the end of this section.

(the power of Test 3) $= 1 - \beta = .5$

Although the decision rules of Test 2 and Test 3 in this example may seem absurd in terms of the hypotheses under consideration, they, nevertheless, define bonafide tests of the hypotheses. The calculations of the example have shown that all three tests have the same probability of rejecting the null hypothesis when the null hypothesis is true. The respective probabilities of accepting the null hypothesis when the alternative hypothesis is true are .001, 1, and .5, and the respective powers (i.e., probabilities of rejecting the null hypothesis when the alternative hypothesis is true) are .999, 0, and .5. Of these three tests, Test 1 has the smallest probability of accepting the null hypothesis when the alternative hypothesis is true. Consequently, Test 1 has the greatest power (i.e., the greatest probability of rejecting the null hypothesis when the alternative hypothesis is true). Therefore, it is obvious that Test 1 is the best one of the three. In the formal language of statistical inference a best test is defined as follows.

Definition 12.2   In the testing of a simple null hypothesis against a simple alternative hypothesis, a test of size α is called a **best test** if, among all tests of size α, none has greater power (or, equivalently, none has a smaller value for β).

Note: we say that the size of a test is α if the probability of type I errors for the test is α. Also, we assume that all tests under consideration are based on random samples of size n.

Although it was plain, good common sense that led us to the particular design of the test in Example 12.5, it can be shown—although it is beyond the scope of this text to do so—that the test is in fact a best test.

The following example considers the problem of comparing tests of simple null hypotheses versus composite alternative hypotheses.

EXAMPLE 12.11

Consider the testing of the hypotheses

$$H_0: \mu = 0 \quad \text{versus} \quad H_1: \mu \neq 0$$

where μ is the mean of a normal population with variance $\sigma^2 = 4$. Find α, the power of the test at $\mu = .5$, and the power of the test at $\mu = -.5$ for each of the tests defined below. In each test, let $\bar{x}$ denote the observed value of the mean of a random sample of size 25 from the population.

Test 1:   Reject H_0 when $|\bar{x}| > .784$.

Test 2:   Reject H_0 when $\bar{x} > .658$.

Test 3:   Reject H_0 when $\bar{x} < -.658$.

SOLUTION

The reader will be asked to verify the numbers given below.

For Test 1:

$$\alpha = .05$$

(the power of the test at $\mu = .5$) = .239

(the power of the test at $\mu = -.5$) = .239

For Test 2:

$$\alpha = .05$$

(the power of the test at $\mu = .5$) $= .347$

(the power of the test at $\mu = -.5$) $= .002$

For Test 3:

$$\alpha = .05$$

(the power of the test at $\mu = .5$) $= .002$

(the power of the test at $\mu = -.5$) $= .347$

All three tests in this example have the same probability of rejecting the null hypothesis when the null hypothesis is true, namely, $\alpha = .5$ for all three tests. The powers of the respective tests at $\mu = .5$ are .239, .347, and .002. Thus, if we were testing the null hypothesis $\mu = 0$ versus the alternative hypothesis $\mu = .5$, then the best test is Test 2, the second-best test is Test 1, and the worst test is Test 3. However, the powers of the respective tests at $\mu = -.5$ are .239, .002, and .347. Therefore, if we were testing the null hypothesis $\mu = 0$ versus the alternative hypothesis $\mu = -.5$, then the best test is Test 3, the second-best test is Test 1, and the worst test is Test 2. Since the power of a test at each value of the parameter provides an evaluation of the test at only that value of the parameter, it is necessary to compute the power of the test at each possible value of the parameter to obtain a complete evaluation of the test. This leads us to the definition of the operating characteristic curve of a test and the power function of a test.

Definition 12.3   Consider a test of a simple null hypothesis concerning a parameter versus a composite alternative hypothesis concerning the same parameter. Let $\beta(\theta)$ denote the probability of accepting the null hypothesis when the value of the parameter is θ. For each value θ of the parameter, the value of $\beta(\theta)$ is called the **operating characteristic** of the test at that value of the parameter. The curve defined by $\beta(\theta)$, taken as a function of θ, is called the **operating characteristic curve** (or the **OC curve**) of the test.

Note: observe that if we applied these notations to the testing of a simple null hypothesis (that $\theta = \theta_0$) versus a simple alternative hypothesis (that $\theta = \theta_1$), then $\beta(\theta_1) = \beta$, where β is the probability of a type II error.

Definition 12.4   Consider a test of a simple null hypothesis concerning a parameter versus a composite alternative hypothesis concerning the same parameter. Let $\not{p}(\theta)$ denote the probability of rejecting the null hypothesis when the value of the parameter is θ. For each value θ of the parameter, the value of $\not{p}(\theta)$ is called the **power** of the test at that value of the parameter. Taken as a function of θ, $\not{p}(\theta)$ is called the **power function** of the test. The power function and the operating characteristic curve of a test are related by the following equation:

$$\not{p}(\theta) = 1 - \beta(\theta)$$

Note: observe that if the null hypothesis is H_0: $\theta = \theta_0$, then $\not{p}(\theta_0) = \alpha$, where α is the probability of a type I error.

Operating characteristic curves and power functions are used to evaluate tests of simple null hypotheses versus composite alternative hypotheses. For example, if the hypotheses are

$$H_0: \theta = \theta_0 \quad \text{versus} \quad H_1: \theta \neq \theta_1$$

then it is desirable to have a high probability of accepting the null hypothesis when $\theta = \theta_0$, and a low probability of accepting the null hypothesis when $\theta \neq \theta_0$. That is, it is desirable to have a large value for $\beta(\theta_0)$, but, when $\theta \neq \theta_0$, then it is desirable to have small values for $\beta(\theta)$. In other words, a desirable test has an operating characteristic curve that is high at the value $\theta = \theta_0$, and low at all other values of θ. Thus, if we are to use these criteria to evaluate the test whose characteristic curves are given in Figure 12.5, we would say that Test 1 is better than Test 2.

In terms of power functions, an equivalent statement of the criteria given above is the following. It is desirable to have a low probability of rejecting the null hypothesis when $\theta = \theta_0$, and a high probability of rejecting the null hypothesis when $\theta \neq \theta_0$. Therefore, it is desirable to have a small value for $\not{p}(\theta_0)$, but when $\theta \neq \theta_0$, then it is desirable to have large values for $\not{p}(\theta)$. In other words, a desirable test has a power function that has a small value at

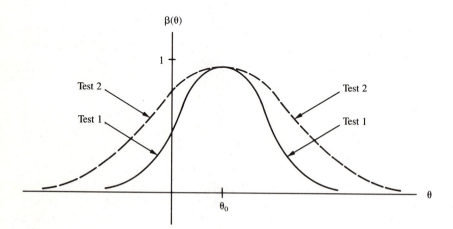

FIGURE 12.5

OC Curves for Testing H_0: $\theta = \theta_0$ versus H_1: $\theta \neq \theta_0$

(Test 1 is better than Test 2)

$\theta = \theta_0$ and large values at all other values of θ. Thus, if we are to compare two tests whose respective power functions are $\not{p}_1(\theta)$ and $\not{p}_2(\theta)$, and whose significance levels are identical (i.e., $\not{p}_1(\theta_0) = \not{p}_2(\theta_0) = \alpha$), then we would say that the first test is at least as good (at least as powerful) as the second test if for each value of θ that is different from θ_0, we have $\not{p}_1(\theta) \geq \not{p}_2(\theta)$. Furthermore, we would say that the first test is better (more powerful) than the second test if the first test is at least as good as the second test and, in addition, there is at least one value of θ that is different from θ_0 for which $\not{p}_1(\theta) > \not{p}_2(\theta)$. For example, if we were to compare the tests whose power functions are given in Figure 12.6, we would say that Test 1 is better (more powerful) than Test 2. Among all tests at the same level of significance, the most desirable test is the one with the greatest power at each value of θ that is different from θ_0. Such a test is called a uniformly most powerful test.

Definition 12.5 In testing a simple null hypothesis concerning a parameter θ against a composite alternative hypothesis concerning the same parameter, a test at the α level of significance is called a **uniformly most powerful test** if among all tests at the same level of significance, none has greater power at any value of θ specified by the alternative hypothesis.

Note: we assume that all tests under consideration are based on random samples of size n.

In many applied situations, uniformly most powerful tests serve only as ideals against which one may evaluate an available test, because uniformly most powerful tests do not exist for the testing of many hypotheses. For example, although the test of Example 12.8 is generally considered to be the most adequate in the given situation, it is nevertheless not a uniformly most

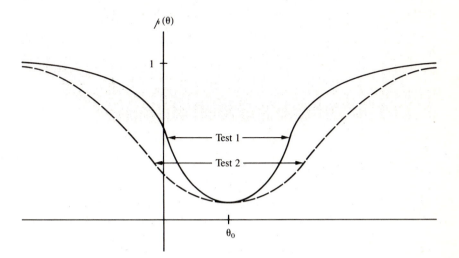

FIGURE 12.6
Power Functions for Tests of $H_0: \theta = \theta_0$ versus $H_1: \theta \neq \theta_0$

(Test 1 is better than Test 2)

powerful test. In fact, it can be shown (reference 4) that there is no uniformly most powerful test for testing the null hypothesis $\mu = \mu_0$ against the alternative hypothesis $\mu \neq \mu_0$ where μ is the mean of a normal population with known variance. The test of Example 12.6 is, however, a uniformly most powerful test.

Exercises

1. Verify the values for α, β and the power of Test 3 in the solution to Example 12.10.

2. Verify the values for α, the power of each of the tests at $\mu = .5$, and the power of each of the tests at $\mu = -.5$ in the solution to Example 12.11.

3. Consider the test given in the solution of Example 12.6. Find the operating characteristic and the power of the test at $\mu = 2$; at $\mu = 3$. Assume that $\mu_0 = 4$, $\sigma = 2$ and $n = 25$.

4. Consider the test given in the solution of Example 12.8. Find the operating characteristic and the power of the test at $\mu = 1$; at $\mu = -1$; at $\mu = .5$; at $\mu = -.5$. Assume that $\alpha = .01$, $\mu_0 = 0$, $\sigma = 1$, and $n = 16$.

References

The topic of testing of statistical hypotheses is discussed in almost every introductory text on mathematical statistics. We list a few such texts below:

1. De Groot, M. H. 1975. *Probability and statistics*, Menlo Park, CA: Addison-Wesley Publishing Company.
2. Freund, J. E., and R. E. Walpole, 1980. *Mathematical statistics*, 3d ed, Englewood Cliffs, NJ: Prentice-Hall, Inc.
3. Hogg, R. V., and A. T. Craig, 1970. *Introduction to mathematical statistics*, 3d ed, London: The Macmillan Company.
4. Mendenhall, W., and R. L. Scheaffer, 1973. *Mathematical statistics with applications*, North Scituate, MA: Duxbury Press.
5. Mood, A. M., F. A. Graybill, and D. C. Boes, 1974. *Introduction to the theory of statistics*, 3d ed, New York: McGraw-Hill Book Company.

Chapter 13

SOME STANDARD TESTS OF STATISTICAL HYPOTHESES

13.1   Introduction

This chapter presents some of the most widely used standard tests of statistical hypotheses. These tests are generally accepted to be the most adequate under the given conditions. We shall see that in each of these tests the decision rule is stated in terms of a test statistic whose distribution is completely determined if the null hypothesis is assumed to be true. We can imagine that these tests were designed according to the method outlined in Section 12.3 of Chapter 12.

All of the tests defined in this chapter may be interpreted as significance tests if the alternative to rejecting the null hypothesis is a decision to withhold judgment concerning the null hypothesis. (Significance tests are defined in Section 12.2.4, Chapter 12.)

13.2   Tests for the Means of Populations

A frequent use of tests for the means of populations is in situations where we wish to compare the average performance of one group with an established or standard performance. This was the case in Example II, Section 12.1.2, Chapter 12, where we wished to compare the mean life of TV tubes manufactured by a new process with the known mean for TV tubes manufactured by the old process. In Example 12.7, we compared the average performance of a certain coffee dispensing machine with a set standard and, in Example 12.9, we compared the average performance of second-graders with the national average. The most commonly used standard tests for the means of populations are given in Sections 13.2.1–13.2.3. The variations in these tests are a result of variations in properties of the populations and in the sample size.

13.2.1   Tests for the Means of Normal Populations with Known Variances

Let $\mathscr{P}$ be a population whose distribution is normal, whose mean is an unknown quantity μ, and whose variance is a known quantity σ^2. The standard test based on a random sample of size n of the hypotheses $H_0: \mu = \mu_0$ versus $H_1: \mu \neq \mu_0$, at the α level of significance, is the test designed in Example 12.8 of Chapter 12. The test statistic is the random variable

$$Z = \frac{\bar{X} - \mu_0}{\sigma/\sqrt{n}}$$

which is standard normal when $\mu = \mu_0$. In terms of observed values z of the test statistic Z, the decision rule is:

$$\text{Reject } H_0 \text{ if } |z| > z_{\alpha/2}$$

Note: the value of z_α is determined by the equation

$$\int_{z_\alpha}^{\infty} \frac{1}{\sqrt{2\pi}} e^{-(x^2/2)} \, dx = \alpha$$

In other words, z_α is the value such that the area to its right under the curve of the standard normal density function is α. (See Figure 10.4 in Chapter 10.) If the alternative hypothesis is changed to $\mu > \mu_0$ or $\mu < \mu_0$, then the decision rule is changed, appropriately, to reject H_0 if $z > z_\alpha$ and $z < -z_\alpha$, respectively (see Example 12.6 and Exercise 3, Section 3, in Chapter 12). Test Format 1 summarizes the test procedure for tests for the mean of a normal population with known variance σ^2 at the α level of significance.

Normal population; σ^2 known; H_0: $\mu = \mu_0$; α level of significance

Test statistic: $Z = \dfrac{\bar{X} - \mu_0}{\sigma/\sqrt{n}}$ (standard normal when $\mu = \mu_0$)

Alternative hypothesis H_1	Decision rule: Reject H_0 if
$\mu \neq \mu_0$	$\|z\| > z_{\alpha/2}$
$\mu > \mu_0$	$z > z_\alpha$
$\mu < \mu_0$	$z < -z_\alpha$

Test Format 1

EXAMPLE 13.1

A producer of chicken feed assures a farmer that their new and less expensive feed is just as effective as the old feed in promoting weight gain in chickens. The farmer knows from past experience that the weight of chickens on a diet of the old feed from birth until the age of 3 months is normally distributed with a mean of 3 pounds and a standard deviation of .25 pounds. To test the feed producer's claim, the farmer selects 10 newborn chickens at random and places them on a diet of the new feed for 3 months. At the end of that period he finds that the average weight of the 10 chickens is $\bar{x} = 2.8$ pounds. Set up a test of hypotheses at the .05 level of significance to test the feed producer's claim. Assume that the standard deviation σ of the weight of chickens on the new feed is also .25 pounds.

SOLUTION

1. *Choice of hypotheses*   Let μ be the average weight of chickens after 3 months on a diet of the new feed. Since the farmer observes an average of 2.8 pounds for 10 chickens on the new feed, naturally he is suspicious of the feed producer's claim that $\mu = 3$. One way of checking the claim is to

test the hypotheses

$$H_0: \mu = 3 \quad \text{versus} \quad H_1: \mu < 3$$

2. *Choice of test statistic and decision rule*   We are testing for the mean of a normal population whose variance is the known quantity $\sigma^2 = (.25)^2$. According to Test Format 1, the appropriate test statistic is

$$Z = \frac{\bar{X} - \mu_0}{\sigma/\sqrt{n}} = \frac{\bar{X} - 3}{.25/\sqrt{10}}$$

and the appropriate decision rule, in terms of observed values z of the test statistic Z is

$$\text{Reject } H_0 \text{ if } z < -z_\alpha$$

3. *Application of decision rule to data*   The data for this test are:

$$\bar{x} = 2.8, \quad \mu_0 = 3, \quad \sigma = .25, \quad n = 10, \quad \alpha = .05, \quad -z_\alpha = -1.645$$

The observed value of the test statistic Z is

$$z = \frac{\bar{x} - \mu_0}{\sigma/\sqrt{n}} = \frac{2.8 - 3}{.25/\sqrt{10}} = -2.5298$$

Therefore, on application of the decision rule to the data we should reject H_0 because

$$z = -2.5298 < -1.645 = -z_\alpha$$

At the .05 level of significance, the farmer should reject the feed producer's claim that the new feed is just as effective as the old feed in promoting weight gain in chickens, and conclude that the new feed, in fact, is less effective than the old feed.

EXAMPLE 13.2

Consider the situation of Example 13.1. The reader may feel that we cannot justify the assumption that the standard deviation σ of the weight of chickens on new feed remains the same as the standard deviation of the weight of chickens on old feed. Had we not made such an assumption, then, how might we have solved the problem?

SOLUTION

Although the farmer does not know the standard deviation σ of the weight of chickens on the new feed, he does know that if the feed producer's claim is true, then the value of σ is .25. Therefore, if the null hypothesis of Example 13.1 is expanded to include the producer's claim concerning σ, then H_0 becomes the hypothesis that $\mu = 3$ *and* $\sigma = .25$. In that case, if the null hypothesis is true, then the test statistic

$$Z = \frac{\bar{X} - 3}{.25/\sqrt{10}}$$

is the standard normal random variable

$$\frac{\bar{X} - \mu}{\sigma/\sqrt{n}}$$

This means that we can use the same test statistic here as in Example 13.1; and the solution to the problem here is identical to the solution in Example 13.1.

In summary, even if the value σ is known only when the null hypothesis is true, we can use Test Format 1 provided that H_0 includes a statement concerning σ.

13.2.2 Tests for the Means of Normal Populations with Unknown Variances

Let $\mathscr{P}$ be a population whose distribution is normal with mean μ and unknown variance. If the null hypothesis is H_0: $\mu = \mu_0$, then the standard test uses the test statistic

$$T = \frac{\bar{X} - \mu_0}{S/\sqrt{n}}$$

The distribution of T is the t distribution with $n - 1$ degrees of freedom if $\mu = \mu_0$. (See Section 10.4.3 of Chapter 10.) Recall that for $n > 29$ the t distribution with n degrees of freedom may be approximated by the standard normal distribution. Test Format 2 summarizes the test procedure for tests for the means of normal populations with unknown variance. See Exercise 6, Section 3, in Chapter 12 for the derivation of these tests.

Normal population; σ^2 unknown; H_0: $\mu = \mu_0$; α level of significance.

Test statistic: $T = \dfrac{\bar{X} - \mu_0}{S/\sqrt{n}}$ (t distribution with $n - 1$ degrees of freedom when $\mu = \mu_0$)

Alternative hypothesis H_1	Decision rule: Reject H_0 if
$\mu \neq \mu_0$	$|t| > t_{\alpha/2, n-1}$
$\mu > \mu_0$	$t > t_{\alpha, n-1}$
$\mu < \mu_0$	$t < -t_{\alpha, n-1}$

Test Format 2

EXAMPLE 13.3

The owner of a restaurant that is up for sale informs a prospective buyer that, not counting holidays, the average number of customers per Saturday evening is 100. To check the owner's claim, the prospective buyer counts the number of customers on 9 randomly chosen Saturday evenings and finds a sample mean of $\bar{x} = 95$ customers and a sample standard deviation of $s = 10$.

(a)   Should the prospective buyer reject the owner's claim at the .05 level of significance? (Assume that the distribution of number of customers is normal.)

(b)   Would the answer be the same if $s = 4$?

(c)   If the answer to (b) is different from the answer to (a), how would you justify the difference?

SOLUTION

(a)

1. *Choice of hypotheses*   With $\bar{x} = 95$, the prospective buyer suspects that the actual average μ of number of customers is less than 100. Therefore, the appropriate choice of hypotheses is

$$H_0: \mu = 100 \quad \text{versus} \quad H_1: \mu < 100$$

2. *Choice of test statistic and decision rule*   This is a case of testing for the mean of a normal population with unknown variance; therefore we should choose Test Format 2. The appropriate test statistic is

$$T = \frac{\bar{X} - \mu_0}{S/\sqrt{n}}$$

and the appropriate decision rule in terms of observed values t of the test statistic T is

$$\text{Reject } H_0 \text{ if } t < -t_{\alpha, n-1}$$

3. *Application of decision rule to data*   The data for this test are:

$$\bar{x} = 95, \quad \mu_0 = 100, \quad s = 10, \quad n = 9, \quad \alpha = .05,$$
$$-t_{\alpha, n-1} = -t_{.05,8} = -1.860$$

The observed value of the test statistic is

$$t = \frac{\bar{x} - \mu_0}{s/\sqrt{n}} = \frac{95 - 100}{10/\sqrt{9}} = -1.5$$

Therefore, on application of the decision rule to the data, we cannot reject H_0 because

$$t = -1.5 \not< -1.860 = -t_{\alpha, n-1}$$

At the .05 level of significance, the prospective buyer cannot reject the owner's claim that the average number of customers is 100.

   (b) If $s = 4$, then the value of the test statistic is

$$t = \frac{\bar{x} - \mu_0}{s/\sqrt{n}} = \frac{95 - 100}{4/\sqrt{9}} = -3.75$$

Therefore, on application of the decision rule, we should reject H_0 because

$$t = -3.75 < -1.860 = -t_{\alpha, n-1}$$

At the .05 level of significance the null hypothesis should be rejected.

   (c) The sample standard deviations in (a) and (b) are $s = 10$ and $s = 4$, respectively. This indicates the possibility of more variability in the number of customers in (a) than in (b). Therefore, under the assumption that $\mu = 100$, the observed value of $\bar{x} = 95$ is more likely to occur in (a) than in (b).

13.2.3 Large Sample Tests for the Means of Populations

First we will consider the case where σ^2 is known. Let $\mathscr{P}$ be a population whose distribution is unknown (hence not necessarily normal). Let μ and σ^2 be the mean and variance of the population. Suppose we wish to test the null hypothesis $H_0: \mu = \mu_0$ against some suitable alternative hypothesis. If σ^2 is known, and if the sample size n is large, we may use the test statistic

$$Z = \frac{\bar{X} - \mu_0}{\sigma/\sqrt{n}}$$

Observe that if $\mu = \mu_0$ and n is large, then Z is approximately standard normal (see Section 10.4.1 in Chapter 10). Test Format 3 summarizes the test procedure for large sample tests for the means of populations with known variances. See Exercise 7, Section 3, in Chapter 12 for the derivation of these tests.

Any population; known σ^2; $H_0: \mu = \mu_0$; large sample size n; α level of significance

Test statistic: $Z = \dfrac{\bar{X} - \mu_0}{\sigma/\sqrt{n}}$ (approximately standard normal if $\mu = \mu_0$).

Alternative Hypothesis H_1	Decision rule: Reject H_0 if
$\mu \neq \mu_0$	$\|z\| > z_{\alpha/2}$
$\mu > \mu_0$	$z > z_{\alpha}$
$\mu < \mu_0$	$z < -z_{\alpha}$

Test Format 3

If the variance σ^2 of the population is unknown, we should use the test statistic

$$Z = \frac{\bar{X} - \mu_0}{S/\sqrt{n}}$$

Observe that if n is large, then Z is approximately standard normal (see Section 10.4.1, Chapter 10). Thus, the tests are those given in Test Format 4 below (see Section 3, Exercise 8 in Chapter 10, for derivation of these tests).

Any population; σ^2 unknown; $H_0: \mu = \mu_0$; large sample size n; α level of significance

Test statistic: $Z = \dfrac{\bar{X} - \mu_0}{S/\sqrt{n}}$ (approximately standard normal if $\mu = \mu_0$)

Alternative hypothesis H_1	Decision rule: Reject H_0 if
$\mu \neq \mu_0$	$\lvert z \rvert > z_{\alpha/2}$
$\mu > \mu_0$	$z > z_\alpha$
$\mu < \mu_0$	$z < -z_\alpha$

Test Format 4

EXAMPLE 13.4

A superintendent of schools has read in a journal that primary-school children watch 15 hours of television per week, on the average. She feels that children in her school district watch less. To test this hypothesis, she randomly selects from the school files the names of 49 primary-school children and queries the parents on their children's television-watching habits.

(a)   If the sample mean of television time is $\bar{x} = 10$ hours, and the sample standard deviation is $s = 5$ hours, should the superintendent reject the null hypothesis of a mean television time of $\mu = 15$ in her school district, at the .05 level of significance, and accept the alternative hypothesis of $\mu < 15$?

(b)   On the basis of her sample data, can the superintendent make the stronger statement, at the .05 level of significance, that primary-school children in her district on the average watch less than 11 hours of television per week?

SOLUTION

(a)

1.   *Choice of hypotheses*

$$H_0: \mu = 15 \quad \text{versus} \quad H_1: \mu < 15$$

2. *Choice of test statistic and decision rule* Although σ^2 is unknown, n is large; therefore, we should use Test Format 4. The appropriate test statistic is

$$Z = \frac{\bar{X} - \mu_0}{S/\sqrt{n}}$$

The appropriate decision rule, in terms of observed values z of the test statistic Z is:

$$\text{Reject } H_0 \text{ if } z < -z_\alpha$$

3. *Application of decision rule to data* The data are:

$$\bar{x} = 10, \quad \mu_0 = 15, \quad s = 5, \quad n = 49, \quad \alpha = .05, \quad -z_\alpha = -z_{.05} = -1.645$$

The observed value of the test statistic is

$$z = \frac{\bar{x} - \mu_0}{s/\sqrt{n}} = \frac{10 - 15}{5/7} = -7$$

Therefore, on application of the decision rule, we should reject H_0 because

$$z = -7 < -1.645 = -z_\alpha$$

In conclusion, the answer to (a) is yes. At the .05 level of significance she should reject the null hypothesis $\mu = 15$ and accept the alternative hypothesis $\mu < 15$.

  (b) The relevant hypotheses are

$$H_0: \mu = 11 \quad \text{versus} \quad H_1: \mu < 11$$

The appropriate decision rule is:

$$\text{Reject } H_0 \text{ if } z < -z_\alpha$$

In this case

$$z = \frac{\bar{x} - \mu_0}{s/\sqrt{n}} = \frac{10 - 11}{5/7} = -1.4, \quad \text{and} \quad -z_\alpha = -z_{.05} = -1.645$$

Therefore, on application of the decision rule to the data, we cannot reject H_0 because

$$z = -1.4 \nless -1.645 = -z_\alpha$$

In conclusion, the answer is no. She cannot reject the null hypothesis that primary-school children in her district on the average watch 11 hours of television per week.

Exercises

In these exercises, a sample size of 49 or greater will be considered sufficiently large for all test formats requiring large sample sizes.

1. Let $\mathscr{P}$ be a population whose distribution is normal with unknown mean μ and known variance σ^2. Test the following hypotheses at the given level of significance α if the sample size n, sample mean $\bar{x}$, and population variance σ^2 have the values given below.
   (a) H_0: $\mu = 3$ versus H_1: $\mu \neq 3$, with $\sigma^2 = 4$, $\alpha = .05$, $n = 9$, $\bar{x} = 4.2$.
   (b) Assume the data of (a) but change the hypotheses to H_0: $\mu = 3$ versus H_1: $\mu > 3$. Compare with the answer in (a).
   (c) H_0: $\mu = -2$ versus H_1: $\mu \neq -2$, with $\sigma^2 = 1$, $\alpha = .01$, $n = 16$, $\bar{x} = -2.6$.
   (d) Assume the data of (c) but change the hypotheses to H_0: $\mu = -2$ versus H_1: $\mu < -2$. Compare with the answer in (c).
   (e) If you are anxious to reject the null hypothesis, and if you know the values in the random sample, would it be better to use a one-tailed or a two-tailed test?

2. Let $\mathscr{P}$ be a population whose distribution is normal with unknown mean μ. Test the following hypotheses at the given level of significance α if the sample size n, sample mean $\bar{x}$, and the sample standard deviation s have the values given below.
   (a) $\alpha = .05$, $n = 9$, $\bar{x} = 22.2$, $s = 3.3$

   $$H_0: \mu = 20 \quad \text{versus} \quad H_1: \mu \neq 20$$

   (b) Assume the data of (a) but change the hypotheses to

   $$H_0: \mu = 20 \quad \text{versus} \quad H_1: \mu > 20$$

   Compare with the answer in (a).
   (c) $\alpha = .01$, $n = 16$, $\bar{x} = 3.9$, $s = 1.6$

   $$H_0: \mu = 5 \quad \text{versus} \quad H_1: \mu < 5$$

   (d) Assume the data of (c) but change the hypotheses to

   $$H_0: \mu = 5 \quad \text{versus} \quad H_1: \mu \neq 5$$

   Compare with the answer in (c).
   (e) If you are anxious to reject the null hypothesis, and if you know the values in the random sample, would it be better to use a one-tailed or a two-tailed test?

3. Let $\mathscr{P}$ be a population with unknown mean μ. Test the following hypotheses at the given level of significance α if the sample size n, sample mean $\bar{x}$, and sample standard deviation s are the values given.

(a) $\alpha = .04$, $n = 100$, $\bar{x} = -3.2$, $s = 1.2$

$$H_0: \mu = -3 \quad \text{versus} \quad H_1: \mu \neq -3$$

(b) $\alpha = .02$, $n = 64$, $\bar{x} = 29.6$, $s = 5.1$

$$H_0: \mu = 28 \quad \text{versus} \quad H_1: \mu > 28$$

(c) $\alpha = .06$, $n = 81$, $\bar{x} = 12.9$, $s = .4$

$$H_0: \mu = 13 \quad \text{versus} \quad H_1: \mu < 13$$

4.   The inventor of a raisin-packing machine claims that the mean weight of boxes of raisins packed by the machine is 3 ounces and that the standard deviation is .1 ounces. Assume a normal distribution.
(a)   Set up a test of hypotheses at the .01 level of significance to check the inventor's claim concerning the mean. Assume a random sample of size $n = 10$.
(b)   If a random sample of 10 boxes of raisins packed by the machine showed a mean of 3.1 ounces, should we, according to the test designed in (a), reject the inventor's claim?
(c)   What are the answers to (a) and (b) if the inventor modifies his claim concerning the standard deviation to .15 ounces?

5.   Incomes of middle-class families in a certain area can be considered to be normally distributed with a mean of $15,000 and a standard deviation of $2,000. Five families claim that, in terms of incomes, they can be considered to be a random sample of size 5 from the middle-class families in the area. Set up the appropriate hypotheses and test their claim at the .05 level of significance if the average family income of the 5 families is $18,000.

6.   If manufactured according to specifications, the shot range of a certain type of BB gun should average 100 feet. A sample of 9 shots with one of these guns yield ranges of 98, 97, 101, 99, 100, 100, 95, 102, and 99 feet, respectively. Does this gun satisfy the manufacturer's specifications? Choose the appropriate hypotheses and test at the .01 level of significance. Assume a normal distribution.

7.   Before the introduction of one-way traffic it took a mean time of 25 minutes to drive from 108th Street to 42nd Street along 5th Avenue. After one-way traffic was established, a random sample of 10 trips along the same route yielded a mean of 23 minutes and a standard deviation of 4 minutes. Has the mean time decreased? Test at the .05 level of significance. Assume a normal distribution.

8.   A company chemist has been directed to maintain the level of a certain toxic chemical in a pesticide at an average of .06%. A random sample of 16 cans of the pesticide yields a mean of .05% and a standard deviation of .01%. Is the level of the toxic chemical being maintained at the desired

average level? Test at the .01 level of significance. Assume a normal distribution.

9. Suppose that the average income of 35-year-olds in the U.S. is $10,000. A random sample of 100 35-year-olds from Arizona shows their average income to be $9,000 with a standard deviation of $2000. At the .05 level of significance, should we conclude that 35-year-olds in Arizona have lower average incomes than the national average?

10. If produced according to specifications, the mean and standard deviation in milligrams of certain vitamin tablets should be 500 mg and 10 mg, respectively. A random sample of 100 such tablets showed a sample mean of 505 mg. At the .02 level of significance, should we conclude that, on the average, the tablets are larger than expected?

11. According to specifications, the average diameter of ball bearings produced by a certain process should be .20 cm. A random sample of 20 ball bearings from the process showed a sample mean of .21 cm and a sample standard deviation of .05 cm. Is there a significant difference between the average size of the actual ball bearings produced and the specified average size? Test at the .05 level of significance. Assume a normal distribution.

12. A manufacturer claims that its brand of light bulbs on the average lasts for 700 hours of continuous use. A random sample of 100 such bulbs showed a mean life of 701 hours and a standard deviation of 50 hours. Does the data support the manufacturer's claim? Test at the .03 level of significance.

13. The average annual income of members of a certain union is $15,000. A random sample of 100 women members showed an average income of $10,800 and a standard deviation of $1,000. Should we conclude that women members, on the average, make $4,000 less per year than the average of the total membership? Test at the .05 level of significance.

14. The distribution of IQs of children in a certain large metropolitan area can be assumed to be normal with a mean of 100 and a standard deviation of 15. A random sample of 10 students from a certain private school in the same area had a mean IQ of 105 and a standard deviation of 20. Should we conclude that the average IQ of children from the private school is significantly higher than the average in the whole area? Test at the .05 level of significance. Assume a normal distribution for IQs of children from the private school.

13.3   Tests for the Difference Between Means

One use of the tests for the difference between means is in comparisons of unknown average performance of two distinct groups. For example, management may wish to compare the average productivity of a group of workers that are given 15-minute coffee breaks in the afternoon with another group of workers that are not given this privilege but instead are compensated with a

15-minute earlier dismissal time. If we let μ_1 and μ_2 be the average productivities of the two groups, then we may wish to test the null hypothesis $H_0: \mu_1 - \mu_2 = 0$ against some suitable alternative (see Example 13.6).

The most common standard tests for the difference between means that are based on random samples from normal populations, or else from random samples that are large, are given in this section. We must caution the reader that if neither conditions of normal populations nor large sample sizes can be reasonably assumed, then the tests given here are not applicable. There are, however, nonparametric tests that deal with such situations. Several of these tests are given in Chapter 15 on nonparametric methods.

Let $\mathscr{P}_1, \mathscr{P}_2$ be distinct and normal populations with means μ_1, μ_2, and variances σ_1^2, σ_2^2, respectively. Suppose we wish to test the null hypothesis $H_0: \mu_1 - \mu_2 = d$, where d is a constant, against some suitable alternative hypothesis. Furthermore, suppose we wish to base the test on a random sample $(X_1^{(1)}, X_2^{(1)}, \ldots, X_{n_1}^{(1)})$ of size n_1 from $\mathscr{P}_1$ and a random sample $(X_1^{(2)}, X_2^{(2)}, \ldots, X_{n_2}^{(2)})$ of size n_2 from $\mathscr{P}_2$. First of all consider the case where σ_1^2 and σ_2^2 are known quantities. Then, we may use the test statistic

$$Z = \frac{\bar{X}_1 - \bar{X}_2 - d}{\sqrt{\dfrac{\sigma_1^2}{n_1} + \dfrac{\sigma_2^2}{n_2}}}$$

where

$$\bar{X}_1 = \frac{X_1^{(1)} + X_2^{(1)} + \cdots + X_{n_1}^{(1)}}{n_1}$$

and

$$\bar{X}_2 = \frac{X_1^{(2)} + X_2^{(2)} + \cdots + X_{n_2}^{(2)}}{n_2}$$

It can be shown (see reference 2) that Z is standard normal if the null hypothesis $H_0: \mu_1 - \mu_2 = d$ is true. If σ_1^2 and σ_2^2 are unknown, but the sample size n_1 and n_2 are large, then we can use the same test statistic if we estimate σ_1^2 and σ_2^2 with the sample variance s_1^2 and s_2^2, respectively. See Test Format 5.

Normal population $\mathscr{P}_1, \mathscr{P}_2$; known σ_1^2, σ_2^2; $H_0: \mu_1 - \mu_2 = d$; α level of significance

Test statistic: $Z = \dfrac{\bar{X}_1 - \bar{X}_2 - d}{\sqrt{(\sigma_1^2/n_1) + (\sigma_2^2/n_2)}}$ (standard normal if $\mu_1 - \mu_2 = d$)

Alternative hypothesis H_1	Decision rule: Reject H_0 if
$\mu_1 - \mu_2 \neq d$	$\|z\| > z_{\alpha/2}$
$\mu_1 - \mu_2 > d$	$z > z_\alpha$
$\mu_1 - \mu_2 < d$	$z < -z_\alpha$

Test Format 5

EXAMPLE 13.5

In comparing the average protein content μ_1 and μ_2 of two brands of dog food, a consumer testing service finds that fifty 5-pound packages of brand A dog food had an average protein content of $\bar{x}_1 = 11$ ounces per package and a standard deviation of $s_1 = 1$ ounce, while sixty 5-pound packages of brand B dog food had an average protein content of $\bar{x}_2 = 9$ ounces per package and a standard deviation of $s_1 = .5$. A difference of .5 ounces is considered to be not sufficiently important to report as a consumer issue. Therefore, a decision was made to test the hypotheses

$$H_0: \mu_1 - \mu_2 = .5 \quad \text{versus} \quad H_1: \mu_1 - \mu_2 > .5$$

Use the observed data to test these hypotheses at the .01 level of significance.

SOLUTION

1. *Choice of hypotheses*

$$H_0: \mu_1 - \mu_2 = .5 \quad \text{versus} \quad H_1: \mu_1 - \mu_2 > .5$$

2. *Choice of test statistic and decision rule*   Following Test Format 5, the appropriate test statistic is

$$Z = \frac{\bar{X}_1 - \bar{X}_2 - d}{\sqrt{\dfrac{\sigma_1^2}{n_1} + \dfrac{\sigma_2^2}{n_2}}}$$

and the decision rule, in terms of observed values z of the test statistic Z, is

$$\text{Reject } H_0 \text{ if } z > z_\alpha$$

3. *Application of decision rule to data*   The data are:

$$\bar{x}_1 = 11, \quad \sigma_1 \approx s_1 = 1, \quad n_1 = 50, \quad d = .5$$
$$\bar{x}_2 = 9, \quad \sigma_2 \approx s_2 = .5, \quad n_2 = 60, \quad \alpha = .01$$

Therefore, on application of the decision rule to the data, we should reject H_0 because

$$z = \frac{\bar{x}_1 - \bar{x}_2 - d}{\sqrt{\dfrac{\sigma_1^2}{n_1} + \dfrac{\sigma_2^2}{n_2}}} = \frac{11 - 9 - .5}{\sqrt{.02 + .00417}} = 9.649 > 2.327 = z_\alpha$$

The conclusion is that at the .01 level of significance we should reject the null hypothesis in favor of the alternative hypothesis that the per-package content of protein in brand A dog food exceeds that in brand B dog food by more than .5 ounces.

In the case where the variances are unknown but equal (i.e., $\sigma_1 = \sigma_2$ but unknown), and the populations are normal, the appropriate test statistic is

$$T = \frac{\bar{X}_1 - \bar{X}_2 - d}{\sqrt{\dfrac{(n_1 - 1)S_1^2 + (n_2 - 1)S_2^2}{n_1 + n_2 - 2}} \cdot \sqrt{\dfrac{1}{n_1} + \dfrac{1}{n_2}}}$$

where S_1^2 and S_2^2 are the sample variances of random samples from $\mathcal{P}_1$ and $\mathcal{P}_2$, respectively. If the null hypothesis $H_0 : \mu_1 - \mu_2 = d$ is assumed to be true, then the distribution of T is the t distribution with $n_1 + n_2 - 2$ degrees of freedom. See Test Format 6.

Normal population $\mathcal{P}_1, \mathcal{P}_2$; $\sigma_1 = \sigma_2$ but unknown; H_0: $\mu_1 - \mu_2 = d$; α level of significance
Test statistic:

$$T = \frac{\bar{X}_1 - \bar{X}_2 - d}{\sqrt{\dfrac{(n_1 - 1)S_1^2 + (n_2 - 1)S_2^2}{n_1 + n_2 - 2}} \cdot \sqrt{\dfrac{1}{n_1} + \dfrac{1}{n_2}}} \qquad \begin{array}{l} (t \text{ distribution with } n_1 + n_2 - 2 \\ \text{ degrees of freedom if } \mu_1 - \mu_2 = d) \end{array}$$

Alternative hypothesis H_1	Decision rule: Reject H_0 if
$\mu_1 - \mu_2 \neq d$	$\lvert t \rvert > t_{\alpha/2,\, n_1 + n_2 - 2}$
$\mu_1 - \mu_2 > d$	$t > t_{\alpha,\, n_1 + n_2 - 2}$
$\mu_1 - \mu_2 < d$	$t < -t_{\alpha,\, n_1 + n_2 - 2}$

Test Format 6

EXAMPLE 13.6

Management must make a decision concerning the adoption of either plan I or plan II for workers. Under plan I, workers will be given a 15-minute coffee break in the afternoon; whereas, under plan II, workers are dismissed 15 minutes earlier in the day. As input to the decision-making process, a test of hypotheses was set up to compare the average productivity μ_1 of workers under plan I versus the average productivity μ_2 of workers under plan II. A random sample of $n_1 = 10$ workers was assigned to plan I, and a random sample of $n_2 = 12$ workers was assigned to plan II. Both groups were assigned to identical tasks on the last stages of an assembly line that produces ball bearings. Productivity is measured in terms of the quantity of ball bearings produced. It was found that the sample means of productivity for the first group $\bar{x}_1 = 100$ units of ball bearings, and the sample standard deviation was $s_1 = 5$. For the second group the sample mean was $\bar{x}_2 = 98$ and the sample standard deviation was $s_2 = 4$. Test the hypotheses

$$H_0: \mu_1 - \mu_2 = 0 \quad \text{versus} \quad H_1: \mu_1 - \mu_2 \neq 0$$

at the .01 level of significance, assuming that the two populations have equal though unknown variances. Assume, also, that the relevant distributions are normal.

SOLUTION

1. *Choice of hypotheses*

$$H_0: \mu_1 - \mu_2 = 0 \quad \text{versus} \quad H_1: \mu_1 - \mu_2 \neq 0$$

2. *Choice of test statistic and decision rule*   Following Test Format 6, the appropriate test statistic is

$$T = \frac{\bar{X}_1 - \bar{X}_2 - d}{\sqrt{\dfrac{(n_1 - 1)S_1^2 + (n_2 - 1)S_2^2}{n_1 + n_2 - 2}} \cdot \sqrt{\dfrac{1}{n_1} + \dfrac{1}{n_2}}}$$

and the appropriate decision rule in terms of observed values t of the test statistic T is:

$$\text{Reject } H_0 \text{ if } |t| > t_{\alpha/2,\, n_1 + n_2 - 2}$$

3. *Application of data to decision rule*   The data are:

$$\bar{x}_1 = 100, \quad s_1 = 5, \quad n_1 = 10, \quad \alpha = .01$$
$$\bar{x}_2 = 98, \quad s_2 = 4, \quad n_2 = 12$$

$$t_{\alpha/2,\, n_1 + n_2 - 2} = t_{.005, 20} = 2.845$$

Therefore, on application of the decision rule to the data we cannot reject H_0 because

$$|t| = \left| \frac{100 - 98 - 0}{\sqrt{\dfrac{(10 - 1)5^2 + (12 - 1)4^2}{10 + 12 - 2}} \cdot \sqrt{\dfrac{1}{10} + \dfrac{1}{12}}} \right| = 1.04 \not> 2.845$$

$$= t_{\alpha/2,\, n_1 + n_2 - 2}$$

In conclusion, we cannot reject the null hypothesis $\mu_1 - \mu_2 = 0$ at the .01 level of significance. In other words, we cannot assume that average productivity is identical under the two plans.

In the case where the variance σ_1^2 and σ_2^2 are unknown, and where the populations $\mathscr{P}_1$ and $\mathscr{P}_2$ are not necessarily normal, but the sample sizes n_1 and n_2 are large, then the appropriate test statistic is

$$Z = \frac{\bar{X}_1 - \bar{X}_2 - d}{\sqrt{\dfrac{S_1^2}{n_1} + \dfrac{S_2^2}{n_2}}}$$

If the null hypothesis $H_0: \mu_1 - \mu_2 = d$ is assumed to be true, then Z is approximately standard normal if n_1 and n_2 are both large. See Test Format 7.

All populations $\mathscr{P}_1, \mathscr{P}_2$; σ_1^2, σ_2^2 unknown; large sample sizes n_1 and n_2; α level of significance
$H_0: \mu_1 - \mu_2 = d$
Test statistic:

$$Z = \frac{\bar{X}_1 - \bar{X}_2 - d}{\sqrt{\dfrac{S_1^2}{n_1} + \dfrac{S_2^2}{n_2}}} \quad \text{(approx. standard normal if } \mu_1 - \mu_2 = d\text{)}$$

Alternative hypothesis H_1	Decision rule: Reject H_0 if
$\mu_1 - \mu_2 \neq d$	$|z| > z_{\alpha/2}$
$\mu_1 - \mu_2 > d$	$z > z_\alpha$
$\mu_1 - \mu_2 < d$	$z < -z_\alpha$

Test Format 7

EXAMPLE 13.7

At a certain large university a sociologist speculates that male students spend considerably more money on junk food than do female students. She guesses that, on the average, male students spend more than $10 more on junk food per week than do their female counterparts. To test her hypothesis, the sociologist randomly selects from the registrar's records the names of 200 students. Of these, 125 are men and 75 are women. She then finds that the sample mean $\bar{x}_1$ of the average amount spent on junk food per week by the 125 men is $\bar{x}_1 = \$20$ and that the sample standard deviation for the men is $s_1 = \$12$. The sample mean and sample standard deviation for the women are $\bar{x}_2 = \$9$ and $s_2 = \$3$. Test the sociologist's hypothesis at the .05 level of significance.

SOLUTION

1. *Choice of hypotheses*

$$H_0: \mu_1 - \mu_2 = 10 \quad \text{versus} \quad H_1: \mu_1 - \mu_2 > 10$$

2. *Choice of test statistic and decision rule*   Following Test Format 7, the appropriate test statistic is

$$Z = \frac{\bar{X}_1 - \bar{X}_2 - d}{\sqrt{\dfrac{S_1^2}{n_1} + \dfrac{S_2^2}{n_2}}}$$

and the decision rule, in terms of observed values z of the test statistic Z, is

$$\text{Reject } H_0 \text{ if } z > z_\alpha$$

3. *Application of decision rule to data*   The data are:

$$\bar{x}_1 = 20, \quad s_1 = 12, \quad d = 10$$
$$\bar{x}_2 = 9, \quad s_2 = 3, \quad \alpha = .05, \quad z_\alpha = 1.645$$

Therefore, on application of the decision rule to the data we cannot reject H_0 because

$$z = \frac{\bar{x}_1 - \bar{x}_2 - d}{\sqrt{\dfrac{s_1^2}{n_1} + \dfrac{s_2^2}{n_2}}} = \frac{20 - 9 - 10}{\sqrt{\dfrac{12^2}{125} + \dfrac{3^2}{75}}} = .8866 \not> 1.645 = z_\alpha$$

In conclusion, the sociologist cannot assume, at the .05 level of significance, that on the average male students at the university spend more than $10 more per week on junk food than do female students.

EXAMPLE 13.8

With reference to Example 13.7, although the results were not significant at the .05 level for $d = 10$, the data certainly seem to indicate that male students spend more on junk food than do female students. Find the value d for which the results would be significant at the .05 level of significance.

SOLUTION

1. *Choice of hypotheses*

$$H_0: \mu_1 - \mu_2 = d \quad \text{versus} \quad H_1: \mu_1 - \mu_2 > d$$

2. *Choice of test statistic and decision rule*   The test statistic is

$$Z = \frac{\bar{X}_1 - \bar{X}_2 - d}{\sqrt{\dfrac{S_1^2}{n_1} + \dfrac{S_2^2}{n_2}}}$$

The decision rule, in terms of observed values z of the test statistic Z, is:

$$\text{Reject } H_0 \text{ if } z > z_\alpha$$

3. *Find d so that we can reject H_0*   In other words, we want to find d so that

$$z = \frac{\bar{x}_1 - \bar{x}_2 - d}{\sqrt{\dfrac{s_1^2}{n_1} + \dfrac{s_2^2}{n_2}}} > z_\alpha$$

The data are: $\alpha = .05$, $z_\alpha = z_{.05} = 1.645$, $\bar{x}_1 = 20$, $\bar{x}_2 = 9$, $s_1 = 12$, $s_2 = 3$, $n_1 = 125$, and $n_2 = 75$. Therefore, we want to find d so that

$$\frac{20 - 9 - d}{\sqrt{\dfrac{12^2}{125} + \dfrac{3^2}{75}}} > 1.645$$

The solution is $d = 9$.

We can conclude, at the .05 level of significance, that college men at this university on the average spend more than $9 more per week on junk food than do college women.

Finally, let us consider the case where σ_1 and σ_2 are unknown, $\sigma_1 \neq \sigma_2$ or $\sigma_1 = \sigma_2, n_1 = n_2 = n$, and $\mathcal{P}_1$ and $\mathcal{P}_2$ are normal. What we can do here is to pair the values obtained in the two samples. That is, define a new random variable X_i as

$$X_i = X_i^{(1)} - X_i^{(2)}$$

for each $i = 1, 2, \ldots, n$. Let $\mu = \mu_1 - \mu_2$.

Then $(X_1, X_2, \ldots, X_n)$ is a random sample of size n from a normal population with mean μ and unknown variance $\sigma^2 = \sigma_1^2 + \sigma_2^2$. The null hypothesis $H_0 \colon \mu_1 - \mu_2 = d$ becomes $H_0 \colon \mu = d$, and the appropriate test statistic and decision rule are given in Test Format 2. The test statistic, then, is

$$T = \frac{\bar{X} - d}{S/\sqrt{n}}$$

The format for this test is given in Test Format 8. If $n_1 \neq n_2$, we can still use this format by defining n to be the smaller of the two values n_1 and n_2 and simply reduce the sample size of the large sample to accommodate the test format.

Normal populations $\mathscr{P}_1, \mathscr{P}_2$; σ_1^2, σ_2^2 unknown, $\sigma_1^2 \neq \sigma_2^2$ or $\sigma_1^2 = \sigma_2^2$, $n_1 = n_2 = n$; α level of significance; H_0: $\mu_1 - \mu_2 = d$

Pair data from the two samples by defining

$$X_i = X_i^{(1)} - X_i^{(2)} \qquad \text{for} \quad i = 1, \ldots, n$$

$$\mu = \mu_1 - \mu_2$$

$$\sigma^2 = \sigma_1^2 + \sigma_2^2$$

Use Test Format 2 to test the null hypothesis H_0: $\mu = d$ against any of the suitable alternative hypotheses.

Test Format 8

EXAMPLE 13.9

The times in minutes it takes the mice in two random samples of 10 mice each (group A and group B) to complete a maze successfully are given in Table 13.1.

Let μ_1 and μ_2 be the mean time for mice of groups A and B, respectively. Test the hypotheses

$$H_0: \mu_1 - \mu_2 = 0 \quad \text{versus} \quad H_1: \mu_1 - \mu_2 \neq 0$$

at the .01 level of significance.

SOLUTION

1. *Choice of hypotheses*   Applying Test Format 8, we define $\mu = \mu_1 - \mu_2$ and $d = 0$ so that the appropriate hypotheses are

$$H_0: \mu = 0 \quad \text{versus} \quad H_1: \mu \neq 0$$

TABLE 13.1

Random sample, group A		Random sample, group B	
Mouse number	Minutes	Mouse number	Minutes
1	10	1	14
2	15	2	20
3	21	3	13
4	13	4	12
5	12	5	18
6	11	6	17
7	15	7	9
8	16	8	12
9	18	9	10
10	14	10	11

2. *Choice of test statistic and decision rule*   Following Test Format 8, we should refer to Test Format 2 for the appropriate test statistic. From Test Format 2, we obtain the test statistic

$$T = \frac{\bar{X} - 0}{S/\sqrt{n}}$$

The appropriate decision rule in terms of observed values t of the test statistic T is:

$$\text{Reject } H_0 \text{ if } |t| > t_{\alpha/2, n-1}$$

3. *Application of decision rule to data*   By pairing the data in the table we obtain:

$$x_1 = 10 - 14 = -4$$
$$x_2 = 15 - 20 = -5$$
$$x_3 = 21 - 13 = 8$$
$$x_4 = 13 - 12 = 1$$
$$x_5 = 12 - 18 = -6$$
$$x_6 = 11 - 17 = -6$$
$$x_7 = 15 - 9 = 6$$
$$x_8 = 16 - 12 = 4$$
$$x_9 = 18 - 10 = 8$$
$$x_{10} = 14 - 11 = 3$$

Then

$$\bar{x} = \frac{x_1 + x_2 + \cdots + x_{10}}{10} = .9$$

$$s^2 = \frac{1}{n-1} \sum_{i=1}^{n} (x_i - \bar{x})^2 = 32.767$$

$$t_{\alpha/2, n-1} = t_{.005, 9} = 3.250$$

Therefore, on application of the decision rule to the data we cannot reject the null hypotheses $H_0: \mu_1 - \mu_2 = 0$ because

$$|t| = \left| \frac{\bar{x} - 0}{s/\sqrt{n}} \right| = \left| \frac{.9 - 0}{5.7242/\sqrt{10}} \right| = .497 \ngtr 3.250$$

$$= t_{\alpha/2, n-1}$$

Exercises

In these exercises, a sample size of 49 or greater will be considered sufficiently large for all test formats requiring large sample sizes.

1. A large group of students were given a French examination. The mean score was 65 and the standard deviation was 15. A random sample of 9 of these students were then given a series of lessons in Spanish. Another random sample of 11 of these students were given a series of lessons in German. Both groups of students were given the French examination again. The mean score for the first group was 68 and the mean score for the second group was 65. Should we conclude that Spanish lessons are more helpful than German lessons in the learning of French? Test at the .05 level of significance. Use a test for the difference between means. Assume normal distributions and assume that the standard deviation remains the same for both groups.

2. The average mature height of a certain type of plant is known to be 3 feet with a standard deviation of 2 inches. Ten seedlings of plants of this type were given treatment A and another 14 seedlings were given treatment B. At maturity the mean height of the first group was 3 feet 2 inches and the mean height of the second group was 3 feet 1 inch. Should we conclude, at the .05 level of significance, that plants of this type at maturity will be taller, on the average, if grown under treatment A than under treatment B? Assume normal distributions and assume that the standard deviation remains the same under both treatments.

3. Twelve students randomly selected from the student body of a school were given a 2-hour review session before a certain arithmetic test. Another 13 students, also randomly selected, watched a football game on television before the same test. The average score for the first group was 75 and the standard deviation was 15. The average score for the second group was 73 and the standard deviation was 17. Should we conclude, at the .05 level of significance, that the two different pretest activities produce a difference in average test performance? Assume normal distributions and equal variances.

4. In an effort to compare the durability of two different types of sandpaper, 10 pieces of type A sandpaper were subjected to treatment by a machine that measures abrasive wear. Eleven pieces of type B sandpaper were subjected to the same treatment. The average amounts of abrasive wear, in coded units, were 27.4 for type A and 24.1 for type B, respectively. The sample standard deviations were 2 and 3, respectively. Test the hypothesis of no difference in average abrasive wear. Use a .01 level of significance. Assume normal distributions and equal variances.

5. In a random sample of 50 fifth-grade boys, the mean time to finish a standardized reading test was 25 minutes. The sample standard deviation was 4 minutes. The mean time for a random sample of 50 fifth-grade girls

was 24 minutes, and the sample standard deviation was 3 minutes. At a .04 level of significance, is there a difference in the mean reading time for fifth-grade boys and girls?

✓ 6.   In 1980 the mean income of workers in Rockbottom County was $1,000 lower than the mean income of workers elsewhere in the state. This year a random sample of 100 workers from Rockbottom County had a mean income of $11,000 and a sample standard deviation of $400, whereas a random sample of 110 workers in other parts of the state had a mean income of $12,200 and a sample standard deviation of $500. At the .05 level of significance, has the difference in income between Rockbottom County and the rest of the state increased since 1980?

✓ 7.   The mature heights of 10 tomato plants treated with Gro-Food once a week were 24, 25, 24, 26, 27, 24.5, 25.5, 26.2, 27.3, 27 inches, respectively; the mature heights of 10 tomato plants treated with Gro-Food twice a week were 23, 28.7, 26, 24, 27.5, 25.5, 26.5, 27.5, 25, 25.8, respectively. Is there a significant difference in the average height under the two treatments? Test at the .01 level of significance. Assume that the distribution of growth is normal.

✓ 8.   Six rabbits on diet A showed weight gains of .5, .2, .4, .2, .5, and .3 pounds, respectively. Another 6 rabbits on diet B showed weight gains of .3, .7, .6, .4, .9, and .5 pounds, respectively. Assume normal distributions.
(a)   Is there a significant difference in average weight gain under the two different diets? Test at the .05 level of significance.
(b)   Is the average weight gain under diet B significantly greater than the average weight gain under diet A? Test at the .05 level of significance.

✓ 9.   A standardized math test was given in the morning to 100 randomly selected second-graders. The same test was given in the afternoon to another 100 randomly selected second-graders. The sample mean and sample standard deviation of scores from the two random samples were $\bar{x}_1 = 2.1, s_1 = .3, \bar{x}_2 = 2.0, s_2 = .4$. Should we conclude that performance in this math test is not affected by the time of day the test is given? Choose the appropriate hypotheses and test at the .05 level of significance.

10.   A sociologist found that the average number of children in a random sample of 98 families with incomes over $35,000 was 2 with a standard deviation of 1, whereas the average number of children in a random sample of 102 families with incomes at or under $35,000 was 2.5 with a standard deviation of 1. Is there a significant difference in the average number of children for these two income groups? Test at the .01 level of significance.

11.   An educator believes that family income can be used as an indication of the reading readiness of kindergarten children. To check this theory he tested a random sample of 64 kindergarten children from families with incomes over $40,000 and found a mean reading readiness score of 9 and a sample standard deviation of 2. He also tested a random sample of 75 kindergarten children from families with incomes under $20,000 and

found a mean reading readiness score of 10 and a sample standard deviation of 4. Should the educator conclude that there is a significant difference in reading readiness between these two income groups? Test at the .03 level of significance.

13.4   Tests for Proportions

A research physician may wish to test for the rate (that is, proportion) of cure for a certain disease after treatment with a given drug. A politician may wish to conduct a poll to test for the proportion of voters that will vote for him on election day. These and many other situations demand tests for proportions.

A test for proportions, through appropriate labeling of the data, can always be interpreted as a test for the proportion of successes. For the doctor, a cure may be interpreted as a success. For the politician, a vote is definitely a success. Let us consider, then, a population $\mathscr{P}$ whose distribution is a Bernoulli distribution. Let p denote the probability of success. Suppose we wish to test the null hypothesis $H_0: p = p_0$ against some suitable alternative hypothesis. Furthermore, suppose we wish to test at the α level of significance and with the use of a random sample of size n. Consider first the case where n is relatively small. If we let S_n be the number of successes in the random sample of size n, then S_n is an appropriate test statistic. The distribution of S_n, if H_0 is true, is the binomial distribution with parameter $p = p_0$. Because S_n is a discrete random variable, it is not always possible to find a critical region whose size is exactly α. The next-best thing is to find a critical region whose size is as close to α as possible but not greater than α. The standard decision rule for testing $H_0: p = p_0$ against $H_1: p > p_0$ is

$$\text{Reject } H_0 \text{ if } s_n \geq k_\alpha$$

where k_α is the smallest integer for which

$$\sum_{x=k_\alpha}^{n} b(x; n, p_0) \leq \alpha$$

The corresponding decision rule for testing $H_0: p = p_0$ against $H_1: p < p_0$ is

$$\text{Reject } H_0 \text{ if } s_n \leq k_\alpha^*$$

where k_α^* is the largest integer for which

$$\sum_{x=0}^{k_\alpha^*} b(x; n, p_0) \leq \alpha$$

If we are testing $H_0: p = p_0$ against $H_1: p \neq p_0$, then the decision rule is

$$\text{Reject } H_0 \text{ if } s_n \geq k_{\alpha/2} \text{ or } s_n \leq k_{\alpha/2}^*$$

See Test Format 9.

Test for proportion; H_0: $p = p_0$; α level of significance; small sample size
Test statistic: S_n (binomial)
Let k_α be smallest integer for which $\sum_{x=k_\alpha}^{n} b(x; n, p_0) \leq \alpha$.

Let k_α^* be largest integer for which $\sum_{x=0}^{k_\alpha^*} b(x; n, p_0) \leq \alpha$.

Alternative hypothesis H_1	Decision rule: Reject H_0 if
$p \neq p_0$	$s_n \geq k_{\alpha/2}$ or $s_n \leq k_{\alpha/2}^*$
$p > p_0$	$s_n \geq k_\alpha$
$p < p_0$	$s_n \leq k_\alpha^*$

Test Format 9

If the sample size n is large, we may use the test statistic

$$Z = \frac{S_n - np_0}{\sqrt{np_0(1 - p_0)}}$$

The distribution of Z is approximately normal if $p = p_0$ and n is large. (See Statement 1, Section 8.6.1, Chapter 8, and Test Format 10.)

Test for proportion; H_0: $p = p_0$; α level of significance; large sample size n

Test statistic: $Z = \dfrac{S_n - np_0}{np_0(1 - p_0)}$ (approximately standard normal for n large and $p = p_0$)

Alternative hypothesis H_1	Decision rule: Reject H_0 if
$p \neq p_0$	$|z| > z_{\alpha/2}$
$p > p_0$	$z > z_\alpha$
$p < p_0$	$z < -z_\alpha$

Test Format 10

EXAMPLE 13.10

A cancer specialist believes that the 5-year survival rate of certain cancer patients, which presently is .3, can be improved if chemotherapy treatment is extended from 1 year to $1\frac{1}{2}$ years. To test her claim, she treated 10 such patients for $1\frac{1}{2}$ years and found that all but one of them survived the 5-year period. At the .01 level of significance, is she justified in claiming that the longer treatment period improves the 5-year survival rate? Assume that the 10 patients may be looked upon as a random sample of size 10 from the set of all such cancer patients. If we feel this assumption to be untenable, the test discussed in this section is not applicable.)

SOLUTION

The appropriate hypotheses are

$$H_0: p = .3 \quad \text{versus} \quad H_1: p > .3$$

where p = proportion of patients that survive 5 years.

At the .01 level of significance the decision rule is:

$$\text{Reject } H_0 \text{ if } s_n > 7$$

In this case $s_n = 9$; therefore, we should reject the null hypothesis and conclude that the treatment does improve the 5-year survival rate.

EXAMPLE 13.11

A quality-control engineer suspects that the proportion of defective units among certain manufactured items has increased from the set limit of .01. To test his claim, he randomly selected 100 of these items and found that the proportion of defective units in the sample was .02. Test the engineer's hypothesis at the .05 level of significance.

SOLUTION

The appropriate hypotheses are

$$H_0: p = .01 \quad \text{versus} \quad H_1: p > .01$$

The appropriate decision rule is:

$$\text{Reject } H_0 \text{ if } \frac{S_n - np_0}{\sqrt{np_0(1 - p_0)}} > z_\alpha$$

In this case, $z_\alpha = z_{.05} = 1.645$, and

$$\frac{S_n - np_0}{\sqrt{np_0(1 - p_0)}} = \frac{(.02)(100) - (100)(.01)}{\sqrt{100(.01)(1 - .01)}} = 1.005$$

Therefore, we cannot reject the null hypothesis.

Exercises

In these exercises, a sample size of 49 or greater will be considered sufficiently large for all test formats requiring large sample sizes.

1. A manufacturer claims that the proportion of consumers who recognize his brand's name is .8. A random sample of 10 consumers included 7 who

recognized the brand name. Should we conclude that the manufacturer overestimated the popularity of his brand's name? Choose the appropriate hypotheses and test at the .05 level of significance.

2. Suppose that in Exercise 1 a random sample of 100 consumers had included 70 who recognized the brand name. Test the same hypotheses at the same level of significance.

3. A random sample of 100 voters in Evanston showed that 49 voted for Proposition A. If the statewide vote was 51% for Proposition A, should we conclude that the vote in Evanston was significantly different from the statewide vote? Choose the appropriate hypotheses and test at the .04 level of significance.

4. Thirty percent of all laundry detergent sold last year at Greatway's supermarkets was Clean-All. Of 300 boxes of laundry detergent sold so far this year, 120 were Clean-All. Has the percentage of sales of Clean-All brand increased significantly? Test at the .05 level of significance.

13.5   Tests for Variances

Often we are interested in testing for the variability of products or operations. A psychologist may wish to test for the variability in subjects' time of response to a certain stimulus. A consumer organization may wish to test for the size of eggs commercially labeled as "large." These and many other situations may lead to tests for the variance of a population.

On the basis of a random sample of size n, consider first the testing of a null hypothesis that the variance σ^2 of a normal population is a given value σ_0^2. The suitable test statistic is the random variable

$$X^2 = \frac{(n-1)S^2}{\sigma_0^2}$$

The distribution of X^2 is the chi-square distribution with $n-1$ degrees of freedom (see Section 10.4.2, Chapter 10). The appropriate tests for the usual alternative hypotheses are given in Test Format 11.

Normal population; $H_0: \sigma^2 = \sigma_0^2$; α level of significance

Test statistic: $X^2 = \dfrac{(n-1)S^2}{\sigma_0^2}$ (chi-square with $n-1$ degrees of freedom if $\sigma^2 = \sigma_0^2$)

Alternative hypothesis H_1	Decision rule: Reject H_0 if
$\sigma^2 \neq \sigma_0^2$	$\chi^2 > \chi_{\alpha/2,\,n-1}^2$ or $\chi^2 < \chi_{1-\alpha/2,\,n-1}^2$
$\sigma^2 > \sigma_0^2$	$\chi^2 > \chi_{\alpha,\,n-1}^2$
$\sigma^2 < \sigma_0^2$	$\chi^2 < \chi_{1-\alpha,\,n-1}^2$

Test Format 11

EXAMPLE 13.12

According to the manufacturer's specifications, the cooling cycle of a certain type of equipment—when set at a fixed mean length of time—should have a standard deviation of, at most, 2 seconds. To check the manufacturer's claim, an engineer ran the equipment through 10 cooling cycles and obtained a sample standard deviation of 3 seconds. At the .01 level of significance test the hypotheses

$$H_0: \sigma = 2 \quad \text{versus} \quad H_1: \sigma > 2$$

Assume a normal distribution.

SOLUTION

1. *Choice of hypotheses*

$$H_0: \sigma = 2 \quad \text{versus} \quad H_1: \sigma > 2$$

2. *Choice of test statistic and decision rule*   Following Test Format 11, the appropriate test statistic is

$$X^2 = \frac{(n-1)S^2}{\sigma_0^2}$$

and the appropriate decision rule, in terms of observed values χ^2 of the test statistic X^2, is:

$$\text{Reject } H_0 \text{ if } \chi^2 > \chi^2_{\alpha, n-1}$$

3. *Application of decision rule to data*   The data are: $n = 10$, $\sigma_0^2 = 4$, $\alpha = .01$, $s^2 = 9$.

Therefore, on application of the decision rule to the data, we cannot reject the null hypothesis at the .01 level of significance because

$$\chi^2 = \frac{(n-1)s^2}{\sigma_0^2} = \frac{9 \cdot 9}{4} = 20.25 \ngtr 21.666 = \chi^2_{.01,9} = \chi^2_{\alpha, n-1}$$

In the case of testing for the equality of the variance σ_1^2 and σ_2^2 of two distinct normal populations $\mathscr{P}_1$ and $\mathscr{P}_2$, the suitable test statistic for the hypotheses $H_0: \sigma_1^2 = \sigma_2^2$ versus $H_1: \sigma_1^2 > \sigma_2^2$ is

$$F = \frac{S_1^2}{S_2^2}$$

whereas the suitable test statistic for the hypotheses $H_0: \sigma_1^2 = \sigma_2^2$ versus $H_1: \sigma_1^2 < \sigma_2^2$ is

$$F = \frac{S_2^2}{S_1^2}$$

In both instances S_1^2 and S_2^2 are the sample variance of random samples of sizes n_1 and n_2 from the populations $\mathscr{P}_1$ and $\mathscr{P}_2$, respectively. The distribution of S_1^2/S_2^2 is the F distribution with $n_1 - 1, n_2 - 1$ degrees of freedom, and the distribution of S_2^2/S_1^2 is the F distribution with $n_2 - 1, n_1 - 1$ degrees of freedom (see Section 10.4.5, Chapter 10).

If the alternative hypothesis is $H_1: \sigma_1^2 \neq \sigma_2^2$, then the appropriate test statistic is

$$F = S_1^2/S_2^2 \qquad \text{if} \quad s_1^2 > s_2^2$$

and

$$F = S_2^2/S_1^2 \qquad \text{if} \quad s_1^2 < s_2^2$$

See Test Format 12.

Normal populations $\mathscr{P}_1, \mathscr{P}_2$; $H_0: \sigma_1^2 = \sigma_2^2$; α level of significance; sample size n_1 from $\mathscr{P}_1$ and n_2 from $\mathscr{P}_2$

Test statistic: $F = S_1^2/S_2^2$ or $F = S_2^2/S_1^2$.

Alternative hypothesis H_1	Decision rule: Reject H_0 if	
$\sigma_1^2 > \sigma_2^2$	$s_1^2/s_2^2 > F_{\alpha, n_1 - 1, n_2 - 1}$	
$\sigma_1^2 < \sigma_2^2$	$s_2^2/s_1^2 > F_{\alpha, n_2 - 1, n_1 - 1}$	
$\sigma_1^2 \neq \sigma_2^2$	$s_1^2/s_2^2 > F_{\alpha/2, n_1 - 1, n_2 - 1}$	if $s_1^2 > s_2^2$
	$s_2^2/s_1^2 > F_{\alpha/2, n_2 - 1, n_1 - 1}$	if $s_1^2 < s_2^2$

Test Format 12

EXAMPLE 13.13

In comparing the variability of family income in two metropolitan areas, a survey yielded the following data: $n_1 = 100$, $n_2 = 110$, $s_1^2 = 25$, and $s_2^2 = 10$, where n_1, n_2 are the sample sizes and s_1^2, s_2^2 are the sample variance of incomes in thousands of dollars for the two areas, respectively. Assuming that the populations are normal, test the hypotheses

$$H_0: \sigma_1^2 = \sigma_2^2 \quad \text{versus} \quad H_1: \sigma_1^2 > \sigma_2^2$$

at the $\alpha = .05$ level of significance.

SOLUTION

1. *Choice of hypotheses*

$$H_0: \sigma_1^2 = \sigma_2^2 \quad \text{versus} \quad H_1: \sigma_1^2 > \sigma_2^2$$

2. *Choice of test statistic and decision rule*   Following Test Format 12, the appropriate test statistic is

$$F = \frac{S_1^2}{S_2^2}$$

and the appropriate decision rule in terms of observed values s_1^2/s_2^2 of the test statistic S_1^2/S_2^2 is

Reject H_0 if $s_1^2/s_2^2 > F_{\alpha, n_1 - 1, n_2 - 1}$

3. *Application of decision rule to data*   The data are: $n_1 = 100$, $n_2 = 110$, $s_1^2 = 25, s_2^2 = 10$. Therefore, on application of the decision rule to the data we should reject H_0 because $s_1^2/s_2^2 = 25/10 = 2.5 > 1.35 = F_{.05, 120, 120} > F_{.05, 99, 109} = F_{\alpha, n_1 - 1, n_2 - 1}$.

In conclusion, we should reject the null hypothesis that $\sigma_1^2 = \sigma_2^2$.

Exercises

In these exercises, a sample size of 49 or greater will be considered sufficiently large for all test formats requiring large sample sizes.

1. The standard deviation of IQs for elementary-school children in the U.S. is 15. A random sample of 15 inner-city elementary-school children had a sample standard deviation of 20. At the .05 level of significance should we conclude that the standard deviation of IQs of inner-city children is greater than 15?

2. If in Exercise 6 of Section 2 the weights of the 10 boxes of raisins in the random sample were 2.9, 3.1, 3.4, 2.8, 2.9, 2.6, 2.5, 3.4, 3.5, and 3.9 ounces, should we conclude, at the .01 level of significance, that the standard deviation of the weight of boxes of raisins is significantly higher than the .1 ounces claimed by the inventor?

3. Does the data of Exercise 7 of Section 3 indicate a significant difference in the variance of mature heights of tomato plants under the two treatments? Test at the .02 level of significance.

4. Does the data of Exercise 8 in Section 3 show a significant difference in the variance of weight gains of rabbits under the two diets? Test at the .10 level of significance.

5. If in Exercise 11 of Section 2 the specification also indicates that the standard deviation of the diameter of the ball bearings should be no more than .04 cm, does the result of the random sample indicate a significant departure from the specified value for the standard deviation? Test at the .01 level of significance.

13.6   Finding the Required Sample Size for a Given Power for a Test

We do not want to reject the null hypothesis H_0 when it is true. Therefore, we favor tests that have small values for α (the probability of rejecting H_0 when H_0 is true). We do, however, want large values for probabilities of rejecting H_0 when H_0 is false. For instance, in medical research related to heretofore incurable diseases, we are anxious to reject the null hypothesis that a new drug is ineffective if it is, in fact, effective; therefore, we want a test that has a high probability of rejecting the null hypothesis when the null hypothesis is false. The ability of a test to reject H_0 is sometimes referred to as the power of the test. The precise definition follows.

Note: the power of a test is also discussed in Section 12.4 of Chapter 12.

Definition 13.1   Consider a test of a null hypothesis H_0: $\theta = \theta_0$ against some suitable alternative hypothesis H_1. If $\theta = \theta_1$ is a viable alternative to the null hypothesis $\theta = \theta_0$, then

$$P(\text{reject } H_0 \mid \theta = \theta_1)$$

is called the **power of the test** when $\theta = \theta_1$.

Observe that if we are testing a simple hypothesis H_0: $\theta = \theta_0$ against a simple alternative hypothesis H_1: $\theta = \theta_1$, then

$$P(\text{reject } H_0 \mid \theta = \theta_1)$$
$$= 1 - P(\text{accept } H_0 \mid \theta = \theta_1)$$
$$= 1 - \beta$$

where β is the probability of a type II error.

Therefore, the usual notation for the power of a test is $1 - \beta$, with the understanding that the value of β is dependent on the value of the alternative parameter θ_1. If H_1 is a composite alternative hypothesis, then the power $1 - \beta$ becomes a function of θ_1 and is called the power function of the test.

An important question concerning the power of a test is the following: How can we design a test at a given α level of significance that will have a high probability (power) of rejecting H_0 when H_0 is false? One answer, of course, is to increase the sample size. However, in most situations there is a practical problem of cost and time that puts a constraint on sample size. A problem of considerable importance, then, is the problem of finding the minimum sample size that will both enable us to design a test at a given α level of significance and also give us a sufficiently high probability (power) of rejecting H_0 when H_0 is false. Let us consider an example.

EXAMPLE 13.14

Let $\mathscr{P}$ be a normal population with unknown mean μ and known variance σ^2. Suppose we wish to test, at the α level of significance, the hypotheses

$$H_0: \mu = \mu_0 \quad \text{versus} \quad H_1: \mu > \mu_0$$

Following the procedure of Test Format 1, Section 13.2.1 the standard decision rule is

$$\text{Reject } H_0 \text{ if } \frac{\bar{X} - \mu_0}{\sigma/\sqrt{n}} > z_\alpha$$

Now, suppose it is quite important for us to be able to reject with great probability, say with a power of at least $1 - \beta$, the null hypothesis $\mu = \mu_0$ when in fact μ is $\mu_0 + d$ or greater (d is a fixed positive constant). Can we find the minimum sample size n so this requirement is met? Let us try.

We want to find n so that the power of the test, when $\mu = \mu_0 + d$, is

$$P(\text{reject } H_0 \mid \mu = \mu_0 + d) = 1 - \beta$$

Observe that

$$1 - \beta = P(\text{reject } H_0 \mid \mu = \mu_0 + d)$$

$$= P\left(\frac{\bar{X} - \mu_0}{\sigma/\sqrt{n}} > z_\alpha \,\middle|\, \mu = \mu_0 + d\right)$$

$$= P\left(\frac{\bar{X} - \mu_0 - d + d}{\sigma/\sqrt{n}} > z_\alpha \,\middle|\, \mu = \mu_0 + d\right)$$

$$= P\left(\frac{\bar{X} - (\mu_0 + d)}{\sigma/\sqrt{n}} + \frac{d}{\sigma/\sqrt{n}} > z_\alpha \,\middle|\, \mu = \mu_0 + d\right)$$

$$= P\left(\frac{\bar{X} - (\mu_0 + d)}{\sigma/\sqrt{n}} > z_\alpha - \frac{d}{\sigma/\sqrt{n}} \,\middle|\, \mu = \mu_0 + d\right)$$

From Section 10.4.1, Chapter 10, we recall that if $\mu = \mu_0 + d$, then $Z = [\bar{X} - (\mu_0 + d)]/(\sigma/n^{1/2})$ is standard normal. Therefore, we now have

$$1 - \beta = P(\text{reject } H_0 \mid \mu = \mu_0 + d)$$

$$= P\left(\frac{\bar{X} - (\mu_0 + d)}{\sigma/\sqrt{n}} > z_\alpha - \frac{d}{\sigma/\sqrt{n}} \,\middle|\, \mu = \mu_0 + d\right)$$

$$= P\left(Z > z_\alpha - \frac{d}{\sigma/\sqrt{n}}\right)$$

The problem is now reduced to the problem of finding n so that

$$1 - \beta = P\left(Z > z_\alpha - \frac{d}{\sigma/\sqrt{n}}\right)$$

This is equivalent to finding n so that

$$-z_\beta = z_\alpha - \frac{d}{\sigma/\sqrt{n}}$$

A little algebra yields the result

$$n = \frac{(z_\alpha + z_\beta)^2 \sigma^2}{d^2}$$

Because of the nature of the decision rule, it is easy to see that if μ is, in fact, even greater than $\mu_0 + d$, then the same value of n will suffice to produce a power of $1 - \beta$ or greater.

EXAMPLE 13.15

Suppose we wish to test, at the $\alpha = .05$ level of significance, the hypothesis

$$H_0: \mu = 100 \quad \text{versus} \quad H_1: \mu > 100$$

of a normal population $\mathscr{P}$ with mean μ and known standard deviation $\sigma = 15$. Find the minimum sample size required if the power of the test is to be .95 when $\mu = 110$.

SOLUTION

Using the notations of Example 11.14, we have

$$\alpha = .05, \quad 1 - \beta = .95, \quad d = 10, \quad \sigma = 15$$

Therefore

$$z_\alpha = z_{.05} = 1.645$$

$$z_\beta = z_{.05} = 1.645$$

and

$$n = \frac{(z_\alpha + z_\beta)^2 \sigma^2}{d^2} = \frac{(1.645 + 1.645)^2 \, 15^2}{10^2}$$

$$= 24.35$$

Since the sample size has to be an integer, the correct answer is

$$n = 25$$

The method used in Example 13.15 to find the required sample size for testing a null hypothesis H_0: $\mu = \mu_0$ against an alternative H_1: $\mu > \mu_0$ may be suitably modified for an alternative hypothesis of $H_1 = \mu < \mu_0$ or H_1: $\mu \neq \mu_0$. The resulting sample size n is given in Table 13.2.

In comparing the means of two normal populations when σ_1 and σ_2 are known, a similar procedure yields the result of Table 13.3.

If the variance of a population is unknown or cannot be estimated, then the computation for the required sample size for a specified power of a test becomes very complex. However, tables exist for determining the required

---

TABLE 13.2

Required Sample Size for Specific Level of Significance and Power

Normal population; σ^2 known; H_0: $\mu = \mu_0$; α level of significance

Test statistic: $Z - \dfrac{\bar{X} - \mu_0}{\sigma/\sqrt{n}}$ (standard normal when $\mu = \mu_0$)

Alternative hypothesis	Decision rule: Reject H_0 if	Required sample size	Power of test: $1 - \beta$ when
$\mu \neq \mu_0$	$\lvert z \rvert > z_{\alpha/2}$	$n = \dfrac{(z_{\alpha/2} + z_\beta)^2 \sigma^2}{d^2}$	$\mu = \mu_0 + d$, or $\mu = \mu_0 - d$
$\mu > \mu_0$	$z > z_\alpha$	$n = \dfrac{(z_\alpha + z_\beta)^2 \sigma^2}{d^2}$	$\mu = \mu_0 + d$
$\mu < \mu_0$	$z < -z_\alpha$	$n = \dfrac{(z_\alpha + z_\beta)^2 \sigma^2}{d^2}$	$\mu = \mu_0 - d$

TABLE 13.3

Required Sample Size for Specific Level of Significance and Power

Normal population $\mathscr{P}_1, \mathscr{P}_2$; known σ_1^2, σ_2^2; H_0: $\mu_1 = \mu_2$; α level of significance; sample sizes $n = n_1 = n_2$

Test statistic: $Z = \dfrac{\bar{X}_1 - \bar{X}_2}{\sqrt{(\sigma_1^2/n) + (\sigma_2^2/n)}}$ (standard normal when $\mu_1 = \mu_2$)

Alternative hypothesis	Decision rule: Reject H_0 if	Required sample size	Power of test: $1 - \beta$ when
$\mu_1 \neq \mu_2$	$\lvert z \rvert > z_{\alpha/2}$	$n = \dfrac{(z_{\alpha/2} + z_\beta)^2(\sigma_1^2 + \sigma_2^2)}{d^2}$	$\mu_1 - \mu_2 = d$, or $\mu_2 - \mu_1 = d$
$\mu_1 > \mu_2$	$z > z_\alpha$	$n = \dfrac{(z_\alpha + z_\beta)^2(\sigma_1^2 + \sigma_2^2)}{d^2}$	$\mu_1 - \mu_2 = d$
$\mu_1 < \mu_2$	$z < -z_\alpha$	$n = \dfrac{(z_\alpha + z_\beta)^2(\sigma_1^2 + \sigma_2^2)}{d^2}$	$\mu_2 - \mu_1 = d$

sample size. Interested readers are referred to the book *Experimental Statistics* by M. G. Natrella, National Bureau of Standards Handbook 91, issued August 1, 1963, by the United States Department of Commerce.

Exercises

In these exercises, a sample size of 49 or greater will be considered sufficiently large for all test formats requiring large sample sizes.

1. Consider a test, at the $\alpha = .01$ level of significance, of the hypotheses

$$H_0: \mu = 2 \quad \text{versus} \quad H_1: \mu > 2$$

of a normal population with mean μ and standard deviation $\sigma = 1$. Find the minimum sample size required if the power of the test is to be .95 when $\mu = 2.5$.

2. Consider a test, at the $\alpha = .05$ level of significance, of the hypotheses

$$H_0: \mu = -3 \quad \text{versus} \quad H_1: \mu < -3$$

of a normal population with mean μ and standard deviation $\sigma = 1.5$. Find the minimum sample size required if the power of the test is to be .90 when $\mu = -4$.

3. Consider a test, at the $\alpha = .05$ level of significance, of the hypotheses

$$H_0: \mu = 70 \quad \text{versus} \quad H_1: \mu \neq 70$$

of a normal population with mean μ and standard deviation $\sigma = 10$. Find the minimum sample size required if the power of the test is to be .85 when $\mu = 73$.

4. Consider a test, at the $\alpha = .01$ level of significance, of the hypotheses

$$H_0: \mu_1 = \mu_2 \quad \text{versus} \quad H_1: \mu_1 \neq \mu_2$$

of the normal populations $\mathscr{P}_1$ and $\mathscr{P}_2$ with means μ_1 and μ_2, respectively, and standard deviations $\sigma_1 = 1$ and $\sigma_2 = 1.5$, respectively. Find the minimum sample size required if the power of the test is to be .95 when $\mu_1 - \mu_2 = 3$.

5. Repeat Exercise 4 with the hypotheses changed to

$$H_0: \mu_1 = \mu_2 \quad \text{versus} \quad H_1: \mu_1 > \mu_2$$

6. Find the significance probability for each of the test results of Examples 13.1, 13.3, 13.4, 13.5, 13.6, 13.7, and 13.9.

13.7 Reports of Tests of Statistical Hypotheses, Significance Probabilities

Suppose that a test of statistical hypotheses at the .01 level of significance is conducted and that the experimental data indicate that the null hypothesis cannot be rejected. How should this result be reported? If it is merely stated that the null hypothesis cannot be rejected at the .01 level of significance, a reader of the report may wonder if it could be rejected at the .05 level of significance, or — indeed — at what minimum level of α it could be rejected. To accommodate legitimate queries of this sort, a general practice in the reporting of statistical results is to include a statement of the lowest level α for which the null hypothesis can be rejected. This smallest value for α is called the significance probability and is denoted by p. It is also called the p-value of the test.

EXAMPLE 13.16

Let $\mathscr{P}$ be a normal population with unknown mean μ and known variance $\sigma^2 = 1$. Test the hypotheses

$$H_0: \mu = 2 \quad \text{versus} \quad H_1: \mu \neq 2$$

at the .01 level of significance if a random sample of size 9 from the population yields a sample mean of $\bar{x} = 2.7$. Find the significance probability p for the test.

SOLUTION

Using Test Format 1, we have

$$|z| = \left| \frac{\bar{x} - \mu_0}{\sigma/\sqrt{n}} \right| = \left| \frac{2.7 - 2}{1/3} \right| = 2.1 \not> 2.575 = z_{\alpha/2}$$

Therefore, we cannot reject H_0 at the .01 level of significance.

To find the significance probability of the test, notice that the observed value of the test statistic is

$$|z| = \left| \frac{\bar{x} - \mu_0}{\sigma/\sqrt{n}} \right| = \left| \frac{2.7 - 2}{1/3} \right| = 2.1 = z_{.0179} = z_{.0358/2}$$

In other words

$$|z| \geq z_{.0358/2}$$

so that the test result is significant at the .0358 level of significance. Therefore the p-value, or significance probability, is

$$p = .0358$$

The p-value is often rounded off to 2 decimal places. In this case the usual practice is to report that

$$p < .04$$

Along with the p-value of a test, it is customary to report the observed value of the test statistic. For instance, in the above example the following statement may be included in a report of the test: "The experimental data indicates that, at the .01 level of significance, we cannot reject the null hypothesis $\mu = 2$ in favor of the alternative hypothesis $\mu \neq 2$ ($z = 2.1$, $p < .04$)." The statement $z = 2.1$ tells the reader that the test statistic was a random variable whose distribution is standard normal and that the value of the test statistic was observed to be 2.1; the statement $p < .04$ tells the reader that the p-value is some number less than .04. If the test statistic had been a random variable whose distribution is the t distribution, or the chi-square distribution, then, instead of the letter z, we would have used the letter t or χ^2, respectively. If the test statistic includes a specified degree of freedom, then the degree of freedom would also be reported.

EXAMPLE 13.17

Let $\mathscr{P}$ be a normal population with unknown mean μ and unknown variance. Test the hypotheses

$$H_0: \mu = 1 \quad \text{versus} \quad H_1: \mu < 1$$

at the .05 level of significance if a random sample of size 16 from the population yields a sample mean of $\bar{x} = .9$ and a sample standard deviation of $s = .15$. Find the significance probability p for the test

SOLUTION

Using Test Format 2 we have

$$t = \frac{\bar{x} - \mu_0}{s/\sqrt{n}} = \frac{.9 - 1}{.15/\sqrt{16}} = -2.667 < -1.753 = -t_{.05,15}$$

At the .05 level of significance we can conclude that μ is significantly less than 1.

To find the p-value of the test, note that

$$t = -2.667 < -t_{.01,15}$$

Therefore

$$p < .01$$

The standard format for reporting this result is: "The test results indicate that the mean is significantly less than 1 ($t = -2.667$, d.f. $= 15$, $p < .01$)."

EXAMPLE 13.18

What is the standard format for reporting the results of the test in Example 13.3(a)?

SOLUTION

Let us find the p-value for the test.

The observed value of the test statistic T was

$$t = -1.5$$

The degrees of freedom for the test statistic T was 8. From the table of values for the t distribution (Table V in Appendix C) we obtain the following bounds for the value $t = -1.5$.

$$-t_{.05,8} = -1.860 < -1.5 < -1.397 = -t_{.10,8}$$

Therefore, the p-value is

$$p < .10$$

The standard format for reporting the result is: "The test results indicate that the number of customers per Saturday evening is not significantly less than 100 ($t = -1.5$, d.f. $= 8$, $.05 < p < .10$)."

References

Derivations of the test statistics discussed in this chapter may be found in advanced texts on mathematical statistics, including the following:

1. De Groot, M. H. 1975. *Probability and statistics*, Menlo Park, CA: Addison-Wesley Publishing Co.
2. Hogg, R. V., and A. T. Craig. 1970. *Introduction to mathematical statistics*, 3d. ed. London: The Macmillan Co.
3. Lindgren, B. W. 1968. *Statistical theory*, 2d. ed. London: The Macmillan Co.

Chapter 14

REGRESSION AND CORRELATION

14.1   Bivariate Linear Regression

Bivariate, or simple, linear regression analysis is concerned with the problem of predicting, or estimating, the value of a random variable Y on the basis of a measurement x when certain conditions of linearity can be assumed. The object of the analysis is to find constants a and b so we can predict—or estimate—with great confidence an approximate value of a measurement y in terms of the value of a measurement x with the linear equation $y = a + bx$. For example, the object of a linear regression analysis of data on prices and earnings of shares of a given group of common stock may be to find constants a and b so that we can predict, with great confidence, the value of y (a stock's earnings per share) in terms of the value of x (its price per share) by using the linear equation $y = a + bx$. Another example might stem from the wish to determine a formula for estimating the level of air pollution in terms of the temperature and humidity index. The object of a linear regression analysis in this case would be to find constants a and b so that we can assert with great confidence that an approximate value of y (the level of pollution at any given time) may be obtained in terms of the value of x (the temperature and humidity index at the same time) by using the linear equation $y = a + bx$. In many applied problems of this kind there are three steps in the analysis. They are:

1. Preliminary examination of sample data to determine the validity of an assumption of linear dependence.

2. Estimation of a regression line with the method of least squares.

3. Computation of confidence limits to evaluate the goodness of the estimation.

A discussion of each of these steps will follow. However, before proceeding with it, we shall elaborate on the exact nature of the "conditions of linearity" that must be present before a linear regression analysis becomes meaningful. We have stated that the object of a bivariate linear regression analysis is to find constants a and b so that we can predict an approximate value of y in terms of x with the linear equation $y = a + bx$. (*Note:* any equation of the form $y = a + bx$, where a and b are constants, is called a linear equation. We say that there is a linear relationship between a variable y and a variable x, or that a variable y is linearly dependent on a variable x, if there are constants a and b such that $y = a + bx$. Furthermore, when the relationship is expressed in this form, x is called the independent variable and y is called the dependent variable.) The inclusion of the word "approximate" implies that we do not expect to predict at all times the correct value of y. We do hope, however, that—on the average—the equation $y = a + bx$ will provide the correct answer. Therefore, it is more accurate to state that the object of bivariate linear regression analysis concerning a random variable Y and a measurement x is to find constants A and B such that

[the expected value of Y, given the measurement x] $= A + Bx$

For example, if the income of a college graduate one year after graduation is expected to predict his income 10 years after graduation, then the object of a linear regression analysis would be to find constants A and B such that

$$\left[\begin{array}{l}\text{the expected income of a college graduate 10 years after} \\ \text{graduation, given that his income 1 year after graduation is } x\end{array}\right] = A + Bx$$

Or, if there are reasons to believe that the amounts of insurance held by families are determined by their incomes, then the object of a linear regression analysis would be to find constants A and B such that

$$\left[\begin{array}{l}\text{the expected amount of insurance held} \\ \text{by a family whose income is } x\end{array}\right] = A + Bx$$

Of course, in each of these cases, the objective can be met only if, in fact, there are constants A and B that satisfy the "condition of linearity" defined by the respective equations. In other words, there is an underlying assumption concerning the random variable Y and the measurement x in a linear regression analysis where values of x are expected to predict values of Y. The assumption is that the expected value of Y given the measurement x is linearly dependent on x, so that there are, in fact, constants A and B such that

$$[\text{the expected value of } Y, \text{ given the measurement } x] = A + Bx$$

Stated in terms of standard notations of mathematical statistics, the underlying assumption concerning the random variable Y and the variable x in a linear regression analysis, where values of x are to predict values of Y, is that there are constants A and B such that

$$E(Y|x) = A + Bx$$

or, equivalently, that

$$\mu_{Y|x} = A + Bx$$

Note: both notations $E(Y|x)$ and $\mu_{Y|x}$ are used to represent the expected value of the random variable Y, given the variable x. Each of them is also called the conditional expectation of Y, given x. The line defined by these equations is called the **regression line** of Y on x, and each of the equations is called the **regression equation** of Y on x.

Observe that if $E(Y|x) = A + Bx$, we can say that, once we have fixed the value of x, the random variable Y can be represented in the form

$$Y = A + Bx + \varepsilon$$

where ε is a random variable that has a zero expectation. This method of representing the linear regression of Y on x allows for an analysis of the error of the prediction in terms of the random variable ε. (See Section 14.1.3 on the standard error of the estimate.)

Note that throughout these introductory remarks we have refrained from calling the measurement x a value of a random variable X. The reason is that in many instances of regression analysis the measurement x is not treated as such a value. Often, it is a value that is preselected by the investigator and therefore is not subject to random fluctuation. For example, in an investigation on the solution of chemicals in water at various temperatures, the investigator may predetermine the temperature at which the experiment is to be conducted. This having been said, we must add that the omission of a reference to x as a value of a random variable X does not preclude the possibility of its being exactly that. In other words, at this point the reader should be aware that the entire discussion on regression analysis remains completely valid if x is a value of a random variable X. However, then we have to be careful to state that we are assuming the observed value of X to be x when we write $Y = A + Bx + \varepsilon$.

14.1.1 Scatter Diagram of Sample Data

As stated in the introductory remarks, the object of a bivariate, or simple, linear regression analysis is to find constants a and b so that we can predict, with great confidence, an approximate value of a measurement y in terms of the value of a measurement x with the linear equation $y = a + bx$. However, before we proceed with the calculation of these constants in any instance of linear regression analysis, it would be sensible for us to conduct a preliminary examination of sample data to determine if such an end result is even remotely possible. If the data consists of a set $\{(x_1, y_1), (x_2, y_2), \ldots, (x_n, y_n)\}$ of n paired observations of measurements x and measurements y, then the simplest method of examining the data for this purpose is to construct a scatter diagram of the data. This consists of plotting the coordinates $(x_1, y_1), (x_2, y_2), \ldots, (x_n, y_n)$ of the n paired measurements.

EXAMPLE 14.1

Nine college graduates were randomly selected from the class of 1970. The incomes, in thousands of dollars, of these graduates one year after graduation and ten years after graduation are given below. Construct a scatter diagram of these data.

College graduate	1	2	3	4	5	6	7	8	9
Income after 1 year	10	10	12	14	14.5	17	18	22	19
Income after 10 years	25	26	28.5	30	28	36	34	39	35

SOLUTION

A scatter diagram of the data is obtained by plotting the coordinates of the 9 paired measurements $(10, 25)$, $(10, 26)$, $(12, 28.5)$, $(14, 30)$, $(14.5, 28)$, $(17, 36)$, $(18, 34)$, $(22, 39)$, and $(19, 35)$:

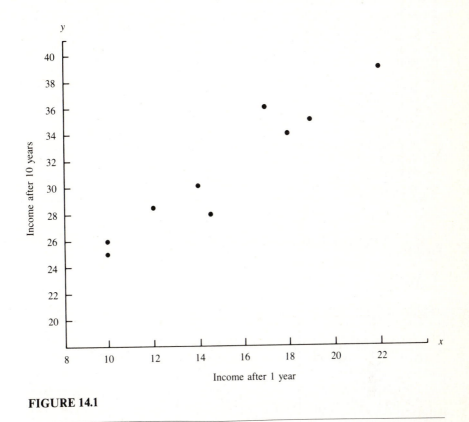

FIGURE 14.1

As can be seen from Figure 14.1, the scatter diagram provides a visual display of the relationship of the paired measurements. In this case, it appears that as the value of x (income after 1 year) increases, the corresponding value of y (income after 10 years) also increases. Furthermore, the points in the scatter diagram seem to fall along a line such as the one marked $y = 9.8 + 1.4x$, drawn onto the scatter diagram in Figure 14.2. This is an indication that, on the average, values of y are linearly dependent on values of x. Therefore the data are appropriate for a linear regression analysis.

Figure 14.3 shows a variety of scatter diagrams; (a), (c), and (e) contain data suitable for bivariate linear regression analysis. Diagrams (b), (d), and (f) contain data unsuitable for bivariate linear regression analysis (although the data in diagrams (b) and (f) are suitable for bivariate curvilinear regression analysis). The data of diagram (d) show no correlation between the x and y values of the paired measurements.

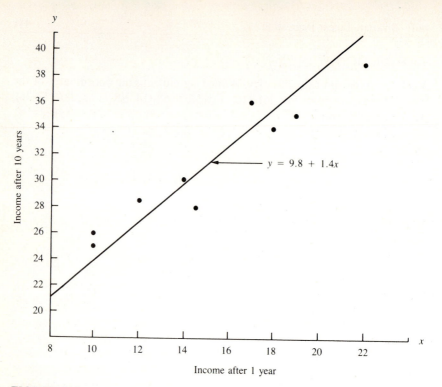

FIGURE 14.2

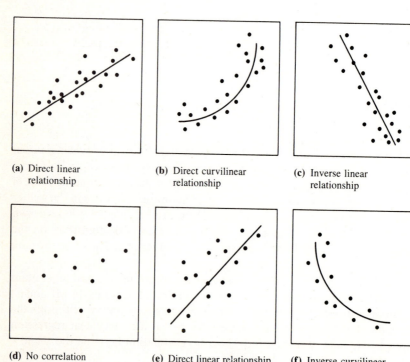

(a) Direct linear
relationship

(b) Direct curvilinear
relationship

(c) Inverse linear
relationship

(d) No correlation

(e) Direct linear relationship
(wider scatter than in (**a**).

(f) Inverse curvilinear
relationship

FIGURE 14.3

14.1.2 Estimation of Regression Line: Method of Least Squares

If a preliminary examination of sample data shows it is reasonable to assume a linear dependence of Y on x, in the sense that there are constants A and B such that

$$[\text{the expected value of } Y \text{ given } x] = A + Bx$$

or, in other words, that a regression line

$$\mu_{Y|x} = A + Bx$$

exists, then the next step in a linear regression analysis of the data is to estimate the regression line. (*Note:* since sample data offer, at best, an incomplete description of the relationship of the random variable Y and the variable x, we can only obtain an estimate of the regression line and hope that it is not too different from the true regression line.) The most frequently used method for estimating the regression line is called the method of least squares.

The object of the method of least squares is to fit a straight line to a set $\{(x_1, y_1), (x_2, y_2), \ldots, (x_n, y_n)\}$ of paired measurements in such a way that the line provides, in a sense, a best fit. But, what do we mean by a best fit? Indeed, how do we evaluate the fit of a line? The least squares method of evaluating the fit of a line is as follows: let $\{(x_1, y_1), (x_2, y_2), \ldots, (x_n, y_n)\}$ be a set of paired measurements. Let $y = a + bx$ be the equation of a straight line (see Figure 14.4). Let e_i be the vertical distance from the point (x_i, y_i) to the

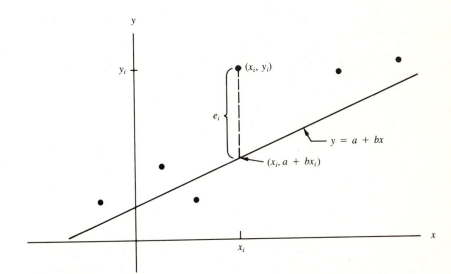

FIGURE 14.4

line $y = a + bx$. Call it the error for the point (x_i, y_i) in the fit of $y = a + bx$ to the given points.

Now, how do we evaluate e_i? Consider the coordinates of the points on the line $y = a + bx$. If the x coordinate of a point on the line is x_i, then the y coordinate is $y_i = a + bx_i$. In other words, $(x_i, a + bx_i)$ are the coordinates of a point on the line $y = a + bx$ that is directly above the point (x_i, y_i), or directly below the point (x_i, y_i), or perhaps is exactly the point (x_i, y_i). Therefore, the vertical distance from the point (x_i, y_i) to the line $y = a + bx$ is

$$e_i = |y_i - (a + bx_i)| = |y_i - a - bx_i|$$

Now, let us compute the sum of the squares of the distances from the points $(x_1, y_1), (x_2, y_2), \ldots, (x_n, y_n)$ to the line $y = a + bx$, denote the sum by S_{e^2}, and call it the sum of the squares of the errors. That is, let

$$[\text{sum of squares of errors}] = S_{e^2} = \sum_{i=1}^{n} e_i^2 = \sum_{i=1}^{n} (y_i - a - bx_i)^2$$

The least squares method of evaluating the fit of a line to a set of points assumes that if S_1 is the sum of the squares of errors for the line L_1, and S_2 is the sum of the squares of errors for the line L_2, then L_1 provides a better fit than L_2 if the sum S_1 is smaller than the sum S_2. Furthermore, it assumes that the line that provides the best fit is the line with the least sum of the squares of errors.

EXAMPLE 14.2

Let $(3, 4), (4, 7), (8, 7), (12, 11), (17, 10)$ be a set of paired measurements. Let L_1 be the line defined by the equation $y = 3 + .75x$. Let L_2 be the line defined by the equation $y = 3 + .5x$. Which line provides a better fit of the data?

SOLUTION

The following table of values contains all the information needed to calculate the sum of the squares of errors for L_1.

i	x_i	y_i	$(y_i - 3 - .75x_i)$	e_i^2
1	3	4	-1.25	1.56
2	4	7	1.00	1.00
3	8	7	-2.00	4.00
4	12	11	-1.00	1.00
5	17	10	-5.75	33.06

$$\sum_{i=1}^{n} e_i^2 = 40.62$$

The sum of the squares of errors for L_1 is $S_1 = S_{e^2} = \sum_{i=1}^{n} e_i^2 = 40.62$.

The following table of values contains all the information needed to calculate the sum of the squares of errors for L_2.

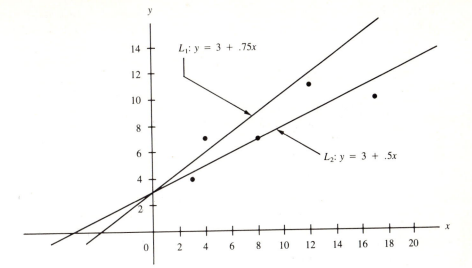

FIGURE 14.5

i	x_i	y_i	$(y_i - 3 - .5x_i)$	e_i^2
1	3	4	$-.5$	.25
2	4	7	2.0	4.00
3	8	7	.0	0.00
4	12	11	2.0	4.00
5	17	10	-1.5	2.25

$$\sum_{i=1}^{n} e_i^2 = 10.50$$

The sum of the squares of errors for L_2 is $S_2 = S_{e^2} = \sum_{i=1}^{n} e_i^2 = 10.50$.
Since $S_2 < S_1$, the line L_2 provides a better fit of the data than the line L_1.
See Figure 14.5 for a scatter diagram of the data and graphs of L_1 and L_2.

Summary

(a) The square of the distance from the point (x_i, y_i) to the line $y = a + bx$ is

$$e_i^2 = (y_i - a - bx_i)^2$$

(b) The sum of the squares of the distances from the points $(x_1, y_1), (x_2, y_2), \ldots, (x_n, y_n)$ to the line $y = a + bx$ is

$$S_{e^2} = \sum_{i=1}^{n} e_i^2 = \sum_{i=1}^{n} (y_i - a - bx_i)^2$$

(c) The least squares method of evaluating the fit of a line to a set of points assumes that if the sum of the squares of errors for line L_1 is less than the sum of the squares of errors for line L_2, then L_1 provides a better fit than L_2. Furthermore, it assumes that the line that provides the best fit is the line with the least sum of the squares of errors.

To find the line with the least sum of the squares of errors, consider the following. The sum of the squares of errors for the line $y = a + bx$ is

$$S_{e^2} = \sum_{i=1}^{n} e_i^2 = \sum_{i=1}^{n} (y_i - a - bx_i)^2$$

We wish to find the values of a and b that will minimize the value of S_{e^2}. Now, by visually checking the scatter diagram of any set of data that is suitable for bivariate linear regression analysis, one can readily see that if we fix the value a of a line with equation $y = a + bx$ (i.e., fix the y intercept of the line) and let the value of b vary (i.e., let the slope of the line vary), then there is a relative minimum but no relative maximum for the value of S_{e^2}. Similarly, if we fix the value of b (i.e., fix the slope of the line) and let the value of a vary (i.e., let the y intercept of the line vary), then there is also a relative minimum but no relative maximum for the value of S_{e^2}. This means that we can use the standard methods of calculus to find the values of a and b that will minimize the value of S_{e^2}. That is, we can find these values by differentiating S_{e^2} (considered as a function of a and b) with respect to a and with respect to b, setting the partial derivatives equal to zero, then solving the simultaneous equations for a and b. The resulting values of a and b will be the values that will produce a line with the equation $y = a + bx$ that will have the least sum of the squares of errors. That line will be the least squares estimate of the regression line.

Following these instructions, we differentiate S_{e^2} with respect to a and with respect to b and obtain

$$\frac{\partial S_{e^2}}{\partial a} = -2 \sum_{i=1}^{n} (y_i - a - bx_i)$$

and

$$\frac{\partial S_{e^2}}{\partial b} = -2 \sum_{i=1}^{n} (y_i - a - bx_i)x_i$$

Setting the partial derivatives equal to zero, dividing through by -2, and separating terms, we obtain

$$\sum_{i=1}^{n} y_i - na - b \sum_{i=1}^{n} x_i = 0$$

and

$$\sum_{i=1}^{n} x_i y_i - a \sum_{i=1}^{n} x_i - b \sum_{i=1}^{n} x_i^2 = 0$$

From these two equations (called the **normal equations**) we solve for a and b and obtain

$$b = \frac{n \sum_{i=1}^{n} x_i y_i - \left(\sum_{i=1}^{n} x_i\right)\left(\sum_{i=1}^{n} y_i\right)}{n \sum_{i=1}^{n} x_i^2 - \left(\sum_{i=1}^{n} x_i\right)^2}$$

and

$$a = \bar{y} - b\bar{x}$$

where

$$\bar{x} = \frac{\sum_{i=1}^{n} x_i}{n} \quad \text{and} \quad \bar{y} = \frac{\sum_{i=1}^{n} y_i}{n}$$

Summary   Least Squares Estimate of Regression Line

Given a sample $\{(x_1, y_1), (x_2, y_2), \ldots, (x_n, y_n)\}$ of paired values from the random variables Y and the variable x, the least squares estimate of the regression line

$$\mu_{Y|x} = A + Bx$$

(provided that it exists) is the line

$$\hat{y} = a + bx$$

where

$$b = \frac{n \sum_{i=1}^{n} x_i y_i - \left(\sum_{i=1}^{n} x_i\right)\left(\sum_{i=1}^{n} y_i\right)}{n \sum_{i=1}^{n} x_i^2 - \left(\sum_{i=1}^{n} x_i\right)^2}$$

$$a = \bar{y} - b\bar{x}$$

$$\bar{x} = \frac{\sum_{i=1}^{n} x_i}{n} \quad \text{and} \quad \bar{y} = \frac{\sum_{i=1}^{n} y_i}{n}$$

EXAMPLE 14.3

Find the least squares estimate of the regression line for the data of Example 14.2. Draw this line onto a scatter diagram of the data. Include drawings of the lines L_1 and L_2 given in the example. Compare the fits of these lines to the given data.

SOLUTION

(a)   An orderly way of finding the least squares estimate of the regression line is to begin by constructing a table for values of x_i, y_i, x_i^2, and $x_i y_i$. Fill in the values for x_i and y_i. Then, compute and fill in the values for x_i^2 and $x_1 y_1$. The table of values for this example follows:

i	x_i	y_i	x_i^2	$x_i y_i$
1	3	4	9	12
2	4	7	16	28
3	8	7	64	56
4	12	11	144	132
5	17	10	289	170

The next step is to use values from this table to compute and fill in the values for the one that follows, as we have done here.

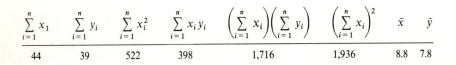

$\sum_{i=1}^{n} x_1$	$\sum_{i=1}^{n} y_i$	$\sum_{i=1}^{n} x_i^2$	$\sum_{i=1}^{n} x_i y_i$	$\left(\sum_{i=1}^{n} x_i\right)\left(\sum_{i=1}^{n} y_i\right)$	$\left(\sum_{i=1}^{n} x_i\right)^2$	$\bar{x}$	$\bar{y}$
44	39	522	398	1,716	1,936	8.8	7.8

Now we substitute these values into the formulas for a and b to obtain

$$b = \frac{n\sum_{i=1}^{n} x_i y_i - \left(\sum_{i=1}^{n} x_i\right)\left(\sum_{i=1}^{n} y_i\right)}{n\sum_{i=1}^{n} x_i^2 - \left(\sum_{i=1}^{n} x_i\right)^2} = \frac{(5)(398) - (44)(39)}{(5)(522) - (1936)} = .407$$

and

$$a = \bar{y} - b\bar{x} = (7.8) - (.407)(8.8) = 4.218$$

Therefore, the least squares estimate of the regression line is

$$\hat{y} = 4.218 + .407x$$

(b)   See Figure 14.6. To compare the fit of these lines to the given data, let us also compute the sum of the squares of errors for the least squares estimate of the regression line. The following table of values shows that the value of S_{e^2} for the least squares estimate of the regression line is 8.523. Therefore, of the lines L_1, L_2, and the least squares estimate of the regression line, the latter offers the best fit.

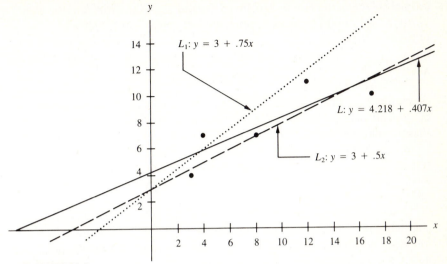

FIGURE 14.6
L: $y = 4.218 + .407x$, the least squares estimate of the regression line
$L_1: y = 3 + .75x$
$L_2: y = 3 + .5x$

i	x_i	y_i	$(y_i - 4.218 - .407x_i)$	e_i^2
1	3	4	-1.439	2.071
2	4	7	1.154	1.332
3	8	7	$-.474$	.225
4	12	11	1.898	3.602
5	17	10	-1.137	1.293

$$S_{e^2} = \sum_{i=1}^{n} e_i^2 = 8.523$$

14.1.3 Confidence Limits for Least Squares Estimates

Having obtained a least squares estimate of the regression line, the next step of a bivariate linear regression analysis of data is the computation of confidence limits that are useful in evaluating the goodness of the estimate. This step of the analysis is very important because we have to come to terms with the fact that the line

$$\hat{y} = a + bx$$

obtained by the least squares method is only an estimate of the true regression line

$$\mu_{Y|x} = A + Bx$$

In other words, the constant a in the estimated regression line is only an estimate of the parameter A in the true regression line; the constant b is only an estimate of the parameter B; and, for each value x_0 of the variable x, the value $\hat{y}_0$ obtained from the estimated regression line $\hat{y} = a + bx$ is only an estimate of the conditional expectation $\mu_{Y|x_0}$. Furthermore, if the estimated regression line $\hat{y} = a + bx$ is used as a formula for predicting a value y_0 for a given value x_0, then $\hat{y}_0$ is only an estimate of y_0.

This section outlines the procedures for obtaining the confidence limits for estimating the constants A and B in the regression line, and for estimating the error in predicting, for each value of x, the values of $\mu_{Y|x}$ and y. The validity of these confidence limits, however, depends on an assumption that for each value of x, the distribution of Y is normal and the mean of Y is $A + Bx$. In terms of the random variable ε in the regression equation $Y = A + Bx + \varepsilon$, the assumption is equivalent to the statement that the distribution of ε is normal and the mean of ε is zero.

The statements of the formulas for the confidence intervals, as well as the computations of the confidence limits, are greatly simplified if we introduce the following notations. In each of these expressions, we assume that $x_1, x_2, \ldots,$ and x_n are n selected values of x, that $y_1, y_2, \ldots,$ and y_n are the corresponding observed values of the random variable Y, and that the constants a and b are least squares estimates of the respective constants A and B of the regression equation $Y = A + Bx + \varepsilon$.

$$S_{xx} = \sum_{i=1}^{n} (x_i - \bar{x})^2 = \sum_{i=1}^{n} x_i^2 - \left[\left(\sum_{i=1}^{n} x_i\right)^2 \bigg/ n\right]$$

$$S_{yy} = \sum_{i=1}^{n} (y_i - \bar{y})^2 = \sum_{i=1}^{n} y_i^2 - \left[\left(\sum_{i=1}^{n} y_i\right)^2 \bigg/ n\right]$$

$$S_{xy} = \sum_{i=1}^{n} (x_i - \bar{x})(y_i - \bar{y}) = \sum_{i=1}^{n} x_i y_i - \left[\left(\sum_{i=1}^{n} x_i\right)\left(\sum_{i=1}^{n} y_i\right) \bigg/ n\right]$$

$$S_{e^2} = \sum_{i=1}^{n} (y_i - a - bx_i)^2$$

$$S^* = \sqrt{S_{e^2}/(n-2)}$$

It follows (see Exercise 11) that

$$b = S_{xy}/S_{xx}$$

$$S_{e^2} = S_{yy} - bS_{xy}$$

and

$$S^* = \sqrt{(S_{yy} - bS_{xy})/(n-2)}$$

The following comments and observations may contribute to an understanding of the meaning of these expressions.

1. Consider the random variable ε in the regression equation $Y = A + Bx + \varepsilon$. We assume that ε is normally distributed with a mean of zero.

Thus, if we write $\varepsilon = Y - A - Bx$, and substitute a and b for the respective values of A and B, then

$$S^* = \sqrt{\frac{\sum_{i=1}^{n} (y_i - a - bx_i)^2}{n - 2}}$$

is an unbiased estimate of the standard deviation of the random variable ε. (We divide by $n - 2$ because the substitution of a and b for A and B removes 2 degrees of freedom.) For this reason, we call S^* the *standard error of the estimate*. Since the mean of ε is assumed to be zero, a small value for S^* would mean that we should expect observed values of Y to be close to the predicted value of $A + Bx$, whereas a large value for S^* would mean that we cannot expect observed values of Y to be consistently close to the predicted value of $A + Bx$.

2. If we let s_x^2 and s_y^2 be the respective sample variances of the x measurements and the y measurements, then

$$s_x^2 = \frac{S_{xx}}{n - 1} \quad \text{and} \quad s_y^2 = \frac{S_{yy}}{n - 1}$$

If we let s_{xy}^2 be the sample covariance of the s and y measurements, then

$$s_{xy}^2 = \frac{S_{xy}}{n - 1}$$

The $1 - \alpha$ confidence intervals for A, B, $\mu_{Y|x_0}$, and y_0, respectively, are given below in terms of the sample data and the values a, b, y_0, and $\hat{y}_0$.

Confidence intervals for least squares estimates   Let $(x_1, y_1), (x_2, y_2), \ldots,$ (x_n, y_n) be a sample of paired values of the variable x and the random variable Y. Let $\hat{y} = a + bx$ be a least squares estimate (computed from sample data) of the regression line $\mu_{Y|x} = A + Bx$.

1. A $1 - \alpha$ confidence interval for A is

$$a - \frac{t_{\alpha/2, n-2}\, S^* \sqrt{\sum_{i=1}^{n} x_i^2}}{\sqrt{nS_{xx}}} < A < a + \frac{t_{\alpha/2, n-2}\, S^* \sqrt{\sum_{i=1}^{n} x_i^2}}{\sqrt{nS_{xx}}}$$

2. A $1 - \alpha$ confidence interval for B is

$$b - \frac{t_{\alpha/2, n-2}\, S^*}{\sqrt{S_{xx}}} < B < b + \frac{t_{\alpha/2, n-2}\, S^*}{\sqrt{S_{xx}}}$$

3. For each value x_0, let y_0 be the corresponding value of Y, and let $\hat{y}_0 = a + bx_0$. A $1 - \alpha$ confidence interval for $\mu_{Y|x_0}$ is

$$\hat{y}_0 - t_{\alpha/2, n-2}\, S^* \sqrt{\frac{1}{n} + \frac{(x_0 - \bar{x})^2}{S_{xx}}} < \mu_{Y|x_0} < \hat{y}_0 + t_{\alpha|2, n-2}\, S^* \sqrt{\frac{1}{n} + \frac{(x_0 - \bar{x})^2}{S_{xx}}}$$

4. A $1 - \alpha$ confidence interval for predicting the value of y_0 in terms of $\hat{y}_0$ is

$$\hat{y}_0 - t_{\alpha/2, n-2}\, S^* \sqrt{1 + \frac{1}{n} + \frac{(x_0 - \bar{x})^2}{S_{xx}}} < y_0 < \hat{y}_0$$
$$+ t_{\alpha/2, n-2}\, S^* \sqrt{1 + \frac{1}{n} + \frac{(x_0 - \bar{x})^2}{S_{xx}}}$$

The derivations of these confidence intervals are considered in Exercises 12, 13, and 14.

EXAMPLE 14.4

The prices per share of 9 stocks from a certain selected group of "growth" stocks were as follows on January 1, 1981, and on January 1, 1982:

i Stock	x_i Price per share Jan. 1981	y_i Price per share Jan. 1982
1	22.40	28.15
2	45.70	54.75
3	15.65	20.05
4	48.25	56.95
5	32.31	40.06
6	55.14	65.45
7	17.65	21.75
8	25.78	30.95
9	13.17	18.10

Let x be the price per share in January, 1981, and let Y be the price per share in January, 1982. Also, assume that there is a linear regression of Y on x and that the regression line is $\mu_{Y|x} = A + Bx$. Furthermore, assume that for each fixed value of x the distribution of Y is normal.

(a) Use the data in the table to find a least squares estimate of the regression line.

(b) Find a .90 confidence interval for A.

(c) Find a .90 confidence interval for B.

(d) If in January, 1981, the price per share of a certain stock of this type was \$30, what could we assert with a confidence of .90 about the expected price per share of this stock in January, 1982?

(e) If in January, 1981, the price per share of a certain stock of this type was \$30, what could we assert with a confidence of .90 about the price per share of this stock in January, 1982?

SOLUTION

A simple and orderly way of solving a problem of this type is to compute the values of x_i^2, y_i^2, and $x_i y_i$ for $i = 1, 2, \ldots, n$ (in this case $n = 9$) and place them in a table as we have done.

i	x_i	y_i	x_i^2	y_i^2	$x_i y_i$
1	22.40	28.15	501.76	792.42	630.56
2	45.70	54.75	2088.49	2997.56	2502.08
3	15.65	20.05	244.92	402.00	313.78
4	48.25	56.95	2328.06	3243.30	2747.84
5	32.31	40.06	1043.94	1604.80	1294.34
6	55.14	65.45	3040.42	4383.70	3608.91
7	17.65	21.75	311.52	473.06	383.89
8	25.78	30.95	664.61	957.90	797.89
9	13.17	18.10	173.45	327.61	238.38
	276.05	336.21	10397.17	15182.35	12517.67
	$\sum_{i=1}^{n} x_i$	$\sum_{i=1}^{n} y_i$	$\sum_{i=1}^{n} x_i^2$	$\sum_{i=1}^{n} y_i^2$	$\sum_{i=1}^{n} x_i y_i$

Next, compute the values of $\sum_{i=1}^{n} x_i$, $\sum_{i=1}^{n} y_i$, $\sum_{i=1}^{n} x_i^2$, $\sum_{i=1}^{n} y_i^2$, and $\sum_{i=1}^{n} x_i y_i$, and place them at the appropriate places at the bottom of the table. Then

$$b = \frac{n \sum_{i=1}^{n} x_i y_i - \left(\sum_{i=1}^{n} x_i\right)\left(\sum_{i=1}^{n} y_i\right)}{n \sum_{i=1}^{n} x_i^2 - \left(\sum_{i=1}^{n} x_i\right)^2} = \frac{(9)(12517.67) - (92810.77)}{(9)(10397.17) - (276.05)^2}$$

$$= 1.1426$$

$$a = \bar{y} - b\bar{x} = \frac{336.21}{9} - 1.1426 \cdot \frac{276.05}{9}$$

$$= 2.31$$

$$S_{xx} = \sum_{i=1}^{n} x_i^2 - \left[\left(\sum_{i=1}^{n} x_i\right)^2 / n\right] = 10397.17 - \frac{(276.05)^2}{9}$$

$$= 1930.10$$

$$S_{yy} = \sum_{i=1}^{n} y_i^2 - \left[\left(\sum_{i=1}^{n} y_i\right)^2 / n\right] = 15082.35 - \frac{(336.21)^2}{9}$$

$$= 2622.67$$

$$S_{xy} = \sum_{i=1}^{n} x_i y_i - \left[\left(\sum_{i=1}^{n} x_i \right) \left(\sum_{i=1}^{n} y_i \right) \bigg/ n \right] = 12517.67 - \frac{(276.05)(336.21)}{9}$$

$$= 2205.36$$

$$S_{e^2} = S_{yy} - bS_{xy} = 2622.67 - (1.1426)(2205.36) = 102.83$$

$$S^* = \sqrt{S_{e^2}/(n-2)} = 3.83$$

$$t_{\alpha/2, n-2} = t_{.05, 7} = 1.895$$

(a)   We have already computed the value of a and b above; therefore, the least squares estimate of the regression line is

$$\hat{y} = 2.31 + 1.14x$$

(b)   By substituting the above values into the formula for a $1 - \alpha$ confidence interval for A, we obtain

$$2.31 - \frac{(1.895)(3.83)\sqrt{10397.17}}{\sqrt{(9)(1930.10)}} < A < .86 + \frac{(1.895)(3.83)10397.17}{\sqrt{(9)(1930.10)}}$$

or

$$-3.31 < A < 7.93$$

as a .90 confidence interval for A.

(c)   By substituting the preceding values into the formula for a $1 - \alpha$ confidence interval for B, we obtain

$$1.14 - \frac{(1.895)(3.83)}{\sqrt{1930.10}} < B < 1.14 + \frac{(1.895)(3.83)}{\sqrt{1930.10}}$$

or

$$.97 < B < 1.31$$

as a .90 confidence interval for B.

(d)   With $x_0 = 30$, the value of $\hat{y}_0$ is $\hat{y}_0 = 2.31 + 1.14x_0 = 36.59$. Therefore, we could have asserted with a confidence of .90 that

$$36.59 - (1.895)(3.83)\sqrt{\frac{1}{9} + \frac{(30 - 30.67)^2}{1039.10}} < \mu_{Y|x_0}$$

$$< 36.59 + (1.895)(3.83)\sqrt{\frac{1}{9} + \frac{(30 - 30.67)^2}{1039.10}}$$

or

$$34.16 < \mu_{Y|x_0} < 39.01$$

In other words, we could have asserted, with a confidence of .90, that if the price per share of a stock of this type was \$30 in January, 1981, then the expected price per share in January, 1982, would be between \$34.16 and \$39.01.

(e)   By substituting the preceding values into the formula for a $1 - \alpha$ confidence interval for y_0, we obtain

$$36.59 - (1.895)(3.83)\sqrt{1 + \frac{1}{9} + \frac{(30 - 30.67)^2}{1039.10}} < y_0$$

$$< 36.59 + (1.895)(3.83)\sqrt{1 + \frac{1}{9} + \frac{(30 - 30.67)^2}{1039.10}}$$

or

$$28.93 < y_0 < 44.25$$

as a .90 confidence interval for y_0. In other words, we could have asserted with a confidence of .90 that if the price per share of a stock of this type was \$30 on January 1, 1981, then the price per share on January 1, 1982, would be between \$28.93 and \$44.25.

Exercises

1.   The level of pollution (measurement y in coded units) at a certain midtown street corner and the corresponding temperature and humidity index (measurement x) were recorded at noon on 10 different days as shown in the table that follows. Construct a scatter diagram of these data. Is this set of data suitable for a linear regression analysis?

Day	1	2	3	4	5	6	7	8	9	10
Temperature and humidity index	77	95	30	45	85	50	65	60	63	82
Level of pollution	1.5	4.0	.5	1.4	2.0	.8	2.5	2.0	1.7	2.8

2.   Seven measurements were taken of the maximum amounts in grams (measurement y) of a certain chemical compound that dissolved in 1 liter of water at various temperatures measured in degrees Celsius (measurement x). Construct a scatter diagram of these data, shown in the table that follows. Is this set of data suitable for a linear regression analysis?

Measurement	1	2	3	4	5	6	7
Temperature	15	30	45	60	70	85	90
Grams of chemical	.3	1.0	1.3	1.9	3.5	7.0	8.1

3. A savings bank wished to examine the relationship between incomes (measurements x) and amounts of savings deposited in a savings bank (measurements y) by heads of families. A survey of 8 randomly selected depositors who are family heads yielded the following results. Construct a scatter diagram of these data. Is this set of data suitable for a linear regression analysis?

Family	1	2	3	4	5	6	7	8
Income (in thousands)	30	20	25	15	35	40	37	27
Savings (in thousands)	3.0	1.5	1.5	1.0	4.0	6.0	5.5	3.5

4. The names of 11 persons who took the SAT examinations in 1967 were randomly selected from file records. Each person's combined score on the examination (measurement x in units of 100) along with his or her income in 1981 (measurement y in units of $1,000) are shown in the following table:

Person	1	2	3	4	5	6	7	8	9	10	11
SAT score (1967)	7.5	9	15	10	6	7	13	5	8	14.5	10.5
Income (1981)	36	16	10	22	12	14	8	4	24	15	40

Construct a scatter diagram of the data. Does the diagram indicate the likelihood of a linear relationship between the variables?

5. Find the least squares estimate of the regression line for the data of Exercise 1. Use the estimated regression line to estimate the expected level of pollution at noon on a day when the temperature and humidity index is (a) 75;   (b) 50.

6. Find the least squares estimate of the regression line for the data of Exercise 3. Use the estimated regression line to predict the expected amount of savings in savings banks of the head of a family whose income is (a) $30,000;   (b) $45,000.

7. (a)   Find the standard error (i.e., S^*) for the estimated regression line of Exercise 5.
(b)   Do the same for the regression line of Exercise 6.

8. (a)   Assuming that a regression line $\mu_{Y|x} = A + Bx$ exists for the random variable Y and the variable x of Exercise 1, find .90 confidence intervals for A, B, $\mu_{Y|x_0}$, and y_0 in terms of the data obtained in Exercise 5. Assume $x_0 = 75$.
(b)   Do the same for the random variable Y and the variable x of Exercise 3. Assume $x_0 = 30$.

9. If the least squares estimate of the regression line of Exercise 5 is used to estimate the level of pollution when the temperature and humidity index is 75, what can we assume, with a confidence of .90, about the maximum

size of an error in the estimate? (*Hint:* the maximum error in the estimate is the maximum value of $|\hat{y}_0 - y_0|$.)

10. If the estimated regression line of Exercise 6 is used to make a prediction concerning the savings of the head of a family whose income is $30,000, what can we assert, with a confidence of .90, about the maximum error of the prediction?

11. (a) Show that $b = S_{xy}/S_{xx}$.
    (b) Show that $\sum_{i=1}^{n} (y_i - a - bx_i)^2 = S_{yy} - bS_{xy}$.

    (*Hint:* replace a with $\bar{y} - b\bar{x}$. Then, regroup the terms so that we have $(y_i - a - bx_i)^2 = (y_i - \bar{y} + b\bar{x} - bx_i)^2 = [(y_i - \bar{y}) - b(x_i - \bar{x})]^2$. Next, square the binomial and use the fact that $b = S_{xy}/S_{xx}$.)

12. Suppose that $\mu_{Y|x} = A + Bx$ is the true regression line and that $\hat{y} = a + bx$ is the least squares estimate of the regression line. Then, $b = S_{xy}/S_{xx}$ and $a = \bar{y} - b\bar{x}$. Since b and a are computed from sample data, we may assume that they are values of random variables B^* and A^*, respectively. It can be shown (see listed references for advanced texts in mathematical statistics) that both A^* and B^* are normally distributed random variables with means $E(A^*) = A$ and $E(B^*) = B$, respectively. Furthermore, the statistic

$$\frac{A^* - A}{S^* \sqrt{\sum_{i=1}^{n} x_i^2 / nS_{xx}}}$$

has a t distribution with $n - 2$ degrees of freedom, and the statistic

$$\frac{B^* - B}{S^*/\sqrt{S_{xx}}}$$

has a t distribution with $n - 2$ degrees of freedom. Use these statistics to derive the confidence intervals for A and B.

13. (Continuation of Exercise 12.) The confidence interval for $\mu_{Y|x_0}$ uses $\hat{y}_0$ as an estimate. Since $\hat{y}_0$ is computed from sample data, it is the value of a random variable $\hat{Y}_0$. It can be shown that the statistic

$$\frac{\hat{Y}_0 - \mu_{Y|x_0}}{S^*\sqrt{(1/n) + [(x_0 - \bar{x})^2/S_{xx}]}}$$

has a t distribution with $n - 2$ degrees of freedom. Use this statistic to derive the confidence interval for $\mu_{Y|x_0}$.

14. (Continuation of Exercise 13.) The confidence interval for y_0 uses $\hat{y}_0$ as an estimate. Since $\hat{y}_0$ is computed from sample data, it is the value of a random variable $\hat{Y}_0$. The value y_0 is itself a value of a random variable Y_0. Therefore, $\hat{Y}_0 - Y_0$ is a random variable. It can be shown that the mean of

$\hat{Y}_0 - Y_0$ is zero. Furthermore, the statistic

$$\frac{\hat{Y}_0 - Y_0}{S^*\sqrt{1 + (1/n) + [(x_0 - \bar{x})^2/S_{xx}]}}$$

has t distribution with $n - 2$ degrees of freedom. Use this statistic to derive the confidence interval for y_0.

14.2   Multivariate Linear Regression

In many applied problems, the value of a measurement y depends not on one, but on several measurements $x_1, x_2, \ldots,$ and x_m. For example, the growth of certain plants may depend on the mineral composition of the soil, the temperature, and the moisture. When this is the case, and when sample data indicate a linear relationship between the measurements (in the sense that we can reasonably hope to predict an approximate value of y with a linear equation $y = a + b_1 x_1 + b_2 x_2 + \cdots + b_m x_m$), then the object of a linear regression analysis (in this case called **multivariate,** or **multiple, linear regression analysis**) is to find an estimate of this linear equation. As in the bivariate case, there are certain underlying theoretical assumptions connected with a multivariate linear regression analysis. We assume that the measurement y is a value of a random variable Y, and—furthermore—that the conditional expectation of Y is linearly dependent on the values $x_1, x_2, \ldots,$ and x_m. In other words, we assume that there are constants $A, B_1, \ldots,$ and B_m such that

$$\begin{bmatrix} \text{the expected value of } Y, \text{ given} \\ \text{the values } x_1, x_2, \ldots, \text{ and } x_m, \end{bmatrix} = A + B_1 x_1 + B_2 x_2 + \cdots + B_m x_m$$

or, stated equivalently in the notations of mathematical statistics, that

$$\mu_{Y \mid x_1, x_2, \ldots, x_m} = A + B_1 x_1 + B_2 x_2 + \cdots + B_m x_m$$

Operating within this framework, the object of a multivariate linear regression analysis is to find a linear equation

$$\hat{y} = a + b_1 x_1 + b_2 x_2 + \cdots + b_n x_n$$

that is an estimate of the linear regression equation

$$\mu_{Y \mid x_1, x_2, \ldots, x_n} = A + B_1 x_1 + B_2 x_2 + \cdots + B_n x_n$$

The estimated regression equation can then be used to predict values of y with values of $\hat{y}$. As in the bivariate case of two variables, this is often accomplished through the use of a least squares method. In the two-variable case, the least squares method fits a line to a set of points in two-dimensional space. The multivariate case requires the fitting of a hyperplane to a set $\{(y_1, x_{11}, x_{21}, \ldots, x_{m1}), (y_2, x_{12}, x_{22}, \ldots, x_{m2}), \ldots, (y_n, x_{1n}, x_{2n}, \ldots, x_{mn})\}$ of points in $(m + 1)$-dimensional space. The method, however, remains essentially the same. Given a point $(y_i, x_{1i}, x_{2i}, \ldots, x_{mi})$ and a hyperplane $y = a + b_1 x_1 + b_2 x_2 + \cdots + b_m x_m$, the error in the fit of the hyperplane to the point is defined to be the distance measured along the line parallel to the y-axis from the point to the hyperplane. That is

$$e_i = |y_i - a - b_1 x_{1i} - b_2 x_{2i} - \cdots - b_m x_{mi}|$$

The sum of the squares of errors is then defined to be

$$S_{e^2} = \sum_{i=1}^{n} e_i^2 = \sum_{i=1}^{n} (y_i - a - b_i x_{1i} - b_2 x_{2i} - \cdots - b_m x_{mi})^2$$

The least squares estimate of the regression equation is the equation

$$\hat{y} = a + b_1 x_1 + b_2 x_2 + \cdots + b_m x_m$$

where $a, b_1, b_2, \ldots,$ and b_m are the values that minimize the value of S_{e^2}. To find the values of $a, b_1, b_2, \ldots,$ and b_m that would minimize the value of S_{e^2}, we would proceed as in the bivariate case. That is, think of S_{e^2} as a function of the variables $a, b_1, b_2, \ldots,$ and b_m. Then, differentiate the function S_{e^2} with respect to each of the variables $a, b_1, \ldots,$ and b_m, respectively, and set the derivatives equal to zero. This produces a system of $m + 1$ equations in $m + 1$ unknowns. By solving this system of equations we would obtain the desired values of a, $b_1, b_2, \ldots,$ and b_m. In practice, these values are usually obtained through the use of computer programs that are available to researchers and statisticians.

EXAMPLE 14.5

Set up a system of equations for finding the values of a, b_1, and b_2 for a least squares estimate of a regression equation for the following sets of data:

(a)

$$\{(y_1, x_{11}, x_{21}), (y_2, x_{12}, x_{22}), \ldots, (y_n, x_{1n}, x_{2n})\}$$

(b)

i	y_i	x_{1i}	x_{2i}
1	14	2	3
2	19	6	2
3	23	4	5
4	11	3	1
5	16	1	4

SOLUTION

(a) Following the multivariate linear regression formula, we have

$$S_{e^2} = \sum_{i=1}^{n} (y_i - a - b_1 x_{1i} - b_2 x_{2i})^2$$

Therefore, we have:

$$\frac{\partial S_{e^2}}{\partial a} = -2 \sum_{i=1}^{n} (y_i - a - b_1 x_{1i} - b_2 x_{2i})$$

$$\frac{\partial S_{e^2}}{\partial b_1} = -2 \sum_{i=1}^{n} (y_i - a - b_1 x_{1i} - b_2 x_{2i}) x_{1i}$$

$$\frac{\partial S_{e^2}}{\partial b_2} = -2 \sum_{i=1}^{n} (y_i - a - b_1 x_{1i} - b_2 x_{2i}) x_{2i}$$

By setting the above derivatives equal to zero and rearranging terms we obtain the following system of equations (called the normal equations):

$$na + b_1 \sum_{i=1}^{n} x_{1i} + b_2 \sum_{i=1}^{n} x_{2i} = \sum_{i=1}^{n} y_i$$

$$a \sum_{i=1}^{n} x_{1i} + b_1 \sum_{i=1}^{n} x_{1i}^2 + b_2 \sum_{i=1}^{n} x_{1i} x_{2i} = \sum_{i=1}^{n} x_{1i} y_i$$

$$a \sum_{i=1}^{n} x_{2i} + b_1 \sum_{i=1}^{n} x_{1i} x_{2i} + b_2 \sum_{i=1}^{n} x_{2i}^2 = \sum_{i=1}^{n} x_{2i} y_i$$

(b) From the table of values we obtain

$$\sum_{i=1}^{5} x_{1i} = 16, \quad \sum_{i=1}^{5} x_{2i} = 15, \quad \sum_{i=1}^{5} y_i = 83, \quad \sum_{i=1}^{5} x_{1i}^2 = 66,$$

$$\sum_{i=1}^{5} x_{2i}^2 = 55, \quad \sum_{i=1}^{5} x_{1i} x_{2i} = 45, \quad \sum_{i=1}^{5} x_{1i} y_i = 283,$$

$$\sum_{i=1}^{5} x_{2i} y_i = 270, \quad n = 5$$

Therefore, the system of equations for this set of data is

$$5a + 16b_1 + 15b_2 = 83$$
$$16a + 66b_1 + 45b_2 = 283$$
$$15a + 45b_1 + 55b_2 = 270$$

Exercises

1. A matrix method for solving the system

$$a_{11}x_1 + a_{12}x_2 + a_{13}x_3 = c_1$$
$$a_{21}x_1 + a_{22}x_2 + a_{23}x_3 = c_2$$
$$a_{31}x_1 + a_{32}x_2 + a_{33}x_3 = c_3$$

of equations states that the solutions are

$$x_1 = \frac{\begin{vmatrix} c_1 & a_{12} & a_{13} \\ c_2 & a_{22} & a_{23} \\ c_3 & a_{32} & a_{33} \end{vmatrix}}{|A|}, \qquad x_2 = \frac{\begin{vmatrix} a_{11} & c_1 & a_{13} \\ a_{21} & c_2 & a_{23} \\ a_{31} & c_3 & a_{33} \end{vmatrix}}{|A|},$$

$$x_3 = \frac{\begin{vmatrix} a_{11} & a_{12} & c_1 \\ a_{21} & a_{22} & c_2 \\ a_{31} & a_{32} & c_3 \end{vmatrix}}{|A|}$$

where

$$|A| = \text{the determinant of } A = \begin{vmatrix} a_{11} & a_{12} & a_{13} \\ a_{21} & a_{22} & a_{23} \\ a_{31} & a_{32} & a_{33} \end{vmatrix}$$

and where for any matrix

$$D = \begin{bmatrix} d_{11} & d_{12} & d_{13} \\ d_{21} & d_{22} & d_{23} \\ d_{31} & d_{32} & d_{33} \end{bmatrix}$$

the determinant of D is

$$|D| = \begin{vmatrix} d_{11} & d_{12} & d_{13} \\ d_{21} & d_{22} & d_{23} \\ d_{31} & d_{32} & d_{33} \end{vmatrix}$$

$$= d_{11}(d_{22}d_{33} - d_{32}d_{23}) - d_{21}(d_{12}d_{33} - d_{32}d_{13})$$
$$+ d_{31}(d_{12}d_{23} - d_{22}d_{13})$$

Use this method to find the least squares estimate of a regression equation for Example 14.5.

2. The soils of 9 plots of seedlings were treated with varying mixes of nutrient A and nutrient B. After three weeks the seedlings were measured

and their average heights recorded. The data are shown in the table that follows.

Plot	Height (inches)	Nutrient A (grams)	Nutrient B (grams)
1	2.4	1	1
2	3.5	2	1
3	4.6	3	1
4	3.0	1	2
5	3.9	2	2
6	5.1	3	2
7	3.5	1	3
8	4.5	2	3
9	5.5	3	3

Assuming that the expected average height of seedlings is linearly dependent on the amounts of nutrients A and B, find a least squares estimate of the regression equation.

14.3   Curvilinear Regression

If a scatter diagram of sample data of paired measurements shows that the points tend to cluster about a curve rather than a straight line (for example, diagrams (b) and (f) of Figure 14.3), then, of course, it is more appropriate to fit a curve, rather than a straight line, to the data. Depending on the nature of the curve, one may choose to fit a second-degree polynomial (i.e., $y = a + bx + cx^2$), or a third-degree polynomial (i.e., $y = a + bx + cx^2 + dx^3$), or, indeed, any kind of function $[y = f(x)]$ to the points. A least squares method applies here just as it did in the case of a straight line. The procedure is to compute the error e_i (defined as the distance measured along the line parallel to the y-axis from the point (x_i, y_i) to the curve) and find the function that minimizes the sum of the squares of errors. In practice, this feat generally is accomplished through the use of a computer program. A similar procedure applies to the multivariate case of sample data of groups of m measurements.

EXAMPLE 14.6

Show that the normal equations for a least squares method of fitting a second degree polynomial, $y = a + bx + cx^2$, to a set $\{(x_1, y_1), (x_2, y_2), \ldots, (x_n, y_n)\}$ of points are

$$na + b \sum_{i=1}^{n} x_i + c \sum_{i=1}^{n} x_i^2 = \sum_{i=1}^{n} y_i$$

$$a \sum_{i=1}^{n} x_i + b \sum_{i=1}^{n} x_i^2 + c \sum_{i=1}^{n} x_i^3 = \sum_{i=1}^{n} x_i y_i$$

$$a \sum_{i=1}^{n} x_i^2 + b \sum_{i=1}^{n} x_i^3 + c \sum_{i=1}^{n} x_i^4 = \sum_{i=1}^{n} x_i^2 y_i$$

that is, show that if the constants a, b, and c satisfy the above set of equations, then $y = a + bx + cx^2$ is the second-degree polynomial that provides the best fit (in the sense of least squares fit) to the given set of points.

SOLUTION

The error for the point (x_i, y_i) in fitting the curve $y = a + bx + cx^2$ to a set of points is

$$e_i = |y_i - (a + bx_i + cx_i^2)| = |y_i - a - bx_i - cx_i^2|$$

Therefore, the sum of the squares of errors is

$$S_{e^2} = \sum_{i=1}^{n} e_i^2 = \sum_{i=1}^{n} (y_i - a - bx_i - cx_i^2)^2$$

From this point on, the problem is the same as the problem of Example 14.5(a) if we make the following replacement:

Example 14.5(a)		Example 14.6
x_{1i}	$\longrightarrow$	x_i
x_{2i}	$\longrightarrow$	x_i^2
a	$\longrightarrow$	a
b_1	$\longrightarrow$	b
b_2	$\longrightarrow$	c

Therefore, the normal equations are the ones given.

Exercises

1. Find the normal equations for a least squares method of fitting a second-degree curve $y = a + bx + cx^2$ to the following set of points: $\{(0, 3), (1, 3.1), (2, 3.7), (3, 4.3), (4, 7)\}$.

2. Use the matrix method described in Exercise 1 of Section 14.2 to find the least squares regression equation for Exercise 1 of Section 14.3.

14.4   Correlation Analysis

While bivariate linear regression analysis assumes—in the sense described in Section 14.1—a linear relationship between corresponding values of paired measurements and proceeds from that point to estimate the regression line, correlation analysis is concerned with measuring the degree to which we can even assume that there is a linear relationship between corresponding values of paired measurements. This analysis is based on the value of a quantity that is called the correlation coefficient of two random variables. The balance of this chapter includes a definition of the correlation coefficient of two random variables, a discussion of how it is used to measure the degree of association between two random variables, a method by which we may estimate the correlation coefficient from sample data, and a significance test concerning the correlation coefficient that may also be used to test for the independence of bivariate normal random variables.

14.4.1   The Correlation Coefficient

Definition 14.1   Let X and Y be any two random variables with means μ_x and μ_y, respectively, and variances σ_x^2 and σ_y^2, respectively. The **correlation coefficient** of X and Y is the quantity

$$\rho(X, Y) = E\left(\frac{X - \mu_x}{\sigma_x} \cdot \frac{Y - \mu_y}{\sigma_y}\right)$$

Note: the covariance of X and Y is defined to be the quantity

$$\text{Cov}(X, Y) = E[(X - \mu_x)(Y - \mu_y)] = E(XY) - \mu_x\mu_y$$

Therefore, we can also define the correlation coefficient of X and Y as

$$\rho(X, Y) = \frac{\text{Cov}(X, Y)}{\sigma_x\sigma_y} = \frac{E(XY) - \mu_x\mu_y}{\sigma_x\sigma_y}$$

See Section 8.7 of Chapter 8 for a definition of $E(XY)$.

Theorem 14.1 is concerned with the properties of the correlation coefficient that are used to measure the degree of association between random variables

Theorem 14.1
    (a)   *For all random variables X and Y, we have*

$$|\rho(X, Y)| \leq 1$$

and if $\rho(X, Y) = \pm 1$, then there are constants a and b such that

$$P(Y = a + bX) = 1$$

    (b)   *If X and Y are independent random variables, then*

$$\rho(X, Y) = 0$$

(c)   If X and Y are bivariate normal random variables and $\rho(X, Y) = 0$, then X and Y are independent.

Proof
(a)   Consider the standardized random variables

$$X^* = \frac{X - \mu_x}{\sigma_x} \quad \text{and} \quad Y^* = \frac{Y - \mu_y}{\sigma_y}$$

Then, we have

$$E(X^*) = 0, \text{Var}(X^*) = E[(X^*)^2] = 1, E(Y^*) = 0, \text{Var}(Y^*) = E[(Y^*)^2] = 1,$$
$$E(X^* + Y^*) = E(X^*) + E(Y^*) = 0, \quad \text{and} \quad \rho(X, Y) = E(X^*Y^*)$$

Also

$$\begin{aligned}
\text{Var}(X^* + Y^*) &= E\{[(X^* + Y^*) - E(X^* + Y^*)]^2\} \\
&= E[(X^* + Y^*)^2] \\
&= E[(X^*)^2 + 2X^*Y^* + (Y^*)^2] \\
&= E[(X^*)^2] + 2E(X^*Y^*) + E[(Y^*)^2] \\
&= 1 + 2\rho(X, Y) + 1 \\
&= 2 + 2\rho(X, Y)
\end{aligned}$$

By a similar argument, we have

$$\text{Var}(X^* - Y^*) = 2 - 2\rho(X, Y)$$

Now, if the variance of any random variable is defined, it is a nonnegative number. Therefore

$$\text{Var}(X^* + Y^*) \geq 0 \quad \text{and} \quad \text{Var}(X^* - Y^*) \geq 0$$

It follows that

$$2 + 2\rho(X, Y) \geq 0, \quad 2 - 2\rho(X, Y) \geq 0$$

and we have

$$-1 \leq \rho(X, Y) \geq +1$$

or

$$|\rho(X, Y)| \leq 1$$

The rest of (a) is considered in Exercise 1.

(b)   From Theorem 8.4 of Chapter 8, we saw that if X and Y are independent random variables, then $E(XY) = E(X)E(Y)$. Therefore, it follows that if X and Y are independent random variables, then

$$\rho(X, Y) = \frac{E(XY) - \mu_x\mu_y}{\sigma_x\sigma_y} = \frac{\mu_x\mu_y - \mu_x\mu_y}{\sigma_x\sigma_y} = 0$$

(c)   The parameter ρ in the bivariate normal density function is the correlation coefficient of X and Y. In Section 7.2.7 of Chapter 7 we discussed the fact that bivariate normal random variables X and Y are independent if and only if $\rho = 0$.

With this theorem, it is understandable why the correlation coefficient is used to measure the association between random variables. According to the theorem, if the correlation coefficient is 1, the association between the random variables—except, perhaps, for values in a set of probability zero—is a perfect linear relationship. At the other end of the scale—that is, if the correlation coefficient is 0 and if the random variables are bivariate normal random variables—the random variables are independent. However, if the random variables are not bivariate normal random variables, it is possible for the mean of one random variable to be completely determined as a nonlinear function of the other random variable even though the correlation coefficient may be zero.

Going beyond this theorem, it can also be shown that if X and Y are bivariate normal random variables, the correlation coefficient also provides a measure of the degree to which values of X and Y are concentrated about the regression line. *Note:* if X and Y are bivariate normal random variables with means μ_x and μ_y, respectively, variances σ_x^2 and σ_y^2, respectively, and correlation coefficient ρ, then

$$\mu_{Y|x} = \mu_y + \rho\left(\frac{\sigma_y}{\sigma_x}\right)(x - \mu_x)$$

This implies the existence of a regression line, namely, the line

$$y = \mu_y + \rho\left(\frac{\sigma_y}{\sigma_x}\right)(x - \mu_x)$$

If the correlation coefficient is 1, the values of X and Y will be right on the regression line (except perhaps for a few points belonging to a set with probability zero). If the correlation coefficient is 0, then there will be no detectable concentration of values on the regression line. In between these two extreme values, the correlation coefficient will vary with the degree to which the values are concentrated about the regression line. For these reasons, correlation analysis is most meaningful for bivariate normal random variables.

14.4.2   Estimating the Correlation Coefficient

Suppose that $[(x_1, y_1), (x_2, y_2), \ldots, (x_n, y_n)]$ is a random sample of n paired values of the random variables X and Y. The **sample correlation coefficient** is defined to be the quantity

$$r = \frac{n \sum\limits_{i=1}^{n} x_i y_i - \sum\limits_{i=1}^{n} x_i \sum\limits_{i=1}^{n} y_i}{\sqrt{\left(n \sum\limits_{i=1}^{n} x_i^2 - \sum\limits_{i=1}^{n} x_i \sum\limits_{i=1}^{n} x_i \right) \left(n \sum\limits_{i=1}^{n} y_i^2 - \sum\limits_{i=1}^{n} y_i \sum\limits_{i=1}^{n} y_i \right)}}$$

If we use the notations already defined in the previous sections, then the sample correlation coefficient may also be defined as

$$r = \frac{S_{xy}}{\sqrt{S_{xx} S_{yy}}} = b \sqrt{\frac{S_{xx}}{S_{yy}}}$$

If the random variables X and Y are bivariate normal random variables, then the sample correlation coefficient is a good estimate of the correlation coefficient of the random variables. Confidence intervals for estimating the correlation coefficient with the sample correlation coefficient may be computed by using the fact that for observations from the bivariate normal distribution, the quantity

$$(1/2)\ln[(1 + r)/(1 - r)]$$

is a value of a random variable whose distribution is approximately normal with mean equal to $(1/2)\ln[(1 + \rho)/(1 - \rho)]$ (where ρ is the correlation coefficient of the random variables) and variance equal to $1/(n - 3)$. (See Exercise 4.)

EXAMPLE 14.7

The following table shows the average rating given by students to 12 calculus teachers and the average grade given by these teachers:

Teacher	1	2	3	4	5	6	7	8	9	10	11	12
Rating	3.9	1.5	3.1	2.4	1.7	3.7	2.7	3.1	2.9	3.6	3.9	3.4
Grade	3.5	2.0	3.2	3.0	2.3	3.2	3.0	2.9	3.0	3.3	3.3	3.0

Find the sample correlation coefficient for this set of data.

SOLUTION

The sample correlation coefficient is easily computed from the data entered into the following table of values.

i	x_i	y_i	$x_i y_i$	x_i^2	y_i^2
1	3.9	3.5	13.65	15.21	12.25
2	1.5	2.0	3.00	2.25	4.00
3	3.1	3.2	9.92	9.61	10.24
4	2.4	3.0	7.20	5.76	9.00
5	1.7	2.3	3.91	2.89	5.29
6	3.7	3.2	11.84	13.69	10.24
7	2.7	3.0	8.10	7.29	9.00
8	3.1	2.9	8.99	9.61	8.41
9	2.9	3.0	8.70	8.41	9.00
10	3.6	3.3	11.88	12.96	10.89
11	3.9	3.3	12.87	15.21	10.89
12	3.4	3.0	10.20	11.56	9.00
Sum	35.9	35.7	110.26	114.45	108.21
	$\sum\limits_{i=1}^{n} x_i$	$\sum\limits_{i=1}^{n} y_i$	$\sum\limits_{i=1}^{n} x_i y_i$	$\sum\limits_{i=1}^{n} x_i^2$	$\sum\limits_{i=1}^{n} y_i^2$

$$r = \frac{n \sum\limits_{i=1}^{n} x_i y_i - \sum\limits_{i=1}^{n} x_i \sum\limits_{i=1}^{n} y_i}{\sqrt{\left(n \sum\limits_{i=1}^{n} x_i^2 - \sum\limits_{i=1}^{n} x_i \sum\limits_{i=1}^{n} x_i\right)\left(n \sum\limits_{i=1}^{n} y_i^2 - \sum\limits_{i=1}^{n} y_i \sum\limits_{i=1}^{n} y_i\right)}}$$

$$= \frac{(12)(110.26) - (35.9)(35.7)}{\sqrt{[(12)(114.45) - (35.9)^2][(12)(108.21) - (35.7)^2]}}$$

$$= .92$$

14.4.3   Hypotheses Concerning the Correlation Coefficient: Testing for Independence

Given a random sample from a bivariate normal distribution, it is possible to design a significance test at the α level of significance of the hypotheses

$$H_0: \rho = \rho_0 \quad \text{versus} \quad H_1: \rho \neq \rho_0$$

(where ρ is the correlation coefficient of the joint distribution). Since the quantity

$$(1/2) \ln [(1 + r)/(1 - r)]$$

(where r is the sample correlation coefficient) is a value of a random variable whose distribution is approximately normal with mean equal

$$(1/2) \ln [(1 + \rho)/(1 - \rho)]$$

and variance equal to $1/(n - 3)$, the test procedure is to compute

$$z = \frac{(1/2) \ln [(1 + r)/(1 - r)] - (1/2) \ln [(1 + \rho_0)/(1 - \rho_0)]}{\sqrt{1/(n - 3)}}$$

and reject H_0 if $|z| > z_{\alpha/2}$. (Observe that the value z above may be considered a value of a random variable whose distribution is standard normal. Therefore, the decision rule given here conforms to those outlined in Chapter 13.)

Because a correlation coefficient of zero for bivariate normal random variables implies that the random variables are independent, the procedure given above may also be used to test for the independence of bivariate normal random variables. The appropriate hypotheses would be

$$H_0: \rho = 0 \quad \text{versus} \quad H_1: \rho \neq 0$$

(In other words, let $\rho_0 = 0$.) Rejection of H_0 would be equivalent to rejection of a hypothesis of independence of the random variables.

EXAMPLE 14.8

In considering the answer to Example 14.7, one may feel that a sample correlation coefficient of .92 is rather convincing evidence that there is a high correlation between student rating of teachers and average grades given by the teachers. Nevertheless, one may have reservations about making such a judgment; after all, a sample size of 12 is rather small. Is it possible, then, that student ratings are independent of average grades in spite of the large sample correlation coefficient? To answer this question, use a .005 level of significance to test the null hypothesis that student rating and average grade are independent random variables. Assume a bivariate normal distribution for these random variables.

SOLUTION

We are testing the hypotheses

$$H_0: \rho = 0 \quad \text{versus} \quad H_1: \rho \neq 0$$

Therefore

$$|z| = \left| \frac{(1/2) \ln [(1 + r)/(1 - r)] - (1/2) \ln [(1 + \rho_0)/(1 - \rho_0)]}{\sqrt{1/(n - 3)}} \right|$$

$$= \left| \frac{(1/2) \ln (1.92/.08)}{(1/3)} \right| = 4.767 > 2.81 = z_{.0025} = z_{\alpha/2}$$

The result of the test is that at the .005 level of significance we should reject the null hypothesis that student ratings are independent of average grades given by teachers.

Exercises

1. Prove that if $\rho(X, Y) = \pm 1$, then there are constants a and b such that $P(Y = a + bX) = 1$. (*Hint:* first show that if W is a random variable with $E(W) = \mu$ and $\text{Var}(W) = 0$, then $P(W = \mu) = 1$.)

2. Find the sample correlation coefficient of the data of these exercises in Section 14.1: (a) Exercise 1; (b) Exercise 2; (c) Exercise 3; (d) Exercise 4.

3. Unless X and Y are bivariate normal random variables, a zero correlation coefficient does not necessarily imply independence of the random variables. For example, suppose that X is a discrete random variable with probability function

$$f(x) = \begin{cases} 1/2 & \text{for} \quad x = -1 \\ 1/2 & \qquad\quad x = +1 \end{cases}$$

Let $Y = X^2$. Show that $\rho(X, Y) = 0$. (In other words, here is an example of two dependent random variables with a zero correlation coefficient.)

4. For bivariate normal random variables, the quantity $(1/2)\ln[(1 + r)/(1 - r)]$ is a value of a random variable whose distribution is approximately normal with mean equal to $(1/2)\ln[(1 + \rho)/(1 - \rho)]$ and variance equal to $1/(n - 3)$. Use this fact to show that a $1 - \alpha$ confidence interval for ρ is given by the inequality

$$\frac{a - 1}{a + 1} < \rho < \frac{b - 1}{b + 1}$$

where

$$a = \frac{(1 + r)}{(1 - r)}\, e^{-(2z_{\alpha/2})/\sqrt{n - 3}}$$

and

$$b = \frac{(1 + r)}{(1 - r)}\, e^{+(2z_{\alpha/2})/\sqrt{n - 3}}$$

Hint: consider the inequality

$$-z_{\alpha/2} < \frac{(1/2)\ln[(1 + r)/(1 - r)] - (1/2)\ln[(1 + \rho)/(1 - \rho)]}{\sqrt{1/(n - 3)}} < +z_{\alpha/2}$$

5. Use the formula given in Exercise 4 to find a .95 confidence interval for the correlation coefficient of the random variables of Example 14.7. Assume that the random variables are bivariate normal.

6. Assuming that the random variables of Exercises 1 and 3 of Section 14.1 are bivariate normal, find .90 confidence intervals for their respective

correlation coefficients. (The sample correlation coefficients were computed in Exercise 2 of Section 14.4.)

7. Assume that the random variables of Exercise 2 of Section 14.1 are bivariate normal.
    (a) Find a .95 confidence interval for their correlation coefficient.
    (b) Test for independence of the random variables at the .10 level of significance.

8. Assume that the random variables of Exercise 4 of Section 14.1 are bivariate normal.
    (a) Find a .95 confidence interval for their correlation coefficient.
    (b) Test for independence of the random variables at the .01 level of significance.

Chapter **15**

NONPARAMETRIC METHODS

15.1   Introduction

Most of the tests that we have studied thus far have decision rules that make reference to specific parameters of the population under consideration, or that are valid only if the distribution of the population is of a certain specific type (such as normal, binomial, or Poisson). For instance, in some tests there are requirements concerning the variance of the distribution. In many tests the distributions of the populations are assumed to be normal. Failing that, sample sizes are increased so that the central limit theorem can be invoked, and in the end we are again testing for a parameter of a normal distribution.

In contrast to these types of tests, there are tests whose decision rules do not depend on parameters of the population nor on narrowly specific requirements concerning the distributions of the populations. These are called **nonparametric** or **distribution-free** tests. Some of them were originally developed to test for the median and other quartiles of continuous distributions. Of course, they can be—and in fact often are—used to test for the mean of a population if the mean happens to coincide with the median. Others were designed to test a null hypothesis that two or more distributions are identical (without focusing on either the parameters of the distributions or the types of distributions), or that two or more random variables are independent, or that a population has a given distribution. There are generally no requirements for the distributions of the populations other than, perhaps, a broad requirement that the distributions can be reasonably assumed to be of the continuous type, or that they conform to certain requirements of symmetry. The appeal of nonparametric tests is that they tend to be easy to understand and easy to use. There is one disadvantage, however; because such tests assume little concerning the distribution of the population, they are not in a position to infer as much as tests that make use of known properties of a distribution.

15.2   The Sign Test

15.2.1   For the Mean and Median

This test was originally developed to test for the median of any continuous distribution. If, however, the mean of a distribution is also the median of the distribution, the test applies equally well to the testing of the mean. We shall outline the test procedure in terms of testing for the mean, because we tend to assign to the mean a more prominent role than to the median.

Let μ be the mean of a population whose distribution can be reasonably assumed to be of the continuous type and whose median is the same as the mean. Suppose we wish to test the null hypothesis $\mu = \mu_0$ against a suitable alternative on the basis of a random sample $(x_1, x_2, \ldots, x_n)$ of size n from the population. The method is as follows.

Step 1:   Compare each value x_i of the sample with the value μ_0. Assign to it a plus sign $(+)$ if $x_i > \mu_0$, and a minus sign $(-)$ if $x_i < \mu_0$. No signs are to be given if $x_i = \mu_0$. (See Figure 15.1.)

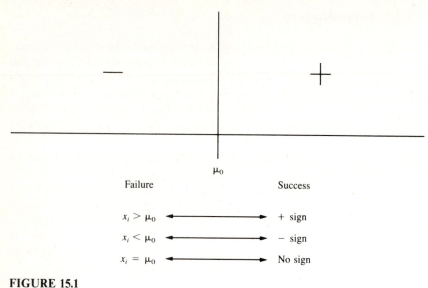

FIGURE 15.1

Step 2:   Count the number of signs given and let

m = the number of signs given

s_m = the number of plus signs ($+$) among the m signs

Now, think of each sign as the result of a Bernoulli trial. That is, think of the experiment of selecting an element at random from the population. If the element is a number that is greater than μ_0, we consider the experiment a success and assign to it the sign ($+$). Then, s_m represents the number of successes in m Bernoulli trials, and the probability p of success is equivalent to the probability of obtaining a plus sign. Thus, $p = 1/2$ if $\mu = \mu_0$ (because if $\mu = \mu_0$, then μ_0 is the median of the distribution and the probability of a plus sign or a value greater than μ_0 is equal to the probability of a minus sign or a value less than μ_0). By the same type of analysis, $p > 1/2$ if $\mu > \mu_0$, and $p < 1/2$ if $\mu < \mu_0$. This means that the hypothesis $\mu = \mu_0$ is equivalent to the hypothesis $p = 1/2$, the hypothesis $\mu > \mu_0$ is equivalent to the hypothesis $p > 1/2$, and the hypothesis $\mu < \mu_0$ is equivalent to the hypothesis $p < 1/2$. We have already considered the testing of these hypotheses concerning p in Chapter 13, Section 13.4. Following the procedure outlined there, the next step is:

Step 3:   If the hypotheses are

$$H_0: \mu = \mu_0 \quad \text{versus} \quad H_1: \mu > \mu_0$$

then reject H_0 if $s_m \geq k_\alpha$.

If the hypotheses are

$$H_0: \mu = \mu_0 \quad \text{versus} \quad H_1: \mu < \mu_0$$

then reject H_0 if $s_m \leq k_\alpha^*$.

If the hypotheses are

$$H_0: \mu = \mu_0 \quad \text{versus} \quad H_1: \mu \neq \mu_0$$

then reject H_0 if $s_m \geq k_{\alpha/2}$ or if $s_m \leq k_{\alpha/2}^*$. (Remember that k_α is the smallest integer for which

$$\sum_{x=k_\alpha}^{m} b(x; m, p_0) \leq \alpha$$

and k_α^* is the largest integer for which

$$\sum_{x=0}^{k_\alpha^*} b(x; m, p_0) \leq \alpha$$

For these particular sign tests, the value of p_0 is 1/2.)

EXAMPLE 15.1

The caffeine content of 16 randomly selected 12-ounce bottles of K&K cola are: .60, .66, .67, .59, .72, .61, .64, .57, .71, .69, .65, .78, .74, .64, .75, .77. At the .10 level of significance, test the null hypothesis that the mean caffeine content for K&K cola is .65 grains against the alternative hypothesis that it is greater than .65 grains.

   SOLUTION

Subjecting the random sample to a sign test, we obtain the following results:

.60	.66	.67	.59	.72	.61	.64	.57	.71	.69	.65	.78	.74	.64	.75	.77
−	+	+	−	+	−	−	−	+	+	none	+	+	−	+	+

There are 15 signs so $m = 15$, and the number of plus signs is

$$s_m = 9$$

We are testing the hypotheses

$$H_0: \mu = .65 \quad \text{and} \quad H_1: \mu > .65$$

Therefore (by referring to Step 3 of the procedure) we should reject the null hypothesis if

$$s_m \geq k_{.10}$$

Referring to the table of binomial probabilities, we obtain for this case

$$k_{.10} = 11$$

Now, since

$$s_m = 9 \not\geq 11 = k_{.10}$$

the conclusion of the test is that we should not reject, at the .10 level of significance, the null hypothesis that the mean caffeine content is .65 grains.

We should remember that this sign test for the mean of a population is valid only if the mean of the population is also the median of the population. Failing this, the above procedure can be used to test for the median of the population.

15.2.2   For the Mean Differences of Paired Measurements

The sign test for the mean can be used for testing the mean of differences between paired measurements, provided that the distribution of differences satisfies the conditions for the test.

EXAMPLE 15.2

A remedial program was given to 16 students who failed a mathematics test. Their scores on this test and on a similar test, taken after the program, are given here.

Student	1	2	3	4	5	6	7	8	9	10	11	12	13	14	15	16
Score after	54	60	65	71	42	53	59	75	80	62	57	69	55	64	72	61
Score before	20	35	40	50	37	48	52	51	57	50	39	45	30	60	57	50

Test the null hypothesis that, on the average, a failing student will improve his or her score by 20 points as a result of the program against the alternative hypothesis that the improvement will be less than 20 points. Use a .11 level of significance.

SOLUTION

Let μ be the mean of the differences in scores for all failing students who complete the program. Then, we wish to test the hypotheses

$$H_0: \mu = 20 \quad \text{versus} \quad H_1: \mu < 20$$

The differences in scores for the 16 students and the signs assigned to these scores are:

Difference	34	25	25	21	5	5	7	24	23	12	18	24	25	4	15	11
Sign	+	+	+	+	−	−	−	+	+	−	−	+	+	−	−	−

Therefore, $m = 16$, $s_m = 8$, $k^*_{.11} = 5$. We should not reject the null hypothesis, because in this case

$$s_m \not\leq k^*_\alpha$$

15.2.3 For the Difference Between Means

Since the difference between two population means is the mean of the differences between paired elements of the two populations (i.e., $E(X_1) - E(X_2) = E(X_1 - X_2)$), the sign test of Section 15.2.1 can be used for testing the difference between means, provided that the same number of elements are obtained from both populations. (This enables us to pair the measurements.) This is the procedure:

Let μ_1 be the mean of the first population. Let μ_2 be the mean of the second population. Suppose we wish to test the null hypothesis $\mu_1 - \mu_2 = d$ against a suitable alternative on the basis of a random sample $(x_{11}, x_{12}, \ldots, x_{1n})$ of size n from the first population and a random sample $(x_{21}, x_{22}, \ldots, x_{2n})$ of size n from the second population. The procedure is to compare each value $x_{1i} - x_{2i}$ with the value d. Assign to it a plus sign $(+)$ if $x_{1i} - x_{2i} > d$ and a minus sign $(-)$ if $x_{1i} - x_{2i} < d$. Let m be the number of signs given, and let s_m be the number of plus signs given. From this point on the procedure is exactly the same as that for the tests described in Section 15.2.1. Of course, we must be reasonably sure that both μ_1 and μ_2 are also medians of their respective populations. Otherwise the test is invalid.

EXAMPLE 15.3

The cash register receipts (in dollars) from two concessions for 9 randomly selected days are given below.

A	98.65	89.77	78.77	90.15	75.82	85.67	69.98	94.55	84.10
B	82.60	91.27	85.16	83.75	70.15	87.98	65.20	85.99	82.75

Test the null hypothesis (at the .10 level of significance) that there is no difference between the two concessions against the alternative hypothesis that the average receipt from concession A is greater than the average receipt from concession B.

SOLUTION

Let μ_1 be the mean receipt for concession A, and let μ_2 be the mean receipt for concession B. We are testing the hypotheses

$$H_0: \mu_1 - \mu_2 = 0 \quad \text{versus} \quad H_1: \mu_1 - \mu_2 > 0$$

Subtracting the receipts of B from the receipts of A and comparing the differences with value of $d = 0$, we obtain the following signs

$$+ \; - \; - \; + \; + \; - \; + \; + \; +$$

Therefore, $m = 9$, $s_m = 6$, $k_{.10} = 7$, and we should not reject the null hypothesis because, in this case

$$s_m \not\geq k_\alpha$$

15.2.4   For the Comparison of Two Distributions

Suppose we wish to test the null hypothesis that the distributions of two populations are the same against the alternative hypothesis that they are different. The following is a sign test for this purpose.

Let $(x_{11}, x_{12}, \ldots, x_{1n})$ be a random sample of size n from the first population, and let $(x_{21}, x_{22}, \ldots, x_{2n})$ be a random sample of size n from the second population. The method is as follows.

Step 1:   Compare each value x_{1i} of the first sample with the corresponding value x_{2i} of the second sample. Assign to it a plus sign $(+)$ if $x_{1i} > x_{2i}$, and a minus sign $(-)$ if $x_{1i} < x_{2i}$. No signs are to be given if $x_{1i} = x_{2i}$.

Step 2:   Count the number of signs, and the number of plus signs given. Let $m =$ the number of signs given, and let $s_m =$ the number of plus signs $(+)$ given.

Now, if the two distributions are identical we should expect the same number of pluses and minuses. Therefore, if the null hypothesis is true, then s_m is the value of a binomial random variable with $p = 1/2$. We can now follow the procedure for testing a binomial probability (see Section 15.2.1 or Chapter 13 Section 13.4). Therefore, the next step is:

Step 3:   Reject the null hypothesis that the two distributions are the same if

$$s_m \geq k_{\alpha/2} \quad \text{or} \quad s_m \leq k_{\alpha/2}^*$$

EXAMPLE 15.4

The annual salaries of men and women in a random selection of employees performing identical tasks for a certain large company are shown in the list that follows. Test the null hypothesis that the distribution of salaries for men and women is identical. Use a .03 level of significance.

Annual Salaries

Men	Women
15,000	14,300
13,500	12,170
12,600	15,000
14,750	10,000
15,600	11,160
13,750	10,750
13,460	12,240
12,750	11,115
14,175	12,680
13,108	12,750

SOLUTION

Let x_{1i} be the salary of the ith man on the list, and let x_{2i} be the salary of the ith woman on the list. Following Steps 1 and 2 we obtain 10 signs and 9 plus signs. Therefore

$$m = 10 \quad \text{and} \quad s_m = 9$$

With $\alpha = .03$, we obtain $k_{\alpha/2} = k_{.015} = 9$ and $k^*_{\alpha/2} = k^*_{.015} = 1$. Therefore, we should reject the null hypothesis that the distribution of salaries for men and women is identical because

$$s_m = 9 \geq 9 = k_{.015}$$

Before we leave this section, we must include a word of caution concerning what one may and may not conclude from a sign test for the comparison of two distributions. The problem is this: In the event that a test indicates that we cannot reject the null hypothesis that two distributions are identical, can we conclude that they are identical? The answer is no. First of all, remember that the sign test is a significance test, and in any significance test the alternative to rejecting the null hypothesis is reserving judgment concerning the null hypothesis. (This is because in a significance test we cannot compute the probability of an error in accepting the null hypothesis when the null hypothesis is false. Therefore, there is no probabilistic evaluation of the consequence of accepting the null hypothesis.) The implication of these remarks for the sign test in the comparison of two distributions is that if we cannot reject the null hypothesis that two distributions are identical, then we should merely reserve judgment concerning the equality of the two distributions—we should not conclude that they are identical. An even more compelling reason for adopting this procedure is the following. The sign test for the comparison of two distributions is based entirely on the assumption that if the distributions are identical, we should expect the same number of plus signs and minus signs. It it easy to see that—even if the distributions are such that the same number of plus signs and minus signs is very likely to be in the sample data—the distributions may still be very different. For example, two such distributions are defined by the probability density functions $f(x)$ and $g(x)$ where

$$f(x) = \begin{cases} 1/2 & \text{for} \quad 0 < x < 1 \\ 1/4 & \quad\quad\quad 1 \le x < 3 \end{cases} \qquad g(x) = \begin{cases} 1/4 & \text{for} \quad 0 < x < 2 \\ 1/2 & \quad\quad\quad 2 \le x < 3 \end{cases}$$

Exercises

In each of the following exercises assume that the population mean is identical to the population median.

1. To test the flammability of a certain type of synthetic fiber, 17 batches of the fiber were subjected to a flame. The following are the ignition times recorded to the nearest tenth of a second: 3.2, 2.1, 3.5, 3.7, 3.0, 2.3, 3.3, 3.9, 3.6, 1.8, 2.7, 3.4, 2.6, 3.4, 2.8, 1.0, 2.4.

    Use the sign test the null hypothesis that the mean ignition time is 3.2 seconds against the alternative that it is less than 3.2 seconds.
    (a) Use a .05 level of significance.
    (b) Use a .10 level of significance.
    (c) Find the smallest (approximate) value of α for which the null hypothesis may be rejected.

2. To measure the response of customers to a new flavor of ice cream, complimentary samples were given to 18 randomly selected customers. Their responses—in terms of ratings from 0 to 10, with 10 being the most favorable—were: 5.9, 5.0, 5.5, 6.7, 9.0, 8.8, 8.7, 7.5, 6.1, 5.3, 4.3, 4.7, 6.2, 7.1, 9.7, 7.1, 6.3, 5.2.

    Use the sign test to test the null hypothesis that the mean rating by all customers for the new flavor is 6.0, against the alernative hypothesis that it is greater than 6.0.
    (a) Use a .05 level of significance.
    (b) Use a .10 level of significance.

3. To determine the effectiveness of a certain work incentive program the following data were collected on the average daily productivity (measured in coded units) of 16 workers one year before the program and one year after the program.

Worker	1	2	3	4	5	6	7	8	9	10	11	12	13	14	15	16
Before	3.1	3.5	4.1	2.8	3.3	4.2	5.4	6.1	3.4	4.3	5.1	6.0	3.7	3.8	4.4	5.2
After	3.9	4.6	5.7	4.0	5.7	5.5	7.7	7.5	5.6	5.8	7.2	6.9	5.6	4.5	6.2	5.8

Use the sign test to test the null hypothesis that the mean increase in productivity is 1.5 units against the alternative hypothesis that the mean increase is less than 1.5 units. Assume that the mean increase is the same as the median. Use a .05 level of significance.

4. The following noise levels (in decibels) were recorded at noon at 18 midtown intersections one month before and one month after a city-wide campaign to reduce noise in public places.

Place	1	2	3	4	5	6	7	8	9	10	11	12	13	14	15	16	17	18
Noise before	55	74	68	80	77	69	57	72	63	52	85	66	71	48	83	78	51	67
Noise after	41	64	61	70	75	60	53	59	61	48	79	65	70	47	81	69	50	62

Can we conclude, at the .01 level of significance, that there has been a decrease of more than 5 decibels in the mean noise level in the city since the campaign? Use the sign test.

5. Twenty randomly selected piglets were put on diet A and another twenty randomly selected piglets were put on diet B. These forty piglets were comparable in age and weight at the start of the diet programs. The piglets' weights in pounds after one month on the respective diets were as follows:

Weights of piglets on diet A

15.1	17.5	16.2	14.7	15.4	17.3	16.5	15.3	17.4	18.0
14.9	15.7	17.3	16.1	17.9	16.8	15.5	16.7	15.9	15.2

Weights of piglets on diet B

15.0	15.2	15.6	17.7	17.2	16.4	16.2	14.8	17.8	15.6
16.7	17.8	15.4	16.9	16.6	15.8	15.7	17.9	15.3	18.3

Use the sign test to test the null hypothesis that there is no difference in the mean weights of piglets on the two different diets against the alternative hypothesis that there is a difference. Use a .05 level of significance.

6. The list that follows shows the scores on a certain chemistry test for 18 randomly chosen chemistry students who had been given a briefing session on test taking and the scores on the same chemistry test for another 18 randomly chosen chemistry students who had not been given the briefing session.

Scores with briefing session

88.3	69.4	75.5	98.0	99.3	75.0	68.7	79.3	81.0
68.0	64.5	57.8	45.0	83.7	77.9	79.3	82.0	82.0

Scores without briefing session

68.7	75.1	99.8	97.5	86.4	65.0	44.0	55.1	30.1
88.3	75.6	79.3	74.9	68.8	55.6	53.8	60.0	61.6

Can we conclude, at the .10 level of significance, that the distributions of test scores are different for students who were briefed from those for students who were not? Use the sign test.

15.3   The Wilcoxon Signed-Rank Test

15.3.1   For the Mean

Although the sign test for the mean is easy to use, it wastes information—only the signs of the differences between the sample values and the value of the null hypothesis concerning the mean are used. A nonparametric test that makes better use of the sample data is the Wilcoxon signed-rank test, which takes into consideration the magnitudes of the differences. This test applies to any population whose distribution is continuous and whose probability density function $f(x)$ has a graph which is symmetric about the vertical line through the mean.

Let μ be the mean of the population. Suppose we wish to test the null hypothesis $\mu = \mu_0$ against a suitable alternative on the basis of a sample $(x_1, x_2, \ldots, x_n)$ of size n from the population. The method is as follows.

Step 1:   Compute $(x_i - \mu_0)$ for $i = 1, 2, \ldots, n$.

Step 2:   Consider the absolute values $|x_i - \mu_0|$ for $i = 1, 2, \ldots, n$. Discard all zero values, and rank the remaining values in order of size with rank 1 going to the smallest value and rank m (where m is the number of values to be ranked) going to the largest value. If two or more values are the same, assign to each of these values the average of the ranks that would have been assigned if they were not the same.

Step 3:   Now, consider only those values of $(x_i - \mu_0)$ that are greater than zero. Let T^+ be the sum of the ranks assigned to the magnitudes of these values.

It can be shown (see listed references for advanced texts in mathematical statistics) that if $m \geq 15$, then T^+ is the value of a random variable whose distribution is approximately normal with mean equal to $m(m + 1)/4$ and variance equal to $m(m + 1)(2m + 1)/24$. Therefore, the next step (if $m \geq 15$) is:

Step 4:   If the hypotheses are

$$H_0: \mu = \mu_0 \quad \text{versus} \quad H_1: \mu > \mu_0$$

then reject H_0 if

$$\frac{T^+ - m(m + 1)/4}{\sqrt{m(m + 1)(2m + 1)/24}} > z_\alpha$$

If the hypotheses are

$$H_0: \mu = \mu_0 \quad \text{versus} \quad H_1: \mu < \mu_0$$

then reject H_0 if

$$\frac{T^+ - m(m + 1)/4}{\sqrt{m(m + 1)(2m + 1)/24}} < -z_\alpha$$

If the hypotheses are

$$H_0: \mu = \mu_0 \quad \text{versus} \quad H_1: \mu \neq \mu_0$$

then reject H_0 if

$$\left| \frac{T^+ - m(m + 1)/4}{\sqrt{m(m + 1)(2m + 1)/24}} \right| > z_{\alpha/2}$$

EXAMPLE 15.5

Use the Wilcoxon signed-rank test for the problem of Example 15.1.

SOLUTION

Following the instructions given, we construct the following table of values.

i	x_i	$x_i - \mu_0$	$|x_i - \mu_0|$	Rank
1	.60	−.05	.05	7
2	.66	.01	.01	2
3	.67	.02	.02	4
4	.59	−.06	.06	8.5
5	.72	.07	.07	10
6	.61	−.04	.04	5.5
7	.64	−.01	.01	2
8	.57	−.08	.08	11
9	.71	.06	.06	8.5
10	.69	.04	.04	5.5
11	.65	.00	.00	none
12	.78	.13	.13	15
13	.74	.09	.09	12
14	.64	−.01	.01	2
15	.75	.10	.10	13
16	.77	.12	.12	14

Therefore

$$T^+ = 2 + 4 + 10 + 8.5 + 5.5 + 15 + 12 + 13 + 14 = 84$$

and

$$\frac{T^+ - m(m + 1)/4}{\sqrt{m(m + 1)(2m + 1)/24}} = 1.36$$

whereas

$$z_{.10} = 1.28$$

Therefore, we should reject H_0 since

$$1.36 > 1.28$$

The procedure outlined is for random samples of sizes greater than 14. If the sample size is smaller, we may refer to a special table of values of the distribution of T^+. (See reference 1.)

15.3.2 For the Mean Differences of Paired Measurements

The Wilcoxon signed-rank test for the mean can be used for testing the mean of differences between paired measurements, provided that the distribution of differences satisfies the conditions for the test.

EXAMPLE 15.6

Use the Wilcoxon signed-rank test for the problem of Example 15.2.

SOLUTION

Let x_{1i} be the score after the program for the ith student. Let x_{2i} be the score before the program for the ith student. Let x_i be the difference in the two scores. That is, $x_i = x_{1i} - x_{2i}$. We are testing the hypotheses

$$H_0: \mu = 20 \quad \text{versus} \quad H_1: \mu < 20$$

where μ is the mean difference in scores. Following the instructions for a Wilcoxon signed-rank test, we would construct the following table of values:

i (student)	x_{1i} (score after)	x_{2i} (score before)	x_i (difference)	$x_i - \mu_0$	$\lvert x_i - \mu_0 \rvert$	Rank
1	54	20	34	14	14	13
2	60	35	25	5	5	7.5
3	65	40	25	5	5	7.5
4	71	50	21	1	1	1
5	42	37	5	−15	15	14.5
6	53	48	5	−15	15	14.5
7	59	52	7	−13	13	12
8	75	51	24	4	4	4.5
9	80	57	23	3	3	3
10	62	50	12	−8	8	10
11	57	39	18	−2	2	2
12	69	45	24	4	4	4.5
13	55	30	25	5	5	7.5
14	64	60	4	−16	16	16
15	72	57	15	−5	5	7.5
16	61	50	11	−9	9	11

From these data we obtain

$$m = 16$$

and

$$T^+ = 13 + 7.5 + 7.5 + 1 + 4.5 + 3 + 4.5 + 7.5 = 48.5$$

and

$$\frac{T^+ - m(m + 1)/4}{\sqrt{m(m + 1)(2m + 1)/24}} = -1.008$$

Testing at the .11 level of significance, we have

$$-z_{.11} = -1.22$$

Therefore, we should not reject the null hypothesis since

$$-1.008 \not< -1.22$$

15.3.3 For the Difference Between Means

Since the difference between two population means is the mean of the differences between paired elements of the two populations (i.e., $E(X_1) - E(X_2) = E(X_1 - X_2)$), the Wilcoxon signed-rank test can be used for testing the difference between means, provided that the same number of elements are obtained from both populations. (This enables us to pair the measurements.) A description of the procedure follows.

Let μ_1 be the mean of the first population. Let μ_2 be the mean of the second population. Suppose we wish to test the null hypothesis $\mu_1 - \mu_2 = d$ against a suitable alternative on the basis of a random sample $(x_{11}, x_{12}, \ldots, x_{1n})$ of size n from the first population and a random sample $(x_{21}, x_{22}, \ldots, x_{2n})$ of size n from the second population. The procedure is to let $x_i = x_{1i} - x_{2i}$ and let $\mu = \mu_1 - \mu_2$. Then, we are testing the null hypothesis $\mu = d$ on the basis of the random sample $(x_1, x_2, \ldots, x_n)$. We can now follow the procedure for testing the mean (Section 15.3.1).

15.3.4 For the Comparison of Two Distributions

The Wilcoxon signed-rank test for the equality of distributions is based on the premise that if two distributions are identical, they must have the same mean. Therefore, the test becomes a test of the null hypothesis that the two means are equal against the alternative hypothesis that the two means are not equal. The tests of these hypotheses were discussed in Section 15.3.3. As in the sign test, if the Wilcoxon signed-rank test indicates that we cannot reject the null hypothesis that two distributions are identical, then we should merely reserve judgment concerning the two distributions—we should not conclude that they are identical (see Section 15.2.4). One reason for this is that although identical distributions must have equal means, distributions with equal means need not be identical.

Exercises

In each of the following exercises, assume that the population has a distribution that is continuous and a probability density function that is symmetric about the vertical line through the mean.

1. Use the Wilcoxon signed-rank test to test for the mean in
   (a) Exercise 1 of Section 15.2, and
   (b) Exercise 2 of Section 15.2.
   (c) Compare the answers in those exercises and the answers here.

2. Use the Wilcoxon signed-rank test to test for the mean difference of the paired measurements of
   (a) Exercise 3 of Section 15.2, and
   (b) Exercise 4 of Section 15.2.
   (c) Compare the answers in those exercises and the answers here.

3. Use the Wilcoxon signed-rank test to test for the difference between means in Exercise 5 of Section 15.2. Compare the answer in that exercise and the answer here.

4. Use the Wilcoxon signed-rank test to compare the two distributions of Exercise 6 of Section 15.2. Compare the answer in that exercise and the answer here.

15.4 The Mann–Whitney–Wilcoxon Rank-Sum Test

15.4.1 For the Difference Between Means

This nonparametric test for the difference between means is also called the **U test** (because of the U statistics). The advantages of this test over the signed-rank test (Section 15.3) for testing the difference between means are that:

1. The samples from the two populations do not have to be of the same size, and

2. the probability density functions do not have to be symmetric about the mean. (The distributions do have to be continuous, however.)

The following is an outline of the procedure for the test.

Let $(x_{11}, x_{12}, \ldots, x_{1n_1})$ and $(x_{21}, x_{22}, \ldots, x_{2n_2})$ be random samples of sizes n_1 and n_2, repectively, from populations with means μ_1 and μ_2, repectively. The procedure is as follows:

Step 1: Combine the elements of both samples into one set and rank them in ascending order so that

$$\text{(the element of rank } r) < \text{(the element of rank } r + 1)$$

If two or more elements have the same value, assign to each of these elements the average of the ranks that would have been assigned if they were not the same.

Step 2:   Let

W_1 = the sum of the ranks of the elements in the first sample
W_2 = the sum of the ranks of the elements in the second sample

and let

$$U_1 = W_1 - \frac{n_1(n_1 + 1)}{2}$$

$$U_2 = W_2 - \frac{n_2(n_2 + 1)}{2}$$

It can be shown (see listed references for advanced texts in mathematical statistics) that if n_1 and n_2 are both greater than 8, it is reasonable to assume that if $\mu_1 = \mu_2$, then both U_1 and U_2 are values of random variables that are normally distributed with the same mean of $(n_1 n_2)/2$ and the same variance of $n_1 n_2(n_1 + n_2 + 1)/12$. Therefore, the next step (if $n_1 > 8$ and $n_2 > 8$) is:

Step 3:   Compute

$$z = \frac{U_1 - (n_1 n_2)/2}{\sqrt{n_1 n_2(n_1 + n_2 + 1)/12}}$$

(Then, for $n_1 > 8$ and $n_2 > 8$, it is reasonable to assume that z is a value of a random variable whose distribution is approximately standard normal.)

Step 4:   If the hypotheses are

$$H_0: \mu_1 = \mu_2 \quad \text{versus} \quad H_1: \mu_1 > \mu_2$$

then reject H_0 if $z > z_\alpha$.
    If the hypotheses are

$$H_0: \mu_1 = \mu_2 \quad \text{versus} \quad H_1: \mu_1 < \mu_2$$

then reject H_0 if $z < -z_\alpha$.
    If the hypotheses are

$$H_0: \mu_1 = \mu_2 \quad \text{versus} \quad H_1: \mu_1 \neq \mu_2$$

then reject H_0 if $|z| > z_{\alpha/2}$.

(In case the reader has doubts about the various inequalities concerning z and z_α, consider the following. The smallest value that W_1 can attain is $1 + 2 + \cdots + n_1 = n_1(n_1 + 1)/2$. This will take place if all values in the sample from the first population are smaller than all values in the sample from the second population. Therefore $U_1 = W_1 - n_1(n_1 + 1)/2$ is always greater than or equal to zero. Furthermore, U_1 will tend to be a larger value if $\mu_1 > \mu_2$ than if $\mu_1 < \mu_2$. Therefore, we should reject the null hypothesis in favor of $\mu_1 \geqslant \mu_2$ if U_1 is large; and we should reject the null hypothesis in favor of $\mu_1 < \mu_2$ if U_1 is small.)

If we had used U_2 instead of U_1 in the computation of z in Step 3 with the result that

$$z = \frac{\left[U_2 - \dfrac{n_1 n_2}{2} \right]}{\sqrt{\dfrac{n_1 n_2 (n_1 + n_2 + 1)}{12}}}$$

then the decision rule would be modified to reject the null hypothesis in favor of $\mu_2 > \mu_1$ if $z > z_\alpha$, reject the null hypothesis in favor of $\mu_2 < \mu_1$ if $z < -z_\alpha$, and reject the null hypothesis in favor of $\mu_2 \neq \mu_1$ if $|z| > z_{\alpha/2}$.

For small sample tests there are special tables of values for the distribution of U_1 and U_2. (See References.)

15.4.2   For the Comparison of Two Distributions

The Mann–Whitney–Wilcoxon rank-sum test for the equality of distributions is based on the premise that if two distributions are identical, they must have the same mean. Therefore, the test becomes a test of the null hypothesis that the two means are equal against the alternative hypothesis that the two means are not equal. The test of these hypotheses was discussed in Section 15.4.1. As in the sign test and the Wilcoxon signed-rank test, if the Mann–Whitney–Wilcoxon rank-sum test indicates that we cannot reject the null hypothesis that two distributions are identical, then we should merely reserve judgment concerning the two distributions—we should not conclude that they are identical (see Sections 15.2.4 and 15.3.4).

Exercises

1. Use the Mann–Whitney–Wilcoxon rank-sum test to test for the difference between means in Exercise 5 of Section 15.2. Compare the answers obtained in Exercise 5 of Section 15.2, in Exercise 3 of Section 15.3 and the answer here.

2. The following table gives the family incomes of 9 randomly selected families from district A and 12 randomly selected families from district B. Use the Mann–Whitney–Wilcoxon rank-sum test to test, at the .05 level of significance, the null hypothesis that the mean family incomes in both districts are identical against the alternative hypothesis that district B has a higher mean income.

Family Income

District A	District B
10,000	7,500
8,000	15,000
9,500	25,000
17,000	35,000
28,000	8,900
23,000	40,000
29,000	7,900
24,200	13,000
16,750	9,750
	8,775
	23,800
	33,960

3. To compare the effectiveness of radical surgery with the effectiveness of modified surgery in prolonging the lives of tumor patients, 17 mice with the same type of tumor and at the same stage of the disease were randomly divided into two groups of 9 and 8 mice, respectively. The mice in the first group of 9 mice received radical surgery, and the mice in the second group of 8 mice received modified surgery. The survival times, in years, from the time of the surgery were:

Radical	3.1	2.6	4.8	3.6	2.9	4.1	3.9	1.6	1.9
Modified	2.4	2.9	3.6	4.1	1.4	3.8	4.9	2.7	

Test the null hypothesis that modified surgery is just as effective as radical surgery in prolonging the lives of mice with this type of tumor against the alternative hypothesis that it is not as effective. Use the Mann–Whitney–Wilcoxon rank-sum test with a .01 level of significance.

4. Use the Mann–Whitney–Wilcoxon rank-sum test to compare the two distributions of Exercise 6 of Section 15.2. Compare the answers obtained in Exercise 6 of Section 15.2 and Exercise 4 of Section 15.3 with the answer here.

5. The silicon content of each of 10 randomly selected buckets of soil from region A and of each of 12 randomly selected buckets of soil from region B are as follows:

Region A	.89	.74	.66	.73	.99	.75	.83	.85	.78	.82		
Region B	.69	.74	.72	.89	.85	.91	.94	.98	.81	.60	.77	.84

Test the null hypothesis that the distributions of silicon in the two regions are identical against the alternative hypothesis that they are not. Use the Mann–Whitney–Wilcoxon rank-sum test with a .05 level of significance.

15.5   Contingency Tables: Chi-Square Test for Independence

When the sample data for two random variables are presented in a contingency (i.e., dependency) table, then a chi-square test for independence of the random variables is applicable. The following is a description of this method of testing for independence.

Suppose we wish to analyze a set of data to determine if success in medical school is related to ability in mathematics. Suppose that 150 randomly selected third-year medical students were rated according to their success in medical school and their ability in mathematics. With respect to each of these characteristics the students were rated as low, average, or high. The number of students in each category is given in the following table.

		Ability in mathematics		
		Low	Average	High
Sucess in medical school	Low	14	8	5
	Average	12	51	11
	High	7	24	18

The table shows, for example, that 14 students were rated as low in terms of success in medical school and low in mathematical ability, 8 students were rated low in terms of success in medical school and average in mathematical ability, and 18 students were rated high in terms of success in medical school and high in mathematical ability. This table is called a 3×3 contingency table for the variable of "success in medical school" (call it the variable A) and the variable of "ability in mathematics" (call it the variable B). In general, an $m \times n$ contingency table has m rows representing the different categories $A_1, A_2, \ldots, A_m$ of the variable A, and n columns representing the different categories $B_1, B_2, \ldots, B_n$ of the variable B. If we let f_{ij} denote the number of items belonging to A_i and B_j, then the value of f_{ij} is placed in the space, called a cell, that is in the ith row and jth column of the table. In addition to these values, a contingency table may also give the values of $f_{i.}$, $f_{.j}$, and f, where

$$f_{i.} = \sum_{j=1}^{n} f_{ij}, \qquad f_{.j} = \sum_{i=1}^{m} f_{ij}, \qquad f = \sum_{j=1}^{n} \sum_{i=1}^{m} f_{ij}$$

(In other words, $f_{i.}$ is the number of items belonging to category A_i and is called the marginal frequency of A_i, $f_{.j}$ is the number of items belonging to category B_j and is called the marginal frequency of B_j, and f is the total number of items represented in the table and is called the grand total.) The following shows the arrangement of values in an $m \times n$ contingency table.

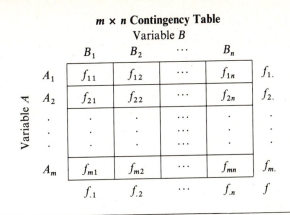

$m \times n$ **Contingency Table**

	Variable B					
	B_1	B_2	$\cdots$	B_n	
A_1	f_{11}	f_{12}	$\cdots$	f_{1n}	$f_{1.}$
A_2	f_{21}	f_{22}	$\cdots$	f_{2n}	$f_{2.}$
A_m	f_{m1}	f_{m2}	$\cdots$	f_{mn}	$f_{m.}$
	$f_{.1}$	$f_{.2}$	$\cdots$	$f_{.n}$	f

Assuming that we can treat the variables A and B as random variables, and assuming that the data in the table are the results of a random sample, then the procedures for a chi-square test of the null hypothesis that the random variables are independent against the alternative hypothesis that they are not independent is as follows.

Step 1:   Compute

$$e_{ij} = (f_{i.}/f)(f_{.j}/f)(f) = \frac{(f_{i.})(f_{.j})}{f}$$

for $i = 1, 2, \ldots, m$, and $j = 1, 2, \ldots, n$.

Note that $(f_{i.}/f)$ is an estimate of the probability of obtaining an item belonging to category A_i, and $(f_{.j}/f)$ is an estimate of the probability of obtaining an item belonging to category B_j. Therefore, if the random variables A and B are independent, the probability of obtaining an item in A_i and B_j would be approximately $(f_{i.}/f)(f_{.j}/f)$, and the expected number of items in A_i and B_j would be approximately $(f_{i.}/f)(f_{.j}/f)(f)$.

Step 2:   Let

$$\chi^2 = \sum_{j=1}^{n} \sum_{i=1}^{m} \frac{(f_{ij} - e_{ij})^2}{e_{ij}}$$

It can be shown that if A and B are independent random variables and if each value of f_{ij} is sufficiently large, then χ^2 is a value of a random variable whose distribution is approximately a chi-square distribution with $(m-1)(n-1)$ degrees of freedom (see listed references for advanced texts in mathematical statistics). In practice, this test should be used only if each value of f_{ij} is no less than 5. This requirement may, in some cases, necessitate the combination of several cells, with a corresponding loss in the number of degrees of freedom.

Step 3:   Reject the null hypothesis that the random variables are independent if

$$\chi^2 \geq \chi^2_{\alpha,(m-1)(n-1)}$$

EXAMPLE 15.7

On the basis of the data given in the previous contingency table, should we conclude that success in medical school is related to ability in mathematics? Test at the .005 level of significance.

SOLUTION

Let "success in medical school" be the variable A, and let "ability in mathematics" be the variable B. We wish to test the null hypothesis that A and B are independent random variables against the alternative hypothesis that they are not independent random variables.

Computing for the marginal frequencies and the grand total from the given contingency table, we obtain

$$f_{1.} = 27 \quad f_{.1} = 33 \quad f = 150$$
$$f_{2.} = 74 \quad f_{.2} = 83$$
$$f_{3.} = 49 \quad f_{.3} = 34$$

Then, following the instruction of Step 1, we obtain the following table of values for e_{ij} (expected cell frequencies).

Values of e_{ij}

		1	*j* 2	3
	1	$\dfrac{(27)(33)}{150} = 5.94$	$\dfrac{(27)(83)}{150} = 14.94$	$\dfrac{(27)(34)}{150} = 6.12$
i	2	$\dfrac{(74)(33)}{150} = 16.28$	$\dfrac{(74)(83)}{150} = 40.95$	$\dfrac{(74)(34)}{150} = 16.77$
	3	$\dfrac{(49)(33)}{150} = 10.78$	$\dfrac{(49)(83)}{150} = 27.11$	$\dfrac{(49)(34)}{150} = 11.11$

Then

$$\chi^2 = \sum_{j=1}^{3} \sum_{i=1}^{3} \frac{(f_{ij} - e_{ij})^2}{e_{ij}}$$

$$= \frac{(14 - 5.94)^2}{5.94} + \frac{(8 - 14.94)^2}{14.94} + \frac{(5 - 6.12)^2}{6.12}$$

$$+ \frac{(12 - 16.28)^2}{16.28} + \cdots + \frac{(18 - 11.11)^2}{11.11}$$

$$= 25.90$$

whereas

$$\chi^2_{.005,(m-1)(n-1)} = \chi^2_{.005,4} = 14.8602$$

Therefore

$$\chi^2 = 25.90 \geq 14.8602 = \chi^2_{.005,4}$$

and, at the .005 level of significance, we should reject the null hypothesis that success in medical school is independent of ability in mathematics, and conclude that A and B are dependent variables.

Exercises

1. An experiment was conducted to determine the effectiveness of a new drug on a certain disease. One hundred patients in the same stages of the disease were randomly divided into three groups. Patients in the first, second, and third groups, respectively, were treated twice daily with a capsule containing 0, 1, and 2 mg of the drug. At the end of the period of treatment the condition of each patient was evaluated as greatly improved, moderately improved, or not at all improved. The number of patients in each category is given in the following contingency table.

	Amount of drug		
	0	1	2
Greatly improved	8	9	10
Moderately improved	12	13	13
No improvement	13	12	10

Do these data provide sufficient evidence, at the .05 level of significance, to indicate that the drug is effective in the treatment of this disease? (*Hint:* test the null hypothesis that the drug is ineffective. Assume that if the drug is ineffective, then the condition of the patient is independent of the amount of drug received. Use the chi-square test with a .05 level of significance.)

2. A randomly selected group of 244 students were interviewed on their opinions concerning a new language requirement. The data are given below.

Opinion of Language Requirement

	Approve	Do not approve	No opinion
Male	50	65	18
Female	48	53	10

Do the data indicate, at the .01 level of significance, that student opinion concerning the language requirement is dependent on the sex of the student?

3. A survey was conducted to determine if there are regional differences among adults concerning the use of vitamins as a dietary supplement. A random sample of adults were interviewed in each of the states of Mississippi, California, North Dakota, and New York. The following data were recorded:

	Miss.	Calif.	N.D.	N.Y.
Take vitamin daily	30	50	25	45
Take vitamin occasionally	55	48	40	60
Never take vitamin	117	92	88	120

Do the data indicate a difference in vitamin intake among adults in the four states?
(a) Test at the .05 level of significance.
(b) Test at the .10 level of significance.

4. A study was made to determine if there is a relationship between the number of workers in a room and the level of fatigue experienced by the workers at the end of the day. One hundred typists were randomly divided into three groups. Each typist in the first group of 10 was assigned to an individual room. The 40 typists in the second group were assigned to 8 rooms with 5 typists to each room. The 50 typists in the third group were assigned to two rooms with 25 typists to each room. All 100 typists were given the same typing assignment for the day. At the end of the day each typist was asked to assess his or her level of fatigue. A count of the number of typists experiencing fatigue at each of three levels follows:

	Group 1	Group 2	Group 3
Very fatigued	1	10	20
Moderately fatigued	7	25	30
Not at all fatigued	2	5	0

Do these data provide sufficient evidence, at the .01 level of significance, to indicate that the level of fatigue experienced by typists is related to the number of typists working together in a room?

15.6   Chi-Square Test for Goodness of Fit

The chi-square test for goodness of fit is a test to determine, on the basis of a random sample from a population, the extent to which it is unreasonable to assume that the population has a given distribution. From another point of view, it is also a test to determine whether a set of data cannot be reasonably assumed to be a random sample from a population having a given distribution. This test is distribution-free in the sense that it applies to all distributions. In the following paragraphs we will outline the procedure for testing for a Poisson distribution (Example 15.8). These instructions apply equally well to any discrete distribution (see Example 15.9). Then, we will outline the procedure for testing for a continuous uniform distribution (Example 15.10). Again, these instructions apply equally well to any continuous distributions.

EXAMPLE 15.8

The following is a record of the number of two-alarm fires that occurred in 1979 in a certain town. On the basis of these data, would it be reasonable to assume that the number of two-alarm fires in any given day in this town is a Poisson random variable?
(a)   Test at the .01 level of significance.
(b)   Test at the .05 level of significance.

Number of fires	0	1	2	3	4	5	6	7
Number of days	151	118	77	19	0	0	0	0

SOLUTION

(a)   To answer the question, we will test the null hypothesis that if X is the number of two-alarm fires in any given day in this town, then the distribution of X is Poisson. The procedure for a chi-square test for this null hypothesis against the alternative that the distribution is not Poisson is as follows: (The procedure may be applied to any discrete distribution.)

Step 1:   Use the sample data to estimate the parameters that are necessary to completely define the distribution of the null hypothesis. In this example, we need to estimate the mean and it is

$$\hat{\lambda} = \frac{(0) \cdot (151) + (1) \cdot (118) + (2) \cdot (77) + (3) \cdot (19)}{365}$$

$$= .90$$

Step 2:   Fill in the values in the following table. In the table, i denotes the value of the random variable X (i.e., the number of fires) and is called the **cell number**; f_i denotes the observed number of values in the random sample that are equal to i (i.e., the observed number of days with i fires) and is called the **observed cell frequency**; p_i is the probability that a value in the random sample will be equal to i (i.e., the probability of i fires in any given day) if the null hypothesis is true and is called the **cell probability under H_0**; and e_i denotes the expected number of values in the random sample that will be equal to i (i.e., the expected number of days with i fires) if the null hypothesis is true and is called the **expected cell frequency under H_0**. (In practice, this test should be used only if each value of f_i is no less than 5. To meet this requirement, we have combined into one cell the number of days having 3 or more fires.)

i	f_i	p_i	e_i
Cell number: Number of fires	Observed cell frequency: Observed number of days with i fires	Cell probability under H_0: Probability of i fires if H_0 is true	Expected cell frequency under H_0: Expected number of days with i fires if H_0 is true
0	151	.4066	148.41
1	118	.3659	133.55
2	77	.1647	60.12
3 or more	19	.0628	22.92

Step 3:   Use the values in the table to compute

$$\chi^2 = \sum_{i=1}^{m} \frac{(f_i - e_i)^2}{e_i}$$

where m is the number of cells (or the number of terms in the summation). In this case, we have

$$\chi^2 = \frac{(151 - 148.41)^2}{148.41} + \frac{(118 - 133.55)^2}{133.55} + \frac{(77 - 60.12)^2}{60.12}$$

$$+ \frac{(19 - 22.92)^2}{22.92}$$

$$= 7.2656$$

It can be shown (see listed references for advanced texts in mathematical statistics) that if the null hypothesis is true (i.e., that the distribution is Poisson) and if the observed cell frequency of each cell is at least 5, then χ^2 is a value of a random variable whose distribution is approximately the chi-square distribution with $m - t - 1$ degrees of freedom, where m is the number of cells and t is the number of independent parameters estimated on the basis of the sample data. In this case, $t = 1$, since the only parameter estimated was the mean, and $m = 4$. Therefore, the next step is:

Step 4:   Reject the null hypothesis that the distribution is Poisson if

$$\chi^2 \geq \chi^2_{\alpha, m-t-1}$$

In this example,

$$\chi^2 = 7.2656, \qquad \chi^2_{\alpha, m-t-1} = \chi^2_{.01, 2} = 9.210$$

Therefore, we should not reject the null hypothesis that the distribution is Poisson, at the .01 level of significance, because

$$\chi^2 = 7.2656 \not\geq 9.210 = \chi^2_{\alpha, m-t-1}$$

(b)   Testing at the .05 level of significance, we have

$$\chi^2 = 7.2656 \quad \text{and} \quad \chi^2_{\alpha, m-t-1} = \chi^2_{.05, 2} = 5.991$$

Therefore, at the .05 level of significance we should reject the null hypothesis that the distribution is Poisson, because

$$\chi^2 = 7.2656 > 5.991 = \chi^2_{\alpha, m-t-1}$$

EXAMPLE 15.9

A theory of Mendel (the geneticist) states that a certain type of pea is round and yellow, wrinkled and yellow, round and green, and wrinkled and green in accordance with the ratio 9:3:3:1. If a random sample of 150 such peas contains 85, 30, 25, and 10 specimens in the respective categories, should we reject Mendel's theory at the
(a)   .05 level of significance;
(b)   .10 level of significance;
(c)   .90 level of significance; or
(d)   .95 level of significance?

SOLUTION

(a)   Suppose we let the numbers 1, 2, 3, and 4 denote the respective categories of round and yellow, wrinkled and yellow, round and green, and wrinkled and green. Furthermore, let X be the category of a randomly selected pea. Then, Mendel's statement that peas are in the categories 1, 2, 3, and 4 according to the ratio 9:3:3:1 is equivalent to the statement that the distribution of X is given by the discrete probability function

$$f(x) = \begin{cases} 9/16 & \text{for} \quad x = 1 \\ 3/16 & x = 2 \\ 3/16 & x = 3 \\ 1/16 & x = 4 \end{cases}$$

Thus, we wish to test the null hypothesis that the distribution of X is the one given here against the alternative hypothesis that it is not. Let us follow the 4 steps outlined in Example 15.8.

Step 1:   There are no parameters to estimate so we can omit this step.

Step 2:   The table of values for a goodness of fit test follows. In this example, i (the cell number) is the category (of the pea), f_i (the observed cell frequency) is the observed number of peas belonging to category i, p_i (the cell probability under H_0) is the probability of belonging to category i, and e_i (the expected cell frequency under H_0) is the expected number of peas belonging to category i.

i	f_i	p_i	e_i
Cell number: Category of pea	Observed cell frequency: Observed number of peas belonging to category i	Cell probability under H_0: Probability of belonging to category i under H_0	Expected cell frequency under H_0: Expected number of peas in category i under H_0
1	85	9/16	84.375
2	30	3/16	28.125
3	25	3/16	28.125
4	10	1/16	9.375

Step 3: Use the values in the table to compute

$$\chi^2 = \sum_{i=1}^{m} \frac{(f_i - e_i)^2}{e_i}$$

where m is the number of terms in the summation. In this case

$$\chi^2 = \frac{(85 - 84.375)^2}{84.375} + \frac{(30 - 28.125)^2}{28.125} + \frac{(25 - 28.125)^2}{28.125}$$

$$+ \frac{(10 - 9.375)^2}{9.375}$$

$$= .5185$$

Note: as stated in Step 3 of Example 15.8, if the null hypothesis is true and if each value of f_i is at least 5, then χ^2 is a value of a random variable whose distribution is approximately the chi-square distribution with $m - t - 1$ degrees of freedom. The value of m is the number of terms in the above summation, and the value of t is the number of parameters that were estimated in Step 1. In this case, $t = 0$.

Step 4:   Reject the null hypothesis that the distribution of X is the one stated in the null hypothesis if

$$\chi^2 \geq \chi^2_{\alpha, m-t-1}$$

In this example, $\chi^2 = .5185$, $\chi^2_{\alpha, m-t-1} = \chi^2_{.05, 3} = 7.815$. Therefore, at the .05 level of significance we should not reject the null hypothesis because

$$\chi^2 = .5185 \not\geq 7.815 = \chi^2_{\alpha, m-t-1}$$

(b)   At the .10 level of significance, we have

$$\chi^2 = .5185 \quad \text{and} \quad \chi^2_{\alpha, m-t-1} = \chi^2_{.10, 3} = 6.25139$$

Therefore, at the .10 level of significance we should not reject Mendel's theory because

$$\chi^2 = .5185 \not\geq 6.25139 = \chi^2_{\alpha, m-t-1}$$

(c)   We cannot reject Mendel's theory at the .90 level of significance because

$$\chi^2 = .5185 \not\geq .584375 = \chi^2_{.90, 3} = \chi^2_{\alpha, m-t-1}$$

(d)   We should reject Mendel's theory at the .95 level of significance because

$$\chi^2 = .5185 > .351846 = \chi^2_{.95, 3} = \chi^2_{\alpha, m-t-1}$$

EXAMPLE 15.10

An automatic packing machine is gauged to discharge a fixed amount of a certain type of cereal into each box in an assembly line. A random sample of 100 boxes provided the following set of data.

Weight in ounces	Number of boxes
15.5–15.6	5
15.6–15.7	9
15.7–15.8	7
15.8–15.9	10
15.9–16.0	12
16.0–16.1	8
16.1–16.2	13
16.2–16.3	15
16.3–16.4	13
16.4–16.5	8

Use a chi-square goodness of fit test to test the null hypothesis that the actual amount of cereal that will be discharged by this machine into a given box is a uniform random variable against the alternative that it is not a uniform random variable. Use
(a)   a .01 level of significance;
(b)   a .10 level of significance;
(c)   a .90 level of significance.

   SOLUTION

Note: the procedure outlined here may be applied to any continuous distribution.

   Step 1:   Choose appropriate intervals of values and count the number of values in the random sample that falls into each of the intervals. Present the data in a table. In this example the values of the random sample were already presented in the proper form.

Step 2:   Use the sample data to estimate the parameters that are necessary to completely define the distribution of the null hypothesis. In this example we need to estimate the parameters α and β in the uniform density function

$$f(x) = \begin{cases} \dfrac{1}{\beta - \alpha} & \text{for} \quad \alpha < x < \beta \\ 0 & \text{elsewhere} \end{cases}$$

To do this, let us first estimate the mean of the distribution. Using the midpoint of each interval as the weight of the boxes, we find that the average weight is

$$\text{average weight} = \frac{(5) \cdot (15.55) + (9) \cdot (15.65) + (7) \cdot (15.75) + \cdots + (8) \cdot (16.45)}{100}$$

$$= 16.05 \approx 16$$

Since 16 is the midpoint of the interval from 15.5 to 16.5 it is reasonable to estimate that $\alpha = 15.5$ and $\beta = 16.5$.

Step 3:   Let f_i (called the observed frequency) be the number of values in the random sample that belong to the ith interval (i.e., the number of boxes in the ith interval of weights). Let p_i be the probability that a value in the random sample will belong to the ith interval if the null hypothesis is true. Let e_i (called the expected frequency) be the expected number of values in the random sample that will belong to the ith interval. In this example, the null hypothesis is that the distribution is uniform, therefore the probability that a value will belong to any one of the 10 intervals of equal length is $1/10$. This implies that the expected number of values in each interval will be 10. Therefore, we obtain the following table of values for this example.

Interval	f_i	p_i	e_i
15.5–15.6	5	.1	10
15.6–15.7	9	.1	10
15.7–15.8	7	.1	10
15.8–15.9	10	.1	10
15.9–16.0	12	.1	10
16.0–16.1	8	.1	10
16.1–16.2	13	.1	10
16.2–16.3	15	.1	10
16.3–16.4	13	.1	10
16.4–16.5	8	.1	10

Step 4:   Use the values in the table in Step 3 to compute

$$\chi^2 = \sum_{i=1}^{m} \frac{(f_i - e_i)^2}{e_i}$$

where m is the number of intervals in the table. In this example

$$\chi^2 = \frac{(5-10)^2}{10} + \frac{(9-10)^2}{10} + \cdots + \frac{(98-10)^2}{10} = 9$$

It can be shown (see listed references for advanced texts in mathematical statistics) that if the null hypothesis is true (i.e., that the distribution is uniform) and if each value of f_i is at least 5, then χ^2 is a value of a random variable whose distribution is approximately a chi-square distribution with $m - t - 1$ degrees of freedom, where m is the number of terms in the summation of χ^2 and t is the number of independent parameters estimated on the basis of the sample data. In this example, $t = 2$ since we estimated both α ans β, and $m = 10$. Therefore, the next step is:

Step 5:   Reject the null hypothesis if

$$\chi^2 \geq \chi^2_{\alpha, m-t-1}$$

In this example, $\chi^2 = 9$, $\chi^2_{\alpha, m-t-1} = \chi^2_{.01, 7} = 18.4753$. Therefore, we should not reject the null hypothesis that the distribution is uniform, because

$$\chi^2 = 9 \ngeq 18.4753 = \chi^2_{\alpha, m-t-1}$$

(b)   At the .10 level of significance, we should not reject the null hypothesis that the distribution is uniform, because

$$\chi^2 = 9 \ngeq 12.0170 = \chi^2_{.10, 7} = \chi^2_{\alpha, m-t-1}$$

(c)   At the .90 level of significance we should reject the null hypothesis that the distribution is uniform, because

$$\chi^2 = 9 > 2.83311 = \chi^2_{.90, 7} = \chi^2_{\alpha, m-t-1}$$

In concluding this discussion on the chi-square test for the goodness of fit let us take a moment to analyze the results of Examples 15.8, 15.9, and 15.10, and determine what these results tell us, in each case, concerning the goodness of fit of the specified distribution to the true distribution of the population. In each of these examples we saw that if we choose a level of significance that is sufficiently small ($\alpha \leq .01$ for Example 15.8, $\alpha \leq .90$ for Example 15.9, and $\alpha \leq .10$ for Example 15.10), then at that level of significance we should not reject the null hypothesis that the data constitute a random sample from a population having the distribution specified by the null hypothesis. However, we also saw that if we choose a level of significance that is sufficiently large ($\alpha \geq .05$ for Example 15.8, $\alpha \geq .95$ for Example 15.9, and $\alpha \geq .90$ for Example 15.10), then at that level of significance we should reject the null hypothesis that the data constitute a random sample from a population having the distribution specified by the null hypothesis.

What do these results tell us, in each case, concerning the goodness of fit of the specified distribution to the true distribution of the population? The answer is simple for Example 15.8. There, the results tell us that, at the .05 level of significance, we should reject the null hypothesis that the data constitute a random sample from a Poisson distribution. This means that we should conclude that the distribution of the population is significantly different (at the .05 level of significance) from a Poisson distribution. In other words, we should conclude that the fit of the Poisson distribution to the distribution of the population is significantly bad (at the .05 level of significance). The answers for Example 15.9 and 15.10 are not so simple. In each of these examples, obviously we cannot conclude, at a level of significance that is reasonably small, that the distribution of the population is significantly different from the specified distribution; but should we conclude that the specified distribution provides a good fit to the distribution of the population? This depends on what we believe a significance test (a chi-square goodness of fit test is such a test) can tell us about acceptance of the null hypothesis (i.e., the hypothesis that the specified distribution is the distribution of the population).

Technically, one should never accept the null hypothesis on the basis of the results of a significance test, because in such tests one cannot conveniently determine the probability of making an error of accepting the null hypothesis when the null hypothesis is false. (See Chapter 12, Understanding Testing of Statistical Hypotheses.) In other words, sound practice in the testing of statistical hypotheses dictates that we use a chi-square goodness of fit test only in cases where we wish to determine the "poorness" of fit (i.e., where we are concerned with the rejection of the null hypothesis). Therefore, if the goodness of fit test tells us that we should not reject the null hypothesis that the specified distribution is the distribution of the population, then we should reserve judgment concerning the fit of the distribution until we have found other means of determining the degree to which we can reasonably assume that the distribution provides a good fit.

One method of assessing the fit of the distribution is to plot the data provided by the random sample. Another method is to compute confidence limits for each of the expected frequencies. It is also a good idea to look at the p-value of the goodness of fit test (the p-value is the minimum level of significance that will lead to rejection of the null hypothesis—see Chapter 13, Section 13.7). For any significance test, a small p-value is statistical evidence that the null hypothesis should be rejected and a large p-value is statistical evidence that the null hypothesis should not be rejected. In terms of the chi-square goodness of fit test, a small p-value is statistical evidence that the fit is poor and a large p-value is statistical evidence that the fit cannot be too poor.

Using these criteria, we might argue that there is evidence that the fit in Example 15.9 cannot be too poor, because the p-value there is a number greater than .90, whereas there is evidence that the fit in Example 15.10 may not be as good, because the p-value there is a number that is somewhere between .10 and .90. Therefore, although the goodness of fit test cannot tell us about the goodness of the fit, in some cases it *can* tell us that the fit cannot be too bad!

Exercises

1. The list that follows shows the number of errors that a word processor made in processing 100 pages of material.

Number of errors per page	Number of pages
0	63
1	28
2	8
3	0
4	1
5	0

   (a)   At the .05 level of significance, should we reject the hypothesis that the distribution of the number of errors per randomly selected page is a Poisson distribution?
   (b)   Test, also, at the .10 level of significance.

2. A certain genetic model concerning random mating states that the number of offspring in three classes should be in the ratio $p^2:2pq:q^2$, where $p + q = 1$ and p is a parameter with $0 \le p \le 1$. A laboratory experiment on a certain type of organism yielded the result of 30, 50, and 30 offspring in the respective classes. What can we say concerning the fit of this type of genetic model to the organism under study?
   (a)   Test at the .05 level of significance.
   (b)   Test at the .10 level of significance.
   (*Hint:* to estimate p, note that if X is the class of the offspring (classes 1, 2, 3 in the ratio $p^2:2pq:q^2$), then $E(X) = 3 - 2p$.)

3. The recommended distribution of grades for a certain course with an enrollment of over 2,000 students is 10% A, 20% B, 50% C, 15% D, and 5% F. A certain section of this course has 100 students and the number of students receiving the grades of A, B, C, D, and F, respectively, were 20, 25, 40, 10, and 5. Can the grades for this section be considered a random sample from a population whose distribution is the recommended distribution of grades?
   (a)   Test at the .05 level of significance.
   (b)   Test at the .10 level of significance.

4. Three months before a certain election an extensive poll showed that 45%, 35%, and 20% of the voters favored candidates A, B, and C, respectively. Then, one month before the election, a random sample of 100 voters showed that 50, 25, and 25 of them favored candidates A, B, and C, respectively.
   (a)   Is there sufficient evidence, at the .05 level of significance, to indicate that there has been a shift in voter preference?
   (b)   What is the answer to this question at the .10 level of significance?

5. The manufacturer of a certain die claims that the die is balanced (i.e., each value has an equal probability of appearing). In checking this claim, a buyer rolled the die 120 times and obtained the following results:

Value of die	1	2	3	4	5	6
Observed frequency	27	22	20	25	16	10

(a) Should the buyer reject the manufacturer's claim at the .05 level of significance?

(b) What is the smallest level of significance at which we should reject the manufacturer's claim?

6. A random sample of 200 college freshmen were given a certain aptitude test; their scores were as follows:

Score x	Number of students
$x < 30$	5
$30 \le x < 40$	15
$40 \le x < 50$	30
$50 \le x < 60$	51
$60 \le x < 70$	60
$70 \le x < 80$	23
$80 \le x < 90$	10
$90 \le x$	6

Conduct a chi-square goodness of fit test for the null hypothesis that the distribution of scores for all college freshmen is a normal distribution with mean equal to 60 and standard deviation equal to 15. Test at the .05, .10, and .90 levels of significance.

7. The burning times of a random sample of 84 60-watt light bulbs from a certain manufacturer are shown in the table that follows. The sample mean was observed to be 4,000 hours.

Burning time (x) in hours	Number of light bulbs
$x < 1,000$	20
$1,000 \le x < 2,000$	15
$2,000 \le x < 3,000$	9
$3,000 \le x < 4,000$	12
$4,000 \le x < 6,000$	8
$6,000 \le x < 8,000$	7
$8,000 \le x < 12,000$	8
$12,000 \le x$	5

Test the hypothesis that the distribution of burning times of 60-watt light bulbs from this manufacturer is an exponential distribution.

(a) Test at the .05 level of significance.

(b) Find the smallest level of significance at which the null hypothesis should be rejected.

Supplementary Exercises

1. Suppose that the distribution of ignition times of Exercise 1 of Section 15.2 can be assumed to be normal. Use the parametric t test of Chapter 13, Section 13.2.2, to test the hypotheses of this exercise. Compare the answers obtained here with the answers to Exercise 1 of Section 15.2 and Exercise 1 of Section 15.3. Which of these three tests is the most powerful for testing the hypotheses under consideration?

2. Suppose that in Exercise 2 of Section 15.2 the distribution of customer responses can be assumed to be normal. Use the parametric t test of Chapter 13, Section 13.2.2, to test the hypotheses of this exercise. Compare the answers obtained here with the answers to Exercise 2 of Section 15.2 and Exercise 1 of Section 15.3. Which of the three tests used in these three exercises is the least powerful for testing the hypotheses under consideration?

References

Tables for small sample tests for various nonparametric methods may be found in:

1. Owen, D. B. 1962. *Handbook of statistical tables*. Menlo Park, CA: Addison-Wesley Publishing Co.

Advanced texts on mathematical statistics include:

2. Bickel, P., and K. Doksum. 1977. *Mathematical statistics*. San Francisco: Holden-Day, Inc.
3. De Groot, M. H. 1975. *Probability and statistics*. Menlo Park, CA: Addison-Wesley Publishing Co.
4. Hogg, R. V., and A. T. Craig. 1970. *Introduction to mathematical statistics*, 3d ed. London: The Macmillan Co.

Texts on nonparametric methods include:

5. Gibbons, J. D. 1971. *Nonparametric statistical inference*. New York: McGraw-Hill Book Co.
6. Lehmann, E. L. 1975. *Nonparametrics: Statistical methods based on ranks*. San Francisco: Holden-Day, Inc.
7. Mosteller, F., and R. E. K. Rourke. 1973. *Sturdy statistics*. Menlo Park, CA: Addison-Wesley Publishing Co., Inc.
8. Noether, G. C. 1976. *Introduction to statistics: A nonparametric approach*, 2d ed. Boston: Houghton Mifflin Co.
9. Siegel, S. 1956. *Nonparametric statistics for the behavioral sciences*. New York: McGraw-Hill Book Co.

APPENDIXES

MOMENT GENERATING FUNCTIONS

Definition of $M_X(t)$

Why $M_X(t)$ Is Called the Moment Generating Function of a Random Variable

Moment Generating Functions of Special Distributions

Properties of Moment Generating Functions

Derivation of Moment Generating Functions of Special Distributions

  Moment generating function of the binomial distribution

  Moment generating function of the Poisson distribution

  Moment generating function of the normal distribution

  Moment generating function of the exponential distribution

  Moment generating function of the gamma distribution

  Moment generating function of the chi-square distribution

A.1 Definition of $M_X(t)$

Definition A.1 The moment generating function of a discrete random variable X is the function

$$M_X(t) = E(e^{tX}) = \sum_x e^{tx} f(x)$$

provided that the series converges for every value t in some interval about $t = 0$.

The moment generating function of a continuous random variable X is the function

$$M_X(t) = E(e^{tX}) = \int_{-\infty}^{\infty} e^{tx} f(x)\, dx$$

provided that the integral exists for every value t in some interval about $t = 0$.

Note: the moment generating function of a random variable is also referred to as the moment generating function of the distribution of the random variable.

EXAMPLE A.1

Find the moment generating function of the random variable X whose probability density function is $f(x) = e^{-x}$ for $x > 0$ and $f(x) = 0$ for $x \leq 0$.

SOLUTION

$$M_X(t) = E(e^{tX}) = \int_{-\infty}^{\infty} e^{tx} f(x)\, dx = \int_0^{\infty} e^{tx} e^{-x}\, dx$$

$$= \int_0^{\infty} e^{-x(1-t)}\, dx = \frac{1}{1-t} \qquad \text{for} \quad |t| < 1$$

EXAMPLE A.2

Find the moment generating function of a binomial random variable.

SOLUTION

$$M_X(t) = E(e^{tX}) = \sum_x e^{tx} f(x) = \sum_{x=0}^{n} e^{tx} b(x; n, p)$$

$$= \sum_{x=0}^{n} e^{tx} \binom{n}{x} p^x (1-p)^{n-x} = \sum_{x=0}^{n} \binom{n}{x} (pe^t)^x (1-p)^{n-x}$$

$$= [pe^t + (1-p)]^n$$

The last step follows from the binomial expansion theorem which states that $(a + b)^n = \sum\limits_{x=0}^{n} \binom{n}{x} a^x b^{n-x}$.

A.2 Why $M_X(t)$ Is Called the Moment Generating Function of a Random Variable

Suppose that X is a discrete random variable. By definition the moment generating function of X is

$$M_X(t) = E(e^{tX}) = e^{tx}f(x)$$

In the preceding series, let us replace e^{tx} with its Maclaurin's series expansion (i.e., the expansion $e^{tx} = 1 + tx + \dfrac{t^2x^2}{2!} + \dfrac{t^3x^3}{3!} + \cdots + \dfrac{t^rx^r}{r!} + \cdots$). Then, we have

$$M_X(t) = \sum_x e^{tx}f(x) = \sum_x \left(1 + tx + \frac{t^2x^2}{2!} + \frac{t^3x^3}{3!} + \cdots + \frac{t^rx^r}{r!} + \cdots\right)f(x)$$

$$= \sum_x f(x) + t \cdot \sum_x xf(x) + \frac{t^2}{2!} \sum_x x^2f(x) + \cdots + \frac{t^r}{r!} \sum_x x^rf(x) + \cdots$$

$$= 1 + tE(X) + \frac{t^2}{2!} E(X^2) + \frac{t^3}{3!} E(X^3) + \cdots + \frac{t^r}{r!} E(X^r) + \cdots$$

Now, recall that for each positive integer r, the quantity $E(X^r)$ is called the rth moment of X. Therefore, it is reasonable to call $M_X(t)$ the moment generating function of X because the function $M_X(t)$, when expanded in powers of t, generates the moments of X. The same result can be obtained for continuous random variables when we replace sums with integrals in the preceding set of equations.

Summary

If $M_X(t)$ is the moment generating function of a random variable X, then

$$M_X(t) = 1 + tE(X) + \frac{t^2}{2!} E(X^2) + \frac{t^3}{3!} E(X^3) + \cdots + \frac{t^r}{r!} E(X^r) + \cdots$$

EXAMPLE A.3

Expand the moment generating function of Example A.1 in powers of t and read off the 3rd and 4th moments of X.

SOLUTION

$$M_X(t) = \frac{1}{1-t} = 1 + t + t^2 + \cdots + t^r + \cdots$$

$$= 1 + t \cdot 1 + \frac{t^2}{2!} \cdot 2! + \frac{t^3}{3!} \cdot 3! + \frac{t^4}{4!} \cdot 4! + \cdots$$

Since $E(X^r)$ is the coefficient of $\dfrac{t^r}{r!}$ in the expansion of $M_X(t)$ in powers of t, it follows that in this example

$$\text{(the 3rd moment of } X) = E(X^3) = 3!$$

and

$$\text{(the 4th moment of } X) = E(X^4) = 4!$$

The following theorem shows that it is possible to obtain the moments of X from $M_X(t)$ without expanding the function in powers of t.

Theorem A.1 *If $M_X(t)$ is the moment generating function of X, then*

$$E(X^r) = \frac{d^r M_X(t)}{dt^r}\bigg|_{t=0}$$

Proof Expanded in powers of t, the moment generating function of X is given by

$$M_X(t) = 1 + tE(X) + \frac{t^2}{2!} E(X^2) + \frac{t^3}{3!} E(X^3) + \cdots + \frac{t^r}{r!} E(X^r) + \cdots$$

Therefore

$$\frac{dM_X(t)}{dt} = E(X) + tE(X^2) + \frac{t^2}{2!} E(X^3) + \cdots + \frac{t^{r-1}}{(r-1)!} E(X^r) + \cdots$$

and

$$\frac{dM_X(t)}{dt}\bigg|_{t=0} = E(X)$$

The second derivative of $M_X(t)$ may be obtained by differentiating $\dfrac{dM_X(t)}{dt}$.
We obtain

$$\frac{d^2 M_X(t)}{dt^2} = E(X^2) + tE(X^3) + \cdots + \frac{t^{r-2}}{(r-2)!} E(X^r) + \cdots$$

and

$$\left.\frac{d^2 M_X(t)}{dt^2}\right|_{t=0} = E(X^2)$$

It is clear, now, that for any positive integer r

$$\left.\frac{d^r M_X(t)}{dt^r}\right|_{t=0} = E(X^r)$$

EXAMPLE A.4

Use Theorem A.1 to find the first and second moments of the random variable X of Example A.1.

SOLUTION

$$M_X(t) = \frac{1}{1-t} \qquad \text{for} \quad |t| < 1$$

Therefore

$$\frac{dM_X(t)}{dt} = \frac{1}{(1-t)^2} \qquad \text{for} \quad |t| < 1$$

and

$$\frac{d^2 M_X(t)}{dt^2} = \frac{2}{(1-t)^2} \qquad \text{for} \quad |t| < 1$$

It follows that

$$E(X) = \left.\frac{dM_X(t)}{dt}\right|_{t=0} = 1$$

and

$$E(X^2) = \left.\frac{d^2 M_X(t)}{dt^2}\right|_{t=0} = 2$$

A.3   Moment Generating Functions of Special Distributions

The moment generating functions of the binomial, Poisson, normal, exponential, gamma, and chi-square distributions are given in Table A.1. The derivation of these moment generating functions is given in Section A.5.

TABLE A.1
Moment Generating Functions of Special Distributions

Distribution	pdf or pf	$M_X(t)$
Binomial	$f(x) = b(x; n, p)$ $= \binom{n}{x} p^x (1-p)^{n-x}$ for $x = 0, 1, \ldots, n$	$M_X(t) = [1 + p(e^t - 1)]^n$
Poisson	$f(x) = p(x; \lambda)$ $= (\lambda^x e^{-\lambda})/x!$ for $x = 0, 1, 2, \ldots$	$M_X(t) = e^{\lambda(e^t - 1)}$
Normal	$f(x) = \dfrac{1}{\sqrt{2\pi}\,\sigma} e^{-\frac{(x-\mu)^2}{2\sigma^2}}$ for $-\infty < x < \infty$	$M_X(t) = e^{\mu t + \frac{t^2}{2\sigma^2}}$
Exponential	$f(x) = \begin{cases} \dfrac{1}{\theta} e^{-x/\theta} & \text{for } x > 0 \\ 0 & \text{elsewhere} \end{cases}$	$M_X(t) = (1 - \theta t)^{-1}$, $\lvert t \rvert < 1/\theta$
Gamma	$f(x) = \begin{cases} \dfrac{1}{\beta^\alpha \Gamma(\alpha)} x^{\alpha-1} e^{-x/\beta} & \\ \text{for } x > 0 & \\ 0 & \text{elsewhere} \end{cases}$	$M_X(t) = (1 - \beta t)^{-\alpha}$, $\lvert t \rvert < 1/\beta$
Chi-Square	$f(x) = \begin{cases} \dfrac{1}{2^{v/2}\Gamma(v/2)} x^{\frac{v-2}{2}} e^{-\frac{x}{2}} & \\ \text{for } x > 0 & \\ 0 & \text{elsewhere} \end{cases}$	$M_X(t) = (1 - 2t)^{-v/2}$, $\lvert t \rvert < 1/2$

A.4   Properties of Moment Generating Functions

Several important properties of moment generating functions are given in the following theorems.

Theorem A.2

    (a) *If X_1 and X_2 are random variables with moment generating functions $M_{X_1}(t)$ and $M_{X_2}(t)$, respectively, then the distributions of X_1 and X_2 are identical if and only if their moment generating functions are identical.*

    (b) *Let X be a random variable with moment generating function $M_X(t)$ and distribution function F. Let $X_1, X_2, X_3, \ldots$ be a sequence of random variables* with respective moment generating functions $M_{X_1}(t), M_{X_2}(t), M_{X_3}(t), \ldots$, and respective distribution functions $F_1, F_2, F_2, \ldots$. Then

$$\lim_{n \to \infty} F_n(x) = F(x) \qquad \text{for} \quad -\infty < x < \infty$$

if and only if

$$\lim_{n \to \infty} M_{X_n}(t) = M_X(t)$$

for each value t in some interval about $t = 0$.

Note: the implication of this statement is that if $\lim_{n \to \infty} M_{X_n}(t) = M_X(t)$, then for n large the distribution of X_n may be approximated by the distribution of X, that is, $P(a < X_n < b) \approx P(a < X < b)$.

The proof of Theorem A.2 is beyond the scope of this book.

EXAMPLE A.5

If the moment generating function of a random variable X is $M_X(t) = e^{(1/2)t^2}$, $-\infty < t < \infty$, then what is the distribution of X?

   SOLUTION

From Section A.3, we saw that if Z is the standard normal random variable, then the moment generating function of Z is

$$M_Z(t) = e^{(1/2)t^2}$$

Since $M_X(t)$ is identical to $M_Z(t)$, it follows from Theorem A.2(a) that the distribution of X is identical to the distribution of Z. In other words, the distribution of X is the standard normal distribution.

EXAMPLE A.6

Suppose that $X_1, X_2, X_3, \ldots$ are random variables with respective moment generating functions given by

$$M_{X_n}(t) = e^{(1/n)t + (1/2)\left(1 + \frac{1}{n}\right)t^2}, \qquad -\infty < t < \infty$$

for $n = 1, 2, 3, \ldots$.
   (a) As $n \to \infty$, the distribution of X_n converges to what known distribution?
   (b) For n large, what is an approximate value for $P(0 < X_n < 1)$?

SOLUTION

(a)   Since $e^{(1/2)t^2}$ is the moment generating function of the standard normal distribution, and since

$$\lim_{n \to \infty} M_{X_n}(t) = \lim_{n \to \infty} e^{(1/n)t + (1/2)\left(1 + \frac{1}{n}\right)t^2} = e^{(1/2)t^2}$$

it follows (from Theorem A.2 (b)) that as $n \to \infty$, the distribution of X_n converges to the standard normal distribution.

(b)   The consequence of (a) is that for n large the distribution of X_n may be approximated by the standard normal distribution. Therefore

$$P(0 < X_n < 1) \approx .3413$$

Theorem A.3   *If a and b are constants, then*

(a)

$$M_{X+a}(t) = E(e^{(X+a)t}) = e^{at} M_X(t)$$

(b)

$$M_{bX}(t) = E(e^{bXt}) = M_X(bt)$$

(c)

$$M_{\frac{X+a}{b}}(t) = E\left[e^{\left(\frac{X+a}{b}\right) \cdot t}\right] = e^{\left(\frac{a}{b}\right) \cdot t} \cdot M_X\left(\frac{t}{b}\right)$$

Proof

(a)

$$M_{X+a}(t) = E(e^{(X+a)t}) = E(e^{Xt} \cdot e^{at}) = e^{at} E(e^{Xt}) = e^{at} M_X(t)$$

(b)

$$M_{bX}(t) = E(e^{bXt}) = E(e^{X(bt)}) = M_X(bt)$$

(c)

$$M_{\frac{X+a}{b}}(t) = M_{\left[\frac{X}{b} + \frac{a}{b}\right]}(t)$$

Applying (a), we obtain

$$M_{\left[\frac{X}{b} + \frac{a}{b}\right]}(t) = e^{\left(\frac{a}{b}\right) \cdot t} M_{\frac{X}{b}}(t)$$

Applying (b), we obtain

$$M_{\frac{X}{b}}(t) = M_X[(1/b)t] = M_X\left(\frac{t}{b}\right)$$

Therefore

$$M_{\frac{X+a}{b}}(t) = e^{\left(\frac{a}{b}\right)\cdot t}\, M_X\!\left(\frac{t}{b}\right)$$

EXAMPLE A.7

Show that if the random variable X is normally distributed with mean μ and variance σ^2, then the random variable $Y = aX + b$ is normally distributed with mean equal to $(a\mu + b)$ and variance equal to $a^2\sigma^2$.

SOLUTION

From Section A.3, we have

$$M_X(t) = e^{\mu t + \frac{1}{2}\sigma^2 t^2} \qquad \text{for} \quad -\infty < t < \infty$$

From Theorem A.3, we have

$$M_Y(t) = M_{aX+b}(t) = e^{bt} M_{aX}(t) = e^{bt} M_X(at) = e^{bt} e^{at + \frac{1}{2}\sigma^2(at)^2}$$

$$= e^{(a+b)t + (1/2)a^2\sigma^2 t^2}$$

The latter is the moment generating function of a random variable that is normally distributed with mean equal to $(a\mu + b)$ and variance equal to $a^2\sigma^2$. We can therefore apply Theorem A.2(a) to conclude that the distribution of Y is normal with mean equal to $(a\mu + b)$ and variance equal to $a^2\sigma^2$.

EXAMPLE A.8   NORMAL APPROXIMATION OF
THE BINOMIAL DISTRIBUTION

Let S_n be a binomial random variable with probability function given by

$$f(x) = b(x; n, p) = \binom{n}{x} p^x q^{n-x} \qquad \text{for} \quad x = 0, 1, \ldots, n$$

Show that for n large, the distribution of $(S_n - np)/(npq)^{1/2}$ is approximately standard normal. (This is Statement 1 of Chapter 8, Section 8.6.1.)

SOLUTION

We will prove the following stronger statement: As $n \to \infty$, the distribution of $\dfrac{S_n - np}{\sqrt{npq}}$ converges to the standard normal distribution. The proof will be in terms of the following 3 steps.

1. Find the moment generating function of $\dfrac{S_n - np}{\sqrt{npq}}$.

2. Show that

$$\lim_{n \to \infty} M_{\frac{S_n - np}{\sqrt{npq}}}(t) = e^{(1/2)t^2}$$

3. Since $e^{(1/2)t^2}$ is the moment generating function of the standard normal random variable, it follows (from Theorem A.2(b)) that as $n \to \infty$, the distribution of $\dfrac{S_n - np}{\sqrt{npq}}$ will converge to the standard normal distribution.

Step 1:   Let us simplify the notations in the proof by adopting the following notations: $X = S_n$, $\mu = np$, and $\sigma^2 = npq$.

Keep in mind the fact that the moment generating function of the binomial random variable is

$$M_X(t) = M_{S_n}(t) = [1 + p(e^t - 1)]^n$$

Then, apply Theorem A.3 to obtain

$$M_{\frac{S_n - np}{\sqrt{npq}}}(t) = M_{\frac{X - \mu}{\sigma}}(t) = e^{-\left(\frac{\mu}{\sigma}\right)\cdot t} M_X\left(\frac{t}{\sigma}\right) = e^{-\left(\frac{\mu}{\sigma}\right)\cdot t}[1 + p(e^{\frac{t}{\sigma}} - 1)]^n$$

Next, take logarithms and substitute the Maclaurin's series of $e^{t/\sigma}$ to obtain

$$\ln M_{\frac{X-\mu}{\sigma}}(t) = -\frac{\mu}{\sigma}\cdot t + n\ln\left[1 + p(e^{\frac{t}{\sigma}} - 1)\right]$$

$$= -\frac{\mu}{\sigma}\cdot t + n\ln\left[1 + p\left\{\frac{t}{\sigma} + \frac{1}{2}\left(\frac{t}{\sigma}\right)^2 + \frac{1}{6}\left(\frac{t}{\sigma}\right)^3 + \cdots\right\}\right]$$

Then, use the infinite series $\ln(1 + x) = x - \frac{1}{2}x^2 + \frac{1}{3}x^3 - \cdots$, which converges for $|x| < 1$, to expand the logarithm in the last term of the preceding equation to obtain

$$\ln M_{\frac{X-\mu}{\sigma}}(t) = -\frac{\mu}{\sigma}\cdot t + np\left[\frac{t}{\sigma} + \frac{1}{2}\left(\frac{t}{\sigma}\right)^2 + \frac{1}{6}\left(\frac{t}{\sigma}\right)^3 + \cdots\right]$$

$$- \frac{np^2}{2}\left[\frac{t}{\sigma} + \frac{1}{2}\left(\frac{t}{\sigma}\right)^2 + \frac{1}{6}\left(\frac{t}{\sigma}\right)^3 + \cdots\right]^2$$

$$+ \frac{np^3}{3}\left[\frac{t}{\sigma} + \frac{1}{2}\left(\frac{t}{\sigma}\right)^2 + \frac{1}{6}\left(\frac{t}{\sigma}\right)^3 + \cdots\right]^3 - \cdots$$

Collecting powers of t, we obtain (with the fact that $\mu = np$)

$$\ln M_{\frac{X-\mu}{\sigma}}(t) = \left(-\frac{\mu}{\sigma} + \frac{np}{\sigma}\right)t + \left(\frac{np}{2\sigma^2} - \frac{np^2}{2\sigma^2}\right)t^2 + \left(\frac{np}{6\sigma^3} - \frac{np^2}{2\sigma^3} + \frac{np^3}{3\sigma^3}\right)t^3 + \cdots$$

$$= \frac{1}{\sigma^2}\left(\frac{np - np^2}{2}\right)t^2 + \frac{n}{\sigma^3}\left(\frac{p - 3p^2 + 2p^3}{6}\right)t^3 + \cdots$$

Then, make the substitution $\sigma = \sqrt{np(1-p)}$ to obtain

$$\ln M_{\frac{X-\mu}{\sigma}}(t) = \frac{1}{2}t^2 + \frac{n}{\sigma^3}\left(\frac{p - 3p^2 + 2p^3}{6}\right)t^3 + \cdots$$

where for $r > 2$, the coefficient of t^r is a constant times n/σ^r.

Step 2:   Since $\sigma = \sqrt{np(1-p)}$, it follows that as n tends to infinity, the expression n/σ^r tends to zero. Therefore, an examination of the last expression for $\ln M_{\frac{X-\mu}{\sigma}}(t)$ in Step 1 shows that

$$\lim_{n \to \infty} \ln M_{\frac{S_n - np}{\sqrt{npq}}}(t) = \lim_{n \to \infty} \ln M_{\frac{X-\mu}{\sigma}}(t) = \frac{1}{2} \cdot t^2$$

Therefore, it follows that

$$\lim_{n \to \infty} M_{\frac{S_n - np}{\sqrt{npq}}}(t) = e^{(1/2)t^2}$$

Step 3:   Since $e^{(1/2)t^2}$ is the moment generating function of the standard normal random variable, it follows (from Theorem A.2(b)) that as n tends to infinity, the distribution of $\dfrac{S_n - np}{\sqrt{npq}}$ will converge to the standard normal distribution. This, in turn, implies that for n large, the distribution of $\dfrac{S_n - np}{\sqrt{npq}}$ is approximately standard normal.

Theorem A.4   *If* $X_1, X_2, \ldots,$ *and* X_n *are independent random variables and* $Y = X_1 + X_2 + \cdots + X_n$, *then the moment generating function of* Y *is given by*

$$M_Y(t) = M_{X_1}(t) \cdot M_{X_2}(t) \cdots M_{X_n}(t)$$

where $M_{X_i}(t)$ *is the moment generating function of* X_i, *for* $i = 1, 2, \ldots, n$.

Proof   Let us consider the case of continuous random variables. Since the random variables are independent, we have

$$f(x_1, x_2, \ldots, x_n) = f_1(x_1) \cdot f_2(x_2) \cdots f_n(x_n)$$

where $f_i(x_i)$ is the probability density function of X_i. It follows that

$$M_Y(t) = E(e^{Yt}) = E(e^{(X_1 + X_2 + \cdots + X_n)t})$$

$$= \int_{-\infty}^{\infty} \cdots \int_{-\infty}^{\infty} e^{(x_1 + x_2 + \cdots + x_n)t} f(x_1, x_2, \ldots, x_n) \, dx_1 dx_2 \cdots dx_n$$

$$= \int_{-\infty}^{\infty} e^{x_1 t} f_1(x_1) \, dx_1 \cdot \int_{-\infty}^{\infty} e^{x_2 t} f_2(x_2) \, dx_2 \cdots \int_{-\infty}^{\infty} e^{x_n t} f_n(x_n) \, dx_n$$

$$= M_{X_1}(t) \cdot M_{X_2}(t) \cdots M_{X_n}(t)$$

The proof for discrete random variables can be obtained by replacing integrals with sums in the preceding equations.

EXAMPLE A.9 SUMS OF INDEPENDENT AND IDENTICALLY DISTRIBUTED NORMAL RANDOM VARIABLES

Prove the following statement: If $X_1, X_2, \ldots, X_n$ are independent and identically distributed normal random variables with means equal to μ and variances equal to σ^2, then $Y = X_1 + X_2 + \cdots + X_n$ is normally distributed with mean equal to $n\mu$ and variance equal to $n\sigma^2$.

SOLUTION

Using Theorem A.4 and the fact that the moment generating function of a normal random variable X with mean μ and variance σ^2 is $M_X(t) = e^{\mu t + \frac{1}{2}\sigma^2 t^2}$, it follows that

$$M_Y(t) = M_{X_1 + X_2 + \cdots + X_n}(t) = M_{X_1}(t) \cdot M_{X_2}(t) \cdots M_{X_n}(t)$$

$$= (e^{\mu t + \frac{1}{2}\sigma^2 t^2}) \cdot (e^{\mu t + \frac{1}{2}\sigma^2 t^2}) \cdots (e^{\mu t + \frac{1}{2}\sigma^2 t^2}) = e^{n\mu t + \frac{1}{2}n\sigma^2 t^2}$$

The last term in the preceding equation is the moment generating function of a normal random variable with mean equal to $n\mu$ and variance equal to $n\sigma^2$. Therefore, it follows from Theorem A.2(a) that the distribution of Y is normal with mean equal to $n\mu$ and variance equal to $n\sigma^2$.

EXAMPLE A.10 THE DISTRIBUTION OF $\bar{X}$

Prove the following statement: If $X_1, X_2, \ldots, X_n$ are independent and identically distributed normal random variables with means equal to μ and variances equal to σ^2, then $\bar{X} = (X_1 + X_2 + \cdots + X_n)/n$ is normally distributed with mean equal to μ and variance equal to σ^2/n.

SOLUTION

From Example A.9, we have

$$M_{X_1 + X_2 + \cdots + X_n}(t) = e^{n\mu t + \frac{1}{2}n\sigma^2 t^2}$$

Therefore, we may apply Theorem A.3(b) to obtain

$$M_{\bar{X}}(t) = M_{\frac{X_1+X_2+\cdots+X_n}{n}}(t) = M_{X_1+X_2+\cdots+X_n}\left(\frac{t}{n}\right) = e^{n\mu(t/n)+\frac{1}{2}n\sigma^2(t/n)^2}$$

$$= e^{\mu t + \left(\frac{\sigma^2}{2n}\right)\cdot t^2}$$

The last term in the preceding equation is the moment generating function of a normal random variable with mean equal to μ and variance equal to σ^2/n. Therefore, it follows from Theorem A.2(a) that the distribution of $\bar{X}$ is normal with mean equal to μ and variance equal to σ^2/n.

EXAMPLE A.11   THE CENTRAL LIMIT THEOREM

Prove the following statement: If $X_1, X_2, \ldots$, and X_n are independent and identically distributed random variables with means equal to μ, variances equal to σ^2, and moment generating functions given by $M_X(t)$, then as $n \to \infty$, the distribution of $\dfrac{\bar{X}-\mu}{\sigma/\sqrt{n}}$ (where $\bar{X} = (X_1 + X_2 + \cdots + X_n)/n$) converges to the standard normal distribution. (This statement is the central limit theorem for random variables with moment generating functions.)

SOLUTION

Applying Theorem A.3, we obtain

$$M_{\frac{\bar{X}-\mu}{\sigma/\sqrt{n}}}(t) = e^{-\sqrt{n}\mu t/\sigma}\cdot M_{\bar{X}}(\sqrt{n}t/\sigma) = e^{-\sqrt{n}\mu t/\sigma}\cdot M_{n\bar{X}}(t/\sigma\sqrt{n})$$

Since $n\bar{X} = X_1 + X_2 + \cdots + X_n$, it follows from Theorem A.4 that

$$M_{\frac{\bar{X}-\mu}{\sigma/\sqrt{n}}}(t) = e^{-\sqrt{n}\mu t/\sigma}\cdot[M_X(t/\sigma\sqrt{n})]^n$$

Therefore

$$\ln M_{\frac{\bar{X}-\mu}{\sigma/\sqrt{n}}}(t) = -\sqrt{n}\mu t/\sigma + n\cdot\ln M_X(t/\sigma\sqrt{n})$$

By expanding $M_X(t/\sqrt{n})$ as a power series in t, we obtain

$$\ln M_{\frac{\bar{X}-\mu}{\sigma/\sqrt{n}}}(t) = -\sqrt{n}\mu t/\sigma + n\cdot\ln\left[1 + \mu_1'\frac{t}{\sigma\sqrt{n}} + \mu_2'\frac{t^2}{2\sigma^2 n}\right.$$

$$\left. + \mu_3'\frac{t^3}{6\sigma^3 n\sqrt{n}} + \cdots\right]$$

where $\mu_1', \mu_2', \mu_3', \ldots$, are the first, second, third, $\ldots$, moments of the distribution that is common to all of the random variables (remember that $X_1, X_2, \ldots, X_n$ are independent and *identically distributed* random variables).

Recall that in Example A.8 we made use of an infinite series expansion for $\ln(1 + x)$ for $|x| < 1$. Now, if n is sufficiently large, we can make use of that same infinite series to obtain

$$\ln M_{\frac{\bar{X}-\mu}{\sigma/\sqrt{n}}}(t) = -\sqrt{n}\,\mu t/\sigma + n\left\{\left[\mu_1'\frac{t}{\sigma\sqrt{n}} + \mu_2'\frac{t^2}{2\sigma^2 n} + \mu_3'\frac{t^3}{6\sigma^3 n\sqrt{n}} + \cdots\right]\right.$$

$$-\frac{1}{2}\left[\mu_1'\frac{t}{\sigma\sqrt{n}} + \mu_2'\frac{t^2}{2\sigma^2 n} + \mu_3'\frac{t^3}{6\sigma^3 n\sqrt{n}} + \cdots\right]^2$$

$$\left.+\frac{1}{3}\left[\mu_1'\frac{t}{\sigma\sqrt{n}} + \mu_2'\frac{t^2}{2\sigma^2 n} + \mu_3'\frac{t^3}{6\sigma^3 n\sqrt{n}} + \cdots\right]^3 - \cdots\right\}$$

Then, collect powers of t to obtain

$$\ln M_{\frac{\bar{X}-\mu}{\sigma/\sqrt{n}}}(t) = \left(-\frac{\sqrt{n}\,\mu}{\sigma} + \frac{\sqrt{n}\,\mu_1'}{\sigma}\right)t + \left(\frac{\mu_2'}{2\sigma^2} - \frac{(\mu_1')^2}{2\sigma^2}\right)t^2$$

$$+\left(\frac{\mu_3'}{6\sigma^3\sqrt{n}} - \frac{\mu_1'\cdot\mu_2'}{2\sigma^3\sqrt{n}} + \frac{(\mu_1')^3}{3\sigma^3\sqrt{n}}\right)t^3 + \cdots$$

Since $\mu_1' = \mu$ and $\mu_2' - (\mu_1')^2 = \sigma^2$, this becomes

$$\ln M_{\frac{\bar{X}-\mu}{\sigma/\sqrt{n}}}(t) = \frac{1}{2}t^2 + \left(\frac{\mu_3'}{6} - \frac{\mu_1'\mu_2'}{2}\frac{(\mu_1')^3}{3}\right)\frac{t^3}{\sigma^3\sqrt{n}} + \cdots$$

It is now clear that as n tends to infinity, all the first term in the preceding series converge to zero, so that

$$\lim_{n\to\infty} \ln M_{\frac{\bar{X}-\mu}{\sigma/\sqrt{n}}}(t) = (1/2)t^2$$

Therefore

$$\lim_{n\to\infty} M_{\frac{\bar{X}-\mu}{\sigma/\sqrt{n}}}(t) = e^{-(1/2)t^2}$$

Since $e^{-(1/2)t^2}$ is the moment generating function of the standard normal distribution, we may apply Theorem A.2(b) to conclude that as $n \to \infty$, the distribution of $\dfrac{\bar{X} - \mu}{\sigma/\sqrt{n}}$ converges to the standard normal distribution.

EXAMPLE A.12   SUMS OF INDEPENDENT POISSON RANDOM VARIABLES

Show that a sum of independent Poisson random variables is a Poisson random variable.

SOLUTION

Let $X_1, X_2, \ldots,$ and X_n be independent Poisson random variables with means equal to $\lambda_1, \lambda_2, \ldots,$ and λ_n, respectively. Let

$$Y = X_1 + X_2 + \cdots + X_n$$

Since $M_{X_i}(t) = e^{\lambda_i(e^t - 1)}$, it follows on application of Theorem A.4 that

$$M_Y(t) = M_{X_1 + X_2 + \cdots + X_n}(t) = M_{X_1}(t) \cdot M_{X_2}(t) \cdots M_{X_n}(t)$$
$$= e^{\lambda_1(e^t - 1)} \cdot e^{\lambda_2(e^t - 1)} \cdots e^{\lambda_n(e^t - 1)} = e^{(\lambda_1 + \lambda_2 + \cdots + \lambda_n)(e^t - 1)}$$

The last term in the preceding set of equations is the moment generating function of a Poisson random variable with mean equal to $\lambda_1 + \lambda_2 + \cdots + \lambda_n$. Therefore (on application of Theorem A.2(a)), Y is a Poisson random variable.

A.5   Derivation of Moment Generating Functions of Special Distributions

A.5.1   Moment Generating Function of the Binomial Distribution

The moment generating function of the binomial distribution is given by

$$M_X(t) = [1 + p(e^t - 1)]^n$$

The derivation was given in Example A.2.

A.5.2   Moment Generating Function of the Poisson Distribution

The moment generating function of the Poisson distribution is given by

$$M_X(t) = e^{\lambda(e^t - 1)}$$

The derivation follows.

$$M_X(t) = E(e^{tX}) = \sum_{x=0}^{\infty} e^{tx} p(x; \lambda) = \sum_{x=0}^{\infty} e^{tx} \cdot \frac{\lambda^x e^{-\lambda}}{x!}$$

$$= \sum_{x=0}^{\infty} \frac{(\lambda e^t)^x e^{-\lambda}}{x!} = e^{-\lambda} \sum_{x=0}^{\infty} \frac{(\lambda e^t)^x}{x!} = e^{-\lambda} e^{\lambda e^t} = e^{\lambda(e^t - 1)}$$

(The preceding equations make use of the fact that $e^r = \sum_{x=0}^{\infty} (r^x/x!)$ is the series expansion of e^r.)

A.5.3   Moment Generating Function of the Normal Distribution

The moment generating function of the normal distribution is given by

$$M_X(t) = e^{\mu t + \frac{1}{2}\sigma^2 t^2}$$

The derivation follows.

$$M_X(t) = E(e^{tX}) = \int_{-\infty}^{\infty} e^{tx} \cdot \frac{1}{\sqrt{2\pi}\,\sigma} e^{-\frac{(x-\mu)^2}{2\sigma^2}}\, dx$$

$$= \int_{-\infty}^{\infty} \frac{1}{\sqrt{2\pi}\,\sigma} e^{-\frac{1}{2\sigma^2}[-2xt\sigma^2 + (x-\mu)^2]}\, dx$$

$$= \int_{-\infty}^{\infty} \frac{1}{\sqrt{2\pi}\,\sigma} e^{-\frac{1}{2\sigma^2}[\{x-(\mu+t\sigma^2)\}^2 - 2\mu t\sigma^2 - t^2\sigma^4]}\, dx$$

$$= e^{\mu t + \frac{1}{2}\sigma^2 t^2} \left\{ \int_{-\infty}^{\infty} \frac{1}{\sqrt{2\pi}\,\sigma} e^{-\frac{1}{2}\left(\frac{x-(\mu+t\sigma^2)}{\sigma}\right)^2}\, dx \right\}$$

when we apply the method of completing the square to the exponent of the fourth term in the preceding equations. Now, observe that the integrand in the last term of the preceding set of equations is the probability density function of a normal random variable with mean equal to $(\mu + t\sigma^2)$ and with variance equal to σ^2. Therefore, the integral must be equal to 1. Consequently

$$M_X(t) = e^{\mu t + \frac{1}{2}\sigma^2 t^2}$$

A.5.4   Moment Generating Function of the Exponential Distribution

The moment generating function of the exponential distribution is given by

$$M_X(t) = (1 - \theta t)^{-1} \qquad \text{for} \quad |t| < 1/\theta$$

The derivation follows.

$$M_X(t) = E(e^{tX}) = \int_0^{\infty} e^{tx} \cdot \frac{1}{\theta} e^{-x/\theta}\, dx$$

$$= \frac{1}{\theta} \int_0^{\infty} e^{-x\left(\frac{1}{\theta} - t\right)}\, dx = \left(\frac{1}{\theta}\right)\left(\frac{-1}{\frac{1}{\theta} - t}\right) e^{-x\left(\frac{1}{\theta} - t\right)} \Bigg|_0^{\infty}$$

$$= \left(\frac{1}{\theta}\right)\left(\frac{-1}{\frac{1}{\theta} - t}\right) = (1 - \theta t)^{-1} \qquad \text{for} \quad |t| < 1/\theta$$

A.5.5 Moment Generating Function of the Gamma Distribution

The moment generating function of the gamma distribution is given by

$$M_X(t) = (1 - \beta t)^{-\alpha} \quad \text{for} \quad |t| < 1/\beta$$

The derivation follows.

$$M_X(t) = E(e^{tX}) = \int_0^\infty e^{tx} \cdot \frac{1}{\beta^\alpha \Gamma(\alpha)} x^{\alpha-1} e^{-x/\beta} \, dx$$

$$= \int_0^\infty \frac{1}{\beta^\alpha \Gamma(\alpha)} x^{\alpha-1} e^{-x\left(\frac{1}{\beta} - t\right)} \, dx$$

Let $y = x(1 - \beta t)$. Then

$$M_X(t) = \int_0^\infty \frac{1}{\beta^\alpha \Gamma(\alpha)} \frac{y^{\alpha-1}}{(1 - \beta t)^{\alpha-1}} e^{-y/\beta} \cdot \frac{1}{(1 - \beta t)} \, dy$$

$$= (1 - \beta t)^{-\alpha} \int_0^\infty \frac{1}{\beta^\alpha \Gamma(\alpha)} y^{\alpha-1} e^{-y/\beta} \, dy \quad \text{for} \quad |t| < 1/\beta$$

The integrand in the last term is the probability density function of a gamma random variable. Therefore, the integral is equal to 1 and it follows that

$$M_X(t) = (1 - \beta t)^{-\alpha} \quad \text{for} \quad |t| < 1/\beta$$

A.5.6 Moment Generating Function of the Chi-Square Distribution

The moment generating function of the chi-square distribution is given by

$$M_X(t) = (1 - 2t)^{-v/2} \quad \text{for} \quad |t| < 1/2$$

Since the chi-square distribution is the gamma distribution with $\alpha = v/2$ and $\beta = 2$, the moment generating function of the chi-square distribution is the moment generating function of the gamma distribution with these values for α and β.

Exercises

1. Show that if $M_X(t)$ is the moment generating function of a random variable, then $M_X(0) = 1$.

2. **Moment Generating Function of the Uniform Distribution.** Let X be a

random variable with probability density function given by

$$f(x) = \begin{cases} \dfrac{1}{\beta - \alpha} & \text{for} \quad \alpha < x < \beta \\ 0 & \text{elsewhere} \end{cases}$$

Show that the moment generating function of X is

$$M_X(t) = \begin{cases} \dfrac{1}{t(\beta - \alpha)} (e^{t\beta} - e^{t\alpha}) & \text{for} \quad t \neq 0 \\ 1 & t = 0 \end{cases}$$

3. **Moment Generating Function of the Bernoulli Distribution.**   Let X be a
    random variable with probability function given by

$$f(x) = \begin{cases} p & \text{for} \quad x = 1 \\ 1 - p & x = 0 \end{cases}$$

Show that the moment generating function of X is

$$M_X(t) = 1 + p(e^t - 1)$$

4. **Moment Generating Function of the Geometric Distribution.**   Let X be a
    random variable with probability function given by

$$f(x) = g(x; p) = q^{x-1} p \qquad \text{for} \quad x = 1, 2, 3, \ldots$$

Show that the moment generating function of X is

$$M_X(t) = \frac{pe^t}{1 - qe^t} \qquad \text{for} \quad t < \ln(q^{-1})$$

5. Use moment generating functions to show that if X is a normal random
    variable with mean μ and variance σ^2, then $(X - \mu)/\sigma$ is standard
    normal.

6. Can the function

$$M(t) = 1/t$$

be the moment generating function of a random variable? Explain.

7. Let X be a random variable with probability density function given by
    $f(x) = 1$ for $0 < x < 1$ and $f(x) = 0$ for all other values of x.
    (a)   Find the moment generating function of X.
    (b)   Expand the moment generating function of X in a power series in t.
    (c)   From the power series of (b), read off the first, second, and third
    moment of X.

8. Let X be a random variable with moment generating function given by

$$M_X(t) = \frac{1}{1-t} \qquad \text{for} \quad |t| < 1$$

Find the mean and variance of X.

9. **Sums of Independent Identically Distributed Exponential Random Variables.** Show that if $X_1, X_2, \ldots,$ and X_n are independent and identically distributed exponential random variables with means equal to θ, than $Y = X_1 + X_2 + \cdots + X_n$ is a gamma random variable with parameters $\beta = \theta$ and $\alpha = n$. (*Hint:* use Theorem A.4 followed by Theorem A.2(a).)

10. Let $X_1, X_2, \ldots,$ and X_n be independent and identically distributed normal random variables with means equal to μ and variances equal to σ^2. Let $\bar{X} = (X_1 + X_2 + \cdots + X_n)/n$. Use moment generating functions to show that the distribution of $\dfrac{\bar{X} - \mu}{\sigma/\sqrt{n}}$ is standard normal.

MONTE CARLO METHOD FOR THE SOLUTION OF DIFFERENTIAL EQUATIONS

The following is an outline of a Monte Carlo method for the solution of a particular type of differential equation. The idea here is to show that if we can define the solution of a problem as the mathematical expectation of some random variable, then we may obtain a good approximation to the solution by taking the average of many observations of the random variable.

Suppose we wish to find the function u defined for values $a \leq x \leq b$ that will satisfy the equation

$$\frac{d^2u}{dx^2} + \gamma(x)\frac{du}{dx} = 0 \tag{1}$$

with boundary conditions $u(a) = A$, $u(b) = B$. Equation (1) is called a differential equation, and u satisfying (1) and the boundary conditions is called a solution of the differential equation.

Let us define functions Δu and $\Delta^2 u$ as follows

$$\Delta u(x) = \frac{u(x + h) - u(x)}{h} \tag{2}$$

$$\Delta^2 u(x) = \frac{\Delta u(x) - \Delta u(x - h)}{h} \tag{3}$$

$$= \frac{u(x + h) - 2u(x) + u(x - h)}{h^2}$$

For h small

$$\frac{du}{dx} \approx \Delta u \qquad \frac{d^2u}{dx^2} \approx \Delta^2 u \tag{4}$$

Substituting (4) into (1), we obtain

$$\Delta^2 u + \gamma(x)\Delta u = 0 \tag{5}$$

Replacing Δu and $\Delta^2 u$ with (2) and (3), we obtain

$$\frac{u(x+h) - 2u(x) + u(x-h)}{h^2} + \gamma(x) \cdot \frac{u(x+h) - u(x)}{h} = 0 \qquad (6)$$

which is equivalent to the difference equation

$$u(x) = \frac{1 + h\gamma(x)}{2 + h\gamma(x)} u(x+h) + \frac{1}{2 + h\gamma(x)} u(x-h) \qquad (7)$$

Let

$$p(x) = \frac{1 + h\gamma(x)}{2 + h\gamma(x)} \quad \text{and} \quad q(x) = \frac{1}{2 + h\gamma(x)} \qquad (8)$$

Observe that for any given x we can choose h sufficiently small so that

$$p(x) > 0 \quad \text{and} \quad q(x) > 0 \qquad (9)$$

Furthermore

$$p(x) + q(x) = 1 \qquad (10)$$

Equation (7) becomes the difference equation

$$u(x) = p(x)u(x+h) + q(x)u(x-h) \qquad (11)$$

with boundary conditions $u(a) = A$, $u(b) = B$.

When h is small, the solution to the differences equation is an approximation of the solution to the differential equation. We propose to solve the difference equation by Monte Carlo method.

Consider the following situation. The interval $[a, b]$ is divided into n equal subintervals of length h. The point x is at one of the division points. A particle is placed at the point x. At each time interval, the particle moves either one point (distance h) to the right or one point to the left. Assume that

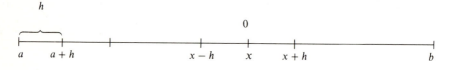

$$P(\text{particle beginning at } x \text{ moves 1 point to right}) = p(x)$$
$$P(\text{particle beginning at } x \text{ moves 1 point to left}) = q(x) \qquad (12)$$

If the particle arrives at position $x + h$, then the probability that it will move to the right by 1 point is $p(x + h)$, and the probability that it will move to the left by 1 point is $q(x + h)$. Similarly, the corresponding probabilities—should the particle arrive at $x - h$—are $p(x - h)$ and $q(x - h)$. The particle continues to move in a random fashion according to the probabilities p and q until it arrives at either a or b. At that point the motion ceases.

Define Z_x as the following random variable.

$$Z_x = \begin{cases} A & \text{(if the particle comes to rest at position } a) \\ B & \text{(if the particle comes to rest at position } b) \end{cases} \tag{13}$$

Let $\mu(x) = E(Z_x)$.
Observe that

$$\mu(x) = E(Z_x) = A \cdot P(Z_x = A) + BP(Z_x = B). \tag{14}$$

Also, observe that

$$P(Z_x = A) = P(Z_{x+h} = A)p(x) + P(Z_{x-h} = A)q(x) \tag{15}$$

$$P(Z_x = B) = P(Z_{x+h} = B)p(x) + P(Z_{x-h} = B)q(x) \tag{16}$$

Therefore

$$\mu(x) = A\{P(Z_{x+h} = A)p(x) + P(Z_{x-h} = A)q(x)\} \tag{17}$$
$$+ B\{P(Z_{x+h} = B)p(x) + P(Z_{x-h} = B)q(x)\}$$
$$= [A \cdot P(Z_{x+h} = A) + BP(Z_{x+h} = B)]p(x)$$
$$+ [AP(Z_{x-h} = A) + BP(Z_{x-h} = B)]q(x)$$
$$= \mu(x + h)p(x) + \mu(x - h)q(x)$$

and

$$\mu(x) = p(x)\mu(x + h) + q(x)\mu(x - h) \tag{18}$$

with

$$\mu(a) = A \quad \text{and} \quad \mu(b) = B$$

Hence, $\mu(x)$ is identical to $u(x)$.

To find $\mu(x)$ recall that $\mu(x) = E(Z_x)$. Therefore, if $Z_{x,i}$ denotes the ith observed value of Z_x, we have

$$\mu(x) \sim \frac{\sum\limits_{i=1}^{n} Z_{x,i}}{n} \tag{19}$$

with great probability, if n is large.

To obtain values of $Z_{x,i}$ we may simulate an experiment, in a computer or by use of a hand calculator, that is equivalent to the example of the particle moving along the line. Substitution of values of $Z_{x,i}$ into equation (19) gives us an approximation to the solution of the differential equation (1).

EXAMPLE B.1

Consider the differential equation

$$\frac{d^2u}{dx^2} - \frac{1}{x} \cdot \frac{du}{dx} = 0$$

with boundary conditions $u(1) = 1$ and $u(2) = 4$. Use the Monte Carlo method described above to find the solution at $x = 1.2$. Compare with the known solution.

SOLUTION

Suppose that the interval $(1, 2)$ is divided into $m = 5$ equal subintervals of length $h = .2$. Then

$$\gamma(x) = -1/x$$

$$p(x) = \frac{1 + (.2)(-1/x)}{2 + (.2)(-1/x)}$$

and

$$q(x) = \frac{1}{2 + (.2)(-1/x)}$$

At the division points 1.2, 1.4, 1.6, and 1.8, we have

$$q(1.2) = .54545$$
$$q(1.4) = .53846$$
$$q(1.6) = .53333$$
$$q(1.8) = .52941$$

To simulate the movement of the particle along the line with end points 1 and 2 we make use of random numbers from Table IV, page 545. Take the first number from Table IV and place a decimal point in front of it. This is equivalent to selecting a number at random from the unit interval $(0, 1)$. Let η denote the number selected. Observe that

$$P[\eta > q(x)] = p(x)$$

and

$$P[\eta \leq q(x)] = q(x)$$

Therefore, we may adopt the following rule: If the particle is initially at

position x, we should
  (a)   move the particle one position to the right if $\eta > q(x)$;
  (b)   move the particle one position to the left if $\eta \leq q(x)$.
After completing this operation, proceed to the next number on Table IV and again follow the rules above. Continue in this fashion until the particle arrives at position 1 or 2.

Let us find $u(1.2)$. For simplicity we set $Z_i = Z_{1.2,i}$. Table B summarizes the motion of the particle and the resulting values of Z_i for $i = 1, \ldots, 21$. The random numbers are taken from Table IV in the order of their appearance in the table. The position of the particle is denoted by x and η is the random number. The initial position of the particle is 1.2; the first random number is .04839; the random number is less than $q(1.2)$; therefore, the particle moves one position to the left to 1.0 and $Z_1 = 1$. The particle is again placed at 1.2; the next random number is .68086; it is greater than $q(1.2)$; therefore, the particle moves to the right to 1.4; the next random number is .39064; it is less than $q(1.4)$; therefore, the particle moves to the left to 1.2; the next random number is .25669; it is less than $q(1.2)$; therefore, the particle moves to the left to 1.0 and $Z_2 = 1$.

Continuing down the table of random numbers, we obtain

$$Z_i = \begin{cases} 4 & \text{for} \quad i = 3, 17, 21, 27 \\ 1 & \text{all other positive integer value of } i \leq 30 \end{cases}$$

Thus, the Monte Carlo estimate for $u(1.2)$ for $n = 30$ is

$$u(1.2) \approx \frac{\sum\limits_{i=1}^{30} Z_i}{30} = 1.33$$

The known solution to the differential equation is $u(x) = x^2$, and

$$u(1.2) = 1.44$$

Exercise

B.1   Use the Monte Carlo method to estimate the solution, at $x = 1.4$, of the differential equation

$$\frac{d^2u}{dx^2} - \frac{2}{2x+1} \cdot \frac{du}{dx} = 0$$

with boundary conditions $u(1) = 2$ and $u(2) = 5$. Divide the interval $(1, 2)$ into $m = 5$ equal parts and use $n = 10$. Use the random numbers of Table IV, page 545.

TABLE B
Position of the Particle

x	η	$q(x)$	Z_i	x	η	$q(x)$	Z_i
1.2	.04839 < q(1.2)			1.2	.47498 < q(1.2)		
1.0			$Z_1 = 1$	1.0			$Z_{11} = 1$
1.2	.68086 > q(1.2)			1.2	.23167 < q(1.2)		
1.4	.39064 < q(1.4)			1.0			$Z_{12} = 1$
1.2	.25669 < q(1.2)			1.2	.23792 < q(1.2)		
1.0			$Z_2 = 1$	1.0			$Z_{13} = 1$
1.2	.64117 > q(1.2)			1.2	.85900 > q(1.2)		
1.4	.87917 > q(1.4)			1.4	.42559 < q(1.4)		
1.6	.62797 > q(1.6)			1.2	.14349 < q(1.2)		
1.8	.95876 > q(1.8)			1.0			$Z_{14} = 1$
2.0			$Z_3 = 4$	1.2	.17403 < q(1.2)		
1.2	.29888 < q(1.2)			1.0			$Z_{15} = 1$
1.0			$Z_4 = 1$	1.2	.23632 < q(1.2)		
1.2	.73577 > q(1.2)			1.0			$Z_{16} = 1$
1.4	.27958 < q(1.4)			1.2	.96423 > q(1.2)		
1.2	.90999 > q(1.2)			1.4	.26432 < q(1.4)		
1.4	.18845 < q(1.4)			1.2	.66432 > q(1.2)		
1.2	.94824 > q(1.2)			1.4	.26422 < q(1.4)		
1.4	.35605 < q(1.4)			1.2	.94305 > q(1.2)		
1.2	.33362 < q(1.2)			1.4	.77341 > q(1.4)		
1.0			$Z_5 = 1$	1.6	.56170 > q(1.6)		
1.2	.88720 > q(1.2)			1.8	.55293 > q(1.8)		
1.4	.39475 < q(1.4)			2.0			$Z_{17} = 4$
1.2	.06990 < q(1.2)			1.2	.88604 > q(1.2)		
1.0			$Z_6 = 1$	1.4	.12908 < q(1.4)		
1.2	.40980 < q(1.2)			1.2	.30134 < q(1.2)		
1.0			$Z_7 = 1$	1.0			$Z_{18} = 1$
1.2	.83974 > q(1.2)			1.2	.49127 < q(1.2)		
1.4	.33339 < q(1.4)			1.0			$Z_{19} = 1$
1.2	.31662 < q(1.2)			1.2	.49618 < q(1.2)		
1.0			$Z_8 = 1$	1.0			$Z_{20} = 1$
1.2	.93546 > q(1.2)			1.2	.78171 > q(1.2)		
1.4	.20492 < q(1.4)			1.4	.81263 > q(1.4)		
1.2	.04153 < q(1.2)			1.6	.64170 > q(1.6)		
1.0			$Z_9 = 1$	1.8	.82765 > q(1.8)		
1.2	.05520 < q(1.2)			2.0			$Z_{21} = 4$
1.0			$Z_{10} = 1$				

STATISTICAL TABLES

TABLE I

Binomial Probabilities: $b(x; n, p) = \binom{n}{x} p^x (1 - p)^{n-x}$

n	x	.05	.10	.15	.20	.25	.30	.35	.40	.45	.50
1	0	.9500	.9000	.8500	.8000	.7500	.7000	.6500	.6000	.5500	.5000
	1	.0500	.1000	.1500	.2000	.2500	.3000	.3500	.4000	.4500	.5000
2	0	.9025	.8100	.7225	.6400	.5625	.4900	.4225	.3600	.3025	.2500
	1	.0950	.1800	.2550	.3200	.3750	.4200	.4550	.4800	.4950	.5000
	2	.0025	.0100	.0225	.0400	.0625	.0900	.1225	.1600	.2025	.2500
3	0	.8574	.7290	.6141	.5120	.4219	.3430	.2746	.2160	.1664	.1250
	1	.1354	.2430	.3251	.3840	.4219	.4410	.4436	.4320	.4084	.3750
	2	.0071	.0270	.0574	.0960	.1406	.1890	.2389	.2880	.3341	.3750
	3	.0001	.0010	.0034	.0080	.0156	.0270	.0429	.0640	.0911	.1250
4	0	.8145	.6561	.5220	.4096	.3164	.2401	.1785	.1296	.0915	.0625
	1	.1715	.2916	.3685	.4096	.4219	.4116	.3845	.3456	.2995	.2500
	2	.0135	.0486	.0975	.1536	.2109	.2646	.3105	.3456	.3675	.3750
	3	.0005	.0036	.0115	.0256	.0469	.0756	.1115	.1536	.2005	.2500
	4	.0000	.0001	.0005	.0016	.0039	.0081	.0150	.0256	.0410	.0625
5	0	.7738	.5905	.4437	.3277	.2373	.1681	.1160	.0778	.0503	.0312
	1	.2036	.3280	.3915	.4096	.3955	.3602	.3124	.2592	.2059	.1562
	2	.0214	.0729	.1382	.2048	.2637	.3087	.3364	.3456	.3369	.3125
	3	.0011	.0081	.0244	.0512	.0879	.1323	.1811	.2304	.2757	.3125
	4	.0000	.0004	.0022	.0064	.0146	.0284	.0488	.0768	.1128	.1562
	5	.0000	.0000	.0001	.0003	.0010	.0024	.0053	.0102	.0185	.0312
6	0	.7351	.5314	.3771	.2621	.1780	.1176	.0754	.0467	.0277	.0156
	1	.2321	.3543	.3993	.3932	.3560	.3025	.2437	.1866	.1359	.0938
	2	.0305	.0984	.1762	.2458	.2966	.3241	.3280	.3110	.2780	.2344
	3	.0021	.0146	.0145	.0819	.1318	.1852	.2355	.2765	.3032	.3125
	4	.0001	.0012	.0055	.0154	.0330	.0595	.0951	.1382	.1861	.2344
	5	.0000	.0001	.0004	.0015	.0044	.0102	.0205	.0369	.0609	.0938
	6	.0000	.0000	.0000	.0001	.0002	.0007	.0018	.0041	.0083	.0156
7	0	.6983	.4783	.3206	.2097	.1335	.0824	.0490	.0280	.0152	.0078
	1	.2573	.3720	.3960	.3670	.3115	.2471	.1848	.1306	.0872	.0547
	2	.0406	.1240	.2097	.2753	.3115	.3177	.2985	.2613	.2140	.1641
	3	.0036	.0230	.0617	.1147	.1730	.2269	.2679	.2903	.2918	.2734
	4	.0002	.0026	.0109	.0287	.0577	.0972	.1442	.1935	.2388	.2734
	5	.0000	.0002	.0012	.0043	.0115	.0250	.0466	.0774	.1172	.1641
	6	.0000	.0000	.0001	.0004	.0013	.0036	.0084	.0172	.0320	.0547
	7	.0000	.0000	.0000	.0000	.0001	.0002	.0006	.0016	.0037	.0078
8	0	.6634	.4305	.2725	.1678	.1001	.0576	.0319	.0168	.0084	.0039
	1	.2793	.3826	.3847	.3355	.2670	.1977	.1373	.0896	.0548	.0312
	2	.0515	.1488	.2376	.2936	.3115	.2965	.2587	.2090	.1569	.1094
	3	.0054	.0331	.0839	.1468	.2076	.2541	.2786	.2787	.2568	.2188
	4	.0004	.0046	.0185	.0459	.0865	.1361	.1875	.2322	.2627	.2734
	5	.0000	.0004	.0026	.0092	.0231	.0467	.0808	.1239	.1719	.2188
	6	.0000	.0000	.0002	.0011	.0038	.0100	.0217	.0413	.0703	.1094
	7	.0000	.0000	.0000	.0001	.0004	.0012	.0033	.0079	.0164	.0312
	8	.0000	.0000	.0000	.0000	.0000	.0001	.0002	.0007	.0017	.0039

Source: *Tables of the Binomial Probability Distribution,* National Bureau of Standards Applied Mathematics Series No. 6. Washington, D.C.: U.S. Government Printing Office, 1950.

TABLE I (continued)

n	x	.05	.10	.15	.20	*p* .25	.30	.35	.40	.45	.50
9	0	.6302	.3874	.2316	.1342	.0751	.0404	.0207	.0101	.0046	.0020
	1	.2985	.3874	.3679	.3020	.2253	.1556	.1004	.0605	.0339	.0176
	2	.0629	.1722	.2597	.3020	.3003	.2668	.2162	.1612	.1110	.0703
	3	.0077	.0446	.1069	.1762	.2336	.2668	.2716	.2508	.2119	.1641
	4	.0006	.0074	.0283	.0661	.1168	.1715	.2194	.2508	.2600	.2461
	5	.0000	.0008	.0050	.0165	.0389	.0735	.1181	.1672	.2128	.2461
	6	.0000	.0001	.0006	.0028	.0087	.0210	.0424	.0743	.1160	.1641
	7	.0000	.0000	.0000	.0003	.0012	.0039	.0098	.0212	.0407	.0703
	8	.0000	.0000	.0000	.0000	.0001	.0004	.0013	.0035	.0083	0176
	9	.0000	.0000	.0000	.0000	.0000	.0000	.0001	.0003	.0008	.0020
10	0	.5987	.3487	.1969	.1074	.0563	.0282	.0135	.0060	.0025	.0010
	1	.3151	.3874	.3474	.2684	.1877	.1211	.0725	.0403	.0207	.0098
	2	.0746	.1937	.2759	.3020	.2816	.2335	.1757	.1209	.0763	.0439
	3	.0105	.0574	.1298	.2013	.2503	.2668	.2522	.2150	.1665	.1172
	4	.0010	.0112	.0401	.0881	.1460	.2001	.2377	.2508	.2384	.2051
	5	.0001	.0015	.0085	.0264	.0584	.1092	.1536	.2007	.2340	.2461
	6	.0000	.0001	.0012	.0055	.0162	.0368	.0689	.1115	.1596	.2051
	7	.0000	.0000	.0001	.0008	.0031	.0090	.0212	.0425	.0746	.1172
	8	.0000	.0000	.0000	.0001	.0004	.0014	.0043	.0106	.0229	.0439
	9	.0000	.0000	.0000	.0000	.0000	.0001	.0005	.0016	.0042	.0098
	10	.0000	.0000	.0000	.0000	.0000	.0000	.0000	.0001	.0003	.0078
11	0	.5688	.3138	.1673	.0859	.0422	.0198	.0088	.0036	.0014	.0005
	1	.3293	.3835	.3248	.2362	.1549	.0932	.0518	.0266	.0125	.0054
	2	.0867	.2131	.2866	.2953	.2581	.1998	.1395	.0887	.0513	.0269
	3	.0137	.0710	.1517	.2215	.2581	.2568	.2254	.1774	.1259	.0806
	4	.0014	.0158	.0536	.1107	.1721	.2201	.2428	.2365	.2060	.1611
	5	.0001	.0025	.0132	.0388	.0803	.1321	.1830	.2207	.2360	.2256
	6	.0000	.0003	.0023	.0097	.0268	.0566	.0985	.1471	.1931	.2255
	7	.0000	.0000	.0003	.0017	.0064	.0173	.0379	.0701	.1128	.1611
	8	.0000	.0000	.0000	.0002	.0011	.0037	.0102	.0234	.0462	.0806
	9	.0000	.0000	.0000	.0000	.0001	.0005	.0018	.0052	.0126	.0269
	10	.0000	.0000	.0000	.0000	.0000	.0000	.0002	.0007	.0021	.0054
	11	.0000	.0000	.0000	.0000	.0000	.0000	.0000	.0000	.0002	.0005
12	0	.5404	.2824	.1422	.0687	.0317	.0138	.0057	.0022	.0008	.0002
	1	.3413	.3766	.3012	.2062	.1267	.0712	.0368	.0174	.0075	.0029
	2	.0988	.2301	.2924	.2835	.2323	.1678	.1088	.0639	.0339	.0121
	3	.0173	.0852	.1720	.2362	.2581	.2397	.1954	.1419	.0923	.0537
	4	.0021	.0213	.0683	.1329	.1936	.2311	.2367	.2128	.1700	.1208
	5	.0002	.0038	.0193	.0532	.1032	.1585	.2039	.2270	.2225	.1934
	6	.0000	.0005	.0040	.0155	.0401	.0792	.1281	.1766	.2124	.2256
	7	.0000	.0000	.0006	.0033	.0115	.0291	.0591	.1009	.1489	.1934
	8	.0000	.0000	.0001	.0005	.0024	.0078	.0199	.0420	.0762	.1208
	9	.0000	.0000	.0000	.0001	.0004	.0015	.0048	.0125	.0277	.0537
	10	.0000	.0000	.0000	.0000	.0000	.0002	.0008	.0025	.0068	.0161
	11	.0000	.0000	.0000	.0000	.0000	.0000	.0001	.0003	.0010	.0029
	12	.0000	.0000	.0000	.0000	.0000	.0000	.0000	.0000	.0001	.0002
13	0	.5133	.2542	.1209	.0550	.0238	.0097	.0037	.0013	.0004	.0001
	1	.3512	.3672	.2774	.1787	.1029	.0540	.0259	.0113	.0045	.0016
	2	.1109	.2448	.2937	.2680	.2059	.1388	.0836	.0453	.0220	.0095

TABLE I (continued)

n	x	.05	.10	.15	.20	.25	.30	.35	.40	.45	.50
13	3	.0214	.0997	.1900	.2457	.2517	.2181	.1651	.1107	.0660	.0349
	4	.0028	.0277	.0838	.1535	.2097	.2337	.2222	.1845	.1350	.0873
	5	.0003	.0055	.0266	.0691	.1258	.1803	.2154	.2214	.1989	.1571
	6	.0000	.0008	.0063	.0230	.0559	.1030	.1546	.1968	.2169	.2095
	7	.0000	.0001	.0011	.0058	.0186	.0442	.0833	.1312	.1775	.2095
	8	.0000	.0000	.0001	.0011	.0047	.0142	.0336	.0656	.1089	.1571
	9	.0000	.0000	.0000	.0001	.0009	.0034	.0101	.0243	.0495	.0873
	10	.0000	.0000	.0000	.0000	.0001	.0006	.0022	.0065	.0162	.0349
	11	.0000	.0000	.0000	.0000	.0000	.0001	.0003	.0012	.0036	.0095
	12	.0000	.0000	.0000	.0000	.0000	.0000	.0000	.0001	.0005	.0016
	13	.0000	.0000	.0000	.0000	.0000	.0000	.0000	.0000	.0000	.0001
14	0	.4877	.2288	.1028	.0440	.0178	.0068	.0024	.0008	.0002	.0001
	1	.3593	.3559	.2539	.1539	.0832	.0407	.0181	.0073	.0027	.0009
	2	.1229	.2570	.2912	.2501	.1802	.1134	.0634	.0317	.0141	.0056
	3	.0259	.1142	.2056	.2501	.2402	.1943	.1366	.0845	.0462	.0222
	4	.0037	.0349	.0998	.1720	.2202	.2290	.2022	.1549	.1040	.0611
	5	.0004	.0078	.0352	.0860	.1468	.1963	.2178	.2066	.1701	.1222
	6	.0000	.0013	.0093	.0322	.0734	.1262	.1759	.2066	.2088	.1833
	7	.0000	.0002	.0019	.0092	.0280	.0618	.1082	.1574	.1952	.2095
	8	.0000	.0000	.0003	.0020	.0082	.0232	.0510	.0918	.1398	.1833
	9	.0000	.0000	.0000	.0003	.0018	.0066	.0183	.0408	.0762	.1222
	10	.0000	.0000	.0000	.0000	.0003	.0014	.0049	.0136	.0312	.0611
	11	.0000	.0000	.0000	.0000	.0000	.0002	.0010	.0033	.0093	.0222
	12	.0000	.0000	.0000	.0000	.0000	.0000	.0001	.0005	.0019	.0056
	13	.0000	.0000	.0000	.0000	.0000	.0000	.0000	.0001	.0002	.0009
	14	.0000	.0000	.0000	.0000	.0000	.0000	.0000	.0000	.0000	.0001
15	0	.4633	.2059	.0874	.0352	.0134	.0047	.0016	.0005	.0001	.0000
	1	.3658	.3432	.2312	.1319	.0668	.0305	.0126	.0047	.0016	.0005
	2	.1348	.2669	.2856	.2309	.1559	.0916	.0476	.0219	.0090	.0032
	3	.0307	.1285	.2184	.2501	.2252	.1700	.1110	.0634	.0318	.0139
	4	.0049	.0428	.1156	.1876	.2252	.2186	.1792	.1268	.0780	.0417
	5	.0006	.0105	.0449	.1032	.1651	.2061	.2123	.1859	.1404	.0916
	6	.0000	.0019	.0132	.0430	.0917	.1472	.1906	.2066	.1914	.1527
	7	.0000	.0003	.0030	.0138	.0393	.0811	.1319	.1771	.2013	.1964
	8	.0000	.0000	.0005	.0035	.0131	.0348	.0710	.1181	.1647	.1964
	9	.0000	.0000	.0001	.0007	.0034	.0116	.0298	.0612	.1048	.1527
	10	.0000	.0000	.0000	.0001	.0007	.0030	.0096	.0245	.0515	.0916
	11	.0000	.0000	.0000	.0000	.0001	.0006	.0024	.0074	.0191	.0417
	12	.0000	.0000	.0000	.0000	.0000	.0001	.0004	.0016	.0052	.0139
	13	.0000	.0000	.0000	.0000	.0000	.0000	.0001	.0003	.0010	.0032
	14	.0000	.0000	.0000	.0000	.0000	.0000	.0000	.0000	.0001	.0005
	15	.0000	.0000	.0000	.0000	.0000	.0000	.0000	.0000	.0000	.0000
16	0	.4401	.1853	.0743	.0281	.0100	.0033	.0010	.0003	.0001	.0000
	1	.3706	.3204	.2097	.1126	.0535	.0228	.0087	.0030	.0009	.0002
	2	.1463	.2745	.2775	.2111	.1336	.0732	.0353	.0150	.0056	.0018
	3	.0359	.1423	.2285	.2463	.2079	.1465	.0888	.0468	.0215	.0085
	4	.0061	.0514	.1311	.2001	.2252	.2040	.1553	.1014	.0572	.0278
	5	.0008	.0137	.0555	.1201	.1802	.2099	.2008	.1623	.1123	.0667
	6	.0001	.0028	.0180	.0550	.1101	.1649	.1982	.1983	.1684	.1222

TABLE I (continued)

n	x	.05	.10	.15	.20	.25	.30	.35	.40	.45	.50
						p					
16	7	.0000	.0004	.0045	.0197	.0524	.1010	.1524	.1889	.1969	.1746
	8	.0000	.0001	.0009	.0055	.0197	.0487	.0923	.1417	.1812	.1964
	9	.0000	.0000	.0001	.0012	.0058	.0185	.0442	.0840	.1318	.1746
	10	.0000	.0000	.0000	.0002	.0014	.0056	.0167	.0392	.0755	.1222
	11	.0000	.0000	.0000	.0000	.0002	.0013	.0049	.0142	.0337	.0667
	12	.0000	.0000	.0000	.0000	.0000	.0002	.0011	.0040	.0115	.0278
	13	.0000	.0000	.0000	.0000	.0000	.0000	.0002	.0008	.0029	.0085
	14	.0000	.0000	.0000	.0000	.0000	.0000	.0000	.0001	.0005	.0018
	15	.0000	.0000	.0000	.0000	.0000	.0000	.0000	.0000	.0001	.0002
	16	.0000	.0000	.0000	.0000	.0000	.0000	.0000	.0000	.0000	.0000
17	0	.4181	.1668	.0631	.0225	.0075	.0023	.0007	.0002	.0000	.0000
	1	.3741	.3150	.1893	.0957	.0426	.0169	.0060	.0019	.0005	.0001
	2	.1575	.2800	.2673	.1914	.1136	.0581	.0260	.0102	.0035	.0010
	3	.0415	.1556	.2359	.2393	.1893	.1245	.0701	.0341	.0144	.0052
	4	.0076	.0605	.1457	.2093	.2209	.1868	.1320	.0796	.0411	.0182
	5	.0010	.0175	.0668	.1361	.1914	.2081	.1849	.1379	.0875	.0472
	6	.0001	.0039	.0236	.0680	.1276	.1784	.1991	.1839	.1432	.0944
	7	.0000	.0007	.0065	.0267	.0668	.1201	.1685	.1927	.1841	.1484
	8	.0000	.0001	.0014	.0084	.0279	.0644	.1134	.1606	.1883	.1855
	9	.0000	.0000	.0003	.0021	.0093	.0276	.0611	.1070	.1540	.1855
	10	.0000	.0000	.0000	.0004	.0025	.0095	.0263	.0571	.1008	.1484
	11	.0000	.0000	.0000	.0001	.0005	.0026	.0090	.0242	0525	.0944
	12	.0000	.0000	.0000	.0000	.0001	.0006	.0024	.0081	.0215	.0472
	13	.0000	.0000	.0000	.0000	.0000	.0001	.0005	.0021	.0068	.0182
	14	.0000	.0000	.0000	.0000	.0000	.0000	.0001	.0004	.0016	.0052
	15	.0000	.0000	.0000	.0000	.0000	.0000	.0000	.0001	.0003	.0010
	16	.0000	.0000	.0000	.0000	.0000	.0000	.0000	.0000	.0000	.0001
	17	.0000	.0000	.0000	.0000	.0000	.0000	.0000	.0000	.0000	.0000
18	0	.3972	.1501	.0536	.0180	.0056	.0016	.0004	.0001	.0000	.0000
	1	.3763	.3002	.1704	.0811	.0338	.0126	.0042	.0012	.0003	.0001
	2	.1683	.2835	.2556	.1723	.0958	.0458	.0190	.0069	.0022	.0006
	3	.0473	.1680	.2406	.2297	.1704	.1046	.0547	.0246	.0095	.0031
	4	.0093	.0700	.1592	.2153	.2130	.1681	.1104	.0614	.0291	.0117
	5	.0014	.0218	.0787	.1507	.1988	.2017	.1664	.1146	.0666	.0327
	6	.0002	.0052	.0301	.0816	.1436	.1873	.1941	.1655	.1181	.0708
	7	.0000	.0010	.0091	.0350	.0820	.1376	.1792	.1892	.1657	.1214
	8	.0000	.0002	.0022	.0120	.0376	.0811	.1327	.1734	.1864	.1669
	9	.0000	.0000	.0004	.0033	.0139	.0386	.0794	.1284	.1694	.1855
	10	.0000	.0000	.0001	.0008	.0042	.0149	.0385	.0771	.1248	.1669
	11	.0000	.0000	.0000	.0001	.0010	.0046	.0151	.0374	.0742	.1214
	12	.0000	.0000	.0000	.0000	.0002	.0012	.0047	.0145	.0354	.0708
	13	.0000	.0000	.0000	.0000	.0000	.0002	.0012	.0045	.0134	.0327
	14	.0000	.0000	.0000	.0000	.0000	.0000	.0002	.0011	.0039	.0117
	15	.0000	.0000	.0000	.0000	.0000	.0000	.0000	.0002	.0009	.0031
	16	.0000	.0000	.0000	.0000	.0000	.0000	.0000	.0000	.0001	.0006
	17	.0000	.0000	.0000	.0000	.0000	.0000	.0000	.0000	.0000	.0001
	18	.0000	.0000	.0000	.0000	.0000	.0000	.0000	.0000	.0000	.0000
19	0	.3774	.1351	.0456	.0144	.0042	.0011	.0003	.0001	.0000	.0000
	1	.3774	.2852	.1529	.0685	.0268	.0093	.0029	.0008	.0002	.0000

TABLE I (continued)

						p					
n	*x*	.05	.10	.15	.20	.25	.30	.35	.40	.45	.50
19	2	.1787	.2852	.2428	.1540	.0803	.0358	.0138	.0046	.0013	.0003
	3	.0533	.1796	.2428	.2182	.1517	.0869	.0422	.0175	.0062	.0018
	4	.0112	.0798	.1714	.2182	.2023	.1491	.0909	.0467	.0203	.0074
	5	.0018	.0266	.0907	.1636	.2023	.1916	.1468	.0933	.0497	.0222
	6	.0002	.0069	.0374	.0955	.1574	.1916	.1844	.1451	.0949	.0518
	7	.0000	.0014	.0122	.0443	.0974	.1525	.1844	.1797	.1443	.0961
	8	.0000	.0002	.0032	.0166	.0487	.0981	.1489	.1797	.1771	.1442
	9	.0000	.0000	.0007	.0051	.0198	.0514	.0980	.1464	.1771	.1762
	10	.0000	.0000	.0001	.0013	.0066	.0220	.0528	.0976	.1449	.1762
	11	.0000	.0000	.0000	.0003	.0018	.0077	.0233	.0532	.0970	.1442
	12	.0000	.0000	.0000	.0000	.0004	.0022	.0083	.0237	.0529	.0961
	13	.0000	.0000	.0000	.0000	.0001	.0005	.0024	.0085	.0233	.0518
	14	.0000	.0000	.0000	.0000	.0000	.0001	.0006	.0024	.0082	.0222
	15	.0000	.0000	.0000	.0000	.0000	.0000	.0001	.0005	.0022	.0074
	16	.0000	.0000	.0000	.0000	.0000	.0000	.0000	.0001	.0005	.0018
	17	.0000	.0000	.0000	.0000	.0000	.0000	.0000	.0000	.0001	.0003
	18	.0000	.0000	.0000	.0000	.0000	.0000	.0000	.0000	.0000	.0000
	19	.0000	.0000	.0000	.0000	.0000	.0000	.0000	.0000	.0000	.0000
20	0	.3585	.1216	.0388	.0115	.0032	.0008	.0002	.0000	.0000	.0000
	1	.3774	.2702	.1368	.0576	.0211	.0068	.0020	.0005	.0001	.0000
	2	.1887	.2852	.2293	.1369	.0669	.0278	.0100	.0031	.0008	.0002
	3	.0596	.1901	.2428	.2054	.1339	.0716	.0323	.0123	.0040	.0011
	4	.0133	.0898	.1821	.2182	.1897	.1304	.0738	.0350	.0139	.0046
	5	.0022	.0319	.1028	.1746	.2023	.1789	.1272	.0746	.0365	.0148
	6	.0003	.0089	.0454	.1091	.1686	.1916	.1712	.1244	.0746	.0370
	7	.0000	.0020	.0160	.0545	.1124	.1643	.1844	.1659	.1221	.0739
	8	.0000	.0004	.0046	.0222	.0609	.1144	.1614	.1797	.1623	.1201
	9	.0000	.0001	.0011	.0074	.0271	.0654	.1158	.1597	.1771	.1602
	10	.0000	.0000	.0002	.0020	.0099	.0308	.0686	.1171	.1593	.1762
	11	.0000	.0000	.0000	.0005	.0030	.0120	.0336	.0710	.1185	.1602
	12	.0000	.0000	.0000	.0001	.0008	.0039	.0136	.0355	.0727	.1201
	13	.0000	.0000	.0000	.0000	.0002	.0010	.0045	.0146	.0366	.0739
	14	.0000	.0000	.0000	.0000	.0000	.0002	.0012	.0049	.0150	.0370
	15	.0000	.0000	.0000	.0000	.0000	.0000	.0003	.0013	.0049	.0148
	16	.0000	.0000	.0000	.0000	.0000	.0000	.0000	.0003	.0013	.0046
	17	.0000	.0000	.0000	.0000	.0000	.0000	.0000	.0000	.0002	.0011
	18	.0000	.0000	.0000	.0000	.0000	.0000	.0000	.0000	.0000	.0002
	19	.0000	.0000	.0000	.0000	.0000	.0000	.0000	.0000	.0000	.0000
	20	.0000	.0000	.0000	.0000	.0000	.0000	.0000	.0000	.0000	.0000

TABLE II

Poisson Probabilities: $p(x; \lambda) = \dfrac{\lambda^x e^{-\lambda}}{x!}$

x	0.1	0.2	0.3	0.4	λ 0.5	0.6	0.7	0.8	0.9	1.0
0	.9048	.8187	.7408	.6703	.6065	.5488	.4966	.4493	.4066	.3679
1	.0905	.1637	.2222	.2681	.3033	.3293	.3476	.3595	.3659	.3679
2	.0045	.0164	.0333	.0536	.0758	.0988	.1217	.1438	.1647	.1839
3	.0002	.0011	.0033	.0072	.0126	.0198	.0284	.0383	.0494	.0613
4	.0000	.0001	.0002	.0007	.0016	.0030	.0050	.0077	.0111	.0153
5	.0000	.0000	.0000	.0001	.0002	.0004	.0007	.0012	.0020	.0031
6	.0000	.0000	.0000	.0000	.0000	.0000	.0000	.0002	.0003	.0005
7	.0000	.0000	.0000	.0000	.0000	.0000	.0000	.0000	.0000	.0001

x	1.1	1.2	1.3	1.4	λ 1.5	1.6	1.7	1.8	1.9	2.0
0	.3329	.3012	.2725	.2466	.2231	.2019	.1827	.1653	.1496	.1353
1	.3662	.3614	.3543	.3452	.3347	.3230	.3106	.2975	.2842	.2707
2	.2014	.2169	.2303	.2417	.2510	.2584	.2640	.2678	.2700	.2707
3	.0738	.0867	.0998	.1128	.1255	.1378	.1496	.1607	.1710	.1804
4	.0203	.0260	.0324	.0395	.0471	.0551	.0636	.0723	.0812	.0902
5	.0045	.0062	.0084	.0111	.0141	.0176	.0216	.0260	.0309	.0361
6	.0008	.0012	.0018	.0026	.0035	.0047	.0061	.0078	.0098	.0120
7	.0001	.0002	.0003	.0005	.0008	.0011	.0015	.0020	.0027	.0034
8	.0000	.0000	.0001	.0001	.0001	.0002	.0003	.0005	.0006	.0009
9	.0000	.0000	.0000	.0000	.0000	.0000	.0001	.0001	.0001	.0002

x	2.1	2.2	2.3	2.4	λ 2.5	2.6	2.7	2.8	2.9	3.0
0	.1225	.1108	.1003	.0907	.0821	.0743	.0672	.0608	.0550	.0498
1	.2572	.2438	.2306	.2177	.2052	.1931	.1815	.1703	.1596	.1494
2	.2700	.2681	.2652	.2613	.2565	.2510	.2450	.2384	.2314	.2240
3	.1890	.1966	.2033	.2090	.2138	.2176	.2205	.2225	.2237	.2240
4	.0992	.1082	.1169	.1254	.1336	.1414	.1488	.1557	.1622	.1680
5	.0417	.0476	.0538	.0602	.0668	.0735	.0804	.0872	.0940	.1008
6	.0146	.0174	.0206	.0241	.0278	.0319	.0362	.0407	.0455	.0504
7	.0044	.0055	.0068	.0083	.0099	.0118	.0139	.0163	.0188	.0216
8	.0011	.0015	.0019	.0025	.0031	.0038	.0047	.0057	.0068	.0081
9	.0003	.0004	.0005	.0007	.0009	.0011	.0014	.0018	.0022	.0027
10	.0001	.0001	.0001	.0002	.0002	.0003	.0004	.0005	.0006	.0008
11	.0000	.0000	.0000	.0000	.0000	.0001	.0001	.0001	.0002	.0002
12	.0000	.0000	.0000	.0000	.0000	.0000	.0000	.0000	.0000	.0001

x	3.1	3.2	3.3	3.4	λ 3.5	3.6	3.7	3.8	3.9	4.0
0	.0450	.0408	.0369	.0334	.0302	.0273	.0247	.0224	.0202	.0183
1	.1397	.1304	.1217	.1135	.1057	.0984	.0915	.0850	.0789	.0733
2	.2165	.2087	.2008	.1929	.1850	.1771	.1692	.1615	.1539	.1465
3	.2237	.2226	.2209	.2186	.2158	.2125	.2087	.2046	.2001	.1954
4	.1734	.1781	.1823	.1858	.1888	.1912	.1931	.1944	.1951	.1954

Source: Reprinted by permission from *Handbook of Probability and Statistics with Tables*, by R. S. Burington and D. C. May, Jr. New York: McGraw-Hill Book Company, 1953.

TABLE II (continued)

x	3.1	3.2	3.3	3.4	3.5	3.6	3.7	3.8	3.9	4.0
5	.1075	.1140	.1203	.1264	.1322	.1377	.1429	.1477	.1522	.1563
6	.0555	.0608	.0662	.0716	.0771	.0826	.0881	.0936	.0989	.1042
7	.0246	.0278	.0312	.0348	.0385	.0425	.0466	.0508	.0551	.0595
8	.0095	.0111	.0129	.0148	.0169	.0191	.0215	.0241	.0269	.0298
9	.0033	.0040	.0047	.0056	.0066	.0076	.0089	.0102	.0116	.0132
10	.0010	.0013	.0016	.0019	.0023	.0028	.0033	.0039	.0045	.0053
11	.0003	.0004	.0005	.0006	.0007	.0009	.0011	.0013	.0016	.0019
12	.0001	.0001	.0001	.0002	.0002	.0003	.0003	.0004	.0005	.0006
13	.0000	.0000	.0000	.0000	.0001	.0001	.0001	.0001	.0002	.0002
14	.0000	.0000	.0000	.0000	.0000	.0000	.0000	.0000	.0000	.0001

λ

x	4.1	4.2	4.3	4.4	4.5	4.6	4.7	4.8	4.9	5.0
0	.0166	.0150	.0136	.0123	.0111	.0101	.0091	.0082	.0074	.0067
1	.0679	.0630	.0583	.0540	.0500	.0462	.0427	.0395	.0365	.0337
2	.1393	.1323	.1254	.1188	.1125	.1063	.1005	.0948	.0894	.0842
3	.1904	.1852	.1798	.1743	.1687	.1631	.1574	.1517	.1460	.1404
4	.1951	.1944	.1933	.1917	.1898	.1875	.1849	.1820	.1789	.1755
5	.1600	.1633	.1662	.1687	.1708	.1725	.1738	.1747	.1753	.1755
6	.1093	.1143	.1191	.1237	.1281	.1323	.1362	.1398	.1432	.1462
7	.0640	.0686	.0732	.0778	.0824	.0869	.0914	.0959	.1002	.1044
8	.0328	.0360	.0393	.0428	.0463	.0500	.0537	.0575	.0614	.0653
9	.0150	.0168	.0188	.0209	.0232	.0255	.0280	.0307	.0334	.0363
10	.0061	.0071	.0081	.0092	.0104	.0118	.0132	.0147	.0164	.0181
11	.0023	.0027	.0032	.0037	.0043	.0049	.0056	.0064	.0073	.0082
12	.0008	.0009	.0011	.0014	.0016	.0019	.0022	.0026	.0030	.0034
13	.0002	.0003	.0004	.0005	.0006	.0007	.0008	.0009	.0011	.0013
14	.0001	.0001	.0001	.0001	.0002	.0002	.0003	.0003	.0004	.0005
15	.0000	.0000	.0000	.0000	.0001	.0001	.0001	.0001	.0001	.0002

λ

x	5.1	5.2	5.3	5.4	5.5	5.6	5.7	5.8	5.9	6.0
0	.0061	.0055	.0050	.0045	.0041	.0037	.0033	.0030	.0027	.0025
1	.0311	.0287	.0265	.0244	.0225	.0207	.0191	.0176	.0162	.0149
2	.0793	.0746	.0701	.0659	.0618	.0580	.0544	.0509	.0477	.0446
3	.1348	.1293	.1239	.1185	.1133	.1082	.1033	.0985	.0938	.0892
4	.1719	.1681	.1641	.1600	.1558	.1515	.1472	.1428	.1383	.1339
5	.1753	.1748	.1740	.1728	.1714	.1697	.1678	.1656	.1632	.1606
6	.1490	.1515	.1537	.1555	.1571	.1584	.1594	.1601	.1605	.1606
7	.1086	.1125	.1163	.1200	.1234	.1267	.1298	.1326	.1353	.1377
8	.0692	.0731	.0771	.0810	.0849	.0887	.0925	.0962	.0998	.1033
9	.0392	.0423	.0454	.0486	.0519	.0552	.0586	.0620	.0654	.0688
10	.0200	.0220	.0241	.0262	.0285	.0309	.0334	.0359	.0386	.0413
11	.0093	.0104	.0116	.0129	.0143	.0157	.0173	.0190	.0207	.0225
12	.0039	.0045	.0051	.0058	.0065	.0073	.0082	.0092	.0102	.0113
13	.0015	.0018	.0021	.0024	.0028	.0032	.0036	.0041	.0046	.0052
14	.0006	.0007	.0008	.0009	.0011	.0013	.0015	.0017	.0019	.0022
15	.0002	.0002	.0003	.0003	.0004	.0005	.0006	.0007	.0008	.0009
16	.0001	.0001	.0001	.0001	.0001	.0002	.0002	.0002	.0003	.0003
17	.0000	.0000	.0000	.0000	.0000	.0001	.0001	.0001	.0001	.0001

TABLE II (continued)

x	6.1	6.2	6.3	6.4	λ 6.5	6.6	6.7	6.8	6.9	7.0
0	.0022	.0020	.0018	.0017	.0015	.0014	.0012	.0011	.0010	.0009
1	.0137	.0126	.0116	.0106	.0098	.0090	.0082	.0076	.0070	.0064
2	.0417	.0390	.0364	.0340	.0318	.0296	.0276	.0258	.0240	.0223
3	.0848	.0806	.0765	.0726	.0688	.0652	.0617	.0584	.0552	.0521
4	.1294	.1249	.1205	.1162	.1118	.1076	.1034	.0992	.0952	.0912
5	.1579	.1549	.1519	.1487	.1454	.1420	.1385	.1349	.1314	.1277
6	.1605	.1601	.1595	.1586	.1575	.1562	.1546	.1529	.1511	.1490
7	.1399	.1418	.1435	.1450	.1462	.1472	.1480	.1486	.1489	.1490
8	.1066	.1099	.1130	.1160	.1188	.1215	.1240	.1263	.1284	.1304
9	.0723	.0757	.0791	.0825	.0858	.0891	.0923	.0954	.0985	.1014
10	.0441	.0469	.0498	.0528	.0558	.0588	.0618	.0649	.0679	.0710
11	.0245	.0265	.0285	.0307	.0330	.0353	.0377	.0401	.0426	.0452
12	.0124	.0137	.0150	.0164	.0179	.0194	.0210	.0227	.0245	.0264
13	.0058	.0065	.0073	.0081	.0089	.0098	.0108	.0119	.0130	.0142
14	.0025	.0029	.0033	.0037	.0041	.0046	.0052	.0058	.0064	.0071
15	.0010	.0012	.0014	.0016	.0018	.0020	.0023	.0026	.0029	.0033
16	.0004	.0005	.0005	.0006	.0007	.0008	.0010	.0011	.0013	.0014
17	.0001	.0002	.0002	.0002	.0003	.0003	.0004	.0004	.0005	.0006
18	.0000	.0001	.0001	.0001	.0001	.0001	.0001	.0002	.0002	.0002
19	.0000	.0000	.0000	.0000	.0000	.0000	.0000	.0001	.0001	.0001

x	7.1	7.2	7.3	7.4	λ 7.5	7.6	7.7	7.8	7.9	8.0
0	.0008	.0007	.0007	.0006	.0006	.0005	.0005	.0004	.0004	.0003
1	.0059	.0054	.0049	.0045	.0041	.0038	.0035	.0032	.0029	.0027
2	.0208	.0198	.0180	.0167	.0156	.0145	.0134	.0125	.0116	.0107
3	.0492	.0464	.0438	.0413	.0389	.0366	.0345	.0324	.0305	.0286
4	.0874	.0836	.0799	.0764	.0729	.0696	.0663	.0632	.0602	.0573
5	.1241	.1204	.1167	.1130	.1094	.1057	.1021	.0986	.0951	.0916
6	.1468	.1445	.1420	.1394	.1367	.1339	.1311	.1282	.1252	.1221
7	.1489	.1486	.1481	.1474	.1465	.1454	.1442	.1428	.1413	.1396
8	.1321	.1337	.1351	.1363	.1373	.1382	.1388	.1392	.1395	.1396
9	.1042	.1070	.1096	.1121	.1144	.1167	.1187	.1207	.1224	.1241
10	.0740	.0770	.0800	.0829	.0858	.0887	.0914	.0941	.0967	.0993
11	.0478	.0504	.0531	.0558	.0585	.0613	.0640	.0667	.0695	.0722
12	.0283	.0303	.0323	.0344	.0366	.0388	.0411	.0434	.0457	.0481
13	.0154	.0168	.0181	.0196	.0211	.0227	.0243	.0260	.0278	.0296
14	.0078	.0086	.0095	.0104	.0113	.0123	.0134	.0145	.0157	.0169
15	.0037	.0041	.0046	.0051	.0057	.0062	.0069	.0075	.0083	.0090
16	.0016	.0019	.0021	.0024	.0026	.0030	.0033	.0037	.0041	.0045
17	.0007	.0008	.0009	.0010	.0012	.0013	.0015	.0017	.0019	.0021
18	.0003	.0003	.0004	.0004	.0005	.0006	.0006	.0007	.0008	.0009
19	.0001	.0001	.0001	.0002	.0002	.0002	.0003	.0003	.0003	.0004
20	.0000	.0000	.0001	.0001	.0001	.0001	.0001	.0001	.0001	.0002
21	.0000	.0000	.0000	.0000	.0000	.0000	.0000	.0000	.0001	.0001

x	8.1	8.2	8.3	8.4	λ 8.5	8.6	8.7	8.8	8.9	9.0
0	.0003	.0003	.0002	.0002	.0002	.0002	.0002	.0002	.0001	.0001
1	.0025	.0023	.0021	.0019	.0017	.0016	.0014	.0013	.0012	.0011

TABLE II (continued)

x	8.1	8.2	8.3	8.4	λ 8.5	8.6	8.7	8.8	8.9	9.0
2	.0100	.0092	.0086	.0079	.0074	.0068	.0063	.0058	.0054	.0050
3	.0269	.0252	.0237	.0222	.0208	.0195	.0183	.0171	.0160	.0150
4	.0544	.0517	.0419	.0466	.0443	.0420	.0398	.0377	.0357	.0337
5	.0882	.0849	.0816	.0784	.0752	.0722	.0692	.0663	.0635	.0607
6	.1191	.1160	.1128	.1097	.1066	.1034	.1003	.0972	.0941	.0911
7	.1378	.1358	.1338	.1317	.1294	.1271	.1247	.1222	.1197	.1171
8	.1395	.1392	.1388	.1382	.1375	.1366	.1356	.1344	.1332	.1318
9	.1256	.1269	.1280	.1290	.1299	.1306	.1311	.1315	.1317	.1318
10	.1017	.1040	.1063	.1084	.1104	.1123	.1140	.1157	.1172	.1186
11	.0749	.0776	.0802	.0828	.0853	.0878	.0902	.0925	.0948	.0970
12	.0505	.0530	.0555	.0579	.0604	.0629	.0654	.0679	.0703	.0728
13	.0315	.0334	.0354	.0374	.0395	.0416	.0438	.0459	.0481	.0504
14	.0182	.0196	.0210	.0225	.0240	.0256	.0272	.0289	.0306	.0324
15	.0098	.0107	.0116	.0126	.0136	.0147	.0158	.0169	.0182	.0194
16	.0050	.0055	.0060	.0066	.0072	.0079	.0086	.0093	.0101	.0109
17	.0024	.0026	.0029	.0033	.0036	.0040	.0044	.0048	.0053	.0058
18	.0011	.0012	.0014	.0015	.0017	.0019	.0021	.0024	.0026	.0029
19	.0005	.0005	.0006	.0007	.0008	.0009	.0010	.0011	.0012	.0014
20	.0002	.0002	.0002	.0003	.0003	.0004	.0004	.0005	.0005	.0006
21	.0001	.0001	.0001	.0001	.0001	.0002	.0002	.0002	.0002	.0003
22	.0000	.0000	.0000	.0000	.0001	.0001	.0001	.0001	.0001	.0001

x	9.1	9.2	9.3	9.4	λ 9.5	9.6	9.7	9.8	9.9	10
0	.0001	.0001	.0001	.0001	.0001	.0001	.0001	.0001	.0001	.0000
1	.0010	.0009	.0009	.0008	.0007	.0007	.0006	.0005	.0005	.0005
2	.0046	.0043	.0040	.0037	.0034	.0031	.0029	.0027	.0025	.0023
3	.0140	.0131	.0123	.0115	.0107	.0100	.0093	.0087	.0081	.0076
4	.0319	.0302	.0285	.0269	.0254	.0240	.0226	.0213	.0201	.0189
5	.0581	.0555	.0530	.0506	.0483	.0460	.0439	.0418	.0398	.0378
6	.0881	.0851	.0822	.0793	.0764	.0736	.0709	.0682	.0656	.0631
7	.1145	.1118	.1091	.1064	.1037	.1010	.0982	.0955	.0928	.0901
8	.1302	.1286	.1269	.1251	.1232	.1212	.1191	.1170	.1148	.1126
9	.1317	.1315	.1311	.1306	.1300	.1293	.1284	.1274	.1263	.1251
10	.1198	.1210	.1219	.1228	.1235	.1241	.1245	.1249	.1250	.1251
11	.0991	.1012	.1031	.1049	.1067	.1083	.1098	.1112	.1125	.1137
12	.0752	.0776	.0799	.0822	.0844	.0866	.0888	.0908	.0928	.0948
13	.0526	.0549	.0572	.0594	.0617	.0640	.0662	.0685	.0707	.0729
14	.0342	.0361	.0380	.0399	.0419	.0439	.0459	.0479	.0500	.0521
15	.0208	.0221	.0235	.0250	.0265	.0281	.0297	.0313	.0330	.0347
16	.0118	.0127	.0137	.0147	.0157	.0168	.0180	.0192	.0204	.0217
17	.0063	.0069	.0075	.0081	.0088	.0095	.0103	.0111	.0119	.0128
18	.0032	.0035	.0039	.0042	.0046	.0051	.0055	.0060	.0065	.0071
19	.0015	.0017	.0019	.0021	.0023	.0026	.0028	.0031	.0034	.0037
20	.0007	.0008	.0009	.0010	.0011	.0012	.0014	.0015	.0017	.0019
21	.0003	.0003	.0004	.0004	.0005	.0006	.0006	.0007	.0008	.0009
22	.0001	.0001	.0002	.0002	.0002	.0002	.0003	.0003	.0004	.0004
23	.0000	.0001	.0001	.0001	.0001	.0001	.0001	.0001	.0002	.0002
24	.0000	.0000	.0000	.0000	.0000	.0000	.0000	.0001	.0001	.0001

TABLE II (continued)

x	11	12	13	14	λ 15	16	17	18	19	20
0	.0000	.0000	.0000	.0000	.0000	.0000	.0000	.0000	.0000	.0000
1	.0002	.0001	.0000	.0000	.0000	.0000	.0000	.0000	.0000	.0000
2	.0010	.0004	.0002	.0001	.0000	.0000	.0000	.0000	.0000	.0000
3	.0037	.0018	.0008	.0004	.0002	.0001	.0000	.0000	.0000	.0000
4	.0102	.0053	.0027	.0013	.0006	.0003	.0001	.0001	.0000	.0000
5	.0224	.0127	.0070	.0037	.0019	.0010	.0005	.0002	.0001	.0001
6	.0411	.0255	.0152	.0087	.0048	.0026	.0014	.0007	.0004	.0002
7	.0646	.0437	.0281	.0174	.0104	.0060	.0034	.0018	.0010	.0005
8	.0888	.0655	.0457	.0304	.0194	.0120	.0072	.0042	.0024	.0013
9	.1085	.0874	.0661	.0473	.0324	.0213	.0135	.0083	.0050	.0029
10	.1194	.1048	.0859	.0663	.0486	.0341	.0230	.0150	.0095	.0058
11	.1194	.1144	.1015	.0844	.0663	.0496	.0355	.0245	.0164	.0106
12	.1094	.1144	.1099	.0984	.0829	.0661	.0504	.0368	.0259	.0176
13	.0926	.1056	.1099	.1060	.0956	.0814	.0658	.0509	.0378	.0271
14	.0728	.0905	.1021	.1060	.1024	.0930	.0800	.0655	.0514	.0387
15	.0534	.0724	.0885	.0989	.1024	.0992	.0906	.0786	.0650	.0516
16	.0367	.0543	.0719	.0866	.0960	.0992	.0963	.0884	.0772	.0646
17	.0237	.0383	.0550	.0713	.0847	.0934	.0963	.0936	.0863	.0760
18	.0145	.0256	.0397	.0554	.0706	.0830	.0909	.0936	.0911	.0844
19	.0084	.0161	.0272	.0409	.0557	.0699	.0814	.0887	.0911	.0888
20	.0046	.0097	.0177	.0286	.0418	.0559	.0692	.0798	.0866	.0888
21	.0024	.0055	.0109	.0191	.0299	.0426	.0560	.0684	.0783	.0846
22	.0012	.0030	.0065	.0121	.0204	.0310	.0433	.0560	.0676	.0769
23	.0006	.0016	.0037	.0074	.0133	.0216	.0320	.0438	.0559	.0669
24	.0003	.0008	.0020	.0043	.0083	.0144	.0226	.0328	.0442	.0557
25	.0001	.0004	.0010	.0024	.0050	.0092	.0154	.0237	.0336	.0446
26	.0000	.0002	.0005	.0013	.0029	.0057	.0101	.0164	.0246	.0343
27	.0000	.0001	.0002	.0007	.0016	.0034	.0063	.0109	.0173	.0254
28	.0000	.0000	.0001	.0003	.0009	.0019	.0038	.0070	.0117	.0181
29	.0000	.0000	.0001	.0002	.0004	.0011	.0023	.0044	.0077	.0125
30	.0000	.0000	.0000	.0001	.0002	.0006	.0013	.0026	.0049	.0083
31	.0000	.0000	.0000	.0000	.0001	.0003	.0007	.0015	.0030	.0054
32	.0000	.0000	.0000	.0000	.0001	.0001	.0004	.0009	.0018	.0034
33	.0000	.0000	.0000	.0000	.0000	.0001	.0002	.0005	.0010	.0020
34	.0000	.0000	.0000	.0000	.0000	.0000	.0001	.0002	.0006	.0012
35	.0000	.0000	.0000	.0000	.0000	.0000	.0000	.0001	.0003	.0007
36	.0000	.0000	.0000	.0000	.0000	.0000	.0000	.0001	.0002	.0004
37	.0000	.0000	.0000	.0000	.0000	.0000	.0000	.0000	.0001	.0002
38	.0000	.0000	.0000	.0000	.0000	.0000	.0000	.0000	.0000	.0001
39	.0000	.0000	.0000	.0000	.0000	.0000	.0000	.0000	.0000	.0001

TABLE III

The Standard Normal Distribution: $P(o < Z < z) = \int_0^z \frac{1}{\sqrt{2\pi}} e^{-x^2/2}\,dx$

z	.00	.01	.02	.03	.04	.05	.06	.07	.08	.09
.0	.0000	.0040	.0080	.0120	.0160	.0199	.0239	.0279	.0319	.0359
.1	.0398	.0438	.0478	.0517	.0557	.0596	.0636	.0675	.0714	.0753
.2	.0793	.0832	.0871	.0910	.0948	.0987	.1026	.1064	.1103	.1141
.3	.1179	.1217	.1255	.1293	.1331	.1368	.1406	.1443	.1480	.1517
.4	.1554	.1591	.1628	.1664	.1700	.1736	.1772	.1808	.1844	.1879
.5	.1915	.1950	.1985	.2019	.2054	.2088	.2123	.2157	.2190	.2224
.6	.2257	.2291	.2324	.2357	.2389	.2422	.2454	.2486	.2517	.2549
.7	.2580	.2611	.2642	.2673	.2704	.2734	.2764	.2794	.2823	.2852
.8	.2881	.2910	.2939	.2967	.2995	.3023	.3051	.3078	.3106	.3133
.9	.3159	.3186	.3212	.3238	.3264	.3289	.3315	.3340	.3365	.3389
1.0	.3413	.3438	.3461	.3485	.3508	.3531	.3554	.3577	.3599	.3621
1.1	.3643	.3665	.3686	.3708	.3729	.3749	.3770	.3790	.3810	.3830
1.2	.3849	.3869	.3888	.3907	.3925	.3944	.3962	.3980	.3997	.4015
1.3	.4032	.4049	.4066	.4082	.4099	.4115	.4131	.4147	.4162	.4177
1.4	.4192	.4207	.4222	.4236	.4251	.4265	.4279	.4292	.4306	.4319
1.5	.4332	.4345	.4357	.4370	.4382	.4394	.4406	.4418	.4429	.4441
1.6	.4452	.4463	.4474	.4484	.4495	.4505	.4515	.4525	.4535	.4545
1.7	.4554	.4564	.4573	.4582	.4591	.4599	.4608	.4616	.4625	.4633
1.8	.4641	.4649	.4656	.4664	.4671	.4678	.4636	.4693	.4699	.4706
1.9	.4713	.4719	.4726	.4732	.4738	.4744	.4750	.4756	.4761	.4767
2.0	.4772	.4778	.4783	.4788	.4793	.4798	.4803	.4808	.4812	.4817
2.1	.4821	.4826	.4830	.4834	.4838	.4842	.4846	.4850	.4854	.4857
2.2	.4861	.4864	.4868	.4871	.4875	.4878	.4881	.4884	.4887	.4890
2.3	.4893	.4896	.4898	.4901	.4904	.4906	.4909	.4911	.4913	.4916
2.4	.4918	.4920	.4922	.4925	.4927	.4929	.4931	.4932	.4934	.4936
2.5	.4938	.4940	.4941	.4943	.4945	.4946	.4948	.4949	.4951	.4952
2.6	.4953	.4955	.4956	.4957	.4959	.4960	.4961	.4962	.4963	.4964
2.7	.4965	.4966	.4967	.4968	.4969	.4970	.4971	.4972	.4973	.4974
2.8	.4974	.4975	.4976	.4977	.4977	.4978	.4979	.4979	.4980	.4981
2.9	.4981	.4982	.4982	.4983	.4984	.4984	.4985	.4985	.4986	.4986
3.0	.4987	.4987	.4987	.4988	.4988	.4989	.4989	.4989	.4990	.4990

Also, for $z = 4.0$, 5.0, and 6.0, the probabilities are .49997, .4999997, and .499999999.

TABLE IV
A Table of Random Numbers

04839	96423	24878	82651	66566	14778	76797	14780	13300	87074
68086	26432	46901	29849	89768	81536	86645	12659	92259	57102
39064	66432	84673	40027	32832	61362	98947	96067	64760	64584
25669	26422	44407	44048	37937	63904	45766	66134	75470	66520
64117	94305	26766	25940	39972	22209	71500	64568	91402	42416
87917	77341	42206	35126	74087	99547	81817	42607	43808	76655
62797	56170	86324	88072	76222	36086	84637	93161	76038	65855
95876	55293	18988	27354	26575	08625	40801	59920	29841	80140
29888	88604	67917	48708	18912	82271	65424	69774	33611	54262
73577	12908	30883	18317	28290	35797	05998	41688	34952	37888
27958	30134	04024	86385	29880	99730	55536	84855	29080	09250
90999	49127	20044	59931	06115	20542	18059	02008	73708	83517
18845	49618	02304	51038	20655	58727	28168	15475	56942	53389
94824	78171	84610	82834	09922	25417	44137	48413	25555	21246
35615	81263	39667	47358	56873	56307	61607	49518	89656	20103
33362	64270	01638	92477	66969	98420	04880	45585	46565	04102
88720	82765	34476	17032	87589	40836	32427	70002	70663	88863
39475	46473	23219	53416	94970	25832	69975	94884	19661	72828
06990	67245	68350	82948	11398	42878	80287	88267	47363	46634
40980	07391	58745	25774	22987	80059	39911	96189	41151	14222
83974	29992	65831	38857	50490	83765	55657	14361	31720	57375
33339	31926	14883	24413	59744	92351	97473	89286	35931	04110
31662	25388	61642	34072	81249	35648	56891	69352	48373	45578
93526	70765	10592	04542	76463	54328	02349	17247	28865	14777
20492	38391	91132	21999	59516	81652	27195	48223	46751	22923
04153	53381	79401	21438	83035	92350	36693	31238	59649	91754
05520	91962	04739	13092	97662	24882	94730	06496	35090	04822
47498	87637	99016	71060	88824	71013	18735	20286	23153	72924
23167	49323	45021	33132	12544	41035	80780	45393	44812	12515
23792	14422	15059	45799	22716	19792	09983	74353	68668	30429
85900	98275	32388	52390	16815	69298	82732	38480	73817	32523
42559	78985	05300	22164	24369	54224	35083	19687	11052	91491
14349	82674	66523	44133	00697	35552	35970	19124	63318	29686
17403	53363	44167	64486	64758	75366	76554	31601	12614	33072
23632	27889	47914	02584	37680	20801	72152	39339	34806	08930

Source: Reprinted by permission from *First Course in Probability* by Sheldon Ross. New York: Macmillan Publishing Co., Inc. (Copyright © 1976 by Sheldon Ross) Page 290.

TABLE V
Values of $t_{\alpha, v}$

v	$\alpha = .10$	$\alpha = .05$	$\alpha = .025$	$\alpha = .01$	$\alpha = .005$	v
1	3.078	6.314	12.706	31.821	63.657	1
2	1.886	2.920	4.303	6.965	9.925	2
3	1.638	2.353	3.182	4.541	5.841	3
4	1.533	2.132	2.776	3.747	4.604	4
5	1.476	2.015	2.571	3.365	4.032	5
6	1.440	1.943	2.447	3.143	3.707	6
7	1.415	1.895	2.365	2.998	3.499	7
8	1.397	1.860	2.306	2.896	3.355	8
9	1.383	1.833	2.262	2.821	3.250	9
10	1.372	1.812	2.228	2.764	3.169	10
11	1.363	1.796	2.201	2.718	3.106	11
12	1.356	1.782	2.179	2.681	3.055	12
13	1.350	1.771	2.160	2.650	3.012	13
14	1.345	1.761	2.145	2.624	2.977	14
15	1.341	1.753	2.131	2.602	2.947	15
16	1.337	1.746	2.120	2.583	2.921	16
17	1.333	1.740	2.110	2.567	2.898	17
18	1.330	1.734	2.101	2.552	2.878	18
19	1.328	1.729	2.093	2.539	2.861	19
20	1.325	1.725	2.086	2.528	2.845	20
21	1.323	1.721	2.080	2.518	2.831	21
22	1.321	1.717	2.074	2.508	2.819	22
23	1.319	1.714	2.069	2.500	2.807	23
24	1.318	1.711	2.064	2.492	2.797	24
25	1.316	1.708	2.060	2.485	2.787	25
26	1.315	1.706	2.056	2.479	2.779	26
27	1.314	1.703	2.052	2.473	2.771	27
28	1.313	1.701	2.048	2.467	2.763	28
29	1.311	1.699	2.045	2.462	2.756	29
inf.	1.282	1.645	1.960	2.326	2.576	inf.

Source: *Biometrika Tables for Statisticians*, Table 12, Vol. I, Cambridge University Press, 1954, by permission of the *Biometrika* trustees.

TABLE VI
Values of $\chi^2_{\alpha, \nu}$

ν	$\alpha = .995$	$\alpha = .99$	$\alpha = .975$	$\alpha = .95$	$\alpha = .05$	$\alpha = .025$	$\alpha = .01$	$\alpha = .005$	ν
1	.0000393	.000157	.000982	.00393	3.841	5.024	6.635	7.879	1
2	.0100	.0201	.0506	.103	5.991	7.378	9.210	10.597	2
3	.0717	.115	.216	.352	7.815	9.348	11.345	12.838	3
4	.207	.297	.484	.711	9.488	11.143	13.277	14.860	4
5	.412	.554	.831	1.145	11.070	12.832	15.086	16.750	5
6	.676	.872	1.237	1.635	12.592	14.449	16.812	18.548	6
7	.989	1.239	1.690	2.167	14.067	16.013	18.475	20.278	7
8	1.344	1.646	2.180	2.733	15.507	17.535	20.090	21.955	8
9	1.735	2.088	2.700	3.325	16.919	19.023	21.666	23.589	9
10	2.156	2.558	3.247	3.940	18.307	20.483	23.209	25.188	10
11	2.603	3.053	3.816	4.575	19.675	21.920	24.725	26.757	11
12	3.074	3.571	4.404	5.226	21.026	23.337	26.217	28.300	12
13	3.565	4.107	5.009	5.892	22.362	24.736	27.688	29.819	13
14	4.075	4.660	5.629	6.571	23.685	26.119	29.141	31.319	14
15	4.601	5.229	6.262	7.261	24.996	27.488	30.578	32.801	15
16	5.142	5.812	6.908	7.962	26.296	28.845	32.000	34.267	16
17	5.697	6.408	7.564	8.672	27.587	30.191	33.409	35.718	17
18	6.265	7.015	8.231	9.390	28.869	31.526	34.805	37.156	18
19	6.844	7.633	8.907	10.117	30.144	32.852	36.191	38.582	19
20	7.434	8.260	9.591	10.851	31.410	34.170	37.566	39.997	20
21	8.034	8.897	10.283	11.591	32.671	35.479	38.932	41.401	21
22	8.643	9.542	10.982	12.338	33.924	36.781	40.289	42.796	22
23	9.260	10.196	11.689	13.091	35.172	38.076	41.638	44.181	23
24	9.886	10.856	12.401	13.848	36.415	39.364	42.980	45.558	24
25	10.520	11.524	13.120	14.611	37.652	40.646	44.314	46.928	25
26	11.160	12.198	13.844	15.379	38.885	41.923	45.642	48.290	26
27	11.808	12.879	14.573	16.151	40.113	43.194	46.963	49.645	27
28	12.461	13.565	15.308	16.928	41.337	44.461	48.278	50.993	28
29	13.121	14.256	16.047	17.708	42.557	45.722	49.588	52.336	29
30	13.787	14.953	16.791	18.493	43.773	46.979	50.892	53.672	30

Source: *Biometrika Tables for Statisticians*, Table 8, Vol. I, Cambridge University Press, 1954, by permission of the *Biometrika* trustees.

v_1 = Degrees of freedom for numerator

v_2 = Degrees of freedom for denominator

v_2 \ v_1	1	2	3	4	5	6	7	8	9	10	12	15	20	24	30	40	60	120	∞
1	161	200	216	225	230	234	237	239	241	242	244	246	248	249	250	251	252	253	254
2	18.5	19.0	19.2	19.2	19.3	19.3	19.4	19.4	19.4	19.4	19.4	19.4	19.4	19.5	19.5	19.5	19.5	19.5	19.5
3	10.1	9.55	9.28	9.12	9.01	8.94	8.89	8.85	8.81	8.79	8.74	8.70	8.66	8.64	8.62	8.59	8.57	8.55	8.53
4	7.71	6.94	6.59	6.39	6.26	6.16	6.09	6.04	6.00	5.96	5.91	5.86	5.80	5.77	5.75	5.72	5.69	5.66	5.63
5	6.61	5.79	5.41	5.19	5.05	4.95	4.88	4.82	4.77	4.74	4.68	4.62	4.56	4.53	4.50	4.46	4.43	4.40	4.37
6	5.99	5.14	4.76	4.53	4.39	4.28	4.21	4.15	4.10	4.06	4.00	3.94	3.87	3.84	3.81	3.77	3.74	3.70	3.67
7	5.59	4.74	4.35	4.12	3.97	3.87	3.79	3.73	3.68	3.64	3.57	3.51	3.44	3.41	3.38	3.34	3.30	3.27	3.23
8	5.32	4.46	4.07	3.84	3.69	3.58	3.50	3.44	3.39	3.35	3.28	3.22	3.15	3.12	3.08	3.04	3.01	2.97	2.93
9	5.12	4.26	3.86	3.63	3.48	3.37	3.29	3.23	3.18	3.14	3.07	3.01	2.94	2.90	2.86	2.83	2.79	2.75	2.71
10	4.96	4.10	3.71	3.48	3.33	3.22	3.14	3.07	3.02	2.98	2.91	2.85	2.77	2.74	2.70	2.66	2.62	2.58	2.54
11	4.84	3.98	3.59	3.36	3.20	3.09	3.01	2.95	2.90	2.85	2.79	2.72	2.65	2.61	2.57	2.53	2.49	2.45	2.40
12	4.75	3.89	3.49	3.26	3.11	3.00	2.91	2.85	2.80	2.75	2.69	2.62	2.54	2.51	2.47	2.43	2.38	2.34	2.30
13	4.67	3.81	3.41	3.18	3.03	2.92	2.83	2.77	2.71	2.67	2.60	2.53	2.46	2.42	2.38	2.34	2.30	2.25	2.21
14	4.60	3.74	3.34	3.11	2.96	2.85	2.76	2.70	2.65	2.60	2.53	2.46	2.39	2.35	2.31	2.27	2.22	2.18	2.13
15	4.54	3.68	3.29	3.06	2.90	2.79	2.71	2.64	2.59	2.54	2.48	2.40	2.33	2.29	2.25	2.20	2.16	2.11	2.07
16	4.49	3.63	3.24	3.01	2.85	2.74	2.66	2.59	2.54	2.49	2.42	2.35	2.28	2.24	2.19	2.15	2.11	2.06	2.01
17	4.45	3.59	3.20	2.96	2.81	2.70	2.61	2.55	2.49	2.45	2.38	2.31	2.23	2.19	2.15	2.10	2.06	2.01	1.96
18	4.41	3.55	3.16	2.93	2.77	2.66	2.58	2.51	2.46	2.41	2.34	2.27	2.19	2.15	2.11	2.06	2.02	1.97	1.92
19	4.38	3.52	3.13	2.90	2.74	2.63	2.54	2.48	2.42	2.38	2.31	2.23	2.16	2.11	2.07	2.03	1.98	1.93	1.88
20	4.35	3.49	3.10	2.87	2.71	2.60	2.51	2.45	2.39	2.35	2.28	2.20	2.12	2.08	2.04	1.99	1.95	1.90	1.84
21	4.32	3.47	3.07	2.84	2.68	2.57	2.49	2.42	2.37	2.32	2.25	2.18	2.10	2.05	2.01	1.96	1.92	1.87	1.81
22	4.30	3.44	3.05	2.82	2.66	2.55	2.46	2.40	2.34	2.30	2.23	2.15	2.07	2.03	1.98	1.94	1.89	1.84	1.78
23	4.28	3.42	3.03	2.80	2.64	2.53	2.44	2.37	2.32	2.27	2.20	2.13	2.05	2.01	1.96	1.91	1.86	1.81	1.76
24	4.26	3.40	3.01	2.78	2.62	2.51	2.42	2.36	2.30	2.25	2.18	2.11	2.03	1.98	1.94	1.89	1.84	1.79	1.73
25	4.24	3.39	2.99	2.76	2.60	2.49	2.40	2.34	2.28	2.24	2.16	2.09	2.01	1.96	1.92	1.87	1.82	1.77	1.71
30	4.17	3.32	2.92	2.69	2.53	2.42	2.33	2.27	2.21	2.16	2.09	2.01	1.93	1.89	1.84	1.79	1.74	1.68	1.62
40	4.08	3.23	2.84	2.61	2.45	2.34	2.25	2.18	2.12	2.08	2.00	1.92	1.84	1.79	1.74	1.69	1.64	1.58	1.51
60	4.00	3.15	2.76	2.53	2.37	2.25	2.17	2.10	2.04	1.99	1.92	1.84	1.75	1.70	1.65	1.59	1.53	1.47	1.39
120	3.92	3.07	2.68	2.45	2.29	2.18	2.09	2.02	1.96	1.91	1.83	1.75	1.66	1.61	1.55	1.50	1.43	1.35	1.25
∞	3.84	3.00	2.60	2.37	2.21	2.10	2.01	1.94	1.88	1.83	1.75	1.67	1.57	1.52	1.46	1.39	1.32	1.22	1.00

Source: M. Merrington and C. M. Thompson, "Tables of percentage points of the inverted beta (F) distribution," Biometrika, Vol. 33 (1943), by permission of the Biometrika trustees.

TABLE VIIb Values of $F_{.01, v_1, v_2}$

v_1 = Degrees of freedom for numerator

v_2	1	2	3	4	5	6	7	8	9	10	12	15	20	24	30	40	60	120	∞
1	4,052	5,000	5,403	5,625	5,764	5,859	5,928	5,982	6,023	6,056	6,106	6,157	6,209	6,235	6,261	6,287	6,313	6,339	6,366
2	98.5	99.0	99.2	99.2	99.3	99.3	99.4	99.4	99.4	99.4	99.4	99.4	99.4	99.5	99.5	99.5	99.5	99.5	99.5
3	34.1	30.8	29.5	28.7	28.2	27.9	27.7	27.5	27.3	27.2	27.1	26.9	26.7	26.6	26.5	26.4	26.3	26.2	26.1
4	21.2	18.0	16.7	16.0	15.5	15.2	15.0	14.8	14.7	14.5	14.4	14.2	14.0	13.9	13.8	13.7	13.7	13.6	13.5
5	16.3	13.3	12.1	11.4	11.0	10.7	10.5	10.3	10.2	10.1	9.89	9.72	9.55	9.47	9.38	9.29	9.20	9.11	9.02
6	13.7	10.9	9.78	9.15	8.75	8.47	8.26	8.10	7.98	7.87	7.72	7.56	7.40	7.31	7.23	7.14	7.06	6.97	6.88
7	12.2	9.55	8.45	7.85	7.46	7.19	6.99	6.84	6.72	6.62	6.47	6.31	6.16	6.07	5.99	5.91	5.82	5.74	5.65
8	11.3	8.65	7.59	7.01	6.63	6.37	6.18	6.03	5.91	5.81	5.67	5.52	5.36	5.28	5.20	5.12	5.03	4.95	4.86
9	10.6	8.02	6.99	6.42	6.06	5.80	5.61	5.47	5.35	5.26	5.11	4.96	4.81	4.73	4.65	4.57	4.48	4.40	4.31
10	10.0	7.56	6.55	5.99	5.64	5.39	5.20	5.06	4.94	4.85	4.71	4.56	4.41	4.33	4.25	4.17	4.08	4.00	3.91
11	9.65	7.21	6.22	5.67	5.32	5.07	4.89	4.74	4.63	4.54	4.40	4.25	4.10	4.02	3.94	3.86	3.78	3.69	3.60
12	9.33	6.93	5.95	5.41	5.06	4.82	4.64	4.50	4.39	4.30	4.16	4.01	3.86	3.78	3.70	3.62	3.54	3.45	3.36
13	9.07	6.70	5.74	5.21	4.86	4.62	4.44	4.30	4.19	4.10	3.96	3.82	3.66	3.59	3.51	3.43	3.34	3.25	3.17
14	8.86	6.51	5.56	5.04	4.70	4.46	4.28	4.14	4.03	3.94	3.80	3.66	3.51	3.43	3.35	3.27	3.18	3.09	3.00
15	8.68	6.36	5.42	4.89	4.56	4.32	4.14	4.00	3.89	3.80	3.67	3.52	3.37	3.29	3.21	3.13	3.05	2.96	2.87
16	8.53	6.23	5.29	4.77	4.44	4.20	4.03	3.89	3.78	3.69	3.55	3.41	3.26	3.18	3.10	3.02	2.93	2.84	2.75
17	8.40	6.11	5.19	4.67	4.34	4.10	3.93	3.79	3.68	3.59	3.46	3.31	3.16	3.08	3.00	2.92	2.83	2.75	2.65
18	8.29	6.01	5.09	4.58	4.25	4.01	3.84	3.71	3.60	3.51	3.37	3.23	3.08	3.00	2.92	2.84	2.75	2.66	2.57
19	8.19	5.93	5.01	4.50	4.17	3.94	3.77	3.63	3.52	3.43	3.30	3.15	3.00	2.92	2.84	2.76	2.67	2.58	2.49
20	8.10	5.85	4.94	4.43	4.10	3.87	3.70	3.56	3.46	3.37	3.23	3.09	2.94	2.86	2.78	2.69	2.61	2.52	2.42
21	8.02	5.78	4.87	4.37	4.04	3.81	3.64	3.51	3.40	3.31	3.17	3.03	2.88	2.80	2.72	2.64	2.55	2.46	2.36
22	7.95	5.72	4.82	4.31	3.99	3.76	3.59	3.45	3.35	3.26	3.12	2.98	2.83	2.75	2.67	2.58	2.50	2.40	2.31
23	7.88	5.66	4.76	4.26	3.94	3.71	3.54	3.41	3.30	3.21	3.07	2.93	2.78	2.70	2.62	2.54	2.45	2.35	2.26
24	7.82	5.61	4.72	4.22	3.90	3.67	3.50	3.36	3.26	3.17	3.03	2.89	2.74	2.66	2.58	2.49	2.40	2.31	2.21
25	7.77	5.57	4.68	4.18	3.86	3.63	3.46	3.32	3.22	3.13	2.99	2.85	2.70	2.62	2.53	2.45	2.36	2.27	2.17
30	7.56	5.39	4.51	4.02	3.70	3.47	3.30	3.17	3.07	2.98	2.84	2.70	2.55	2.47	2.39	2.30	2.21	2.11	2.01
40	7.31	5.18	4.31	3.83	3.51	3.29	3.12	2.99	2.89	2.80	2.66	2.52	2.37	2.29	2.20	2.11	2.02	1.92	1.80
60	7.08	4.98	4.13	3.65	3.34	3.12	2.95	2.82	2.72	2.63	2.50	2.35	2.20	2.12	2.03	1.94	1.84	1.73	1.60
120	6.85	4.79	3.95	3.48	3.17	2.96	2.79	2.66	2.56	2.47	2.34	2.19	2.03	1.95	1.86	1.76	1.66	1.53	1.38
∞	6.63	4.61	3.78	3.32	3.02	2.80	2.64	2.51	2.41	2.32	2.18	2.04	1.88	1.79	1.70	1.59	1.47	1.32	1.00

v_2 = Degrees of freedom for denominator

Reproduced from M. Merrington and C. M. Thompson, "Tables of percentage points of the inverted beta (F) distribution," *Biometrika*, Vol. 33 (1943), by permission of the *Biometrika* trustees.

Factorials

n	$n!$	$\log n!$
0	1	0.0000
1	1	0.0000
2	2	0.3010
3	6	0.7782
4	24	1.3802
5	120	2.0792
6	720	2.8573
7	5,040	3.7024
8	40,320	4.6055
9	362,880	5.5598
10	3,628,800	6.5598
11	39,916,800	7.6012
12	479,001,600	8.6803
13	6,227,020,800	9.7943
14	87,178,291,200	10.9404
15	1,307,674,368,000	12.1165

TABLE VIII
Binomial Coefficients

n	$\binom{n}{0}$	$\binom{n}{1}$	$\binom{n}{2}$	$\binom{n}{3}$	$\binom{n}{4}$	$\binom{n}{5}$	$\binom{n}{6}$	$\binom{n}{7}$	$\binom{n}{8}$	$\binom{n}{9}$	$\binom{n}{10}$
0	1										
1	1	1									
2	1	2	1								
3	1	3	3	1							
4	1	4	6	4	1						
5	1	5	10	10	5	1					
6	1	6	15	20	15	6	1				
7	1	7	21	35	35	21	7	1			
8	1	8	28	56	70	56	28	8	1		
9	1	9	36	84	126	126	84	36	9	1	
10	1	10	45	120	210	252	210	120	45	10	1
11	1	11	55	165	330	462	462	330	165	55	11
12	1	12	66	220	495	792	924	792	495	220	66
13	1	13	78	286	715	1287	1716	1716	1287	715	286
14	1	14	91	364	1001	2002	3003	3432	3003	2002	1001
15	1	15	105	455	1365	3003	5005	6435	6435	5005	3003
16		16	120	560	1820	4368	8008	11440	12870	11440	8008
17		17	136	680	2380	6188	12376	19448	24310	24310	19448
18		18	153	816	3060	8568	18564	31824	43758	48620	43758
19		19	171	969	3876	11628	27132	50388	75582	92378	92378
20	1	20	190	1140	4845	15504	38760	77520	125970	167960	184756

TABLE IX
Values of e^x and e^{-x}

x	e^x	e^{-x}	x	e^x	e^{-x}	x	e^z	e^{-x}	x	e^x	e^{-x}
.0	1.000	1.000	2.5	12.18	.082	5.0	148.4	.0067	7.5	1,808.0	.00055
.1	1.105	.905	2.6	13.46	.074	5.1	164.0	.0061	7.6	1,998.2	.00050
.2	1.221	.819	2.7	14.88	.067	5.2	181.3	.0055	7.7	2,208.3	.00045
.3	1.350	.741	2.8	16.44	.061	5.3	200.3	.0050	7.8	2,440.6	.00041
.4	1.492	.670	2.9	18.17	.055	5.4	221.4	.0045	7.9	2,697.3	.00037
.5	1.649	.607	3.0	20.09	.050	5.5	244.7	.0041	8.0	2,981.0	.00034
.6	1.822	.549	3.1	22.20	.045	5.6	270.4	.0037	8.1	3,294.5	.00030
.7	2.014	.497	3.2	24.53	.041	5.7	298.9	.0033	8.2	3,641.0	.00027
.8	2.226	.449	3.3	27.11	.037	5.8	330.3	.0030	8.3	4,023.9	.00025
.9	2.460	.407	3.4	29.96	.033	5.9	365.0	.0027	8.4	4,447.1	.00022
1.0	2.718	.368	3.5	33.12	.030	6.0	403.4	.0025	8.5	4,914.8	.00020
1.1	3.004	.333	3.6	36.60	.027	6.1	445.9	.0022	8.6	5,431.7	.00018
1.2	3.320	.301	3.7	40.45	.025	6.2	492.8	.0020	8.7	6,002.9	.00017
1.3	3.669	.273	3.8	44.70	.022	6.3	544.6	.0018	8.8	6,634.2	.00015
1.4	4.055	.247	3.9	49.40	.020	6.4	601.8	.0017	8.9	7,332.0	.00014
1.5	4.482	.223	4.0	54.60	.018	6.5	665.1	.0015	9.0	8,103.1	.00012
1.6	4.953	.202	4.1	60.34	.017	6.6	735.1	.0014	9.1	8,955.3	.00011
1.7	5.474	.183	4.2	66.69	.015	6.7	812.4	.0012	9.2	9,897.1	.00010
1.8	6.050	.165	4.3	73.70	.014	6.8	897.8	.0011	9.3	10,938	.00009
1.9	6.686	.150	4.4	81.45	.012	6.9	992.3	.0010	9.4	12,088	.00008
2.0	7.389	.135	4.5	90.02	.011	7.0	1,096.6	.0009	9.5	13,360	.00007
2.1	8.166	.122	4.6	99.48	.010	7.1	1,212.0	.0008	9.6	14,765	.00007
2.2	9.025	.111	4.7	109.95	.009	7.2	1,339.4	.0007	9.7	16,318	.00006
2.3	9.974	.100	4.8	121.51	.008	7.3	1,480.3	.0007	9.8	18,034	.00006
2.4	11.023	.091	4.9	134.29	.007	7.4	1,636.0	.0006	9.9	19,930	.00005

ANSWERS TO EXERCISES

Chapter 1

Section 1.2

classical (a, b, g); empirical (c, e, f); subjective (d, h, i)

Section 1.3

1. (a)

$$S = \begin{cases} (A, B), (A, C), (A, D), (A, E), \\ (B, A), (B, C), (B, D), (B, E), \\ (C, A), (C, B), (C, D), (C, E), \\ (D, A), (D, B), (D, C), (D, E), \\ (E, A), (E, B), (E, C), (E, D) \end{cases}$$

The first letter in each pair denotes the best-tasting brand; the second letter in each pair denotes the worst-tasting brand.

(b) $S = \{atc, act, tac, tca, cat, cta\}$

(c) $S = \{(H, H), (H, P), (H, F), (P, H), (P, P), (P, F), (F, H), (F, P), (F,F)\}$

The first letter in each pair denotes the score of the first student and the second letter denotes the score of the second student.

(d) Let 1 and 0 denote "heads" and "tails", respectively. Let i, j, and k be the faces on the top sides of the coins after the first, second, and third tosses, respectively. The sample space is $S = \{(i, j, k) \mid i = 0$ or 1; $j = 0$ or 1; $k = 0$ or $1\}$

(e) Let n be the number of tosses required. $S = \{n \mid n$ is a positive integer$\}$

(f) Let n be the number of heads. $S = \{n \mid n = 0, 1, 2, 3\}$

(g) $S = \begin{cases} (1, 5), (1, 10), (1, 25), (5, 1), (5, 10), (5, 25) \\ (10, 1), (10, 5), (10, 25), (25, 1), (25, 5), (25, 10) \end{cases}$

The first number in each pair denotes the value of the first coin; the second number is the value of the second coin. **(h)** $S = \{6, 11, 15, 26, 30, 35\}$

(i) Let x, y, and z be the conditions of patients 1, 2 and 3, respectively. $S = \{(x, y, z) \mid x = G, M,$ or N; $y = G, M,$ or N; $z = G, M,$ or $N\}$

2. (a) (A, B), (A, C), (A, D), (A, E);

(b) (B, C), (B,D), (B, E), (C, B), (C, D), (C,E), (D, B), (D, C),(D, E), (E,B), (E, D), (E, C); **(c)** (P, P), (P, F), (F, P), (F, F); **(d)** (1, 1, 1), (1, 1, 0), (1, 0, 1), (0, 1, 1);

(e) 2, 3; **(f)** (1, 5), (5, 1), (1, 10), (10,1), (1, 25), (25, 1); **(g)** (5, 10), (10, 5);

(h) 15; **(i)** (G, M, N), (M, G, N), (N, G, M), (G, N, M), (N, M, G), (M, N, G)

3. (a)

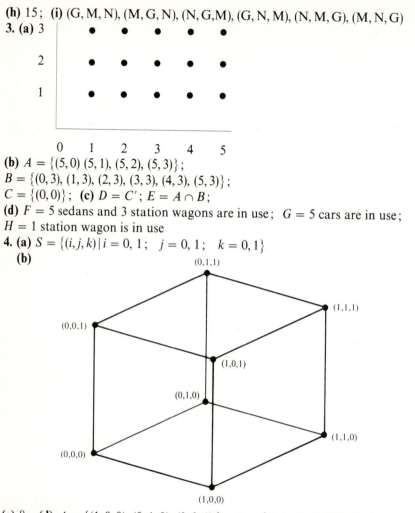

0   1   2   3   4   5

(b) $A = \{(5, 0)\ (5, 1),\ (5, 2),\ (5, 3)\}$;
$B = \{(0, 3),\ (1, 3),\ (2, 3),\ (3, 3),\ (4, 3),\ (5, 3)\}$;
$C = \{(0, 0)\}$; **(c)** $D = C'$; $E = A \cap B$;
(d) $F = 5$ sedans and 3 station wagons are in use; $G = 5$ cars are in use;
$H = 1$ station wagon is in use

4. (a) $S = \{(i, j, k) \mid i = 0, 1;\quad j = 0, 1;\quad k = 0, 1\}$

(b)

(0,1,1)

(1,1,1)

(0,0,1)

(1,0,1)

(0,1,0)

(1,1,0)

(0,0,0)

(1,0,0)

(c) 8; **(d)** $A = \{(1, 0, 0),\ (0, 1, 0),\ (0, 0, 1),\}$; $B = \{(1, 1, 0),\ (1, 0, 1),\ (0, 1, 1)\}$;
$C = \{(1, 0, 0),\ (0, 1, 0),\ (0, 0, 1),\ (1, 1, 0),\ (1, 0, 1),\ (0, 1, 1),\ (1, 1, 1)\}$; **5. (a)** yes;
(b) no; **(c)** no; **6. (a)** A'; **(b)** $A \cap B$; **(c)** $A' \cap B' \cap C'$; **(d)** $A \cup B \cup C$;
(e) $(A \cap B' \cap C') \cup (A' \cap B \cap C') \cup (A' \cap B' \cap C) \cup (A' \cap B' \cap C')$;
(f) $A \subset B$; **(g)** $B \cap C = \varnothing$; **(h)** $A \cup B \cup C$;
7. (a) $A = \{2, 4, 6, \ldots\}$; $B = \{1, 2, 3, 4, 5, 6, 7, 8, 9, 10\}$; $C = \{10, 11, 12, \ldots\}$;
(b) A'; $B \cap C$; B'; **8. (a)** B'; **(b)** B; **(c)** none; **(d)** B'; **10.** b;
11. (a) nondiscrete; **(b)** finite; **(c)** nondiscrete; **(d)** infinite but discrete;
(e) finite; **(f)** infinite but discrete (*Note*: we assume that the Dow Jones
Industrial Average is recorded to two decimal places); **(g)** finite

Section 1.4

1. (a) .5; **(b)** .65; **(c)** 0; **(d)** 1; **2. (a)** .75; **(b)** .5; **(c)** .5; **(d)** .1; **(e)** .25;
3. (a) .55; **(b)** .45; **(c)** .16; **(d)** .46; **(e)** .75; **4. (a)** .2; **(b)** .7; **(c)** .5;
(d) 1; **(e)** 0; **(f)** .5; **(g)** 1; **5. (a)** .2; **(b)** .8; **(c)** .15;

6. 10/25, 5/25, 10/25, 5/25, 21/25, 15/25; **7. (a)** .96; **(b)** .01; **(c)** .03;
(d) .02; **(e)** .01; **8. (a)** 1/26; **(b)** 3/13; **(c)** 1/2; **(d)** 4/52; **(e)** 24/52;
(f) 36/52; **(g)** 1/2; **9.** b, d, g, h;
10. 1/2 for Exercise 2(d); 1/6 for Exercise 2(g); **11.** $P(A) = 4/24$,
$P(B) = 6/24$, $P(C) = 1/24$, $P(D) = 23/24$, $P(E) = 1/24$, $P(F) = 1/24$,
$P(G) = 4/24$, $P(H) = 6/24$; **12.** $P(A) = 3/8$, $P(B) = 3/8$, $P(C) = 7/8$;
14. (a) 25/10,000; **(b)** 30/10,000; **(c)** 9/10,000; **(d)** 5/10,000;
(e) 9953/10,000; **(f)** 9957/10,000; **(g)** 22/10,000; **(h)** 9995/10,000;
(i) 9944/10,000; **(j)** 9/10,000; **(k)** 14/10,000; **(l)** 42/10,000; **15. (a)** yes;
(b) no; **(c)** no; **(d)** no; **16.** $4/5 \le p < 5/6$; **17.** $.1 \le p \le .3$; **18. (a)** .66;
(b) .33; **(c)** .01; **19. (a)** .4; **(b)** .6; **(c)** .75; **(d)** .05; **20. (a)** .5; **(b)** .85;
(c) .35; **22.** no; **24. (a)** Contradicts Theorem 1.2;
(b) The statement implies that the probability of being asked by Jim or Mark
is 1.2;
(c) The statement implies that the probability of the Yankees winning at
least one game is 1.1;
(d) Contradicts the statement in Exercise 21; **26.** none

Chapter 2

Section 2.1

1. (a) Let x_i be the value on the ith die. $i = 1, 2, 3, 4, 5, 6$;
$S = \{(x_1, x_2, x_3, x_5, x_6) | x_i = 1, 2, 3, 4, 5, 6 \text{ for } i = 1, 2, 3, 4, 5, 6\}$;
$N = 6^6 = 46656$; **(b)** Let $x_i = $ h if heads on the ith flip, and let $x_i = $ t if
tails on the ith flip.
$S = \{(x_1, x_2, \ldots, x_{20}) | x_i = \text{h or t}, \quad i = 1, 2, \ldots, 20\}$; $N = 2^{20} = 1,048,576$;
(c) Let $x_i = $ t if question i is marked "true", and let $x_i = $ f if question i is
marked "false". $S = \{(x_1, x_2, \ldots, x_{15}) | x_i = \text{t or f}; i = 1, 2, \ldots, 15\}$;
$N = 2^{15} = 32,768$; **(d)** Let $x_i = 1$ if the ith patient shows improvement.
Let $x_i = 0$ if the ith patient shows no improvement.
$S = \{(x_1, x_2, \ldots, x_{10}) | x_i = 0 \text{ or } 1\}$; $N = 2^{10} = 1,024$; **(e)** Let $x_i = $ d if the
ith building is demolished; $x_i = $ r if it is renovated; and $x_i = $ s if it is
left as a slum. $S = \{(x_1, x_2, x_3, x_4) | x_i = \text{d, r, or s}\}$; $N = 3^4 = 81$;
2. (a) $6/(6^6) = 1/6^5)$; **(b)** $2/(2^{20}) = 1/(2^{19})$; **(c)** $1/(2^{15})$; **(d)** $1/(2^{10})$;
(e) 1/81; **3. (a)** $(6 \cdot 5^5)/(6^6) = (5^5)/(6^5)$; **(b)** $(3^6)/(6^6) = 1/(2^6)$; **(c)** $7/(6^6)$;
(d) $20/(2^{20})$; **(e)** $2/(2^{15})$; **(f)** $15/(2^{15})$; **(g)** $1 - [1/(2^{10})] = 1023/1024$;
(h) 16/81; **4. (a)** Let n be the value on the die. Let $x_i = $ h if heads on
the ith toss and $x_i = $ t if tails on the ith toss.
$S = \{(n, x_1, x_2, x_3, x_4, x_5, x_6) | n = 1, 2, 3, 4, 5, 6; \ x_i = \text{h or t},$
$i = 1, 2, 3, 4, 5, 6\}$; $N = 6 \cdot 2^6 = 384$; **(b)** Assume that the physics, math,
and English books are numbered 1 to 3, 1 to 5, and 1 to 7, respectively.
Let x_1 be the number of the physics book, x_2 be the number of the math
book, and x_3 be the number of the English book.
$S = \{(x_1, x_2, x_3) | x_1 = 1, 2, 3; \ x_2 = 1, 2, 3, 4, 5; \ x_3 = 1, 2, \ldots, 7\}$;

$N = 3 \cdot 5 \cdot 7 = 105$; **(c)** Assume that the returns from groups A, B, and C are numbered from 1 to 100, 1 to 200, and 1 to 150, respectively. Let a, b, and c be the identification numbers of the returns from group A, group B, and group C, respectively.
$S = \{(a,b,c) \mid a = 1,2,\ldots,100; \; b = 1,2,\ldots,200; \; c = 1,2,\ldots,150\}$;
$N = (100) \cdot (200) \cdot (150) = 3,000,000$; **(d)** Assume that the calculators from shipments A, B, and C are numbered from 1 to 100, 1 to 150, and 1 to 125, respectively. Let a, b, and c be the numbers of the calculators from shipment A, shipment B, and shipment C, respectively.
$S = \{(a,b,c) \mid a = 1,2,\ldots,100; \; b = 1,2,\ldots,150; \; c = 1,2,\ldots,125\}$;
$N = (100) \cdot (150) \cdot (125) = 1,875,000$; **5. (a)** $1/(2^6)$; **(b)** $12/35$; **(c)** $.006$; **(d)** $.05$

Section 2.2

1. (a) S = set of permutations of 4 players from the set of 15 players. The 1st, 2nd, 3rd, and 4th players in the permutations are the catcher, pitcher, first baseman, and second baseman respectively.
$N = 15 \cdot 14 \cdot 13 \cdot 12 = 32,760$;
(b) S = set of permutations of 2 professors from the set of 40 professors. The first and second professors in the permutations are the chairman and deputy chairman, respectively. $N = 40 \cdot 39 = 1,560$; **(c)** S = set of permutations of 3 wines from the set of 20 wines.
$N = 20 \cdot 19 \cdot 18 = 6,840$;
(d) S = set of permutations of 3 funds from the set of 20 mutual funds. The first, second, and third mutual funds in the permutation are investments of $3,000, $10,000, and $15,000, respectively.
$N = 20 \cdot 19 \cdot 18 = 6,840$; **2. (a)** $1/32760$; **(b)** $2/1560$;
(c) $(6 \cdot 5 \cdot 4)/(20 \cdot 19 \cdot 18) = .0175$; **(d)** $3!/6840 = .00088$;
3. $(6 \cdot 5 \cdot 4 \cdot 3 \cdot 2 \cdot 1)/(6^6) = .0154$;
4. (a) $N = 52 \cdot 51 \cdot 50 \cdot 48 = 6,497,400$; **(b)** $100 \cdot 99 \cdot 98 = 970,200$;
5. (a) $1/6497400$; **(b)** $(4 \cdot 3 \cdot 2 \cdot 1)/6497400 = 1/(270,725)$;
(c) $(4 \cdot 3 \cdot 2 \cdot 1)(4^4)/6497400 = (4^4)/(270,725)$; **(d)** $(50 \cdot 49 \cdot 48)/970200$;
(e) $(50 \cdot 49 \cdot 48)/970200$; **(f)** $98/970200$; **6. (a)** $6^4 = 1,296$;
(b) $6^4 - 5^4 = 671$

Section 2.3

1. (a) S = set of all combinations of 5 questions from the set of 15 questions;
$N = \binom{15}{5} = 3,003$; **(b)** S = set of all combinations of 5 members from the membership of 20 members; $N = \binom{20}{5} = 15,504$; **(c)** S = set of all combinations of 4 cards from the deck of 52 cards; $N = \binom{52}{4} = 270,725$;
(d) S = set of all combinations of 5 businesses from the group of 25 failing small businesses; $N = \binom{25}{5} = 53,130$; **2. (a)** $\binom{10}{5}/\binom{15}{5} = .084$;
(b) $\binom{8}{5}/\binom{20}{5} = .0036$; **(c)** $\binom{26}{4}/\binom{52}{4} = .055$; **(d)** $\binom{4}{4}\binom{21}{1}/\binom{25}{5} = .00039$;
3. no for 5(a); yes for 5(b) and 5(c); answer to 5(b) is $1/\binom{52}{4}$; answer to 5(c) is $4^4/\binom{52}{4}$

4. (a) $\dfrac{\binom{8}{3}\binom{7}{2}}{\binom{15}{5}}$; **(b)** $\dfrac{\binom{8}{1}\binom{12}{4}}{\binom{20}{5}}$; **(c)** $\dfrac{\binom{13}{3}\binom{13}{1}}{\binom{52}{4}}$; **(d)** $\dfrac{\binom{4}{0}\binom{21}{5} + \binom{4}{1}\binom{21}{4} + \binom{4}{2}\binom{21}{3}}{\binom{25}{5}}$;

5. $\dfrac{\binom{15}{5}}{2^{15}}$; **6.** $\dfrac{\binom{15}{0} + \binom{15}{1} + \binom{15}{2} + \binom{15}{3} + \binom{15}{4}}{2^{15}}$ **11. (a)** $\binom{10}{2} \cdot \binom{15}{2} \cdot \binom{20}{2} = 897750$;

(b) $\dfrac{\binom{3}{2}\binom{4}{2}\binom{7}{2}}{\binom{10}{2}\binom{15}{2}\binom{20}{2}} = .00042$; **(c)** $\dfrac{\binom{7}{2}\binom{11}{2}\binom{13}{2}}{\binom{10}{2}\binom{15}{2}\binom{20}{2}} = .1405$;

12. $\dfrac{\binom{1}{1}\binom{14}{1}\binom{2}{2}\binom{18}{1}\binom{2}{2}\binom{8}{0}}{\binom{15}{2}\binom{20}{3}\binom{10}{2}} = .000047$

Section 2.4

1. $\dfrac{14!}{4!\ 3!\ 7!}$; **2.** $\dfrac{1}{\left(\frac{100!}{30!\ 20!\ 35!\ 15!}\right)}$; **3.** $\dfrac{\left(\frac{6!}{2!\ 3!\ 1!}\right)}{6^6}$; **4.** $\dfrac{4!}{\left(\frac{8!}{2!\ 2!\ 2!\ 2!}\right)}$;

5. $\dfrac{\left(\frac{20!}{10!\ 10!}\right) \cdot \left(\frac{15!}{10!\ 5!}\right)}{\left(\frac{100!}{30!\ 20!\ 35!\ 15!}\right)}$; **6. (a)** $\dfrac{10!}{2!\ 3!\ 5!}$; **(b)** $\dfrac{2 \cdot \binom{5}{2}\binom{3}{3}\binom{5}{2}}{\left(\frac{10!}{2!\ 3!\ 5!}\right)}$;

(c) $\dfrac{\left(\frac{5!}{1!\ 1!\ 3!} \cdot \frac{5!}{1!\ 2!\ 2!} + \frac{5!}{1!\ 2!\ 2!} \cdot \frac{5!}{1!\ 1!\ 3!} + \frac{5!}{1!\ 3!\ 1!} \cdot \frac{5!}{1!\ 0!\ 4!} + \frac{5!}{2!\ 1!\ 2!} \cdot \frac{5!}{0!\ 2!\ 3!} + \frac{5!}{2!\ 2!\ 1!} \cdot \frac{5!}{0!\ 1!\ 4!}\right)}{\left(\frac{10!}{2!\ 3!\ 5!}\right)}$

Section 2.5

1.

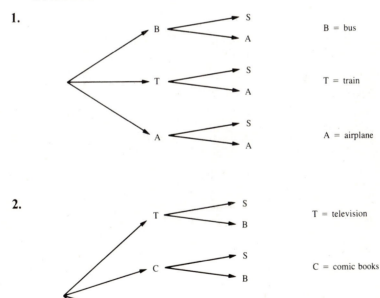

B = bus

T = train

A = airplane

2.

T = television

C = comic books

L = loafing

S = study

3.

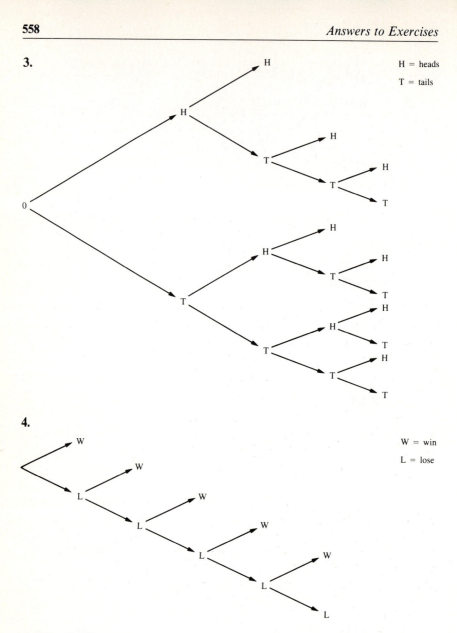

H = heads
T = tails

4.

W = win
L = lose

6. 2/3; **7.** 3/8

Supplementary Exercises

1. $\dfrac{6 \cdot 5 \cdot 4}{6^3} = \frac{5}{9}$; **2.** $1 - \dfrac{\binom{97}{4}}{\binom{100}{4}}, \dfrac{\binom{97}{4}}{\binom{100}{4}}, \dfrac{\binom{3}{1}\binom{97}{3}}{\binom{100}{4}}$; **3.** $\dfrac{1}{4!}$;

4. $1 - \dfrac{\binom{20}{10}}{\binom{25}{10}}, \dfrac{\binom{20}{10}}{\binom{25}{10}}, \dfrac{\binom{5}{2}\binom{20}{8}}{\binom{25}{10}}$; **5. (a)** $\binom{52}{13}$; **(b)** $25 \cdot 24 \cdot 23 \cdot 22 = 303,600$;

(c) $\binom{1000}{2}\binom{980}{2}\binom{850}{2}\binom{840}{2}$; **(d)** $5 \cdot 10 \cdot 6 = 300$; **6. (a)** $\dfrac{4 \cdot \binom{13}{13}}{\binom{52}{13}}$; **(b)** $\dfrac{1}{303,600}$;

(c) $\dfrac{\binom{400}{2}\binom{350}{2}\binom{300}{2}\binom{300}{2}}{\binom{1000}{2}\binom{980}{2}\binom{850}{2}\binom{840}{2}}$; **(d)** $\dfrac{2\cdot 2\cdot 2}{5\cdot 10\cdot 6}=\dfrac{2}{75}$; **7.** $\dfrac{\binom{13}{2}\binom{13}{2}\binom{13}{2}\binom{13}{2}}{\binom{52}{8}}$;

8. $\dfrac{\binom{4}{2}\binom{4}{3}\binom{4}{1}\binom{4}{4}\binom{4}{2}\binom{4}{1}}{\binom{52}{13}}$; **9. (a)** $\dfrac{1}{\left(\frac{14!}{4!\ 3!\ 7!}\right)}$; **(b)** $\dfrac{3!}{\left(\frac{14!}{4!\ 3!\ 7!}\right)}$

Supplementary Exercises on Combinatorial Methods

Permutations
1. $10!$; **2.** $9\cdot 8\cdot 7\cdot 6=3{,}024$; $1\cdot 8\cdot 7\cdot 6=336$; **3.** $9!$;
4. $26\cdot 25\cdot 24=15{,}600$; **5.** $10\cdot 9\cdot 8\cdot 7=5{,}040$; **6.** $6\cdot 5\cdot 4\cdot 3\cdot 2=720$

Combinations
1. $\binom{10}{4}$; **2.** $\binom{15}{5}$; **3.** $\binom{5}{3}$; **4.** $\binom{6}{2}\cdot\binom{8}{3}$; **5.** $\binom{5}{3}\cdot\binom{3}{2}\cdot\binom{6}{4}$

Multinomial coefficients
1. $\dfrac{11!}{5!\ 2!\ 2!\ 1!\ 1!}$; **3.** $\dfrac{12!}{3!\ 2!\ 3!\ 4!}$; **3.** $\dfrac{8!}{3!\ 2!\ 3!}$; **4.** $\dfrac{20!}{7!\ 7!\ 6!}$

Chapter 3

Section 3.1

1. $.5$; **2.** $.5$; **3. (a)** $1/3$; **(b)** $4/5$; **4.** $3/4$; **5.** $1/3$; **6.** $\binom{50}{4}\cdot\binom{46}{4}/\binom{51}{4}\cdot\binom{47}{5}$;
7. $\binom{13}{5}/\binom{47}{5}$; **8. (a)** $30/65$; **(b)** $30/35$

Section 3.2

1. $1/3$; **2.** $7/12$; **3. (a)** $1/6$; **(b)** $5/12$; **4.** $11/50$; **5. (a)** $.2$; **(b)** $.1$;
(c) $1/30$; **(d)** $4/5$; **(e)** $3/5$; **(f)** $1/2$; **(g)** $3/4$; **(h)** $1/40$; **(i)** 1; **(j)** $29/30$;
6. (a) $.021$; **(b)** 23.81%; **7. (a)** $.875$; **(b)** $.2714$; **8. (a)** 88%; **(b)** $15/22$;
9. $2/3$; **11.** prior probabilities are $.6, .3, .1$; posterior probabilities are
$6/29, 21/29, 2/29$; **12. (a)** $.6$

Section 3.3

1. $P(h_1)P(h_2)P(h_3)\cdots P(h_{10})=(1/2)^{10}$; **2.** $P(e_1)P(e_2)\cdots P(e_5)=(1/2)^{5}$;
3. $(1/2)(1/6)(1/13)$; **4. (a)** $P(AB')=P(A)-P(AB)=P(A)-P(A)\cdot P(B)=$
$P(A)[1-P(B)]=P(A)P(B')$; **(b)** follows from (a)
5. Use the results of Exercise 4. If A and B are independent then

$$P(A\mid B')=\frac{P(AB')}{P(B')}=\frac{P(A)P(B')}{P(B')}=P(A)\quad\text{and}$$

$$P(B\mid A')=\frac{P(A'B)}{P(A')}=\frac{P(A')P(B)}{P(A')}=P(B).$$

Whereas, if $P(A\mid B')=P(A)$ then $\dfrac{P(AB')}{P(B')}=P(A)$,

and $P(AB')=P(A)P(B')$. Therefore, A and B are independent.

6. $P(A'B') = P(A')P(B') = (.5)(.2) = .1$;
7. $P(S_1S_2S_3F_4) + P(S_1S_2F_3S_4) + P(S_1F_2S_3S_4) + P(F_1S_2S_3S_4) = 4p^3q$;
$P(S_1S_2S_3S_4) = p^4$; **8.** q^3; $3pq^2$; $3p^2q$; p^3; **9.** the events of (a), (b), (c), (d)

Section 3.5

1. (a)

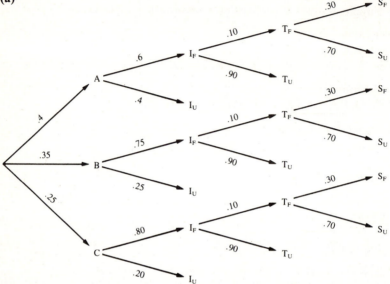

A = interview with personnel officer A; B = interview with personnel officer B; C = interview with personnel officer C; I_F = favorable interview; I_U = unfavorable interview; T_F = favorable test score; T_U = unfavorable test score; S_F = favorable staff member review; S_U = unfavorable staff member review
(b) $[(.4)(.6)(.10)(.30)] + [(.35)(.75)(.10)(.30)] + [(.25)(.80)(.10)(.30)] = .02$;
(c) $(.10)(.30) = .03$

Supplementary Exercises

1. .0476; **2. (a)** 1/3; **(b)** 43.18%; **3. (a)** 1/16; **(b)** 1/8; **(c)** 3/8; **(d)** 1/2;
(e) 15/16; **4.** 6/11; **5. (a)** 36/91; **(b)** 30/91; **6. (a)** 1/57; **(b)** 91/285;
(c) 91/190; **7.** 8/28; 15/28, 5/28; **8. (a)** .0595; **(b)** .0005;
(c) $118/119 = .9916$; **9. (a)** no; **(b)** 1/20; **(c)** 1/10; **(d)** increase;
(e) .029; **(f)** .028

Chapter 4

Section 4.2

1. (a) $X(i, j) = i - j$ for $i = 1, 2, \ldots, 6$; $j = 1, 2, \ldots, 6$;
(b) $Y(i, j) = i \cdot j$ for $i = 1, 2, \ldots, 6$; $j = 1, 2, \ldots, 6$

2.

Sample point	(h, h, h)	(t, h, h)	(h, t, h)	(h, h, t)	(t, t, h)	(t, h, t)	(h, t, t)	(t, t, t)
Value of X	0	2	2	2	2	2	2	0
Value of Y	3	1	1	1	1	1	1	3

3. (a) $X(i, j, k) = \dfrac{i + j + k}{3}$ for $i = 1, 2, 3; \ k = 1, 2, 3$

(b) Sample point	(1, 1, 1)	(2, 1, 1)	(1, 2, 1)	(1, 1, 2)	(1, 2, 2)	$\cdots$	(3, 3, 3)
Value of X	1	4/3	4/3	4/3	5/3	$\cdots$	3

Section 4.3

1.

x	-5	-4	-3	-2	-1	0	1	2	3	4	5
$f(x)$	$\frac{1}{36}$	$\frac{2}{36}$	$\frac{3}{36}$	$\frac{4}{36}$	$\frac{5}{36}$	$\frac{6}{36}$	$\frac{5}{36}$	$\frac{4}{36}$	$\frac{3}{36}$	$\frac{2}{36}$	$\frac{1}{36}$

y	1	2	3	4	5	6	8	9	10	12	15	16	18	20	24	25	30	36
$g(y)$	$\frac{1}{36}$	$\frac{2}{36}$	$\frac{2}{36}$	$\frac{3}{36}$	$\frac{2}{36}$	$\frac{4}{36}$	$\frac{2}{36}$	$\frac{1}{36}$	$\frac{2}{36}$	$\frac{4}{36}$	$\frac{2}{36}$	$\frac{1}{36}$	$\frac{2}{36}$	$\frac{2}{36}$	$\frac{2}{36}$	$\frac{1}{36}$	$\frac{2}{36}$	$\frac{1}{36}$

2.

x	0	2
$f(x)$	$\frac{2}{8}$	$\frac{6}{8}$

y	1	3
$g(y)$	$\frac{6}{8}$	$\frac{2}{8}$

3. $f(x) = \frac{1}{6}$ for $x = 350$, $\frac{1}{6}$ for $x = 600$, $\frac{2}{6}$ for $x = 750$, $\frac{1}{6}$ for $x = 900$, and $\frac{1}{6}$ for $x = 1{,}150$; **4.** $f(x) = \frac{1}{4}$ for $x = 108\frac{1}{3}$, $\frac{1}{4}$ for $x = 116\frac{2}{3}$, and $\frac{1}{2}$ for $x = 125$; **5. (a)** $P(X < 4) = f(2) + f(3) = \frac{3}{36}$;
(b) $P(3 \leqslant X \leqslant 9) = f(3) + f(4) + f(5) + f(6) + f(7) + f(8) + f(9) = \frac{29}{36}$;
(c) $f(3) + f(6) + f(9) + f(12) = \frac{12}{36}$; **6.** $f(x) = \frac{2}{6}$ for $x = 6$,
$\frac{1}{6}$ for $x = 10$, $\frac{1}{6}$ for $x = 11$, and $\frac{2}{6}$ for $x = 15$;
7. $P(-2 \leqslant X < 4) = \frac{3}{10}$; $P(X > 0) = \frac{7}{10}$; $P(X \leqslant 4) = \frac{7}{10}$; **8. (a)** yes;
(b) no, contradicts $\sum_x f(x) = 1$; **(c)** no, contradicts $\sum_x f(x) = 1$;
(d) yes; **9.** $\frac{1}{20}$

Section 4.5

2. $b(x; n, p) = \binom{n}{x} p^x q^{n-x} = \binom{n}{n-x} q^{n-x} p^{n-(n-x)} = b(n - x; n, q)$; **3. (a)** .0090;
(b) .1983; **(c)** .1859; **(d)** .0577; **4.** $b(1; 5, .01) = .048$;

5. (a) $b(1; 15, 105) = .3658$;

(b) $b(0; 15, .05) + b(1; 15, .05) + b(2; 15, .05) = .9639$;

(c) $1 - b(0; 15, .05) - b(1; 15, .05) = .1709$;

6. (a) $b(10; 20, .65) = b(10; 20, .35) = .0686$;

(b) $\displaystyle\sum_{x=10}^{20} b(x; 20, .65) = \sum_{y=0}^{10} b(y; 20, .35) = .9469$;

(c) $\displaystyle\sum_{x=8}^{12} b(x; 20, .65) = \sum_{y=8}^{12} b(y; 20, .35) = .393$; **7. (a)** $b(2; 4, \frac{1}{3}) = .2963$;

(b) $b(0; 4, \frac{1}{3}) = .1975$; **(c)** $1 - b(0; 4, \frac{1}{3}) - b(1; 4, \frac{1}{3}) = .4074$;

8. (a) $\displaystyle\sum_{x=3}^{10} b(x; 10, .01) = .00011$; **(b)** $b(0; 10, .05) = .5987$;

(c) $b(1; 10, .01) + b(2; 10, .01) = .0955$; **9.** $b(0; 10, .1) = .3487$;

10. $1 - b(0; 5, .1) - b(1; 5, .1) = .0815$; **11. (a)** $\dfrac{\binom{5}{0}\binom{19}{3}}{\binom{24}{3}} = .4788$;

(b) $\dfrac{\binom{5}{1}\binom{19}{2}}{\binom{24}{3}} = .4224$; **(c)** $1 - \dfrac{\binom{5}{0}\binom{19}{3}}{\binom{24}{3}} = .5212$; **12. (a)** $\dfrac{\binom{4}{0}\binom{16}{5}}{\binom{20}{5}} = .2817$;

(b) $\dfrac{\binom{4}{4}\binom{16}{1}}{\binom{20}{5}} = .0010$; **(c)** $\dfrac{\binom{4}{1}\binom{16}{4}}{\binom{20}{5}} = .4696$;

13. (a) $\dfrac{\binom{50}{2}\binom{50}{1}}{\binom{100}{3}} = .3788$; **(b)** $b(2; 3, .5) = .3750$;

14. (a) $\dfrac{\binom{30}{3}\binom{120}{1}}{\binom{150}{4}} = .0240$; **(b)** $b(3; 4, .2) = .0256$;

15. (a) $b(3; 20, .01) = .00096$; $p(3; .2) = .0011$; **(b)** $b(5; 100, .05) = .1800$;
$p(5; 5) = .1755$; **17. (a)** $p(5; 5) = .1755$; **(b)** $1 - p(0; 5) = .9933$;
18. (a) $b(15; 30, .5) \approx p(15; 15) = .1024$;

(b) $\displaystyle\sum_{x=0}^{3} b(x; 60, \frac{1}{6}) \approx \sum_{x=0}^{3} p(x; 10) = .0104$; **19. (a)** $p(3, 2) = .1804$;

(b) $p(0; 2) + p(1; 2) + p(2; 2) = .6767$; **(c)** $1 - .6767 = .3233$;

20. $1 - \sum_{x=8}^{15} p(x; 10) = .0487$;

23. (a) $P(\text{no car in 5-minute interval}) = P(X_5 = 0) = e^{-1.5} = .223$;

(b) $P(X_5 > 1) = 1 - P(X_5 = 0) - P(X_5 = 1) =$
$\qquad 1 - e^{-1.5} - (1.5)e^{-1.5} = .442$;

(c) $P(15 \le X_{60} \le 25) = \sum_{x=15}^{25} \dfrac{18^x e^{-18}}{x!}$;

24. (a) $P(X_{10} = 10) = \dfrac{10^{10} e^{-10}}{10!} = .125$;

(b) $P(X_{10} \ge 1) = 1 - P(X < 10) = 1 - \sum_{x=0}^{9} \dfrac{10^x e^{-10}}{x!}$

Section 4.6

3. (a) $P(X \le 3) = F(3) = .4$; $P(X \le 4) = F(4) = .4$;
$P(1.5 < X \le 5.2) = F(5.2) - F(1.5) = .9 - .1 = .8$; $P(X = 3) = .3$;
$P(X = 3.5) = 0$; **(b)** $f(x) = .1$ for $x = 1$, $.3$ for $x = 3$, $.5$ for $x = 5$, and $.1$ for
$x = 5.5$

Supplementary Exercises

1. 4; **2.** 8; **3.** 44; **4.** $P(S_n \geq 17) = .0445$; yes; **5. (a)** .5287; **(b)** 13;
6. (a) $f(x) = P(X = x) = P$(a 6 occurs for the first time on the xth throw) $=$
P(no 6's in the first $(x - 1)$ throws followed by a 6 on the xth throw) $=$
$(\frac{5}{6})^{x-1}(\frac{1}{6}) = q^{x-1}p$ where $p = \frac{1}{6}$; **(b)** $f(4) = (\frac{5}{6})^3(\frac{1}{6}) = .096$; **(c)** .69; **9.** .974;
10. .82

Chapter 5

Section 5.2

1. c, d, f; **2.** $f(x) = 1/3$ for $-1 < x < 2$, $f(x) = 0$ elsewhere;
3. $f(x) = 1/(b - a)$ for $a < x < b$, $f(x) = 0$ elsewhere;
4. (b) $f(x) = 1/6$ for $0 < x < 1$, $f(x) = 5/6$ for $1 < x < 2$, $f(x) = 0$
elsewhere.

Section 5.4

1. .1467, .8533, .38, 0, 0, 0; **2. (b)** .32, .68, .16, .42; **3. (a)** $5/32 = .15625$;
(b) $11/32 = .34375$; **(c)** .21825; **(d)** .1255; **(e)** 0; **4. (a)** .7407; **(b)** .2593;
(c) .4259; **(d)** .5; **5.** a, c, d, e, f, and g are true statements.

Section 5.5

1. (a) $F(x) = 0$ for $x \leq 1$; $F(x) = \frac{1}{3}(t^2 - 1)$ for $1 < x < 2$; $F(x) = 1$ for $x \geq 2$;
(b) $F(x) = 0$ for $x \leq -.5$; $F(x) = 2(.25 - x^2)$ for $-.5 < x \leq 0$;
$F(x) = .5 + 2x^2$ for $0 < x < .5$; $F(x) = 1$ for $x \geq .5$;
(c) $F(x) = 0$ for $x \leq -1$;
$F(x) = \frac{3}{4}(x - \frac{x^3}{3} + \frac{2}{3})$ for $-1 < x < 1$; $F(x) = 1$ for $x \geq 1$; **2.** 0, .25, 1, .84,
.21; **3. (a)** .16; **(b)** .96; **(c)** .078; **4. (a)** 0; **(b)** .125.

Section 5.6

7. b, c, e; **8. (a)** 1/12; **(c)** 15/24; **(d)** $F(x) = 0$ for $x \leq 1$, $\frac{1}{24}(x^2 - 1)$ for
$1 < x < 5$, and 1 for $x \geq 5$; **(g)** yes; **(h)** no; **9.** b; **10. (a)** 9;
(c) 7/16, 81/400, 1/4, 0; **(d)** $f(x) = 18/x^3$ for $x \geq 3$, 0 for $x < 3$

Section 5.7

1. (a) .2; **(b)** .3; **(c)** 0; **(d)** .2; **2. (a)** 1/3; **(b)** $8/81 = .0988$;
3. (a) $e^{-2} = .1353$; **(b)** $e^{-1/2} - e^{-3/2} = .3834$;
(c) $1 - b(0; 5, e^{-2}) = .5167$; **4. (a)** $e^{-3/2} = .223$; **(b)** $1 - e^{-5/2} = .918$;
5. The waiting time (in years) until a miracle is an exponential random
variable with mean equal to 217.14724. Therefore, the probability of at
least 1 miracle in 200 years is .6019. **6. (a)** .996; **(b)** .964;
7. (a) .9332; **(b)** .0162; **(c)** .0506; **(d)** .0934; **(e)** .9941; **(f)** .9436;
(g) .0609; **(h)** 0; **8. (a)** 2.15; **(b)** .5; **(c)** $-.17$; **(d)** -1.13; **(e)** 1.6;

9. (a) .0228; **(b)** .8413; **(c)** .383; **(d)** .1587; **10. (a)** .0228; **(b)** .9332;
(c) .0228; **(d)** .8664; **(e)** 0; **11. (a)** .0228; **(b)** .8413; **(c)** .6826; **(d)** .3446;
12. (a) .8413; **(b)** .0668; **(c)** .1587; **(d)** .3413

Chapter 6

Section 6.1

1. 11/3; **2. (a)** 1/3; **(b)** 1/2; **3. (a)** 7; **(b)** $161/36 = 4.47$; **4.** 1.8;
5. 31/6; **6.** $15,000; **7. (a)** 1; **(b)** 1; **(c)** the strong law of large numbers;
8. (a) 5; **(b)** yes, because the expected amount is 5 oz.; **9.** $85/221 = .385$;
10. 3.25; **11.** 1; **12. (a)** 16/15; **(b)** $1066.67; **(c)** yes; **(d)** yes

Section 6.2

1. $5; **2.** no, B, 89¢; **5.** no, expected cost is $2.50 for illegal parking;
6. 1; **9.** Buy! Expected profit is $1,350,000.

Section 6.3

1. $g(X)(i, j) = (i + j)^{1/2}$, $E[g(X)] = 2.602$; **2.** 47, 2.4, 4.8; **3.** .2, 0, -1, .375;
4. 1.42 π square inches; **5.** $175
6.

x	$g_4(x)$	$g_5(x)$	$g_6(x)$	$g_7(x)$
4	120	135	150	165
5	160	150	165	180
6	200	190	180	195
7	240	230	220	210

$E[g_4(X)] = 180$
$E[g_5(X)] = 176.25$
$E[g_6(X)] = 178.75$
$E[g_7(X)] = 187.50$

7. $E[g_3(X)] = 1.50$; $E[g_4(X)] = 1.58$; $E[g_5(X)] = 1.33$; $E[g_6(X)] = .92$

8.

x	$g_1(x)$	$g_2(x)$
0	-1	-2
1	2	1
2	2	4

$E[g_1(X)] = 1.25$
$E[g_2(X)] = 1$

Section 6.4

1. 44; **2.** $731/108 = 6.7685$; **3.** 45; **5.** 1.63; **6.** $500

Section 6.5

1. *Exercise 1:* $\sigma^2 = 11.556$; *Exer. 2:* (a) $\sigma^2 = .0556$; (b) $\sigma^2 = 1.083$;
Exer. 3: (a) $\sigma^2 = 5.833$; (b) $\sigma^2 = 1.9714$; *Exer. 4:* $\sigma^2 = .56$;
Exer. 5: $\sigma^2 = .4722$; **2.** 48 oz, 48 oz, Group II, .00003, .0228; **3.** the first;
4. the first; **5.** mean height $= 4'4\frac{2}{3}''$; variance $= 2\frac{2}{3}''$; $E(X) = 4'4\frac{2}{3}''$;
$\mathrm{Var}(X) = 2\frac{2}{3}''$

Section 6.6

1. 4, 36; **2.** *Exercise 1:* $\sigma^2 = 11.556$; *Exer. 2:* (a) $\sigma^2 = .0556$;
(b) $\sigma^2 = 1.083$; *Exer. 3:* (a) $\sigma^2 = 5.833$; (b) $\sigma^2 = 1.9714$;
3. $\mathrm{Var}(X) = 30.333$, $\mathrm{Var}[g(X)] = .02075$; **4.** 100

Section 6.7

3. .75, 0; **4.** .479, 0

Section 6.8

1. 5, $5/\sqrt{6}$; **2.** 10.5; **3.** 50; **4.** 5 oz.; .0033 oz.; **5.** \$10,000; \$2000; **6.** 60 min.; 4 min.; **7.** 4206.5; **8.** 5; 175.5; **9.** 9; **10.** 3 hrs.; 3 hrs.; **11.** 15; **12.** 1 min.; **13.** 60

Section 6.9

1. (a) .0228; **(b)** .1587; **(c)** .1587; **(d)** .0228; **(e)** .0228; **2. (a)** .1587; **(b)** .0668; **(c)** .8185; **3. (a)** .1587; **(b)** \$7,440; **4.** $x = -2.24$; **5.** 4.64; **6. (a)** .1359; **(b)** 72.8 min.; **7.** $\sigma = 22.7$; **8.** $K = 4$

Section 6.10

1. Obtain the average income of many residents selected at random from the city. **2.** Roll the die many times. If the die is not loaded, we would expect the number of ones, twos, threes, fours, fives, and sixes to be approximately equal. The method is justified by the strong law of large numbers.

Supplementary Exercises

2. The probabilities of obtaining 0–4, 5, 6, 7, 8, 9, and 10 matches are .9338095365, .0514276877, .0130099805, .0016111431, .0001354194, .0000061206, and .0000001122, respectively. The expected net gain for the player is -24.4 cents. The game is not a fair game. **3.** $16\frac{2}{3}$, 50/9; **4.** 17; **5.** The probabilities that player A, B, or C will win are 5/14, 5/14, and 4/14, respectively. The expected net gain in dollars for players A, B, and C are 5/7, 5/7, and $-10/7$. **6.** $f(x)$ is a pdf, $E(X)$ is not defined; **8.** exact probability is .0475, Chebyshev estimate is .36; **9.** expected profit is 105, standard deviation is 21.21; **10.** 9050, 500; **11.** 20, 10; **12.** $.38 \times 10^{10,000,000}$ years.

Chapter 7

Section 7.2

1. $f(x_1, x_2) = (1/6)b(x_2; 3, .5)$ for $x_1 = 1, 2, \ldots, 6$; $x_2 = 0, 1, 2, 3$.

		x_1					
		1	2	3	4	5	6
x_2	0	1/48	1/48	1/48	1/48	1/48	1/48
	1	3/48	3/48	3/48	3/48	3/48	3/48
	2	3/48	3/48	3/48	3/48	3/48	3/48
	3	1/48	1/48	1/48	1/48	1/48	1/48

$$f(x_1, x_2)$$

2.

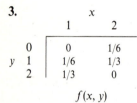

			x			
	1	2	3	4	5	6
−5	0	0	0	0	0	1/36
−4	0	0	0	0	1/36	1/36
−3	0	0	0	1/36	1/36	1/36
−2	0	0	1/36	1/36	1/36	1/36
−1	0	1/36	1/36	1/36	1/36	1/36
y 0	1/36	1/36	1/36	1/36	1/36	1/36
1	1/36	1/36	1/36	1/36	1/36	0
2	1/36	1/36	1/36	1/36	0	0
3	1/36	1/36	1/36	0	0	0
4	1/36	1/36	0	0	0	0
5	1/36	0	0	0	0	0

$f(x, y)$

3.

	x	
	1	2
0	0	1/6
y 1	1/6	1/3
2	1/3	0

$f(x, y)$

4. 4/36, 1/36, 27/36, 0, 6/36, 0, 1/36; **5. (a)** .0625; **(b)** .140625; **(c)** .1914; **(d)** 0; **6.** $f(x, y) = 4xy$ for $0 < x < 1, 0 < y < 1$; $f(x, y) = 0$ elsewhere; **7. (a)** .046875; **(b)** 1/6; **(c)** .1563; **8. (a)** $F(x, y) = 0$ for $x \le 1, -\infty < y < \infty$; $F(x, y) = 0$ for $-\infty < x < \infty, y \le 1$; $F(x, y) = \frac{1}{9}(x^2 - 1)(y^2 - 1)$ for $1 < x < 2, 1 < y < 2$; $F(x, y) = \frac{1}{3}(x^2 - 1)$ for $1 < x < 2, y \ge 2$; $F(x, y) = \frac{1}{3}(y^2 - 1)$ for $x \ge 2, 1 < y < 2$; $F(x, y) = 1$ for $x \ge 2, y \ge 2$; **(b)** 0, .0734, .5803; **(c)** .5; **9. (a)** $f_X(-1) = .35, f_X(0) = .55, f_X(3) = .1$; **(b)** $f_Y(0) = .27, f_Y(1) = .34, f_Y(2) = .39$; **(c)** .9, 1; **10. (a)** $f_X(x) = (2/3)x$ for $1 < x < 2$, 0 elsewhere; **(b)** $f_Y(y) = (2/3)y$ for $1 < y < 2$, 0 elsewhere; **(c)** .4167, .2708; **11. (a)** $f(x \mid y = 1) = .294$ for $x = -1, .588$ for $x = 0$, and .118 for $x = 3$; **(b)** $f(y \mid x = 3) = .2$ for $y = 0, .4$ for $y = 1$, and .4 for $y = 2$; **(c)** .882; **(d)** .909; **12. (a)** $f(x \mid y = 1.5) = (2/3)x$ for $1 < x < 2$, and 0 elsewhere; **(b)** $f(y \mid x = 1.4) = (2/3)y$ for $1 < y < 2$, and 0 elsewhere; **(c)** .4167; **(d)** .2708; **13.** .021; **14.** .0338; **15.** .0154; **16.** .0463; **17.** .2205; **18. (a)** $f_{X_1}(x_1) = x_1/6$ for $x_1 = 1, 2, 3$; **(b)** $f_{X_2, X_3}(x_2, x_3) = (1/27)x_2 x_3$ for $x_2 = 1, 2$ and for $x_3 = 4, 5$; **19. (a)** $f_{X_2}(x_2) = (5 + x_2)/13$ for $x_2 = 1, 2$; **(b)** $f_{X_1, X_3}(x_1, x_3) = (3 + 2x_1 + 2x_3)$ for $x_1 = 1, 2, 3$ and for $x_3 = 2, 4$; **20. (a)** $f_{X_2}(x_2) = 2x_2$ for $0 < x_2 < 1$, and 0 elsewhere; **(b)** $f_{X_1, X_2}(x_1, x_2) = 4x_1 x_2$ for $0 < x_1 < 1, 0 < x_2 < 1$; $f_{X_1, X_2}(x_1, x_2) = 0$ elsewhere; **21. (a)** $f_{X_3}(x_3) = (2 + 2x_3)/3$ for $0 < x_3 < 1$; $f_{X_3}(x_3) = 0$ elsewhere; **(b)** $f_{X_1, X_2}(x_1, x_2) = (2x_1 + 2x_2 + 1)/3$ for $0 < x_1 < 1, 0 < x_2 < 1$; $f_{X_1, X_2}(x_1, x_2) = 0$ elsewhere;

24. (a) for $y > 0$, $f_X(x\,|\,y) = 1$ for $0 < x \le 1$, $f_X(x\,|\,y) = 0$ elsewhere;
for $0 < x < 1$, $f_Y(y\,|\,x) = e^{-y}$ for $y > 0$, $f_Y(y\,|\,x) = 0$ elsewhere;
(b) .5; **(c)** .3496

Section 7.3

1. the random variables are not independent because $f_X(-1)f_Y(0) =$
$(.35)(.27) \ne .2 = f(-1,0)$; **2.** the random variables are independent because
$f_X(x)f_Y(y) = f(x, y)$; **3.** *Exercise 18:* X_1, X_2, X_3 are independent;
Exer. 19: X_1, X_2, X_3 are not independent;
Exer. 20: X_1, X_2, X_3 are independent; *Exer. 21:* X_1, X_2, X_3 are not
independent; **4.** yes, they are independent. See statement (a) of Section
7.3.3; **5.** yes, they are independent. See statement (a) of Section 7.3.3;
6. $f(x_1, x_2) = (1/840)e^{-x_1/35}\,e^{-x_2/24}$ for $x_1 > 0$, $x_2 > 0$; $f(x_1, x_2) = 0$
elsewhere; **7.** $e^{-48/135}\,e^{-48/24} = .034$

Chapter 8

Section 8.2

2. Let $S = \{(i, j)\,|\,i = 1, 2, 3, 4, 5, 6;\ j = 0, 1\}$ where $i =$ outcome of roll of
the die, and $j =$ number of heads. Let $X_1(i, j) = i$ and $X_2(i, j) = j$.
3. (a) Let $X = X_1 + X_2 + X_3$ where $X_i =$ outcome of the ith die in a throw
of 3 dice. **(b)** Let $X = X_1 + X_2$ where $X_1 =$ verbal SAT score of student
selected at random from Queens College and $X_2 =$ mathematics SAT score
of the same student. **(c)** Select a married couple at random from Chicago.
Let X_1 be the income of the wife and let X_2 be the income of the husband.
Let $X = X_1 + X_2$.

Section 8.3

1. $E(X_1 + X_2) = -.55$, $\text{Var}(X_1 + X_2) = 3.6475$; **2.** $E(X_1 + X_2) = 9.333$,
$\text{Var}(X_1 + X_2) = 116.63$; **3.** $E(X_1 + X_2) = 1.2917$, $\text{Var}(X_1 + X_2) = .1177$;
5. (a) 4; **(b)** 14; **(c)** -7; **(d)** 34; **6.** 1950 lbs.; 58.09 lbs.

Section 8.4

1. 100; 10; **2.** .13175; **3.** .1097; **4. (a)** .0228; **(b)** .3446; **5.** .09;
7. standard normal; **8. (a)** 200; **(b)** .0000003; **9.** .0038; **10.** .1082

Section 8.5

1. .2776; **2.** .0228; **3.** .9544; **4.** .9979; **5.** The initial estimate was
probably overly optimistic because the probability of a total contribution of
less than \$80,000 is approximately zero if the initial estimate was correct.

Section 8.6

1. .7286; **2.** .0475 (without correction), .0446 (with correction);
3. .9871 (without correction), .9913 (with correction); **4.** $n = 2571$; **5.** .99;

6. .0054; **7.** .5239; **8.** .0475; **9. (a)** .1742; **(b)** .2514; **(c)** .4972; **10.** .9699; **11.** .1335; **12.** .0342; **13.** .3085; **14.** .0876 (normal approximation); .0888 (exact probability); **15.** .0369 (normal approximation); .0305 (exact probability); **16.** .5; **17.** .6179

Section 8.7

1. .876; **2. (a)** .108; **(b)** .044; **3.** $E(Z) = 9.375$; $\text{Var}(Z) = 52.886$; **4. (a)** 12.25; **(b)** 79.965; **(c)** 0; **(d)** 0; **6. (a)** 4.1; **(b)** 68.29; **(c)** $-.06$; **(d)** $-.0175$ The random variables are not independent. **7. (a)** 1/3; **(b)** 1/18; **(c)** $-.006$; **(d)** $-.08$

Chapter 9

Section 9.2

1. (a)

Score	Frequency	Relative frequency
0	1	.02
1	2	.03
2	2	.03
3	6	.10
4	8	.14
5	10	.17
6	12	.21
7	7	.12
8	5	.09
9	3	.05
10	2	.03

(b)

Score	Cumulative frequency	Cumulative relative frequency
0	1	.02
1	3	.05
2	5	.09
3	11	.19
4	19	.33
5	29	.50
6	41	.71
7	48	.83
8	53	.91
9	56	.97
10	58	1.00

2. (a)

Running time	Frequency	Relative frequency	Percentage
10.00–10.99	4	.100	10.0
11.00–11.99	7	.175	17.5
12.00–12.99	10	.250	25.0
13.00–13.99	9	.225	22.5
14.00–14.99	7	.175	17.5
15.00–15.99	2	.050	5.0
16.00–16.99	1	.025	2.5

(b)

Running time	Cumulative frequency	Cumulative relative frequency
up to 10.99	4	.100
up to 11.99	11	.275
up to 12.99	21	.525
up to 13.99	30	.750
up to 14.99	37	.925
up to 15.99	39	.975
up to 16.99	40	1.000

(c)

Class	Class mark
10.00–10.99	10.495
11.00–11.99	11.495
12.00–12.99	12.495
13.00–13.99	13.495
14.00–14.99	14.495
15.00–15.99	15.495
16.00–16.99	16.495

3. (a)

Age in years	Frequency	Relative frequency	Percentage
20–24	6	.10	10
25–29	11	.18	18
30–34	8	.13	13
35–39	9	.15	15
40–44	8	.13	13
45–49	9	.15	15
50–54	3	.05	5
55–59	3	.05	5
60–64	1	.02	2
65–69	2	.03	3

(b)

Age in years	Cumulative frequency	Cumulative relative frequency
up to 24	6	.10
up to 29	17	.28
up to 34	25	.42
up to 39	34	.57
up to 44	42	.70
up to 49	51	.85
up to 54	54	.90
up to 59	57	.95
up to 64	58	.97
up to 69	60	1.00

(c)

Class	Class mark
20–24	22
25–29	27
30–34	32
35–39	37
40–44	42
45–49	47
50–54	52
55–59	57
60–64	62
65–69	67

4.

Running time (in minutes)	Frequency	Relative frequency
10–under 11	4	.100
11–under 12	7	.175
12–under 13	10	.250
13–under 14	9	.225
14–under 15	7	.175
15–under 16	2	.050
16–under 17	1	.025

5.

Running time	Frequency	Relative frequency
10.0–under 10.5	2	.050
10.5–under 11.0	2	.050
11.0–under 11.5	0	.000
11.5–under 12.0	7	.175
12.0–under 12.5	6	.150
12.5–under 13.0	4	.100
13.0–under 13.5	7	.175
13.5–under 14.0	2	.050
14.0–under 14.5	2	.050
14.5–under 15.0	5	.125
15.0–under 15.5	1	.025
15.5–under 16.0	1	.025
16.0–under 16.5	1	.025

Section 9.4

1. mean = 124.36; median = 124; mode at 120; variance = 30.05;
s.d. = 5.48; 2. mean = 684.46; median = 699.10; variance = 11,178.15;
s.d. = 105.73; 3. (a) mean = 12.923; variance = 2.080; s.d. = 1.442;
(b) mean = 12.945; variance = 2.1475; s.d. = 1.465;
4. (a) mean = 38.433; variance = 126.478; s.d. = 11.246;
(b) mean = 38.333; variance = 132.389; s.d. = 11.506;
5. (a) 1st quartile = 52.393; 2nd quartile = 59.833; 3rd quartile = 67.171;
(b) 80th percentile = 68.635; 95th percentile = 78.084;
(c) .65 fractile = 65th percentile = 64.237

Chapter 10

Section 10.2

2. Consider the experiment of selecting a student at random from Professor James' class. Let X be the grade of the student selected. The distribution of the population is the distribution of X as given by

x	95	87	85	75	70
$f(x)$	.2	.3	.2	.1	.2

3. c is a parameter

Section 10.3

1. (a) (2, 5), (2, 9), (5, 2), (5,9), (9, 2), (9, 5); (b) (2, 2), (2, 5), (2, 9), (5, 5), (5, 2),
(5, 9), (9, 5), (9, 9), (9, 2); 2. (a) (a, b, c), (a, c, b), (b, a, c), (b, c, a), (c, a, b),
(c, b, a); (a, b, d), (a, d, b), (b, a, d), (b, d, a), (d, a, b), (d,b,a); (a, c, d), (a, d, c),
(c, a, d), (c, d, a), (d, a, c), (d, c, a); (b, c, d), (b, d, c), (c, b, d), (c, d, b), (d, b, c),
(d, c, b); (b) Add the following to those in (a): (a, a, a), (b, b, b), (c, c, c),
(d, d, d); (a, a, b), (a, b, a), (b, a, a), (a, a, c), (a, c, a), (c, a, a), (a, a, d), (a, d, a),
(d, a, a); (b, b, a), (b, a, b), (a, b, b), (b, b, c), (b, c, b), (c, b, b), (b, b, d), (b, d, b),
(d, b, b); (d, d, a), (d, a, d), (a, d, d), (d, d, b), (d, b, d), (b, d, d), (d, d, c), (d, c, d),
(c, d, d); (c, c, a), (c, a, c), (a, c, c), (c, c, b), (c, b, c), (b, c, c), (c, c, d), (c, d, c), (d, c, c);
3. sample mean = 2.5; sample variance = 8.333; sample standard
deviation = 2.887; 4. sample mean = .8; sample variance = 4.7,
sample standard deviation = 2.168; 5. $E(\bar{X}) = 1$; $E(S^2) = 9$;
6. $E(\bar{X}) = 2$; $E(S^2) = 4$

Section 10.4

1. $E(\bar{X}) = 1$; $Var(\bar{X}) = 4/9$; the distribution of $\bar{X}$ is normal with mean 1
and variance 4/9; $P(\bar{X} > 1.5) = .2266$; 2. .7486; 3. .3085;

4. The question is, if the sample mean is as low as 1,100 gallons per day, can the average sales per day be as high as 1,200 gallons? If $\mu = 1,200$, $\sigma = 100$, then $P(\bar{X} < 1,100) = P(Z < -4) = .00003$. There certainly is reason to question the owner's claim. **5. (a)** .05; .975; .005; .9; **(b)** $t = 2.160$; $t = -1.895$; **6.** There is reason to question the administration's claim because the chances of getting a sample mean that is greater or equal to 76.1875 is greater than .25 if the true mean is 75 minutes. **7. (a)** .995; .99; .95; **(b)** 38.932; 7.815; 22.362

Chapter 11

Section 11.1

2. No. In Example 11.2 the sample median was a better estimate for 24 of the 64 random samples. **4.** $48.41 < \mu < 52.19$; **5.** $.297 < \mu < .303$; **6. (a)** $48.071 < \mu < 52.529$; **(b)** The length of the confidence interval is shorter when σ^2 is known. **(c)** $s = 2.5$; **(d)** yes; **7.** $2.71 < \mu < 3.63$; **8. (a)** $2.621 < \mu < 7.129$; **(b)** $8.19 < \mu < 12.83$; **9.** $-.002 < \mu < .002$, yes; **10.** error $\leq .46$; **11.** error $\leq .00286$; **12.** $n \geq 97$; **13.** $n \geq 224$

Section 11.2

1. $2.404 < \mu_1 - \mu_2 < 9.596$; **2.** $.4823 < \mu_1 - \mu_2 < 2.1177$; **3.** $5.421 < \mu_1 - \mu_2 < 10.579$; **4.** $3.403 < \mu_1 - \mu_2 < 5.985$; **5.** $3.267 < \mu_1 - \mu_2 < 6.121$; **7.** error $\leq .014$ lbs.;

8. (a) length $= 2 \cdot (t_{\alpha/2, v}) \cdot s^*$; **(b)** length $= 2t_{\alpha/2, v} \cdot \sqrt{\dfrac{s_1^2}{n_1} + \dfrac{s_2^2}{n_2}}$;

(c) length $= 5.25$ if $\sigma_1 = \sigma_2$; length $= 5.338$ if $\sigma_1 \neq \sigma_2$; **(d)** length $= 6.42$ if $\sigma_1^2 = \sigma_2^2$; length $= 5.98$ if $\sigma_1^2 \neq \sigma_2^2$; **(e)** length $= 14.31$ if $\sigma_1^2 = \sigma_2^2$; length $= 11.63$ if $\sigma_1^2 \neq \sigma_2^2$; **(f)** assume $\sigma_1 = \sigma_2$ if $|s_1 - s_2| = 0$; assume $\sigma_1 \neq \sigma_2$ if $|s_1 - s_2|$ small; assume $\sigma_1 \neq \sigma_2$ if $|s_1 - s_2|$ large.

Section 11.3

1. $.073 < p < .126$; **2.** $.152 < p < .514$; **4.** We can assert with a confidence of .95 that between 82% and 98% of patients will be cured. **6.** $n \geq 148$; **7.** $n \geq 3458$

Section 11.4

1. $34.22 < \sigma^2 < 275.23$; **2.** $3.06 < \sigma^2 < 50.06$; **3.** $.30 < \sigma < .46$; **4.** $s = .37$; yes, because $.25 < \sigma < .71$ is a .95 confidence interval for σ.

Section 11.5

1. (a) false; **(b)** true; **2. (a)** $\sigma/\sqrt{n}$; **(b)** $\sqrt{(\sigma_1^2/n_1) + (\sigma_2^2/n_2)}$

Section 11.6

2. $k = 1/\alpha$

Chapter 12

Section 12.1

1. $\alpha = .3135$; $\beta = .0593$; **2.** $\alpha = .0668$

Section 12.2

1. (a) parameter; **(b)** null; **(c)** simple; **(d)** composite;
(e) decision rule; **(f)** type I; α; type II; β; **(g)** significance;
(h) the probability of a type I error; **(i)** test statistics; **(j)** sample;
(k) critical; **2. (a)** $\alpha = .75$; $\beta = .0625$; **(b)** $\Omega = \{(r,r),(r,b),(b,r),(b,b)\}$;
(c) critical region $= \{(r,b),(b,r),(b,b)\}$; **(d)** simple; **(e)** simple

Section 12.3

1. Reject H_0 when $\bar{x} > 6.1635$;
2. (a) no, $z = -1.5 \not< -1.645 = -z_{.05} = -z_\alpha$;
(b) yes, $z = -1.5 < -1.28 = -z_{.10} = -z_\alpha$;
4. yes, $z = 2.5 > 1.881 = z_{.03} = z_\alpha$; **5.** yes, $|z| = 2 > 1.96 = z_{.025} = z_{\alpha/2}$

Section 12.4

3. $\rho(2) = 1$; $\rho(3) = .8037$; **4.** $\beta(1) = .0771$; $\rho(1) = .9229$; $\beta(-1) = .0771$;
$\rho(-1) = .9229$; $\beta(.5) = .7173$; $\rho(.5) = .2827$; $\beta(-.5) = .7173$;
$\rho(-.5) = .2827$

Chapter 13

Section 13.2

1. (a) $|z| = 1.8$; $z_{.025} = 1.96$; cannot reject H_0; **(b)** $z = 1.8$; $z_{.05} = 1.645$;
reject H_0; **(c)** $|z| = 2.4$; $z_{\alpha/2} = 2.575$; cannot reject H_0; **(d)** $z = -2.4$;
$-z_\alpha = -2.327$; reject H_0; **(e)** use one-tail tests; **2. (a)** $|t| = 2$;
$t_{\alpha/2,n-1} = 2.306$; cannot reject H_0; **(b)** $t = 2$; $t_{\alpha,n-1} = 1.86$; reject H_0;
(c) $t = -2.75$; $-t_{\alpha,n-1} = -2.602$; reject H_0;
(d) $|t| = 2.75$; $t_{\alpha/2,n-1} = 2.947$; cannot reject H_0;
(e) use one-tail tests; **3. (a)** $|z| = 1.667$; $z_{\alpha/2} = 2.054$; cannot reject H_0;
(b) $z = 2.5098$; $z_\alpha = 2.054$; reject H_0; **(c)** $z = -2.25$;
$-z_\alpha = -1.555$; reject H_0; **4. (a)** $H_0: \mu = 3$ versus $H_1: \mu \neq 3$;
reject H_0 if $|z| = \left|\dfrac{\bar{x} - 3}{.1/\sqrt{10}}\right| > 2.575$; **(b)** $|z| = 3.162$; reject inventor's claim;
(c) $|z| = 2.107$; cannot reject inventor's claim;

5. H_0: $\mu = 15,000$ versus H_1: $\mu \neq 15,000$; $|z| = 3.354$; $z_{\alpha/2} = 1.96$;
reject the families' claim; **6.** $|t| = 1.1415$; $t_{\alpha/2, n-1} = 3.355$;
cannot reject manufacturer's claim; **7.** $t = -1.58$; $-t_{\alpha, n-1} = -1.833$;
cannot conclude that the mean time has decreased; **8.** $|t| = 4$;
$t_{.005, 15} = 2.947$; cannot conclude that toxic chemical is being maintained
at the desired average amount; **9.** $z = -5$; $-z_{\alpha} = -1.64$; yes;
10. $z = 5$; $z_{\alpha} = 2.05$; yes; **11.** $|t| = .8944$; $t_{\alpha/2, n-1} = 2.093$; no;
12. $|z| = .2$; $z_{\alpha/2,} = 2.17$; cannot reject manufacturer's claim;
13. Let μ = mean income of women; H_0: $\mu = 11,000$ versus H_1: $\mu \neq 11,000$;
$|z| = 2$; $z_{\alpha/2} = 1.96$; cannot reject H_0 that women make on the average
$4,000 less than men; **14.** $t = .79$; $t_{.05, 9} = 1.833$; no

Section 13.3

1. $z = .445$; $z_{\alpha} = 1.645$; no; **2.** $z - 1.2076$; $z_{\alpha} = 1.645$; no;
3. $|t| = .3108$; $t_{\alpha/2, n_1 + n_2 - 2} = 2.069$; no; **4.** $|t| = 3.47$; $t_{.005, 19} = 2.861$;
at the .01 level of significance there is a difference in average
abrasive wear; **5.** $|z| = 1.414$; $z_{\alpha/2} = 2.05$; no; **6.** $z = -19.28$;
$-z_{\alpha} = -1.645$; yes; **7.** $|t| = .499$; $t_{\alpha/2, n-1} = 3.250$; no;
8. (a) $|t| = 2.2117$; $t_{\alpha/2, n-1} = 2.571$; no; **(b)** $t = -2.2117$;
$-t_{\alpha, n-1} = -2.015$; yes; **9.** $|z| = 2$; $z_{\alpha/2} = 1.96$; no; **10.** $|z| = 3.534$;
$z_{\alpha/2} = 2.575$; yes; **11.** $|z| = 1.897$; $z_{\alpha/2} = 2.575$; no

Section 13.4

1. $s_n = 7$; $k^* = 5$; no; **2.** $z = -2.5$; $-z_{\alpha} = -1.645$; yes;
3. $|z| = .40$; $z_{\alpha/2} = 2.05$; no; **4.** $z = 3.779$; $z_{\alpha} = 1.645$; yes

Section 13.5

1. $\chi^2 = 24.89$; $\chi^2_{.05, 14} = 23.6849$; yes; **2.** $\chi^2 = 175.82$;
$\chi^2_{\alpha, n-1} = 21.666$; yes; **3.** $s_2^2/s_1^2 = 1.878$; $F_{\alpha/2, n_2 - 1, n_1 - 1} = 5.35$; no;
4. $s_2^2/s_1^2 = 2.457$; $F_{\alpha/2, n_2 - 1, n_1 - 1} = 5.05$; no; **5.** $\chi^2 = 29.6875$;
$\chi^2_{\alpha/2, n-1} = 38.5822$; $\chi^2_{1 - \alpha/2, n-1} = 6.84398$; no

Section 13.6

1. $n = 64$; **2.** $n = 20$; **3.** $n = 100$; **4.** $n = 7$; **5.** $n = 6$

6. Example	Value of test statistic	p-value
13.1	$z = -2.532$	$.005 < p < .006$
13.3 (a)	$t = -1.5$	$.05 < p < .10$
13.3 (b)	$t = 3.75$	$.10 < p < .005$
13.4 (a)	$z = -7$	$p < .0001$
13.4 (b)	$z = -1.4$	$p = .0808$
13.5	$z = .1036$	$.46 < p < .45$
13.6	$t = 1.04$	$p > .10$
13.7	$z = .8866$	$.18 < p < .19$
13.9	$|t| = .497$	$p > .05$

Chapter 14

Section 14.1

1. yes; **2.** yes, although a curvilinear regression analysis is better;
3. yes; **4.** no; **5.** $\hat{y} = -.805 + .0418x$; **(a)** 2.331; **(b)** 1.285;
6. (a) 3.5321; **(b)** 6.6101; **7. (a)** .6031; **(b)** .6705;
8. (a) $-2.073 < A < .462$; $.023 < B < .060$; $1.9305 < \mu_{Y|x_0} < 2.7295$;
$1.14 < y_0 < 3.52$; **9.** 1.19

Section 14.2

1. $\hat{y} = 3.309 + 1.705x + 2.612x$; **2.** $\hat{y} = .9 + 1.05x_1 + 6.5x_2$ where x_1 is the
amount of nutrient A, x_2 is the amount of nutrient B, and y is the height

Section 14.3

1. $5a + 10b + 30c = 21.1$; $10a + 30b + 100c = 51.4$;
$30a + 100b + 354c = 168.6$; **2.** $\hat{y} = 3.123 - .566x + 83.25x^2$

Section 14.4

2. (a) .07; **(c)** .914; **5.** $.733 < \rho < .978$;
6. *Exercise 1*: $-.50 < \rho < .60$; *Exercise 2*: $.62 < \rho < .98$

Chapter 15

Section 15.2

1. (a) $s_m = 7$; $m = 6$; $k^*_{.05} = 4$; cannot reject H_0; **(b)** $k^*_{.10} = 5$;
cannot reject H_0; **(c)** $k^*_{.40} = 7$; can reject H_0 at $\alpha = .40$; **2. (a)** m = 18;
$s_m = 11$; $k^*_{.05} = 13$; cannot reject H_0; **(b)** $k_{.10} = 13$; cannot reject H_0;
3. $m = 15$; $s_m = 7$; $k^*_{.05} = 3$; cannot reject H_0; **4.** $m = 17$; $s_m = 8$;
$k_{.01} = 13$; no; **5.** $m = 20$; $s_m = 11$; $k_{.025} = 15$; $k^*_{.025} = 1$;
cannot conclude that there is a difference; **6.** $m = 18$; $s_m = 12$;
$k_{.05} = 13$; $k^*_{.05} = 5$; no

Section 15.3

1. (a) $m = 16$; $T^+ = 38$; cannot reject H_0 at $\alpha = .05$; cannot reject
H_0 at $\alpha = .10$, smallest α at which we may reject H_0 is $\alpha = .06$;
(b) $m = 18$; $T^+ = 115$; cannot reject H_0 at $\alpha = .05$; can reject H_0 at
$\alpha = .10$; **(c)** *Exercise 1*: For sign test, cannot reject H_0 at $\alpha = .05$, cannot
reject H_0 at $\alpha = .10$, smallest α to reject H_0 is $\alpha = .40$. For Wilcoxon
sign test, cannot reject H_0 at $\alpha = .05$, can reject at $\alpha = .10$,
smallest α to reject H_0 is $\alpha = .06$. *Exercise 2*: For sign test, cannot
reject H_0 at $\alpha = .05$, cannot reject H_0 at $\alpha = .10$. For Wilcoxon sign test,
cannot reject H_0 at $\alpha = .05$, can reject H_0 at $\alpha = .10$. **2. (a)** $m = 15$;
$T^+ = 58.5$; cannot reject H_0 at $\alpha = .05$; **(b)** $m = 17$; $T^+ = 91.5$;
cannot reject H_0 at $\alpha = .01$; **(c)** In all cases, cannot reject H_0.

3. $m = 20$; $T^+ = 103$; cannot reject H_0. **4.** $m = 18$; $T^+ = 116$; cannot reject H_0.

Section 15.4

1. If we use the test with U_1, then we have $U_1 = 187.5$, $|z| = .388$, $z_{.025} = 1.96$, and we cannot reject H_0 at $\alpha = .05$. But, if we use the test with U_2, then we have $U_2 = 632.5$, $|z| = 11.699$, and we can reject H_0 at $\alpha = .05$. Therefore, we cannot reject H_0 with the sign test, we cannot reject H_0 with the Wilcoxon sign-rank test, but we can reject H_0 with the Mann–Whitney–Wilcoxon rank-sum test. **2.** If we use the test with U_1, then we have $U_1 = 58$, $z = .2843$, $-z_{.05} = -1.645$, and we cannot conclude that district B has higher mean income. If we use the test with U_2, then we have $U_2 = 50$, $z = -.28426$ and we still cannot conclude that district B has higher mean income. **3.** If we use the test with U_1, then we have $U_1 = 35.5$, $z = -.048$, $z_{.01} = 2.327$, and we cannot reject H_0. If we use the test with U_2, then we have $U_2 = 36.5$, $z = .048$ and we still cannot reject H_0. **4.** If we use the test with U_1, then we have $U_1 = 209$, $|z| = 1.202$, $z_{.05} = 1.645$, and we cannot reject H_0. If we use the test with U_2, then we have $U_2 = 115$, $|z| = 1.77$, and we can reject H_0. In comparing the 3 tests, we see that we cannot reject H_0 using the sign test, we cannot reject H_0 using the Wilcoxon sign-rank test, but we can reject H_0 using the Mann–Whitney–Wilcoxon rank-sum test. **5.** If we use the test with U_1, then we have $U_1 = 56.5$, $|z| = .2308$, $z_{.025} = 1.96$, and we cannot reject H_0. If we use the test with U_2, then we have $U_2 = 63.5$, $|z| = .2308$, and we still cannot reject H_0.

Section 15.5

1. $\chi^2 = .6613$; $\chi^2_{.05, 4} = 9.4877$; cannot conclude that the drug is effective; **2.** $\chi^2 = 1.578$; $\chi^2_{.01, 2} = 9.2103$; cannot conclude that student opinion is dependent on sex of student; **3. (a)** $\chi^2 = 9.767$; $\chi^2_{.05, 6} = 12.5916$; cannot conclude that there is a difference at $\alpha = .05$; **(b)** $\chi^2_{.10, 6} = 10.6446$; can conclude that there is a difference at $\alpha = .10$; **4.** $\chi^2 = 10.989$; $\chi^2_{.01, 4} = 13.2767$; yes

Section 15.6

1. (a) $\bar{\lambda} = .48$; $\chi^2 = 15.97$; $\chi^2_{.05, 1} = 3.8415$; cannot reject hypothesis that the distribution is Poisson; **(b)** At $\alpha = .10$, $\chi^2_{.10, 1} = 2.7055$; and we still cannot reject hypothesis that the distribution is Poisson; **2. (a)** $\chi^2 = .909$; $\chi^2_{.05, 1} = 3.8415$; cannot reject hypothesis that the proportions are $p^2:2pq:q^2$; **(b)** At $\alpha = .10$, $\chi^2_{.10, 1} = 2.7055$ and we still cannot reject the hypothesis that the proportions are $p^2:2pq:q^2$; **3. (a)** $\chi^2 = 14.917$; $\chi^2_{.05, 4} = 9.4877$; reject H_0 that the distribution of grades are 10%, 20%, 50%, 15%, and 5%; **(b)** At $\alpha = .10$, $\chi^2_{.10, 4} = 7.7794$; and we should reject H_0; **4. (a)** $\chi^2 = 4.6626$; $\chi^2_{.05, 2} = 5.9915$; cannot reject H_0 that distribution of voter preferences are 45%, 35%, and 20%; **(b)** If $\alpha = .10$,

$\chi^2_{.10,2} = 4.6052$; and we can reject H_0; **5. (a)** $\chi^2 = 9.7$; $\chi^2_{.05,5} = 11.0705$; no; **(b)** At $\alpha = .10$, yes; **6.** $\chi^2 = 6.415$; At $\alpha = .05$ we cannot reject H_0 that distribution is normal; at $\alpha = .10$ we cannot reject H_0; at $\alpha = .90$ we can reject H_0. **7.** $\chi^2 = 3.4614$; At $\alpha = .05$ we cannot reject the null hypothesis that the distribution is exponential. The smallest α for which we can reject H_0 is $\alpha = .75$

Supplementary Exercises

1. $t = -1.798$; **(a)** reject H_0; **(b)** reject H_0; **(c)** smallest α is $\alpha = .05$

		Comparison:	
α	Sign test	Wilcoxon-rank test	t-test
.05	cannot reject	cannot reject	can reject
.10	cannot reject	can reject	can reject
smallest α	.40	.06	.05

The most powerful test is the t test.

2. With respect to Exercise 2 of Section 15.2, $t = 1.641$. We can reject H_0

		Comparison:	
α	Sign test	Wilcoxon sign-rank test	t-test
.05	cannot reject	cannot reject	cannot reject
.10	cannot reject	can reject	can reject

The least powerful test is the sign test.

INDEX